Charged Beam Dynamics, Particle Accelerators and Free Electron Lasers

IOP Plasma Physics Series

Series Editors

About the Series

The IOP Plasma Physics ebook series aims at comprehensive coverage of the physics and applications of natural and laboratory plasmas, across all temperature regimes. Books in the series range from graduate and upper-level undergraduate textbooks, research monographs and reviews.

The conceptual areas of plasma physics addressed in the series include:

- Equilibrium, stability and control
- Waves: fundamental properties, emission, and absorption
- Nonlinear phenomena and turbulence
- Transport theory and phenomenology
- Laser-plasma interactions
- Non-thermal and suprathermal particle populations
- Beams and non-neutral plasmas
- High energy density physics
- Plasma-solid interactions, dusty, complex and non-ideal plasmas
- Diagnostic measurements and techniques for data analysis

The fields of application include:

- Nuclear fusion through magnetic and inertial confinement
- Solar-terrestrial and astrophysical plasma environments and phenomena
- Advanced radiation sources
- Materials processing and functionalisation
- Propulsion, combustion and bulk materials management
- Interaction of plasma with living matter and liquids
- Biological, medical and environmental systems
- Low temperature plasmas, glow discharges and vacuum arcs
- Plasma chemistry and reaction mechanisms
- Plasma production by novel means

Charged Beam Dynamics, Particle Accelerators and Free Electron Lasers

Giuseppe Dattoli, Andrea Doria, Elio Sabia and Marcello Artioli
ENEA Frascati Research Center, Frascati, Rome, Italy

IOP Publishing, Bristol, UK

ISBN 978-0-7503-1239-4 (ebook)
ISBN 978-0-7503-1240-0 (print)
ISBN 978-0-7503-1241-7 (mobi)

DOI 10.1088/978-0-7503-1239-4

Version: 20170901

IOP Expanding Physics
ISSN 2053-2563 (online)
ISSN 2054-7315 (print)

British Library Cataloguing-in-Publication Data: A catalogue record for this book is available from the British Library.

Published by IOP Publishing, wholly owned by The Institute of Physics, London

IOP Publishing, Temple Circus, Temple Way, Bristol, BS1 6HG, UK

US Office: IOP Publishing, Inc., 190 North Independence Mall West, Suite 601, Philadelphia, PA 19106, USA

To Angelo, Marcello, Pier Luigi

Contents

Preface

This book summarizes different topics in the field of accelerators and free electron laser (FEL) devices. It is the result of the working experience of the authors, who have been involved in this field of research for a considerable amount of time.

The scientific location the authors developed in is that of the Frascati Laboratories, which gathers together some of the most significant Italian research institutions. This environment has provided the conditions to merge different areas of expertise, allowing the growth and maturation of FEL technologies in their various aspects, which has been required for the synergic superposition of fields of research such as accelerators, lasers, magnetic materials—which are all historically present in this location.

Even though the technology was proposed some years before, the FEL 'revolution' started in 1975 with the realization, at Stanford University, of the first FEL amplifier. This experiment, conducted by J M J Madey and coworkers, paved the way for the FEL oscillator experiment, which was successfully carried out in 1976. In the 40 years since then, the FEL has undergone spectacular evolution in terms of world-wide distribution, size and financial investment.

For many years, since its discovery, this type of device had been operated in the oscillator configuration. However, on the eve of the 1990s, the demand for increasing x-ray brightness drove the realization of fourth-generation synchrotron radiation sources, based on the so-called self-amplified spontaneous emission device.

The study and development of FELs has provided new conceptions in the construction of accelerators, designed to provide beams with unprecedented high quality, namely high current, low-energy dispersion and low emittance. The effective realization of these goals has also been made possible by an impressive development of sophisticated numerical codes and by 'start to end' simulations, which are capable of including both accelerators and FEL items. The quest for larger and larger FEL devices—which now compete with and in many cases overcome the accelerators dedicated to high-energy physics—has opened the quest for new technologies reducing the relevant size and cost. New types of undulators and high-gradient accelerators have accordingly been proposed.

This book is an account of the different aspects of this type of device. There are, however, authoritative monographs which have accomplished the same task. We have therefore attempted to provide the point of view of a 'manual', in the sense that rather than providing an academic monograph with very rigorous and complete mathematical treatment, we have chosen a pragmatic exposition aimed at explaining how to design an FEL device and the accelerator providing the driving beam.

We have chosen to keep the bibliography to a minimum. Essentially we refer to review papers or books which are reported at the end of each chapter. Apart from chapter 3, all chapters are composed of two parts. The second of these contains the complements and exercises. The complements cover topics which have only been touched upon in the first part, while the exercises have been designed to introduce the reader to actual problems encountered in real-life research.

We have collected further materials concerning the topics treated in this book on a dedicated and continuously upgraded website devoted to FEL physics research and hosted by ENEA at www.fel.enea.it. In particular the reader is directed to 'Booklet for FEL design: a collection of practical formulae with interactive graphs' by Marcello Artioli, Giuseppe Dattoli and Simonetta Pagnutti (www.fel.enea.it/booklet/live/), which provides a tool for the 'quick' design of an FEL device operating in different configurations.

Regarding the chosen system of units, most of this book uses SI (MKSA) units, but chapter 3 employs (for historical reasons) gauss (G) units, and we have largely adopted practical units. In any case, when using gauss units, we give sufficient warning and provide a 'translation to SI counterparts'.

To avoid any misunderstanding, we recall here that the translation from one system of units to the other is realized by means of the following equations:

$$\begin{aligned} B_G &\to \sqrt{\frac{4\pi}{\mu_0}}\, B_{SI}, \\ E_G &\to \frac{1}{\sqrt{4\pi\,\varepsilon_0}} E_{SI}, \\ q_G &\to \sqrt{4\pi\,\varepsilon_0}\, q_{SI}, \\ c &\to \frac{1}{\sqrt{\varepsilon_0 \mu_0}}. \end{aligned}$$

According to the above correspondence, the Lorentz force is (in gauss units)

$$\vec{F}_G = e_G\, \vec{E}_G + e_G \frac{\vec{v} \times \vec{B}_G}{c},$$

while, according to the previous conversion prescription, we find for the SI expression

$$\vec{F}_{SI} = \frac{e_{SI}}{\sqrt{4\pi\,\varepsilon_0}} \left(\sqrt{4\pi\,\varepsilon_0}\, \vec{E}_{SI}\right) + \frac{e_{SI}}{\sqrt{4\pi\,\varepsilon_0}} \left(\sqrt{\varepsilon_0 \mu_0} \sqrt{\frac{4\pi}{\mu_0}}\, \vec{v} \times \vec{B}_{SI}\right) = e_{SI}\left(\vec{E}_{SI} + \vec{v} \times \vec{B}_{SI}\right).$$

In the following chapters we will drop the labels 'G' and 'SI' and advise the reader in the case of a change of units.

During the compilation of the book we have been assisted by many colleagues, who helped us with their advice and experience. We owe them all our gratitude.

Over the course of the years we have had the privilege of working with outstanding scientists, some of whom are no longer with us; we would like to recall John Madey, who began the activity on FELs more than four decades ago, and John Walsh, who pioneered the study of Cerenkov FEL devices.

Finally we would like to dedicate our work to our late friends Pier Luigi Ottaviani, Angelo Marino and Marcello Quattromini, for sharing with us their work and their friendship.

Giuseppe Dattoli, Andrea Doria, Elio Sabia and Marcello Artioli
Frascati, December 2016

Author biographies

Giuseppe Dattoli

Giuseppe Dattoli was born in Lagonegro, Italy, in 1953. He received the PhD degree in physics from 'La Sapienza' University of Rome, Italy, in 1976. He is an ENEA Researcher and has been involved in different research projects, including high energy accelerators, free electron lasers, and applied mathematics networks since 1979. Dr Dattoli has taught in Italian and foreign universities, and has received the FEL Prize Award for his outstanding achievements in the field.

Andrea Doria

Andrea Doria received the Laurea degree in physics from 'La Sapienza' University of Rome, Italy, in 1987. He has been a Permanent Staff Member of the experimental free electron laser (FEL) group of ENEA, Frascati, Italy, since 1988. His research interests are Cerenkov based FELs, compact FEL sources in the millimetre and far IR regions, and coherence and correlation effects on FEL power output. He collaborated in designing and operating user application experiments on THz interaction with biological organisms, solid state systems and cultural heritage samples using the compact FEL sources of ENEA. He has been the Principal Investigator of the combined ENEA, INFN, CNR research project named SPARC/SPARX, for the realization of a VUV and soft x-ray FEL operating in SASE and external seeding regime and funded by the Italian Minister of Research. He is now member of the team for the study of a CARM-type source for electron cyclotron resonance plasma heating.

Elio Sabia

Elio Sabia was born in Naples, Italy, in 1952. He received the PhD degree in physics from the University of Naples Federico II, Italy, in 1978. His dissertation concerned the study of the magnetic structure of a storage ring free electron laser (FEL). He has been a Staff Member with the ENEA Research Center of Frascati since 1981.

Since 1981, he has been involved in theoretical problems connected with particle dynamics in high-energy electron accelerators driven FELs. He has taught for three years at University of Naples as a contract professor. His major areas of interest and research include experimental activities on SPARC/SPARX FELs, and theoretical activities for the development of 250 GHz RF source for tokamak plasma heating.

Marcello Artioli

Marcello Artioli received the Laurea degree in Electronic Engineering from 'Alma Mater' University of Bologna, Italy, in 1996. During and after the PhD in Electrotechnics he did some research on fault diagnosis of electromechanical systems and signal processing. In 2011 he joined the FEL team at ENEA, Frascati, Italy, where he has been working on control systems and computer science. He is now member of the team for the study of a CARM-type source for electron cyclotron resonance plasma heating.

IOP Publishing

Charged Beam Dynamics, Particle Accelerators and Free Electron Lasers

Giuseppe Dattoli, Andrea Doria, Elio Sabia and Marcello Artioli

Chapter 1

Charged particle dynamics and beam transport optics

1.1 Introduction

This chapter deals with the basic tools for understanding the interaction between charged particles and static electric and magnetic fields, and/or electromagnetic waves.

The 'tool' allowing the coupling of a charged particle to the fields is the charge itself, and the feedback on the particle motion is provided by the Lorentz force experienced by a particle, with charge q, moving under the combined action of electric $\vec{E}$ and magnetic $\vec{B}$ field vectors, namely

$$\vec{F} = q(\vec{E} + \vec{v} \times \vec{B}). \tag{1.1}$$

The previous equation holds for static and time-dependent fields, whose spatial and temporal behavior is ruled by the in-vacuum Maxwell's equations (since we are dealing with in-vacuum Maxwell's equations we have not explicitly mentioned the electric displacement and the magnetic intensity vectors $\vec{D} = \varepsilon_0 \vec{E}$, $\vec{B} = \mu_0 \vec{H}$, respectively)

$$\begin{aligned}
&\vec{\nabla} \times \vec{E} = -\frac{\partial \vec{B}}{\partial t}, && \vec{\nabla} \cdot \vec{E} = \frac{\rho}{\varepsilon_0}, \\
&\vec{\nabla} \times \vec{B} = \mu_0 \vec{J} + \frac{1}{c^2}\frac{\partial \vec{E}}{\partial t}, && \vec{\nabla} \cdot \vec{B} = 0, \\
&\varepsilon_0 = 8.85 \cdot 10^{-12}\left[\frac{\mathbb{F}}{\mathbb{m}}\right], && \mu_0 = 4\pi \cdot 10^{-7}\left[\frac{\mathbb{H}}{\mathbb{m}}\right], \\
&c^2 = \frac{1}{\varepsilon_0 \mu_0},
\end{aligned} \tag{1.2}$$

where SI units are used and the electric permittivity ε_0 and magnetic permeability μ_0 are expressed accordingly. The Lorentz force equation holds in the relativistic case as well, provided that

$$\vec{F} = \frac{d}{dt}\vec{P}$$
$$\vec{P} = m_0 \gamma \vec{v}, \tag{1.3}$$
$$\gamma = \frac{1}{\sqrt{1 - |\vec{\beta}|^2}}, \qquad \vec{\beta} = \frac{\vec{v}}{c}.$$

It is evident that for non-relativistic motion ($|\vec{\beta}| \ll 1$) the mass growth due to relativistic factor γ does not affect the equation of motion and, accordingly, for a constant field the above equations can be solved in a straightforward manner.

To clarify the quantitative meaning of relativistic and non-relativistic motion, we consider a charged particle ruled by a static electric field and look for the relevant motion equations.

In the following we will try to learn as much as possible from the previous set of equations.

The first example we analyze is the dynamical behavior of a particle moving under the influence of a static electric field. We proceed by casting the Lorentz equation with $\vec{B} = 0$ in the form

$$\frac{d}{dt}(\gamma \vec{v}) = \vec{a},$$
$$\vec{a} = \frac{q\vec{E}}{m_0}, \tag{1.4}$$
$$\vec{v}_0 = 0.$$

We leave the symbol of the vector intact, even though the motion is one-dimensional and the velocity has the same direction as the acceleration. The integration of (1.4) proceeds as follows:

$$(\gamma \vec{v}) = \vec{a}t, \tag{1.5}$$

and using the third equation of (1.3) and by solving for the velocity we find

$$\vec{v}(t) = \frac{\vec{a}t}{\sqrt{1 + \frac{|\vec{a}t|^2}{c^2}}}. \tag{1.6}$$

Furthermore a second integration with $\vec{r}_0 = 0$ yields

$$r(t) = \frac{c^2}{|\vec{a}|}\left(\sqrt{1 + \frac{|\vec{a}t|^2}{c^2}} - 1\right). \tag{1.7}$$

In figure 1.1 we have reported specific examples regarding charged particles (with different charges and masses) moving under the influence of electric fields of different intensities. The integration of the equations of motion (obtained with not necessarily vanishing initial conditions) can be exploited to understand how and when the particle motion becomes relativistic, namely when the acquired velocity becomes comparable to that of light.

To understand this point in a quantitative way, we examine the case of a light particle such as the electron, whose charge and mass are reported below (we will not care about the sign of the electron charge)

$$\begin{aligned} |q| &= e \cong 1.6 \times 10^{-19}\ \mathrm{C}, \\ m_e &\cong 9.1 \times 10^{-31}\ \mathrm{kg}. \end{aligned} \tag{1.8}$$

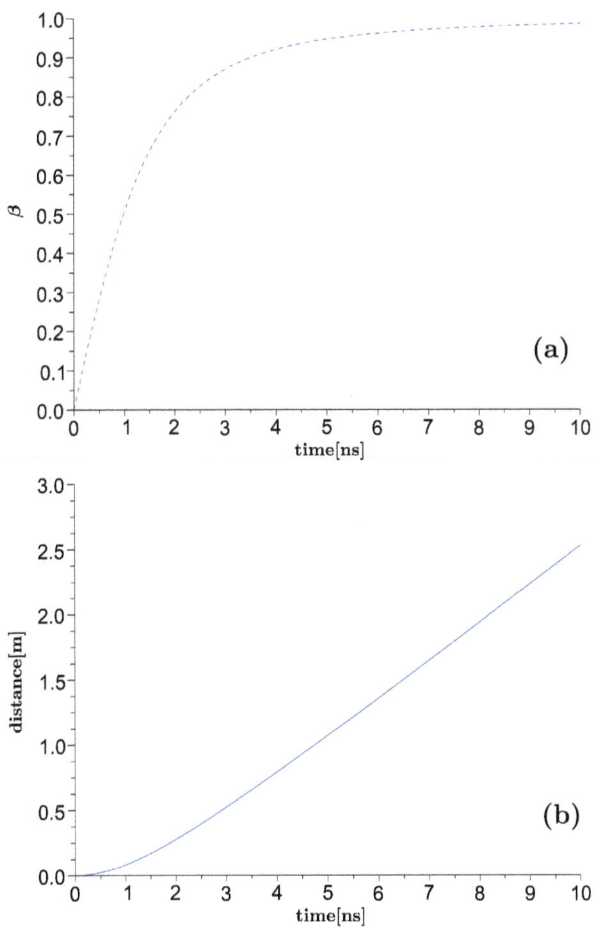

Figure 1.1. (a) Velocity ($\beta = v/c$) acquired by an electron under the influence of an electric field of 10^6 V m^{-1}. (b) Distance traveled.

From (1.6) we find that the time taken to reach a given velocity $|\vec{v}|$ is given by

$$t = \frac{|\vec{\beta}|}{\sqrt{1 - |\vec{\beta}|^2}} \frac{m_e c}{|e\vec{E}|}, \tag{1.9}$$

which, in 'practical' units (in a practical formula the units are explicitly given in square brackets), is written as

$$t[\mathrm{s}] \cong 1.7 \times 10^{-3} \frac{|\vec{\beta}|}{\sqrt{1 - |\vec{\beta}|^2}} \frac{1}{E[\mathrm{V/m}]}. \tag{1.10}$$

An applied electric field of 1 V/m accelerates the electron to 1/10 of the light velocity in about 1.7×10^{-4} s, after which the particle has traveled a distance of around 2.56×10^3 m. Relativistic velocities can therefore be reached in a reasonable distance using significantly larger electric field strengths. Keeping $E[\mathrm{V/m}] = 10^6$ we obtain the results reported in figure 1.1.

The previous example, albeit naïve, yields a feeling of the orders of magnitude involved in an acceleration process of electrons using a static electric field.

According to special relativity, the total energy of an electron moving at velocity $\vec{v}$ is given by

$$\mathcal{E} = m_e \gamma c^2. \tag{1.11}$$

The relevant kinetic energy is obtained from (1.11) by subtracting the rest energy, thus obtaining

$$\mathcal{T} = (\gamma - 1)\, m_e c^2. \tag{1.12}$$

See table 1.1 for a synopsis of the interplay between the different relativistic kinematics quantities.

The theorem of kinetic energy (namely that the variation of kinetic energy of a particle is provided by the work done by an external force on the particle itself) holds in the relativistic case too, we can accordingly cast the energy acquired by the electron during the acceleration in the form

$$\begin{aligned} \mathcal{T} &= \mathcal{E}_0\left(\sqrt{1 + \left(\frac{|\vec{a}|\, t}{c}\right)^2} - 1\right) = e\vec{E} \cdot \vec{r}(t), \\ \vec{E} &= \frac{V_0}{d} \frac{\vec{r}}{|\vec{r}|}, \\ \mathcal{T} &= \frac{eV_0}{d} r(t), \\ \mathcal{E}_0 &= m_e c^2 \cong 8.19 \times 10^{-14}\ \mathrm{J}. \end{aligned} \tag{1.13}$$

Equation (1.13) suggests that, in order to avoid redundant exponential notations, we can adopt a more convenient reference unit for the energy. If an electron crosses a region 1 m in length moving parallel to the line of forces of the corresponding

Table 1.1. Synopsis of relativistic kinematical quantities and the relevant trigonometric picture.

	$\beta = \frac{v}{c}$	P	$\mathcal{E}$	$\mathcal{T}$
γ	$(1-\beta^2)^{-\frac{1}{2}}$	$\sqrt{1+\left(\frac{P}{m_0c}\right)^2}$	$\frac{\mathcal{E}}{m_0c^2}$	$\frac{\mathcal{T}}{m_0c^2}+1$
P	$\frac{m_0\beta c}{\sqrt{1-\beta^2}}$	P	$\frac{\mathcal{E}}{c}\sqrt{1-\left(\frac{m_0c^2}{\mathcal{E}}\right)^2}$	$mc\sqrt{\frac{\mathcal{T}}{m_0c^2}\left(\frac{\mathcal{T}}{m_0c^2}+2\right)}$
$\mathcal{E}$	$\frac{m_0c^2}{\sqrt{1-\beta^2}}$	$\sqrt{(m_0c)^2+P^2}$	$\mathcal{E}$	$\mathcal{T}+m_0c^2$
$\mathcal{T}$	$m_0c^2\left(\frac{1}{\sqrt{1-\beta^2}}-1\right)$	$m_0c^2\left(\sqrt{1+\left(\frac{P}{m_0c}\right)^2}-1\right)$	$\mathcal{E}-m_0c^2$	$\mathcal{T}$
β	β	$\frac{P}{\sqrt{P^2+(m_0c)^2}}$	$\sqrt{1-\left(\frac{\mathcal{E}}{m_0c^2}\right)^{-2}}$	$\frac{\sqrt{\frac{\mathcal{T}}{m_0c^2}\left(\frac{\mathcal{T}}{m_0c^2}+2\right)}}{\frac{\mathcal{T}}{m_0c^2}+1}$

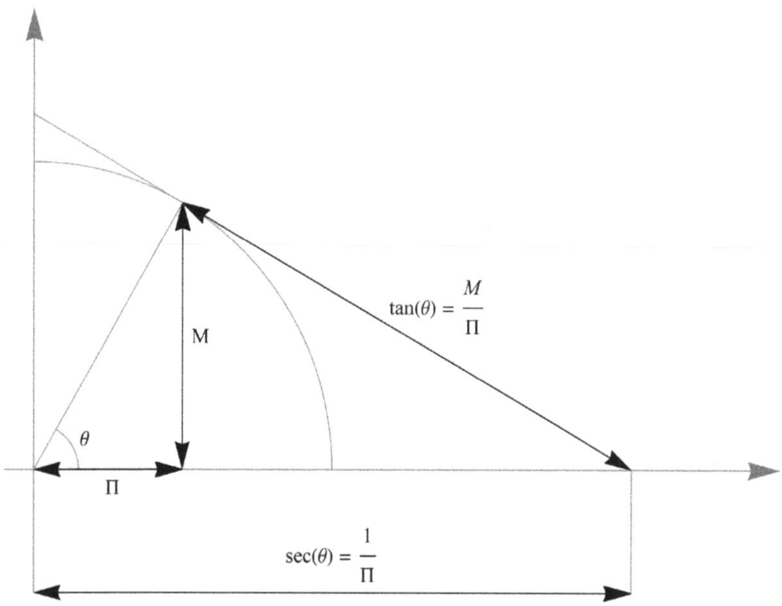

$$\Pi = \frac{pc}{E+mc^2} = \frac{\gamma\beta}{\gamma+1}$$

$$M = \sqrt{\frac{2mc^2}{E+mc^2}} = \sqrt{\frac{2}{\gamma+1}}$$

$$\Pi^2 + M^2 = 1$$

$$\cos(\theta) = \sqrt{\frac{T}{E+mc^2}}$$

$$\sin(\theta) = \sqrt{\frac{2mc^2}{E+mc^2}}$$

$$T = 2mc^2[\cot(\theta)]^2,$$

$$E = mc^2\frac{1+[\cos(\theta)]^2}{[\sin(\theta)]^2}$$

$$\beta = \frac{2\cos(\theta)}{1+[\cos(\theta)]^2}$$

electric field, the energy gained is $\mathcal{T}_0 = 1.6 \times 10^{-19}$ J = 1 eV, and the electron rest energy compared to $\mathcal{T}_0$ is

$$\mathcal{E}_0 \cong 0.511 \times 10^6 \text{ eV} = 0.511 \text{ MeV}. \tag{1.14}$$

MeV $\equiv 10^6$ eV will be the reference unit we will adopt in the following to refer to the mass of particles and to the relevant energy.

Since the first equation of (1.13) is recognized as the relativistic version of the theorem of kinetic energy, we can conclude that in general the variation of the relativistic factor can be associated with the work done on the particle, namely

$$\gamma_2 - \gamma_1 = \frac{e\vec{E}}{\mathcal{E}_0} \cdot (\vec{r}_2 - \vec{r}_1). \tag{1.15}$$

In this introductory section we have provided a very naïve idea of how concepts ranging from classical to relativistic mechanics should be embedded to treat the interaction of a charged particle with an external field. We have solved an elementary problem in mechanics, fully equivalent to that of a particle driven by an external force. Within such a context the concept of the interaction particle field is not transparent. In the following sections we will reach a more convincing picture by considering less idealized model configurations.

1.2 A planar diode and the Child–Langmuir law

In this section we will make a step further toward understanding the field–particle interaction, which we mentioned earlier. The example we have addressed in the introductory section assumes the existence of a static field, without any spatial inhomogeneity. Such a field is that existing between two infinitely extended conducting planes, separated by a distance, d, and brought to different potentials (see figure 1.2). The lower plate (the cathode) is at $V = 0$ and the upper side (the anode) at $V = V_0$. The transverse homogeneity of the field is ensured by the infinite extension of the conducting planes, determining the absence of fringes.

In figure 1.2 we have sketched the geometry of the emission process and the transition from the cathode to the anode. Inside the metal the electrons do not

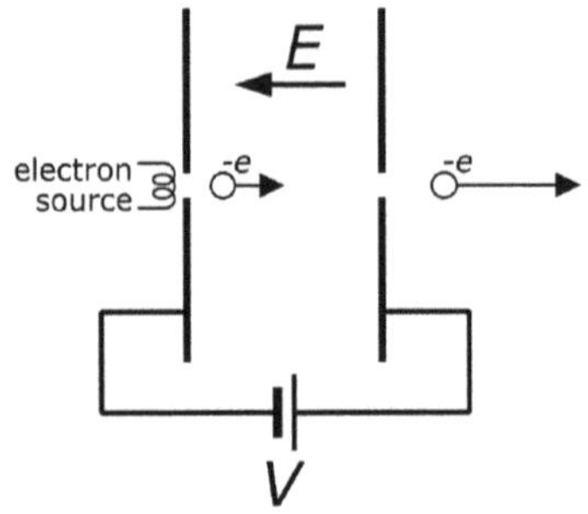

Figure 1.2. Sketch of the geometry of the emission process and the transition from the cathode to the anode. This image has been obtained by the author(s) from the Wikimedia website https://commons.wikimedia.org/wiki/File:Electron_gun.svg, where it is stated to have been released into the public domain. It is included within this article on that basis.

experience any electric field and the forces pulling the electrons inside the conductor are those due to the image charges ('seen' by the electrons outside the conductor), which act as confining forces (for further comments see section 1.15). Thermal agitation can however provide sufficient energy to overcome the potential well and the electrons, expelled via thermionic emission with almost zero velocity at the metal surface, are accelerated toward the anode. Although they are only loosely summarized here, the thermionic emission and the associated laws will be extensively discussed in this and the forthcoming chapters.

In an emission and transport process of the type we have described, the electrons extracted from the cathode move in the forward direction, with velocity determined by energy conservation (we limit ourselves to the non-relativistic case)

$$\frac{1}{2}m\dot{x}^2 - eV_0 = 0. \tag{1.16}$$

The electron current flowing from the cathode is characterized by a current density and our goal is that of understanding the interplay between the diode potential V_0, the charge e and the current density J, whose vector associated with the charge density ρ is

$$J = -\dot{x}\rho, \tag{1.17}$$

and points in the opposite direction of the electron motion. According to (1.16) we find

$$\rho = -\frac{J}{\sqrt{\frac{2e}{m}V}} \tag{1.18}$$

Using the standard relation between the potential and electric field

$$\vec{E} = -\vec{\nabla}V, \tag{1.19}$$

according to the Gauss law, we obtain for the potential satisfying the Poisson equation

$$\vec{\nabla}^2 V = -\frac{\rho}{\varepsilon_0}. \tag{1.20}$$

We can now combine the last equation with the definition of charge density given in (1.18) and as the field is transversally homogeneous we can reduce (1.20) to a one-dimensional problem corresponding to the following second order non-linear equation (the apex denotes the derivative with respect to x)

$$\begin{aligned} V'' &= \alpha V^{-\frac{1}{2}}, \\ \alpha &= \frac{1}{\varepsilon_0}\sqrt{\frac{m}{2e}}J. \end{aligned} \tag{1.21}$$

The initial conditions for (1.21) are imposed by noting that the field and its derivative are zero at the interface of the conductor, namely

$$
\begin{aligned}
V(0) &= 0 \\
V'(0) &= 0.
\end{aligned} \tag{1.22}
$$

We can integrate (1.21) by successive steps. We first multiply both sides by V', thus finding

$$
V'V'' = \alpha V'V^{-\frac{1}{2}}. \tag{1.23}
$$

This can be reduced to quadrature by noting that

$$
\begin{aligned}
V'V'' &= \frac{1}{2}\frac{d}{dx}[V'^2 - V'(0)^2], \\
V'V^{-\frac{1}{2}} &= 2\frac{d}{dx}\left[\sqrt{V} - \sqrt{V(0)}\right].
\end{aligned} \tag{1.24}
$$

Accordingly we find

$$
\frac{1}{2}V'^2 = 2\alpha\sqrt{V}. \tag{1.25}
$$

Thus obtaining the differential equation

$$
V' = 2\sqrt{\alpha}(V)^{\frac{1}{4}}. \tag{1.26}
$$

A second integration finally yields

$$
V = \left(\frac{9}{4}\frac{J}{\varepsilon_0}\sqrt{\frac{m}{2e}}\right)^{\frac{2}{3}} x^{\frac{4}{3}}. \tag{1.27}
$$

This represents the *Child–Langmuir law*, whose physical meaning will be discussed below.

In the exercises we will see how the Child–Langmuir law can be derived from simple dimensional considerations; now before proceeding further we will check whether (1.27) is dimensionally correct. To this aim we recall that, in the chosen units, the dimensions of the electric permittivity are

$$
[\varepsilon_0] = \frac{[Q^2]}{[M\ L^3T^{-2}]}, \qquad [J] = [Q\ T^{-1}L^{-2}], \tag{1.28}
$$

and therefore those of the rhs of (1.27) are

$$
\begin{aligned}
\left[\left(\frac{J}{\varepsilon_0}\sqrt{\frac{m}{2\,e}}\right)^{\frac{2}{3}} x^{\frac{4}{3}}\right] &= \left[Q\ T^{-1}L^{-2}\left(\frac{M\ L^3T^{-2}}{Q^2}\right)M^{\frac{1}{2}}Q^{-\frac{1}{2}}\right]^{\frac{2}{3}}\left[L^{\frac{4}{3}}\right] \\
&= \left[Q^{-\frac{3}{2}}M^{\frac{3}{2}}T^{-3}L\right]^{\frac{2}{3}}\left[L^{\frac{4}{3}}\right] = \frac{[ML^2T^{-2}]}{[Q]} = [V].
\end{aligned} \tag{1.29}
$$

According to (1.27) the potential of the diode is modified by the presence of the inside flowing current. The electron equations of motion which can be obtained from (1.27) are therefore different from those derived in the previous section, indeed by embedding (1.1) and (1.19) we find (the dots denote the derivative with respect to time)

$$\begin{aligned} m\ddot{x} &= e\frac{\partial V}{\partial x} = \frac{4}{3}e\left(\frac{9}{4}\frac{J}{\varepsilon_0}\sqrt{\frac{m}{2e}}\right)^{\frac{2}{3}} x^{\frac{1}{3}}, \\ x_0 &= 0,\ \dot{x}_0 = 0. \end{aligned} \tag{1.30}$$

The integration of the above differential equation proceeds as noted earlier and indeed we obtain

$$\begin{aligned} \dot{x} &= \sqrt{\frac{e}{m}}\left(9\frac{J}{\varepsilon_0}\sqrt{\frac{1}{4}\frac{m}{e}}\right)^{\frac{1}{3}} x^{\frac{2}{3}}, \\ x(t) &= \frac{1}{6}\frac{e}{m}\frac{J}{\varepsilon_0}t^3. \end{aligned} \tag{1.31}$$

If otherwise we solve (1.27) for the current density, we obtain

$$J = \frac{4}{9}\frac{\varepsilon_0}{d^2}\sqrt{\frac{2\,e}{m}}\,V_A^{\frac{3}{2}}, \tag{1.32}$$

yielding the maximum current density, which can be obtained (at fixed geometry and potential) from the cathode.

The previous comments offer a twofold understanding of the physical content of the Child–Langmuir law, further discussion will be provided at the end of this chapter.

We have provided an elementary picture of how field and motion equations should be embedded to obtain a relation which has played a significant role in the theory of vacuum tube devices; in the following section we make a further step toward an actual device by discussing the very first model of the klystron device.

1.3 The klystron concept

The little we have learned so far allows the understanding of the working principle of the klystron and important mechanisms such as current bunching, which underlies all the coherent emission processes we will describe in this book.

In figure 1.3 we show a schematic layout of a klystron in which an electron gun provides a non-relativistic beam of electrons, which are then sent into a cavity where an oscillating electromagnetic field is present. In this cavity the field induces a velocity modulation of the electron beam, which propagates along a straight path leading to a second cavity. In this *drift* section the velocity modulation transforms into a density modulation, causing an amplification of the modulating field. The electron beam is then sent into a collector where it is damped, while the radiation is extracted from the output cavity.

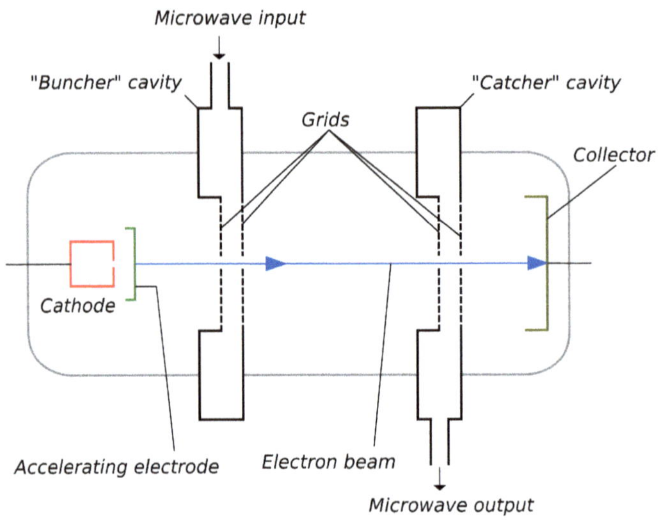

Figure 1.3. Schematic layout of a klystron.

From the energetic point of view we can say that part of the kinetic energy of the electrons is transferred to the electromagnetic field. Within this context an important quantity is the electron-beam power, which is expressible in terms of its current and energy. By noting that the beam kinetic energy can be written as $\mathcal{T} = QV$ with Q being the beam charge, we can express the associated power as

$$\mathcal{P} = \frac{d}{dt}\mathcal{T} = V\frac{d}{dt}Q = VI, \text{ in practical units } \mathcal{P}[\mathbb{W}] = I[\mathbb{A}]\mathcal{T}[e\mathbb{V}], \tag{1.33}$$

where I is the electron-beam current. The relative amount of power lost by the electrons toward the field

$$\eta = \frac{\Delta\mathcal{P}}{\mathcal{P}} \tag{1.34}$$

defines the *efficiency* of the device.

In the gridded region the electrons are brought to the velocity specified by

$$u_0 = \sqrt{\frac{2eV_0}{m_e}} \tag{1.35}$$

Inside the input cavity they interact with the oscillating field $V_1 \sin(\omega t)$. Therefore, by the conservation of energy, the kinetic energy will be provided by the sum of two terms in which the first is the initial kinetic energy and the second is due to the interaction with the field,

$$\frac{1}{2}m_e u^2 = \frac{1}{2}m_e u_0^2 + eV_1 \sin(\omega t), \tag{1.36}$$

where the amplitude of the oscillating field is that effectively coupled in the cavity by the radio-frequency (RF) circuit.

From (1.36) and (1.35) we find for the velocity

$$u = u_0\sqrt{1 + \frac{V_1}{V_0}\sin(\omega t)} \tag{1.37}$$

and, if $V_1 \ll V_0$, by expanding the square root at the lowest order we obtain

$$u \cong u_0\left(1 + \frac{V_1}{2V_0}\sin(\omega t)\right). \tag{1.38}$$

The electron velocity has undergone a modulation and, therefore, in the drift section the electrons that have undergone acceleration tend to overcome those that have been decelerated.

To describe quantitatively what we have just put in words, we assume that the gap of the input cavity is sufficiently small to guarantee the assumption that the time of the arrival time to and exit time from the cavity are the same and will be denoted by t_1. Accordingly, if we indicate by l the drift length before the second cavity, we obtain that the electrons will be at the center of the second gap in a time

$$t_2 = t_1 + \frac{l}{u} = t_1 + \frac{l}{u_0\left(1 + \frac{V_1}{2V_0}\sin(\omega t_1)\right)} \cong t_1 + T - X\frac{\sin(\omega t_1)}{\omega}, \tag{1.39}$$

where

$$T = \frac{l}{u_0},\ X = \frac{\omega T}{2}\frac{V_1}{V_0}. \tag{1.40}$$

It is evident that the modulation in the arrival times produces an analogous modulation in current

$$I_2 = \frac{dQ}{dt_2} = \frac{dQ}{dt_1\left(\frac{dt_2}{dt_1}\right)} \cong \frac{I_0}{1 - X\cos(\omega t_1)}, \tag{1.41}$$

furthermore since

$$\begin{aligned} \omega t_2 &\cong \omega t_1 + \vartheta - X\sin(\omega t_1), \\ \vartheta &= \omega T, \end{aligned} \tag{1.42}$$

we also find that the interaction has induced a phase shift and a phase modulation.

We can understand qualitatively the current modulation by performing a Fourier series expansion of the current in the second cavity, accordingly we find

$$I_2 = I_0 + \sum_n (c_n\cos(n(\omega t_2 - \vartheta)) + s_n\sin(n(\omega t_2 - \vartheta))), \tag{1.43}$$

with the relevant coefficients being provided by

$$c_n = \frac{1}{\pi}\int_{-\pi+\vartheta}^{\pi+\vartheta} I_2 \cos\left[n\omega\left(t_2 - \frac{\vartheta}{\omega}\right)\right]d(\omega t_2),$$
$$s_n = \frac{1}{\pi}\int_{-\pi+\vartheta}^{\pi+\vartheta} I_2 \sin\left[n\omega\left(t_2 - \frac{\vartheta}{\omega}\right)\right]d(\omega t_2). \quad (1.44)$$

The expansion has been done in such a way that the constant phase shift ϑ can be eliminated. Furthermore by substituting (1.42) into (1.44) and since $I_2 \cong I_0 + \beta X$ we find

$$c_n \cong \frac{I_0}{\pi}\int_{-\pi}^{\pi} \cos\left[n\omega(t_1 - X\sin(\omega t_1))\right]d(\omega t_1)$$
$$s_n = \frac{I_0}{\pi}\int_{-\pi}^{\pi} \sin\left[n\omega(t_1 - X\sin(\omega t_1))\right]d(\omega t_1). \quad (1.45)$$

It is now evident that $s_n = 0$, since the relevant integrand is an odd function of time, while the integral representation of Bessel functions ($J_n(x) = \frac{1}{\pi}\int_0^{\pi} \cos[x\sin\vartheta - n\vartheta]d\vartheta$) yields

$$c_n = 2I_0 J_n(nX). \quad (1.46)$$

We obtain therefore

$$I_2 \cong I_0\left(1 + 2\sum_{n=1}^{\infty} J_n(nX)\cos(n(\omega t_2 - \vartheta))\right). \quad (1.47)$$

The current modulation occurs therefore as reported in figure 1.4 where we have reported I_2 versus time for different values of X(called the bunching parameter, see

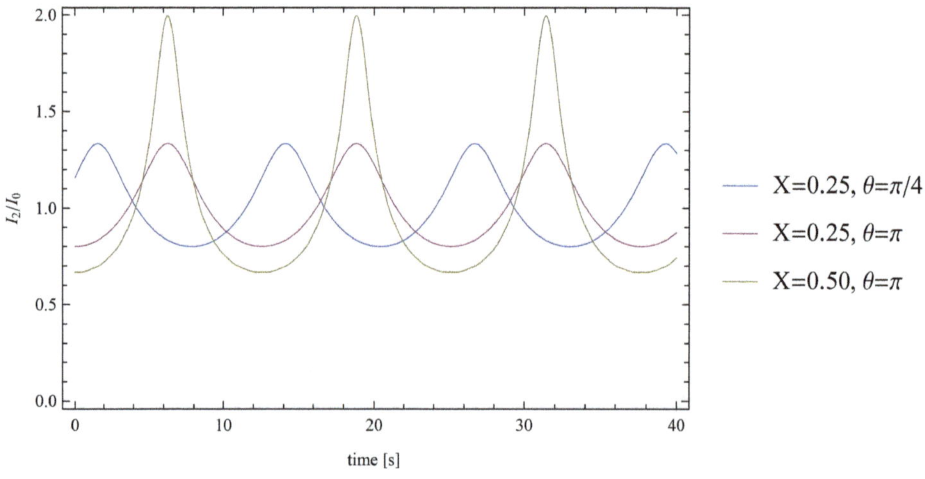

Figure 1.4. Current in the second cavity versus time for different values of the bunching parameter X.

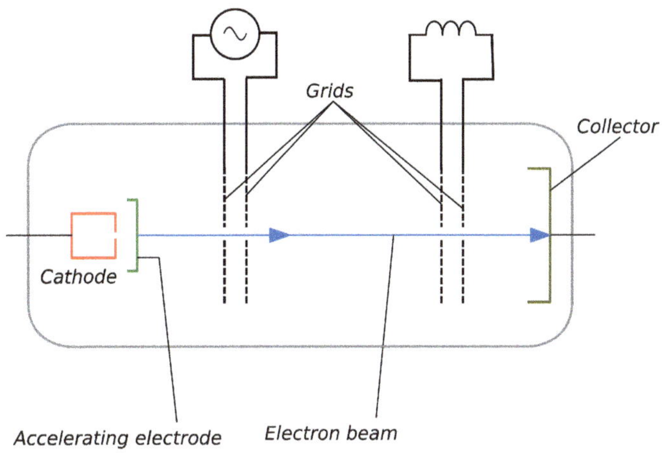

Figure 1.5. Equivalent circuit of a cavity resonator.

(1.40)) and phase ϑ. According to the above result the electrons enter the second cavity in bunches (klystern in German, hence the name of the device).

As further commented on in chapter 2, it is important to keep in mind (see figure 1.5) that a cavity resonator and a conventional circuit are conceptually equivalent and indeed the cavity can be schematized as a capacitor and an inductor. Therefore, the bunched electron current stream excites in the second cavity, called the catcher, a time-dependent field according to the conventional laws of electromagnetic induction.

By keeping now in mind what we have mentioned earlier on the electron-beam power, we can estimate the power delivered by the

$$\mathcal{P}_{\mathcal{K}} \cong \overline{I_2 V_2} = \overline{I_0 V_0 \left(1 + 2\sum_{n=1}^{\infty} J_n(nX) \cos(n(\omega t_2 - \vartheta)) \right) \cos(\omega t_2 - \vartheta)}, \tag{1.48}$$

where the average is taken either over the phase or period, thus finally obtaining

$$\mathcal{P}_k \cong J_1(X)\mathcal{P}_0 \tag{1.49}$$

The efficiency of the power delivered by the electrons to the cavity on the first harmonic is determined by the first order bunching coefficient, linked to the cylindrical Bessel function of order 1, exhibiting the first maximum at $x = 1.84$ where $J_1(1.84) \cong 0.58$, we can accordingly state that the efficiency of a two cavity klystron is 58%.

The high level of efficiency and thus large amount of energy delivered to the cavity can be understood as follows. In the first cavity the incoming electrons are continuously distributed, they are therefore accelerated and decelerated in almost equal number. In the second cavity they are bunched on a spatial extension which is a wavelength of the oscillating field. The decelerated bunch loses all the power associated with the spatial bunching, which as we saw is linked to its bunching coefficient.

We have described so far a klystron operating in the so-called *ballistic regime*, namely the space charge forces are ignored and the faster particles just overcome the slower ones. In the reality, space charge effects counteract the purely ballistic bunching and the efficiency may not grow that large (however, we will see in the following that using multi-cavity configurations and bunching optimization criteria, space charge forces can be exploited to obtain greater efficiency).

The discussion provided in this section is very inaccurate and contains sloppily introduced concepts, which will be more carefully discussed in forthcoming chapters. Here we simply want to see how what we had learned so far can be assembled to describe an actual device.

1.4 Charged particle motion in combined electric and magnetic fields (the non-relativistic case)

So far we have grappled on the surfaces of problems dealing with field charge interaction and have learned something without mounting any particular mathematical artillery. In this section we proceed further following the same strategy, and will deal with the full Lorentz equation by considering the motion of charged particles ruled by their combined action.

Just to fix the elements of the game we cast the non-relativistic version of (1.1) in the form of first order non-homogeneous equation, namely

$$\begin{gathered}\frac{d}{dt}\vec{v} = -\,\vec{\Omega} \times \vec{v} + \vec{\kappa},\\ \vec{\Omega} = \frac{q\vec{B}}{m},\ \vec{\kappa} = \frac{q\vec{E}}{m},\end{gathered} \tag{1.50}$$

and assume that both fields are static and homogeneous.

Equation (1.50) is equivalent to a first order non-homogeneous differential equation and therefore its formal solution reads (note that the solution of $y' + \alpha y = \beta$, $y(0) = y_0$ reads $y = e^{-\alpha x} y_0 + e^{-\alpha x} \int_0^x e^{\alpha x'} \beta\, dx'$)

$$\vec{v} = e^{-[\hat{\Omega}]t}\vec{v}_0 + e^{-[\hat{\Omega}]t} \int_0^t e^{[\hat{\Omega}]t'} \vec{\kappa} dt', \tag{1.51}$$

where $\vec{v}_0$ is the initial velocity and $e^{[\hat{\Omega}]t}$ is a vector evolution operator, whose action on $\vec{v}_0$ should be understood by performing a Taylor expansion of the evolution operator, thus finding

$$\begin{aligned}
&e^{[\hat{\Omega}]t}\vec{v}_0 = \sum_{n=0}^{\infty} t^n \frac{[\hat{\Omega}]^n}{n!} \vec{v}_0,\\
&[\hat{\Omega}]^0 = \hat{1},\\
&[\hat{\Omega}]^1 = \vec{\Omega} \times ,\\
&[\hat{\Omega}]^2 = \vec{\Omega} \times (\vec{\Omega} \times ,\\
&[\hat{\Omega}]^3 = \vec{\Omega} \times (\vec{\Omega} \times (\vec{\Omega} \times ,\\
&\ldots.
\end{aligned} \tag{1.52}$$

The use of the cyclical properties of the vector product allows one to obtain a solution in closed form as discussed later. Equation (1.51) can be handled such as to obtain all the information we need to study the motion of a charged particle under the combined effect of (static) electric and magnetic fields. It is, however, more instructive to decompose the problem, by considering particular solutions of (1.50). We consider first the case of the absence of the electric field ($\vec{\kappa} = 0$) and a magnetic field pointing in the z-direction; in this case we have

$$\frac{d}{dt}\vec{v} = \begin{vmatrix} \vec{i} & \vec{j} & \vec{k} \\ 0 & 0 & -\Omega \\ v_x & v_y & v_z \end{vmatrix}, \qquad \Omega = \frac{qB}{m} \equiv \text{Larmor frequency.} \tag{1.53}$$

The motion is simply two-dimensional and can be reduced to

$$\frac{d}{dt}\begin{pmatrix} v_x \\ v_y \end{pmatrix} = \Omega\hat{S}\begin{pmatrix} v_x \\ v_y \end{pmatrix}, \qquad \hat{S} = \begin{pmatrix} 0 & 1 \\ -1 & 0 \end{pmatrix}, \tag{1.54}$$

where $\hat{S}$ denotes the two-dimensional symplectic matrix.

The use of the evolution operator method applied to (1.54) yields the solution

$$\begin{pmatrix} v_x \\ v_y \end{pmatrix} = e^{\Omega t\,\hat{S}}\begin{pmatrix} v_{x,0} \\ v_{y,0} \end{pmatrix}. \tag{1.55}$$

Furthermore, by exploiting the following property of the matrix (sometimes called *symplectic*)

$$e^{\alpha\hat{S}} = \cos(\alpha)\hat{1} + \sin(\alpha)\hat{S}, \qquad \hat{1} = \begin{pmatrix} 1 & 0 \\ 0 & 1 \end{pmatrix}. \tag{1.56}$$

(see section 1.15 for further details) we end up with

$$\begin{pmatrix} v_x \\ v_y \end{pmatrix} = \begin{pmatrix} \cos(\Omega t) & -\sin(\Omega t) \\ \sin(\Omega t) & \cos(\Omega t) \end{pmatrix}\begin{pmatrix} v_{x,0} \\ v_{y,0} \end{pmatrix}. \tag{1.57}$$

To be consistent with the emerging physics, we redefine the initial condition in a polar form, by introducing a phase δ,

$$\begin{pmatrix} v_{x,0} \\ v_{y,0} \end{pmatrix} = \begin{pmatrix} v_\perp \cos(\delta) \\ v_\perp \sin(\delta) \end{pmatrix}. \tag{1.58}$$

Thus writing the solution of our problem as

$$\begin{aligned} v_x &= v_\perp \cos(\Omega t + \delta), \\ v_y &= v_\perp \sin(\Omega t + \delta), \\ v_z &= v_{z,0}, \\ V^2 &= v_\perp^2 + v_z^2, \end{aligned} \tag{1.59}$$

finally, a further integration provides the particle trajectory as (see figure 1.6 for further comments and trajectory examples)

$$\begin{aligned} r_x - r_{0,x} &= \mathcal{R}_L[\sin(\Omega t + \delta) - \sin(\delta)], \\ r_y - r_{0,y} &= \mathcal{R}_L[\cos(\Omega t + \delta) - \cos(\delta)], \\ z &= z_0 + \sqrt{V^2 - v_\perp^2}\, t, \\ \mathcal{R}_L &= \frac{m v_\perp}{qB} \equiv \text{Larmor radius}. \end{aligned} \tag{1.60}$$

It seems that there is not too much to learn from the previous identities, except that the velocity is conserved during the evolution and that the charged particle executes a helical trajectory of radius $\mathcal{R}_L$ and pitch $h = \frac{2\pi}{\Omega} v_z$ around the axis fixed by the z motion and a guiding center, which in the transverse plane is specified by

$$C \equiv \left(r_{0,x} + \mathcal{R}_L \sin(\delta),\ r_{0,y} + \mathcal{R}_L \cos(\delta)\right). \tag{1.61}$$

We have considered a highly idealized problem in which the magnetic field is static, without any field inhomogeneity, and there is no transition from a region with and without magnetic field.

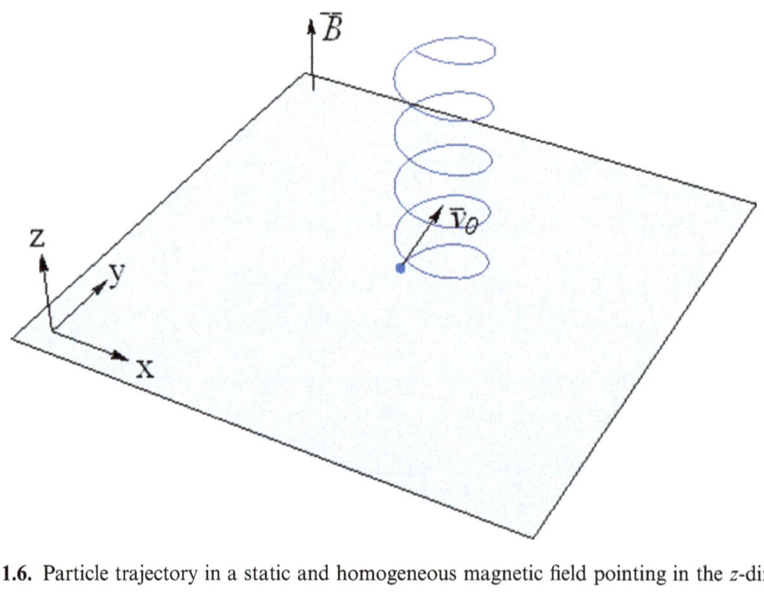

Figure 1.6. Particle trajectory in a static and homogeneous magnetic field pointing in the z-direction.

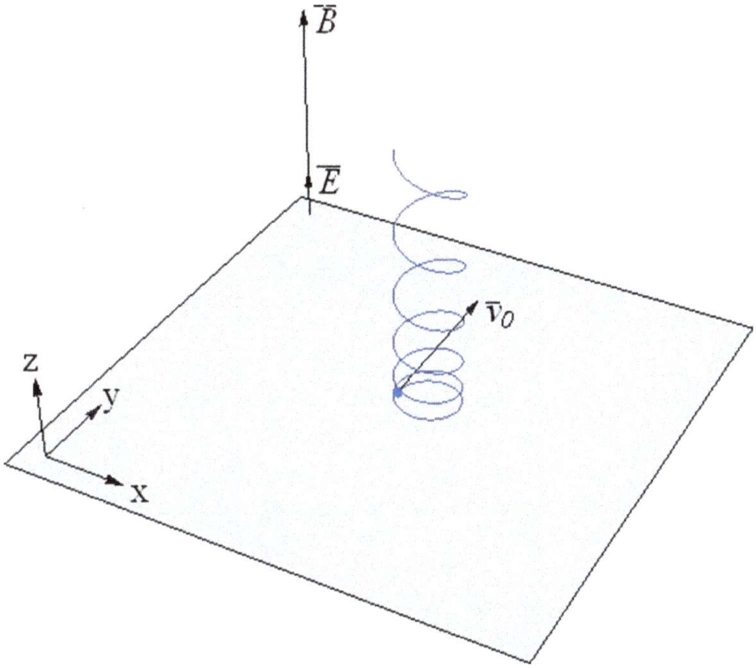

Figure 1.7. Particle trajectory in static and homogeneous magnetic and electric fields pointing in the z-direction.

The second (idealized) condition is that both the electric and magnetic fields are parallel to the z-axis and the charged particle is assumed to have an initial velocity with longitudinal and transverse components (see figure 1.7). In the non-relativistic case the problem is almost straightforward and reduces to the two decoupled equations (the charge has been chosen in such a way that the particle accelerates in the positive z-direction)

$$\begin{aligned} \frac{d}{dt}\begin{pmatrix} v_x \\ v_y \end{pmatrix} &= \Omega \hat{S} \begin{pmatrix} v_x \\ v_y \end{pmatrix}, \\ \frac{d}{dt} v_z &= \kappa, \end{aligned} \tag{1.62}$$

whose parametric solution is provided by (1.60) with the third component replaced by

$$z = z_0 + \sqrt{V^2 - v_\perp^2}\, t + \frac{1}{2}\kappa t^2. \tag{1.63}$$

The motion is again provided by a helical trajectory with a non-constant pitch, as illustrated in figure 1.7.

Let us now consider the case of electric and magnetic fields perpendicular to each other, according to the vector arrangement in figure 1.8. The motion is slightly more complicated than the examples discussed so far.

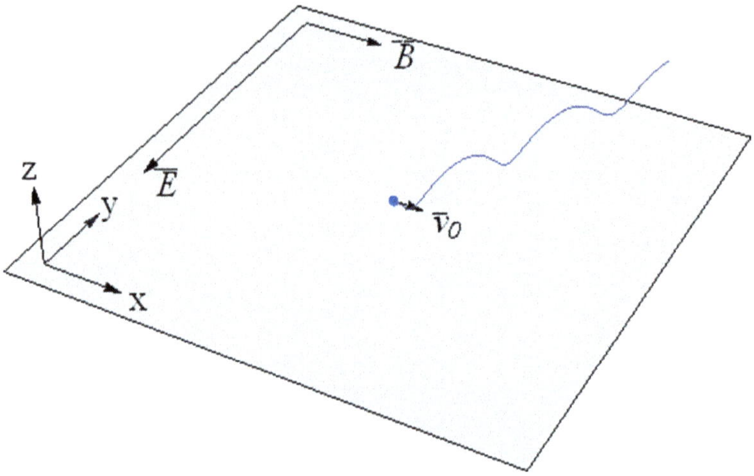

Figure 1.8. Particle trajectory in a static and homogeneous magnetic and electric fields pointing in the x-direction and y-direction, respectively.

The problem can, however, be solved using the same naïve mathematical tools exploited so far.

According to figure 1.8 we write

$$\begin{aligned}\frac{d}{dt}v_x &= 0,\\ \frac{d}{dt}v_y &= \Omega v_z + \kappa,\\ \frac{d}{dt}v_z &= -\Omega v_y.\end{aligned} \tag{1.64}$$

If we introduce the complex velocity

$$\mathcal{V} = v_y - iv_z \tag{1.65}$$

we can combine (1.64) into the single differential equation

$$\frac{d}{dt}\mathcal{V} = i\Omega\mathcal{V} + \kappa, \tag{1.66}$$

which is a first order equation with a non-homogeneous term, and the relevant solution is

$$\mathcal{V} = e^{i\Omega t}\mathcal{V}_0 + \frac{e^{i\Omega t} - 1}{i\Omega}\kappa. \tag{1.67}$$

A further integration yields the trajectory

$$\mathcal{R} = \mathcal{R}_0 + \frac{e^{i\Omega t} - 1}{i\Omega}\mathcal{V}_0 + i\frac{\kappa}{\Omega}t - \frac{\kappa}{\Omega^2}e^{i\Omega t}. \tag{1.68}$$

The additional term

$$\mathcal{V}_d = \frac{\kappa}{\Omega} = \frac{E}{B} \tag{1.69}$$

represents the drift velocity directed along the z-direction (see figure 1.8) and in vector terms is provided by

$$\vec{\mathcal{V}}_d \propto \frac{\vec{E} \times \vec{B}}{B^2}. \tag{1.70}$$

We are now sufficiently confident with the physics of the problem to appreciate the consequences of (1.51), which yields for the velocity (see section 1.15 for further details)

$$\begin{aligned}
&\vec{v} = \cos(\Omega t)\vec{v}_0 + \left[\frac{\sin(\Omega t)}{\Omega}\right]\vec{\kappa} + (\vec{n} \cdot \vec{l})\vec{n} - \vec{n} \times \vec{m}, \\
&\vec{n} = \frac{\vec{B}}{|\vec{B}|}, \\
&\vec{l} = (1 - \cos(\Omega t))\vec{v}_0 + \left[t - \frac{\sin(\Omega t)}{\Omega}\right]\vec{\kappa}, \ \vec{m} = \frac{1}{\Omega}\frac{d}{dt}\vec{l},
\end{aligned} \tag{1.71}$$

and for the trajectory

$$\begin{aligned}
&\vec{r} = \vec{r}_0 + \frac{\sin(\Omega t)}{\Omega}\vec{v}_0 + \left[\frac{1 - \cos(\Omega t)}{\Omega^2}\right]\vec{\kappa} + (\vec{n} \cdot \vec{o})\,\vec{n} - \frac{1}{\Omega}\vec{n} \times \vec{l}, \\
&\vec{o} = \left[t - \frac{\sin(\Omega t)}{\Omega}\right]\vec{v}_0 + \left[\frac{t^2}{2} - \frac{1 - \cos(\Omega t)}{\Omega^2}\right]\vec{\kappa}.
\end{aligned} \tag{1.72}$$

It is worth stressing again that the vector contribution $\vec{n} \times \vec{m}$ arises only when the electric and magnetic fields are not parallel. Such a term appears when the initial velocity and/or the electric field are not parallel to the magnetic field and reads

$$\vec{n} \times \vec{m} = \sin(\Omega t)\frac{\vec{B} \times \vec{v}_0}{|\vec{B}|} - 2\sin^2\left(\frac{\Omega t}{2}\right)\frac{\vec{E} \times \vec{B}}{|\vec{B}|^2} \tag{1.73}$$

Further comments on the role of this vector and examples of trajectories can be found at the end of the chapter.

1.5 Charged particle motion in combined electric and magnetic fields (the relativistic case)

In the previous section we have treated the problem of the motion of a charged particle in static electric and magnetic fields using non-relativistic mechanics. The inclusion of relativistic effects does not introduce any conceptual problem; by taking into account the Lorentz factor, the relativistic version of (1.50) is written

$$\frac{d}{dt}(\gamma\vec{v}) = -\vec{\Omega} \times \vec{v} + \vec{\kappa}, \qquad \gamma = \frac{1}{\sqrt{1 - \frac{|\vec{v}|^2}{c^2}}}. \tag{1.74}$$

It is therefore evident that, unlike the non-relativistic case, it is not amenable for straightforward analytical solutions.

By keeping the scalar product with respect to $\vec{v}$ of both sides of (1.74), we find (note that $-\vec{v} \cdot (\vec{\Omega} \times \vec{v}) = 0$)

$$\vec{v} \cdot \frac{d}{dt}(\gamma\vec{v}) = \vec{v} \cdot \vec{\kappa}, \tag{1.75}$$

which can be further simplified by noting that (see section 1.15)

$$\vec{v} \cdot \frac{d}{dt}(\gamma\vec{v}) = \frac{1}{2\gamma}\frac{d}{dt}(\gamma^2\,|\vec{v}|^2) = c^2\frac{d}{dt}\gamma. \tag{1.76}$$

It is therefore convenient to couple the equation accounting for the variation of the relativistic factor

$$c^2\frac{d}{dt}\gamma = \vec{v} \cdot \vec{\kappa} \tag{1.77}$$

to the vector equation (1.74), which can also be cast as

$$\frac{d}{dt}\vec{v} = -\left(\left[\frac{\vec{\Omega}}{\gamma}\right] + \frac{\dot{\gamma}}{\gamma}\right)\vec{v} + \frac{\vec{\kappa}}{\gamma}. \tag{1.78}$$

(Note that according to the previously stated notation (1.52) $\left[\frac{\vec{\Omega}}{\gamma}\right]\vec{v} = \frac{1}{\gamma}(\vec{\Omega} \times \vec{v})$.)

The solution of (1.78) can be obtained through the usual means, which yield

$$\vec{v} = \hat{U}(t)\left[\vec{v}_0 + \int_0^t \hat{U}^{-1}(t')\frac{\vec{\kappa}}{\gamma(t')}dt'\right], \tag{1.79}$$

where we have defined the evolution operator

$$\hat{U}(t) = e^{-\int_0^t \frac{\dot{\gamma}}{\gamma}dt'}e^{-\int_0^t \frac{dt'}{\gamma(t')}[\vec{\Omega}]} = \frac{\gamma_0}{\gamma}e^{-\int_0^t \frac{dt'}{\gamma(t')}[\vec{\Omega}]}. \tag{1.80}$$

The above result is the best we can do with an analytical procedure, there are however particular cases which are worth discussing since they can disclose interesting physical features.

We consider first the relativistic motion in the presence of the magnetic field only. According to the fact that $|\vec{v}|^2$ is a constant of motion, we can pull out γ from the derivative and obtain the same solution as before by just replacing the Larmor frequency with its relativistic counterpart

$$\Omega_L = \frac{\Omega}{\gamma} \tag{1.81}$$

In the case in which the motion is ruled by parallel electric and magnetic fields, we can cast the relevant equations in the form

$$\begin{aligned} &\frac{d}{dt}(\gamma \mathcal{V}_\perp) = i\Omega\, \mathcal{V}_\perp, \\ &\frac{d}{dt}(\gamma v_z) = \kappa, \\ &\mathcal{V}_\perp = v_x + i v_y. \end{aligned} \tag{1.82}$$

The first equation ensures that the transverse momentum is a conserved quantity, while from the second we find $(\gamma v_z|_{t=0} = 0)$

$$\gamma v_z = \kappa t. \tag{1.83}$$

The total momentum is therefore provided by

$$\begin{aligned} &\frac{p}{m_0} = \sqrt{\frac{p_\perp^2}{m_0^2} + (\kappa t)^2}\,, \\ &m_0 \equiv \text{rest mass}. \end{aligned} \tag{1.84}$$

By the relativistic definition of momentum we find

$$\gamma|v| = \sqrt{\frac{p_\perp^2}{m_0^2} + (\kappa t)^2}\,, \tag{1.85}$$

which yields for the modulus of the velocity

$$\begin{aligned} &|\vec{v}|^2 = V_\perp^2 + v_z^2, \\ &V_\perp = \frac{\dfrac{p_\perp}{m_0}}{\sqrt{1 + \left(\dfrac{p_\perp}{m_0 c}\right)^2 + \left(\dfrac{\kappa t}{c}\right)^2}}, \\ &v_z = \frac{\left(\dfrac{\kappa t}{c}\right)}{\sqrt{1 + \left(\dfrac{p_\perp}{m_0 c}\right)^2 + \left(\dfrac{\kappa t}{c}\right)^2}}. \end{aligned} \tag{1.86}$$

It is accordingly evident that the Lorentz factor is time-dependent and we can write

$$\gamma = \sqrt{1 + \left(\frac{p_\perp}{m_0 c}\right)^2 + \left(\frac{\kappa t}{c}\right)^2}. \tag{1.87}$$

We can now infer some information on the motion from the previous identities. The transverse velocity in the second equation of (1.86) can be exploited to define the centrifugal force balanced by the magnetic part of the Lorenz force as

$$\frac{V_\perp^2}{\mathcal{R}} = \frac{eBV_\perp}{m_0\gamma}, \tag{1.88}$$

where $\mathcal{R}$ is the curvature radius in the transverse plane of the motion and can be expressed as

$$\mathcal{R} = \frac{m_0\gamma V_\perp}{eB} = \frac{p_\perp}{eB}, \tag{1.89}$$

which, on account of (1.86), can be recognized as a constant of motion. The associated angular velocity is easily shown to be Ω_L which is not conserved and decreases with increasing energy.

The motion is accordingly a kind of helix with a constant radius and decreasing angular velocity. With increasing time the transverse velocity decreases, the longitudinal counterpart approaches to that of light and the motion is essentially a straight line.

On the analytical side we note that the relevant equations of motion are

$$\dot{\mathcal{V}} + \frac{\dot{\gamma}}{\gamma}\mathcal{V} = i\frac{\Omega}{\gamma}\mathcal{V}, \tag{1.90}$$

and can be integrated as

$$\begin{aligned} \mathcal{V} &= \frac{\gamma_0}{\gamma} e^{i\Phi(t)}\,\mathcal{V}_0, \\ \Phi(t) &= \Omega\int_0^t \frac{dt'}{\gamma(t')} \\ &= \frac{c\Omega}{\kappa}\sinh^{-1}\left(\frac{\kappa t}{c\sqrt{1 + \left(\frac{p_\perp}{m_0 c}\right)^2}}\right), \end{aligned} \tag{1.91}$$

this last result confirms the conclusions we have previously qualitatively obtained.

The considerations developed in this section are complemented by the further discussion at the end of the chapter.

1.6 Hamiltonian theory of charged particle motion

In the previous sections we have treated the charged particle motion using essentially the Newton equations and their relativistic extensions. In this section we will exploit a more complete mathematical description allowing the introduction of concepts of crucial importance in transport theory, such as the Liouville theorem, the beam emittance and the brightness.

To make the discussion self-consistent we make a quick review of some pivoting concepts of analytical mechanics.

The starting point of the Hamiltonian description of mechanics is the assumption of the existence of a function characterizing the motion of any mechanical system. In the case of N independent particles, such a function, namely the *Lagrangian*, depends on the trajectories $Q_n(t) = (x_n, y_n, z_n)$, $n = 1 \ldots N$ of the particles and on their velocities $\dot{Q}_n(t)$ will be denoted by

$$\begin{gathered} \mathcal{L}(Q, \dot{Q}; t), \\ Q = \{Q_n(t)\}, \ \dot{Q} = \{\dot{Q}_n(t)\}. \end{gathered} \tag{1.92}$$

The function $\mathcal{L}(Q, \dot{Q}; t)$ plays a major role in analytical mechanics and is exploited to characterize the evolution of a dynamical system. It defines the so-called action integral

$$S = \int_{t_1}^{t_2} \mathcal{L}(Q, \dot{Q}; t)dt, \tag{1.93}$$

specifying the motion, along with the relevant evolution, from times t_1 to t_2. The Hamilton–Mapertuis, or *the least action principle*, states that the integral (1.93) is an extremum. The use of the above statement, along with standard rules from variational calculus, allow the conclusion that the Lagrangian satisfies the differential equation

$$\frac{d}{dt}\frac{\partial}{\partial \dot{Q}_n}\mathcal{L} - \frac{\partial}{\partial Q_n}\mathcal{L} = 0, \tag{1.94}$$

which can be viewed as implicitly containing the Newton laws and completely determines the motion of the system, once the Lagrangian is known.

The generalized momenta and forces associated with the Lagrangian are defined through the following derivatives

$$\begin{aligned} P_n &= \frac{\partial}{\partial \dot{Q}_n}\mathcal{L}, \\ \mathcal{F}_n &= \frac{\partial}{\partial Q_n}\mathcal{L}. \end{aligned} \tag{1.95}$$

The Newton equations follow from (1.94, 1.95) as

$$\mathcal{F}_n = \frac{d}{dt}P_n = \frac{\partial}{\partial Q_n}\mathcal{L}. \tag{1.96}$$

The Lagrangian can be derived through general considerations, which will not be reported here. However, considering a system characterized by velocities and position-dependent forces, the most natural choice is

$$\begin{aligned} \mathcal{L} &= T(\dot{Q}^2) - V(Q), \\ T &= \frac{m}{2}\dot{Q}^2. \end{aligned} \tag{1.97}$$

Accordingly from (1.96),

$$m\ddot{Q} = -\frac{\partial}{\partial Q}V. \tag{1.98}$$

We should, however, pay attention, because things are not always that straightforward and the *canonical* momentum (P_n) does not always coincide with the *kinetic* ($m\dot{q}_n$) momentum.

The *Hamiltonian* function is linked to the Lagrangian by the identity (see the exercise at the end of the chapter)

$$\mathcal{H} = P\dot{Q} - \mathcal{L}, \tag{1.99}$$

which, referring to (1.97), is easily recognized as the total energy of the mechanical system. The equation of motion can be expressed in terms of the Hamiltonian as reported below (see section 1.15)

$$\begin{aligned} \frac{d}{dt}P_n &= -\frac{\partial}{\partial Q_n}\mathcal{H}, \\ \frac{d}{dt}Q_n &= \frac{\partial}{\partial P_n}\mathcal{H}, \end{aligned} \tag{1.100}$$

which in vector notation can also be written as

$$\begin{aligned} \frac{d}{dt}\vec{P} &= -\vec{\nabla}_Q\mathcal{H}, \\ \frac{d}{dt}\vec{Q} &= \vec{\nabla}_P\mathcal{H}. \end{aligned} \tag{1.101}$$

We have so far reported a very quick survey of analytical mechanics (with further details provided in section 1.15), let us therefore try to understand how these notions match with the treatment so far of the charged particle dynamics.

We have so far written the Lorentz equation in terms of the electric and magnetic fields. Being interested in the interaction of a charge with the electromagnetic field, we proceed further by introducing the vector $\vec{A}$ and scalar potentials V allowing the definition of the electric and magnetic fields as

$$\begin{aligned} \vec{E} &= -\frac{\partial}{\partial t}\vec{A} - \vec{\nabla}V, \\ \vec{B} &= \vec{\nabla} \times \vec{A}. \end{aligned} \tag{1.102}$$

The Lorentz equation will therefore be

$$m\frac{d}{dt}\vec{v} = -q\frac{\partial}{\partial t}\vec{A} + q\vec{v} \times (\vec{\nabla} \times \vec{A}) - q\vec{\nabla}V, \tag{1.103}$$

and note that since (the vector product identity is worth stressing $\vec{a} \times (\vec{b} \times \vec{c}) = (\vec{a} \cdot \vec{c})\vec{b} - (\vec{a} \cdot \vec{b})\vec{c}$)

$$\begin{aligned} \frac{d}{dt}\vec{A} &= \frac{\partial}{\partial t}\vec{A} + (\vec{v} \cdot \vec{\nabla})\vec{A}, \\ \vec{v} \times (\vec{\nabla} \times \vec{A}) &= -(\vec{v} \cdot \vec{\nabla})\vec{A}, \end{aligned} \tag{1.104}$$

equation (1.103) can also be written as

$$\frac{d}{dt}(m\,\vec{v} + q\,\vec{A}) = -q\,\vec{\nabla}V. \tag{1.105}$$

This last identity is of paramount importance for our discussion and clarifies the difference between kinetic and canonical momentum. Indeed by setting

$$\vec{P} = m\vec{v} + q\vec{A}, \tag{1.106}$$

we easily infer that the corresponding Hamiltonian can be written as (see section 1.15)

$$\mathcal{H} = \frac{1}{2m}(\vec{P} - q\vec{A})^2 + qV. \tag{1.107}$$

The *canonical* and the *kinetic* momenta are linked by the identity (1.106), usually defined as *minimal coupling*.

The relativistic extension of the previous formalism is easily obtained and indeed the Hamiltonian (1.107) is simply replaced by

$$\mathcal{H} = c\sqrt{(m_0c)^2 + (\vec{P} - q\vec{A})^2} + qV. \tag{1.108}$$

We have insisted on the notion of canonical and kinetic momentum because confusions might be extremely dangerous when applied to charged particle transport.

1.7 Beam transport, Liouville theorem and beam emittance

Before going further we should clarify what we mean by a beam of charged particles (in particular of electrons). They are a flow or stream of particles moving in a given direction determined by the forces acting on them (be they external or internal). Such a system is accordingly subject to conservation laws very similar to those occurring in a fluid.

The forces may be either external or internal, and may be responsible for the symmetry properties of the beam itself. The examples we have considered in this section do not include internal and self-consistent effects.

When talking about beams rather than following the behavior of the individual (indistinguishable) particles, it is important to follow the evolution of global quantities (such as the beam envelope, the divergence, the energy spread...) which characterizes the beam as a whole.

The most suitable representation is the beam distribution $F(P, Q, t)$ in phase space, such a quantity is the particle density in phase space, namely

$$F(P, Q, t) = \frac{1}{N} f(P, Q, t), \\ f(P, Q, t) = \left(\frac{dn}{dP\, dQ} \right), \tag{1.109}$$

with N being the total number of particles, accordingly, since

$$\iint f(P, Q, t) dP\, dQ = N, \tag{1.110}$$

we have that $F(P, Q, t)$ is normalized to unity.

If the beam is ruled by external forces only and other effects, such as intra-beam scattering, radiation losses, particle absorption, etc are not included, the number of particles is a constant of motion and therefore we find

$$\frac{d}{dt} F = \dot{P} \frac{\partial}{\partial P} F + \dot{Q} \frac{\partial}{\partial Q} F + \frac{\partial}{\partial t} F = 0. \tag{1.111}$$

It should be noted that a more appropriate notation is

$$\dot{P} \frac{\partial}{\partial P} \equiv \vec{\dot{P}} \cdot \vec{\nabla}_P, \\ \dot{Q} \frac{\partial}{\partial Q} \equiv \vec{\dot{Q}} \cdot \vec{\nabla}_Q, \tag{1.112}$$

consistent with the fact that the phase space is a six-dimensional space.

Equation (1.111) implies that the density distribution function evolution is ruled by the partial differential equation (PDE) (see (1.100))

$$\frac{\partial}{\partial t} F = \hat{L} F, \\ \hat{L} = \frac{\partial \mathcal{H}}{\partial Q} \frac{\partial}{\partial P} - \frac{\partial \mathcal{H}}{\partial P} \frac{\partial}{\partial Q}, \tag{1.113}$$

where $\hat{L}$ is the Liouville operator, which rules the transport of a non-interacting ensemble of particles ruled by the Hamiltonian $\mathcal{H}$ (the interacting case will be discussed later in this chapter).

It is accordingly important to stress that since we are referring to a phase-space variable, Q, P are canonical variables and the arguments of the density distribution function are specified by canonical variables. A distribution of velocity and position for a problem ruled by a Hamiltonian in which canonical and kinetic momenta do not coincide is not representative of the phase-space density function and is not ruled by the Liouville equation.

The mathematical problem, represented by (1.113), belongs to the family of evolution PDEs, which can be treated with a variety of methods. In very rare cases these methods are analytical in nature and in most cases efficient numerical algorithms can be exploited, as discussed in detail in this and forthcoming chapters.

Just to give an idea of the methods that can be employed in the solution of PDEs involving the Liouville operator, we consider the case of a beam undergoing a drift space, namely a region in which no external forces are present. The Hamiltonian ruling the single particle dynamics and the Liouville operator determining the beam distribution evolution are in this case provided, respectively, by

$$\begin{aligned} \mathcal{H} &= \frac{P^2}{2m}, \\ \hat{L} &= -\frac{P}{m}\frac{\partial}{\partial Q}, \end{aligned} \tag{1.114}$$

we find therefore

$$\frac{\partial}{\partial t}F(P,\, Q;\, t) = -\frac{P}{m}\frac{\partial}{\partial Q}F(P,\, Q;\, t). \tag{1.115}$$

The solution of (1.115) requires at least an 'initial condition' that will be assigned as

$$F(P,\, Q;\; 0) = F_0(P,\, Q). \tag{1.116}$$

The variables P, Q are canonical coordinates and within the present context they are provided by the kinetic momenta and by the spatial coordinates. The relevant phase space is accordingly a six-dimensional space (to be more precise, N being the particles constituting the beam, the phase space is $6N$). Since there is no coupling between the different spaces we can consider the distribution as the product of three distributions for the three variables x, y, z.

Let us now consider a region of the x-phase space representative of the beam (see figure 1.9). We recall that in each region of the space we are referring to an individual set of particles in terms of their position and momenta. Referring to figure 1.9, the particles initially occupying region A will drift to region B, which is different from A, but with an identical area (i.e. no particles have indeed been lost in the process). Such a symplectic shape change (see below for further comments) is simply due to the fact that the particles on the upper edge move faster.

This is what we expect and what should be obtained from (1.115), which can be handled by the use of a plethora of techniques, such as the method of characteristics. We prefer, however, the already quoted method of the evolution operator. The use of the evolution operator method (see section 1.15) allows us to cast the solution of (1.115) in the form

$$F(P, Q; t) = e^{-\frac{t}{m}\frac{P}{}\frac{\partial}{\partial Q}} F_0(P, Q),$$
$$\vec{Q} \equiv (x, y, z), \tag{1.117}$$
$$\vec{P} \equiv (m\dot{x}, m\dot{y}, m\dot{z}),$$

which is just a formal solution, obtained by using the analogy with a first order differential equation $\dot{y} = ky$ with solution $y = e^{kt}y_0$, and by making the following correspondence $y \to F$, $\to k\frac{P}{m}\frac{\partial}{\partial Q}$, $y_0 \to F_0$.

The action of the evolution operator on the initial distribution reflects what is shown in figure 1.9 and provides a shift of the coordinate Q (to this aim it is worth mentioning that $e^{\alpha\frac{\partial}{\partial x}}f(x) = f(x + \alpha)$ and for further remarks see section 1.15), yielding the solution

$$F(P, Q; t) = F_0\left(P, Q - \frac{P}{m}t\right). \tag{1.118}$$

Let us now assume that the explicit form of the initial distribution is

$$F_0(P, Q) = \frac{1}{2\pi\Sigma}e^{-\frac{\beta P^2 + \gamma Q^2 + 2\alpha PQ}{2\Sigma}}. \tag{1.119}$$

We will discuss later the meaning of the β, α, γ, Σ coefficients, and we only note that the condition (1.110) requires that

$$\int_{-\infty}^{+\infty}\int_{-\infty}^{+\infty} F_0(P, Q)dQd\,P = 1, \tag{1.120}$$

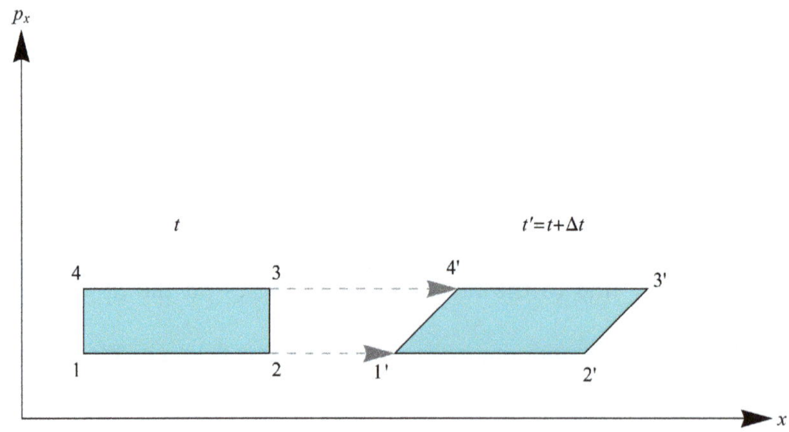

Figure 1.9. x-phase space representative of the beam.

which is ensured (see section 1.15) by

$$\beta\gamma - \alpha^2 = 1. \tag{1.121}$$

According to (1.118) we write the argument of the distribution in (1.119) as

$$\begin{aligned}
&\beta P^2 + 2\alpha P\left(Q - \frac{P}{m}t\right) + \gamma\left(Q - \frac{P}{m}t\right)^2 \to \beta' P^2 + 2\alpha' PQ + \gamma' Q^2,\\
&\beta' = \left(\beta - 2\frac{\alpha}{m}t + \gamma\frac{t^2}{m^2}\right),\\
&\alpha' = \left(\alpha - \frac{\gamma}{m}t\right),\\
&\gamma' = \gamma.
\end{aligned} \tag{1.122}$$

Even though it is straightforward, the result (1.122) is extremely significant from the physical point of view.

Indeed, it states that the quadratic form representing the argument of the distribution remains a quadratic form, namely is left unchanged, by the evolution operator and the normalization condition (1.121) as well ($\beta'\gamma' - \alpha'^2 = 1$). This is not surprising since the evolution operator can be viewed as a canonical transform preserving the number of particles.

The coefficients β, α, γ play a crucial role in the physics of charged beam transport, within this framework they are called Twiss parameters and Σ is called the beam emittance. We will spend a few words specifying their statistical meaning when referring to a distribution.

We define the co-variance matrix

$$\begin{aligned}
\hat{C} &= \begin{pmatrix} \sigma_{Q,Q}^2 & \sigma_{Q,P}^2 \\ \sigma_{P,Q}^2 & \sigma_{P,P}^2 \end{pmatrix},\\
\sigma_{M,N} &= \sqrt{\langle MN\rangle - \langle M\rangle\langle N\rangle},
\end{aligned} \tag{1.123}$$

where $\langle ...\rangle$ means the average taken on the distribution itself. The use of the properties of the Gaussian integrals (see section 1.15) yields

$$\hat{C} = \begin{pmatrix} \beta\Sigma & -\alpha\Sigma \\ -\alpha\Sigma & \gamma\Sigma \end{pmatrix}. \tag{1.124}$$

Let us now consider a beam of particles propagating in a drift space. The relevant quantities are the dimensions of the beam itself (which are evidently associated with the position coordinate Q and thus with $\sigma_{Q,Q}$) while for the other quantities we need to take more care.

Very roughly speaking, $\gamma\Sigma$ could be associated with the rms velocity and $\alpha\,\Sigma$ with velocity position correlation. It is also evident that the square root of the determinant of the covariance matrix is Σ and that this represents the phase-space area associated with the beam distribution itself.

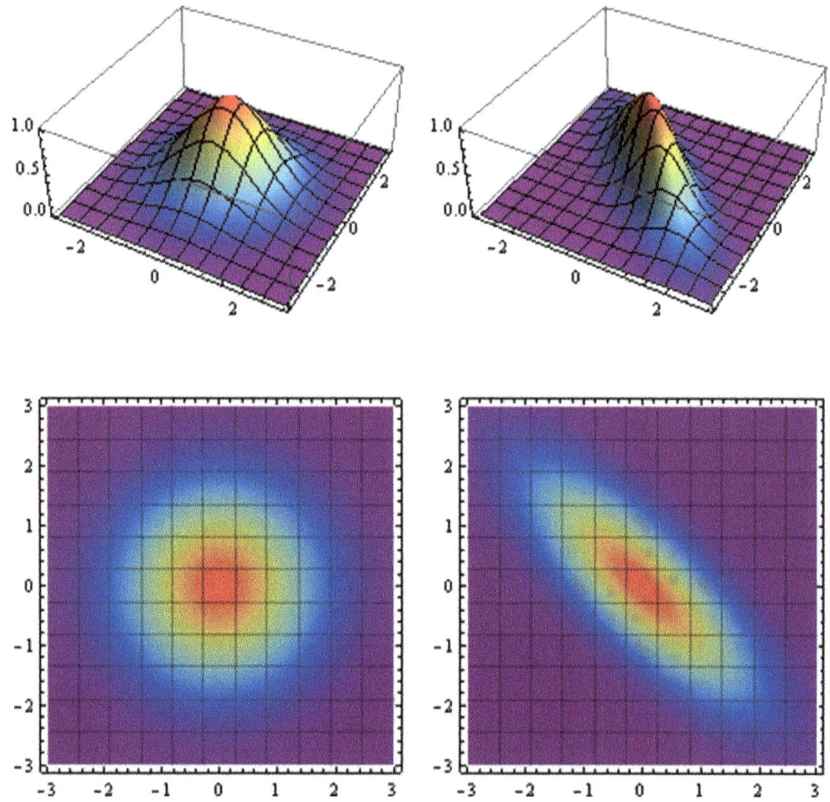

Figure 1.10. Gaussian distribution and the relevant contour plots.

From the statistical point of view the emittance can be treated as an rms value and can be understood from the inspection of figure 1.10, where we have reported the Gaussian distribution and the relevant contour plots. Within such a geometric context the emittance is defined as

$$\Sigma = \frac{A}{\pi}, \tag{1.125}$$

where $3A$ is the area of the contour ellipse containing 84% of the beam.

The use of the contour plots is extremely useful since by following their changes in form during the evolution we may have an idea of the beam evolution itself.

Let us now discuss a slightly more complicated problem, regarding the evolution of a beam ruled by a constant external force $\vec{f}$. In this case the associated Hamiltonian is

$$H = \frac{1}{2m}p^2 - fx, \tag{1.126}$$

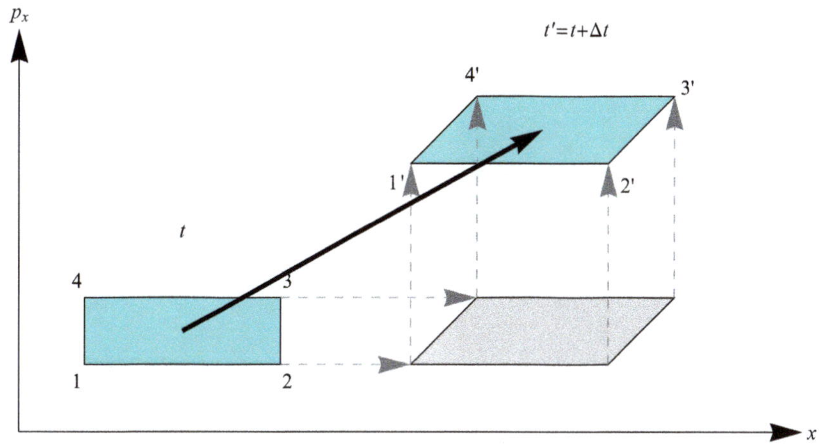

Figure 1.11. x-phase space evolution of a beam ruled by a constant external force.

the consequent Liouville operator is

$$\hat{L} = -\frac{p}{m}\frac{\partial}{\partial x} - \frac{f}{m}\frac{\partial}{\partial p}, \tag{1.127}$$

and its physical meaning should be recognized with reference to figure 1.11.

The solution of the associated Liouville equation is accordingly provided by

$$F(p, x, t) = e^{-t\left(\frac{p}{m}\frac{\partial}{\partial x}+\frac{f}{m}\frac{\partial}{\partial p}\right)}F_0(p, x). \tag{1.128}$$

The action of the exponential operator on the initial distribution yields

$$F(p, x, t) = F_0\left(p - \frac{f}{m}t, x - \frac{p}{m}t + \frac{1}{2}\frac{f}{m}t^2\right) \tag{1.129}$$

and the relevant validity can be verified by a direct check (namely by plugging the solution inside the Liouville equation and taking the relevant derivatives). Equation (1.129) contains a significant amount of physics and the role played by forces, acceleration and velocities on the phase-space coordinates. We invite the reader to use the previous result to study the evolution of the Twiss coefficients and check the relevant conclusion with the contour plot evolutions shown in figure 1.12.

We have given a very initial idea of how beam envelope evolution should be treated using the mean of classical analytical mechanics. The following section will be devoted to a deeper discussion of the problem, in which we will clarify the role of the Vlasov equation in the study of this type of problem.

1.8 The Vlasov equation and beam transport

The last example of the previous section dealt with an example of the Liouville equation regarding the motion of particles in a linear potential, $V(x) = -Fx$ and the consequent distribution evolution in phase space.

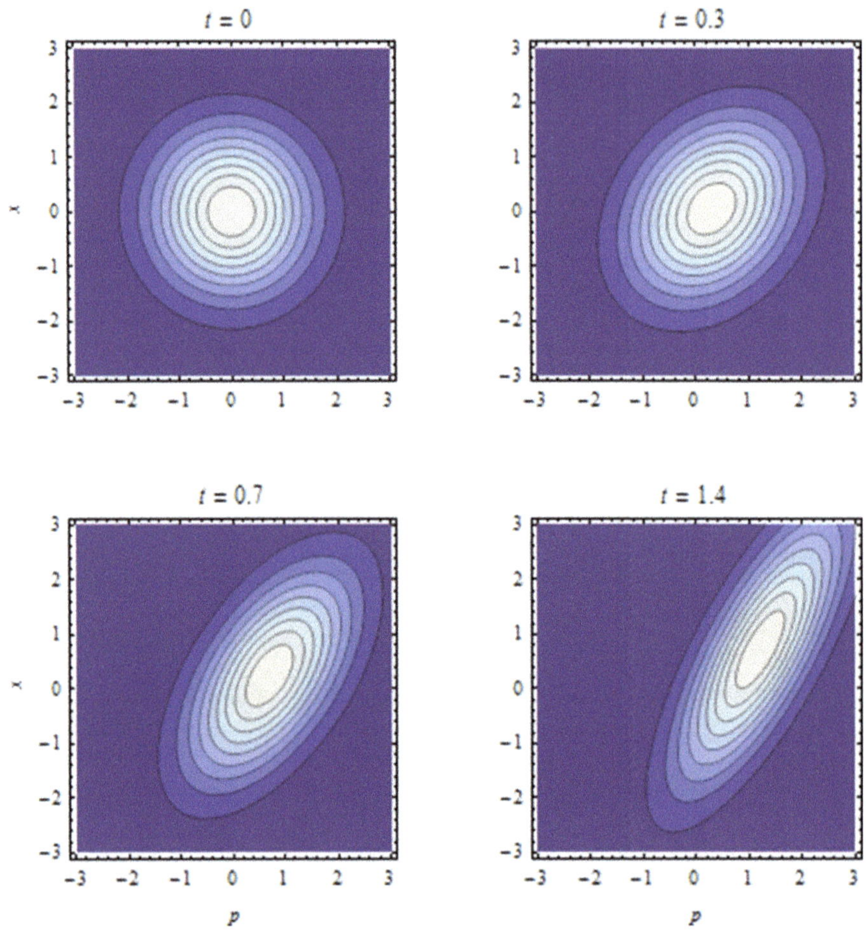

Figure 1.12. Contour plot evolution of the Liouville equation solution (1.129).

To complete the discussion and to provide an idea of the types of mathematical problems we will be faced with, we consider the Liouville equation for ensembles of particles ruled by a Hamiltonian (see (1.113)) with a generic potential $V(x)$.

In this case the evolution equation is

$$\begin{aligned}\frac{\partial}{\partial t}F &= -\frac{p}{m}\frac{\partial}{\partial x}F + V'(x)\frac{\partial}{\partial p}F,\\ V'(x) &= \frac{\partial}{\partial x}V(x),\end{aligned} \tag{1.130}$$

and its solution reads

$$F = e^{-t\left(\frac{p}{m}\frac{\partial}{\partial x} - V'(x)\frac{\partial}{\partial p}\right)}F_0. \tag{1.131}$$

Due to the argument of the exponential operators, we cannot naïvely apply the exponential disentanglement underlying the exponential semi group property

$$e^{\hat{A}+\hat{B}} = e^{\hat{A}}e^{\hat{B}}, \tag{1.132}$$

because

$$\begin{aligned}\hat{A} &= -\frac{pt}{m}\frac{\partial}{\partial x},\\ \hat{B} &= V'(x)t\frac{\partial}{\partial p}.\end{aligned} \tag{1.133}$$

The additional terms (see the final part of the chapter) depend on successive commutators involving the operators themselves. At first order we find, e.g.

$$e^{\hat{A}+\hat{B}} \cong e^{\hat{A}}e^{\hat{B}}e^{-\frac{1}{2}[\hat{A},\hat{B}]}, \tag{1.134}$$

while the higher order corrections involve commutators of the type $[\hat{A}, [\hat{A}, \hat{B}]], [\hat{B}, [\hat{B}, \hat{A}]]\ldots$

If one approximates the problem of the evolution as the successive application of the evolution for small time steps δt, allowing contributions of $(\delta t)^2$ to be neglected, we can write

$$F(p, x) \cong e^{\delta t V'(x)\frac{\partial}{\partial p}}e^{-\delta t\frac{p}{m}\frac{\partial}{\partial x}} \ldots\ldots e^{\delta t V'(x)\frac{\partial}{\partial p}}e^{-\delta t\frac{p}{m}\frac{\partial}{\partial x}}F_0(p, x), \tag{1.135}$$

and, at the first step, the effect of the evolution operator produces the following changes in the phase-space distribution

$$e^{\delta t V'(x)\frac{\partial}{\partial p}}e^{-\delta t\frac{p}{m}\frac{\partial}{\partial x}}F_0(p, x) = F_0\left(p + \delta t V'(x), x - \frac{p\delta t}{m}\right). \tag{1.136}$$

The two distinct shifts (in x- and p-coordinates), associated with the action of the evolution operator in (1.136), are shown in figure 1.11. The full dynamics can be realized through a series of successive steps in the phase space. The contour plots undergo a series of area-preserving transformations, which may also significantly modify their geometrical forms. We have so far looked at evolution dynamics turning ellipses into ellipses. This is not a general rule—as we will see in the following—it applies only to evolution processes ruled by Hamiltonians, which are quadratic forms in the canonical coordinates.

We will discuss all these problems within the context of specific applications in the next section and we go back to the distinction between canonical and non-canonical momenta.

The Liouville equation has been derived in a rather sloppy way. We have simply imposed (see (1.111)) the condition that the total derivative of the phase-space density distribution is vanishing!

As already stressed, the Liouville theorem states that the phase-space density of a certain element in phase space is fixed during its evolution. A different derivation of the same theorem, also touched upon in the previous section, is the fact that the evolution of canonical variables induced by a Hamiltonian is equivalent to a canonical transformation, which leaves invariant the phase-space volume. This point will be more carefully discussed in section 1.15, where we will see that a unitary transformation corresponds to a change of variables in phase-space integrals with unit Jacobian determinant.

From the geometrical point of view, the conservation statement we are dealing with is equivalent to the multi-dimensional extension of the pictures in figures 1.9 and 1.11, and the physical picture we should have in mind is that shown in figure 1.13 reporting a flow of particles crossing a given surface, and accordingly we can define the relevant current density as

$$\begin{aligned} \vec{\mathcal{J}} &= F\vec{\mathcal{G}}, \\ \vec{\mathcal{G}} &\equiv (\dot{P}, \dot{Q}), \end{aligned} \tag{1.137}$$

where $\vec{\mathcal{G}}$ is the six-dimensional phase-space velocity vector. We define the following six-dimensional divergence operator

$$\vec{\mathcal{D}} \equiv \left(\vec{\nabla}_P, \vec{\nabla}_Q\right), \tag{1.138}$$

and, recalling that the continuity equation is written $\vec{\nabla} \cdot \vec{J} + \frac{\partial}{\partial t}\rho = 0$, we introduce the following continuity equation

$$\vec{\mathcal{D}} \cdot \vec{\mathcal{J}} + \frac{\partial}{\partial t} F = 0, \tag{1.139}$$

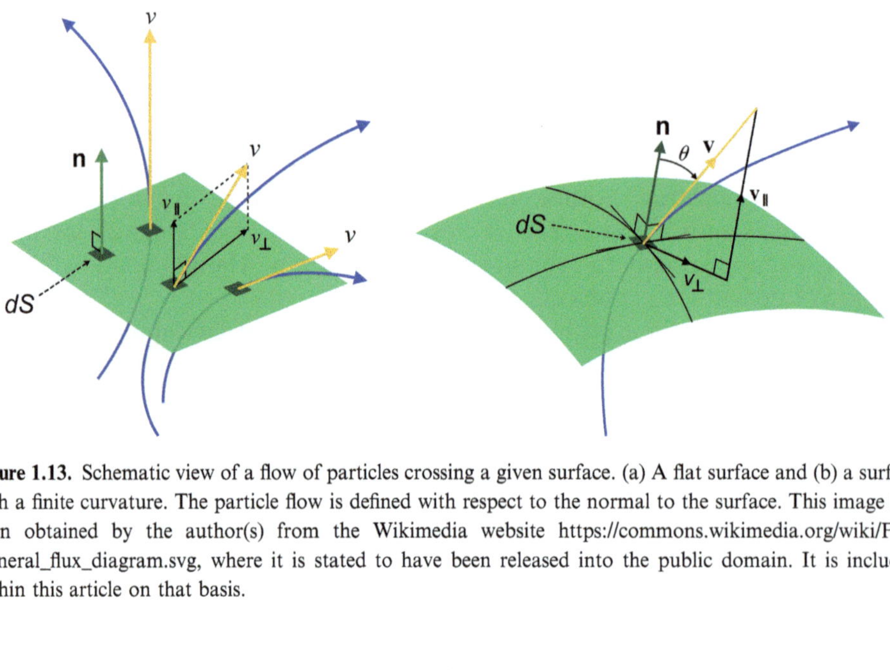

Figure 1.13. Schematic view of a flow of particles crossing a given surface. (a) A flat surface and (b) a surface with a finite curvature. The particle flow is defined with respect to the normal to the surface. This image has been obtained by the author(s) from the Wikimedia website https://commons.wikimedia.org/wiki/File:General_flux_diagram.svg, where it is stated to have been released into the public domain. It is included within this article on that basis.

which, according to the previous prescription, explicitly reads

$$\frac{\partial}{\partial t}F + \vec{\nabla}_P \cdot (F\vec{\mathcal{G}}) + \vec{\nabla}_Q \cdot (F\vec{\mathcal{G}}) = 0. \tag{1.140}$$

By expanding the gradient operators we find

$$\begin{aligned} &\vec{\nabla}_P \cdot (F\vec{\mathcal{G}}) + \vec{\nabla}_Q \cdot (F\vec{\mathcal{G}}) \\ &= (\vec{\nabla}_P \cdot \vec{\mathcal{G}})F + (\vec{\mathcal{G}} \cdot \vec{\nabla}_P)F + (\vec{\nabla}_Q \cdot \vec{\mathcal{G}})F + (\vec{\mathcal{G}} \cdot \vec{\nabla}_Q)F, \end{aligned} \tag{1.141}$$

and note that, with respect to the previous Liouville equation, we have two extra contributions

$$(\vec{\nabla}_P \cdot \vec{\mathcal{G}}) + (\vec{\nabla}_Q \cdot \vec{\mathcal{G}}) = \vec{\nabla}_P \cdot \dot{P} + \vec{\nabla}_Q \cdot \dot{Q}. \tag{1.142}$$

P, Q being canonical variables and considering that we are referring to trajectories ruled by Hamilton equations, we end up with

$$\vec{\nabla}_P \cdot \dot{P} + \vec{\nabla}_Q \cdot \dot{Q} = 0 \tag{1.143}$$

(according to the Hamilton equations we have $\dot{P} = -\vec{\nabla}\mathcal{H}$, $\dot{Q} = \vec{\nabla}\mathcal{H}$, which yields $\vec{\nabla}_P \cdot \dot{P} + \vec{\nabla}_Q \cdot \dot{Q} = 0$), which holds for Hamiltonian systems only (see (1.101)), thus we obtain from (1.140) the Liouville equation.

We complete our discussion by noting that the problem of beam transport can also be discussed by considering the distribution in terms of coordinate and velocities, namely $\mathcal{F}(\vec{v}, \vec{r}, t)$.

Even though we are getting pedantic, we prefer to discuss the problem from this point of view, which is summarized in figure 1.14 as a transition from a state A to a state B in the v, x-plane and conservation of the six-dimensional volume.

By following the same discussion as before, we write the associated continuity equation in the form

$$\begin{aligned} &\vec{D} \cdot \vec{J} + \frac{\partial}{\partial t}\mathcal{F} = 0, \\ &\vec{J} = (\dot{\vec{v}}, \dot{\vec{r}})\,\mathcal{F}, \\ &\vec{D} = (\vec{\nabla}_v, \vec{\nabla}_r), \end{aligned} \tag{1.144}$$

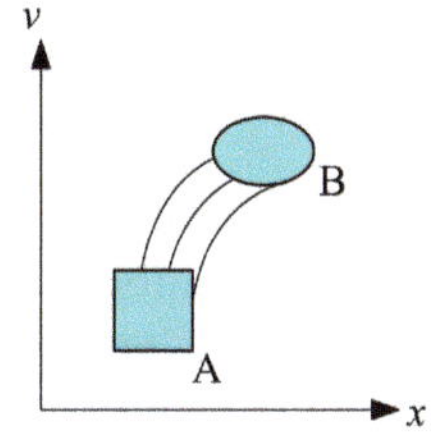

Figure 1.14. Transition from a state A to a state B in the v, x-plane and conservation of the six-dimensional volume.

which finally yields

$$\frac{\partial}{\partial t}\mathcal{F} + (\vec{v} \cdot \vec{\nabla}_r)\mathcal{F} + (\dot{\vec{v}} \cdot \vec{\nabla}_v)\mathcal{F} + (\vec{\nabla}_r \cdot v + \vec{\nabla}_v \cdot \dot{\vec{v}})\mathcal{F} = 0. \tag{1.145}$$

The last term should again be considered carefully. In general we have that $\vec{\nabla}_r \cdot \vec{v} = 0$ while $\vec{\nabla}_v \cdot \dot{\vec{v}}$ may not be zero, as in the case of friction forces depending on velocity. This does not apply to the Lorenz force for which we obtain

$$\vec{\nabla}_v \cdot \dot{\vec{v}} = \frac{q}{m}\vec{\nabla}_v \cdot [\vec{E} + \vec{v} \times \vec{B}] = 0 \tag{1.146}$$

since

$$\begin{aligned} \vec{\nabla}_v \cdot \vec{E} &= 0, \\ \vec{\nabla}_v \cdot (\vec{v} \times \vec{B}) &= 0. \end{aligned} \tag{1.147}$$

Equation (1.145) in the case of a beam of charged particles moving in static electric and magnetic fields can be written as

$$\frac{\partial}{\partial t}\mathcal{F} + (\vec{v} \cdot \vec{\nabla}_r)\mathcal{F} + \left(\frac{q}{m}[\vec{E} + \vec{v} \times \vec{B}] \cdot \vec{\nabla}_v\right)\mathcal{F} = 0. \tag{1.148}$$

The physical content of the Liouville equation and of (1.148) (the Vlasov equation or collisionless Boltzmann equation) is the same, with the difference that the latter can be used also for problems treated in terms of non-canonical variables.

It should accordingly be noted that, by keeping the total derivative with respect to time of the function $\mathcal{F}$, yields

$$\frac{d}{dt}\mathcal{F} = \frac{\partial}{\partial t}\mathcal{F} + (\vec{v} \cdot \vec{\nabla}_r)\mathcal{F} + (\dot{\vec{v}} \cdot \vec{\nabla}_v)\,\mathcal{F}. \tag{1.149}$$

The physical interpretation of the total derivative (also called the *convective, co-moving* or *Lagrangian* derivative denoted by $\frac{\mathcal{D}}{\mathcal{D}t}$) should be understood as a measure of the variation of $\mathcal{F}$ due to the change in the $\vec{v}$, $\vec{r}$ space.

The assumption that the total derivative vanishes is equivalent to the statement that the particle density remains unchanged along the particle orbit.

According to the above discussion the equation $\frac{\mathcal{D}}{\mathcal{D}t}\mathcal{F} = 0$ is equivalent to (1.145), provided that the condition $\vec{\nabla}_r \cdot \vec{v} + \vec{\nabla}_v \cdot \dot{\vec{v}} = 0$ is satisfied. This last assumption amounts to saying, as we will see in the following, that the beam is treated as an incompressible fluid.

Before closing this section we want to emphasize again that the Vlasov equation we have derived holds in the *collisionless* regime, namely when no abrupt event, such as an intra-particle collision, determines a variation of the position and velocity determining a flow out of the particle from the considered phase-space volume. In general the equation holds on a temporal scale involving times less the collision time.

We will touch on these points later in this and forthcoming chapters.

1.9 The Vlasov equation and the method of momenta

The Vlasov equation can be exploited to obtain an idea of the distribution evolution by means of a kind of indirect procedure, by establishing a set of equations yielding the time dependence of the moments of the distribution itself. By defining the kth moment of the velocity distribution as

$$\mathcal{M}_k(\vec{r}, t) = \int v^k \mathcal{F}(\vec{v}, \vec{r}, t) d\vec{v}, \tag{1.150}$$

the equation ruling its evolution can be obtained from (1.149), after setting $\frac{\mathcal{D}}{\mathcal{D}t}\mathcal{F} = 0$, multiplying by v^k and integrating as prescribed in (1.150). Limiting ourselves to the one-dimensional case, we obtain for the zeroth order moment the equation

$$\frac{\partial}{\partial t}\mathcal{M}_0 + \frac{\partial}{\partial x}\mathcal{M}_1 + a\int_{-\infty}^{+\infty} \frac{\partial}{\partial v}\mathcal{F} dv = 0, \tag{1.151}$$

which is derived from (1.149), after noting that $\vec{v} \cdot \vec{\nabla}_r \to v\partial_x$. The physical role of

$$\mathcal{M}_0 = \int_{-\infty}^{+\infty} \mathcal{F}(v, x, t) dv = n(x, t), \tag{1.152}$$

is nothing but that of the particle spatial density.

Regarding $\mathcal{M}_1$, we find

$$\begin{aligned} \mathcal{M}_1(x, t) &= \int_{-\infty}^{+\infty} v\mathcal{F}(v, x, t) dv \\ &= \left(\int_{-\infty}^{+\infty} \mathcal{F}(v, x, t) dv\right)\left(\frac{\int_{-\infty}^{+\infty} v\mathcal{F}(v, x, t) dv}{\int_{-\infty}^{+\infty} \mathcal{F}(v, x, t) dv}\right) \\ &= \langle v(x, t)\rangle n(x, t), \end{aligned} \tag{1.153}$$

and therefore it is understood as the particle flux at the position x, with the further constraint

$$\int \frac{\partial}{\partial v}\mathcal{F} dv = \mathcal{F}\Big|_{-\infty}^{+\infty} = 0. \tag{1.154}$$

Thus obtaining from the zeroth order moment equation

$$\frac{\partial}{\partial t}n(x, t) + \frac{\partial}{\partial x}(\langle v(x, t)\rangle n(x, t)) = 0, \tag{1.155}$$

which is essentially a simplified version of the continuity equation, which was the starting point of our discussion.

A bit more physics can be obtained from the equation for the first order moment, which is written

$$\frac{\partial}{\partial t}\mathcal{M}_1 + \frac{\partial}{\partial x}\mathcal{M}_2 + a\int_{-\infty}^{+\infty} v\,\frac{\partial}{\partial v}\mathcal{F}dv = 0. \tag{1.156}$$

To understand the meaning of (1.156) we should clarify the role of

$$\mathcal{M}_2 = \int_{-\infty}^{+\infty} v^2\mathcal{F}dv = n(x,\,t)\langle v(x,\,t)^2\rangle, \tag{1.157}$$

which can be understood as specified below. After integration by parts, we obtain, by use of the condition (1.154)

$$\int_{-\infty}^{+\infty} v\,\frac{\partial}{\partial v}\mathcal{F}dv = -\int_{-\infty}^{+\infty}\mathcal{F}dv = -n(x,\,t). \tag{1.158}$$

The physical role of $\mathcal{M}_2$ emerges from (1.157) itself, which can be further handled by noting that

$$v(x) = \langle v(x)\rangle + \delta_v(x), \tag{1.159}$$

which means that, in accelerated beams, the velocity is provided by an average value and by a term fluctuating near such a mean and, by definition,

$$\langle \delta_v(x)\rangle = 0. \tag{1.160}$$

Accordingly we obtain

$$\langle v(x)^2\rangle = \langle v(x)\rangle^2 + \langle \delta_v(x)^2\rangle. \tag{1.161}$$

Thus finding for the equation, defining the first order momenta, the expression (which can be derived without much effort)

$$\begin{aligned}
&\frac{\partial}{\partial t}\langle \mathcal{P}(x)\rangle = n(x)f - \frac{\partial}{\partial x}\mathcal{E}(x) - \frac{\partial}{\partial x}\varphi(x)\\
&\quad \mathcal{P}(x) = m_0 n(x)\langle v(x)\rangle,\ \mathcal{E}(x) = m_0 n(x)\langle v(x)\rangle^2,\\
&\quad \langle\varphi(x)\rangle = m_0 n(x)\langle \delta_v(x)^2\rangle,\\
&\quad f = m_0 a.
\end{aligned} \tag{1.162}$$

We recall that the beam is provided by N identical particles of mass m_0.

This last equation should be interpreted as follows; the local average momentum of the beam depends on the applied force (the first term on the rhs of (1.162)), the second term is the kinetic momentum change due to the replacement of particles flowing out of the space region by other particles of the beam entering from a different region, and the third term is due to the effect of fluctuations. In some sense it measures the ‘temperature’ of the beam itself and when this term is negligible, the beam is said to be cold.

We prefer to spend a few more words on this point, by going deeper into the physical meaning of $\mathcal{E}(x)$, which can be recast as

$$\mathcal{E}(x) = [m_0\, n(x)\ \langle v(x)\rangle]\,\langle v(x)\rangle, \tag{1.163}$$

and understood either as a flux of kinetic momentum or as a kind of kinetic pressure. The use of the same point of view leads to the interpretation of $\varphi(x)$as an internal pressure.

It is evident that these contributions are due to the collective nature of the problem under study. They are different from the first term which is 'deterministic' in nature, since it determines the change in momenta induced by the force applied to an individual particle. The other terms specify average processes and are characteristic of the Euler point of view, according to which it is not significant to identify a specific particle with its associated position. Attempting an analogy with laminar fluid transport, (1.162) could be understood as a restatement of the Bernoulli theorem.

1.10 Emittance, normalized emittance and beam brightness

Even though not said explicitly, the discussion of the previous sections apply to classical and relativistic motion as well. However, a few remarks can be given. We consider therefore the relativistic free particle evolution and the associated Liouville equation, describing a beam of relativistic particles with initial phase-space distribution $F_0(P, Q)$ drifting in free space

$$\begin{gathered} \mathcal{H} = c\sqrt{P^2 + (m_0c)^2}, \\ \frac{\partial}{\partial t}F(P, Q, t) = -\frac{Pc}{\sqrt{P^2 + (m_0c)^2}}\frac{\partial}{\partial Q}F(P, Q, t), \end{gathered} \tag{1.164}$$

which can be solved by the use of the of the evolution operator method as

$$F(P, Q, t) = e^{-\frac{Pct}{\sqrt{P^2+(m_0c)^2}}\frac{\partial}{\partial Q}}F_0(P, Q) = F_0\left(P, Q - \frac{Pct}{\sqrt{P^2 + (m_0c)^2}}\right). \tag{1.165}$$

On account of the fact that (see table 1.1)

$$\frac{Pc}{\sqrt{P^2 + (m_0c)^2}} = v, \tag{1.166}$$

equation (1.165), even though it is derived within a relativistic framework, has the same physical content as the non-relativistic case, summarized in (1.117).

Regarding the inclusion of, e.g. a constant force in the relativistic Liouville problem, we modify the relevant equations as follows:

$$\begin{gathered} \mathcal{H} = c\sqrt{P^2 + (m_0c)^2} - fQ, \\ \frac{\partial}{\partial t}F(P, Q, t) = -\frac{Pc}{\sqrt{P^2 + (m_0c)^2}}\frac{\partial}{\partial Q}F(P, Q, t) - f\frac{\partial}{\partial P}F(P, Q, t), \end{gathered} \tag{1.167}$$

and obtain the solution of the problem as

$$F(P, Q, t) = F_0\left(P - m_0at,\ Q + \frac{c^2}{a}\right.$$
$$\left.\times\left[\sqrt{1+\left(\frac{P - m_0at}{m_0c}\right)^2} - \sqrt{1+\left(\frac{P}{m_0c}\right)^2}\right]\right) \tag{1.168}$$
$$a = \frac{f}{m_0},$$

which, in the non-relativistic limit, coincides with the result obtained in the previous sections.

It should, however, be noted that the Hamiltonians in (1.164) and (1.167) are not in quadratic form in the canonical variables. We cannot therefore expect that the associated phase-space contour plots undergo ellipses preserving transformations.

In general the study of the beam evolution in relativistic conditions is a significantly more complex problem than the corresponding non-relativistic case. The relativistic version of the Vlasov equation for a relativistic charged particle in a combined electric and magnetic field reads

$$\frac{\partial}{\partial t}\mathcal{F} + \left(\frac{\vec{p}}{m_0\gamma}\cdot\vec{\nabla}_r\right)\mathcal{F} + q[\vec{E} + \vec{v}\times\vec{B}]\cdot\vec{\nabla}_p\mathcal{F} = 0, \tag{1.169}$$

and the difficulty in handling the problem (such as the derivation of the equation of momenta) is associated with the non-constancy of the relativistic factor γ. We will accordingly avoid the use of the above equation in full relativistic terms by using appropriate disentanglement of longitudinal and transverse motion.

We prefer now to bring the discussion onto more 'practical' ground by discussing an operative (i.e. amenable for an experimental implementation) point of view to the concepts associated with beam emittance and phase space.

We remark therefore that charged particle beams move in a chosen direction called the beam axis, which might be either straight or curved. In this book we will essentially treat the former case. The associated trajectories may be complicated, but in general we can decouple the longitudinal motion (along the axis) from the transverse part. According to the above description, if we identify with z the propagation axis, the beam motion is characterized by two symmetry planes (x–z, y–z). It is indeed often possible to express the problem of transverse motion in the form of Newtonian equations with the rest mass replaced by the relativistic mass.

The validity of such an approximation is analogous to the paraxial propagation in optics. Accordingly, the beam is assumed to be well collimated, and therefore the transverse velocity components are small compared to the longitudinal velocity of the beam particles. We can be less qualitative by considering a particle, moving at relativistic velocity along the z-axis, with a force applied in the transverse direction (say along the x-direction). The relativistic equations of motion of the x-component therefore read

$$\gamma v_x\left(\frac{\dot{\gamma}}{\gamma}+\frac{\dot{v}_x}{v_x}\right)=\frac{F_x}{m_0}. \tag{1.170}$$

Let us now note that if $v_x \ll v_z$ then $|\frac{\dot{\gamma}}{\gamma}| \ll |\frac{\dot{v}_x}{v_x}|$, therefore (1.170) can be safely replaced by

$$m_0\gamma\frac{d\,v_x}{d\,t}=F_x, \tag{1.171}$$

which is simply the Newton equation with m_0 replaced by $m_0\gamma$.

We have stressed that the assumption of non-relativistic transverse motion amounts to the paraxial approximation. It is indeed worth noting that

$$\frac{v_x}{v_z}=\frac{dx}{dz}=x' \ll 1 \tag{1.172}$$

means that the angle of deviation from the propagation axis is always very small. We will see in the following that such an interpretation is a key element to understand the parameters characterizing the beam divergence.

In the following we will refer to x', y', rather than to the velocities, to characterize the transverse motion, because these parameters are more useful for the purposes of defining the beam emittance and brightness, two figures of merit of crucial importance to specify the 'quality' of transported charged beams and to qualify the correctness of the transport system itself.

There are many definitions of emittance given in the literature, which may sometimes cause confusion either regarding the data comparison or the planning of specific measurements. The reason for such a wealth of points of view is certainly due to the fact that the concept of emittance is so general and the experimental conditions to which it applies are so different, either because any beam, produced by a specific accelerator, has its own peculiarities, or because the use of the beam itself varies widely.

A definition of emittance should accordingly capture different features, including the effect induced by acceleration and transport mechanisms. It should furthermore provide a non-ambiguous criterion of the procedure leading to its measure.

To proceed further we need a few remarks, listed below:

a. The phase-space area is conserved in a $6N$-dimensional geometry.
b. The conservation does not imply that the geometrical shape of the volume or hyper-surface remains invariant, filamentation effects may arise when non-linear forces are active (see the following sections).
c. If motions among the different symmetry planes are not coupled then $2N$-dimensional conserved phase spaces can be considered.

We introduce the quantity

$$A_x = \iint_{V_2} dx\, dx', \tag{1.173}$$

to denote the *trace space.* The relevant conservation stems directly from the Liouville theorem and it could be a good candidate for a definition of beam emittance. Even though it is appropriate, it is too general to fit all the requirements, including that of certifying the quality of the beam and that of the transport itself, which are for the moment ill-defined concepts.

Consider therefore a series of translations in phase (or trace) space represented by the successive steps in evolution reported in (1.135), now rewritten as

$$F_0(p_{n+1}, x_{n+1}) \cong e^{\delta t V'(x)\frac{\partial}{\partial p}} e^{-\delta t \frac{p}{m}\frac{\partial}{\partial x}} F_0(p_n, x_n), \tag{1.174}$$

which is an example of a symplectic map (see the next section). It is evident that the conservation of the trace area does not depend on whether the forces acting on the particles of the beam are linear or not (see figure 1.15). On the other side, the shape of the area is strongly affected by the linear or non-linear nature of the force itself, which, in the latter case, determines a significant distortion of the phase-space contour area. We define therefore as good quality transport devices those which do not induce any severe distortion, leaving the beam phase-space contour plots free (or almost free) from filamentation contaminations. A good quality beam is a beam free (or almost free) from these distortions.

We have already noted that, starting from the solution of the Liouville equation it can be 'operatively' correct to link the emittance to the determinant of the covariance matrix which, adapted to the previously discussed trace space, yields the following definition of rms emittance

$$\tilde{\varepsilon}_x = [\langle x^2\rangle\langle x'^2\rangle - \langle xx'\rangle^2]^{\frac{1}{2}}, \tag{1.175}$$

a quantity expressed in m × rad (while the two-dimensional phase-space area has the dimensions of angular momentum). The average $\langle xx'\rangle$ in (1.175) reflects, as noted in the previous sections, the correlation between position and divergence, which may be induced by, e.g. a kick on the trajectory induced by a focusing force.

The differences between $\tilde{\varepsilon}_x$ and A_x are striking and worth stressing. When the phase-space area undergoes filamentations A_x is conserved while $\tilde{\varepsilon}_x$ increases and this fits with our request of having a figure of merit going beyond the Liouville theorem itself.

If we consider an ideal transport system, namely free from non-linear distortions, the two definitions are equivalent (not coincident) and referring to elliptical contour plots we find that the area encircling almost 100% of the particles inside the beam is

$$A_x = \pi\, x_M x'_M, \tag{1.176}$$

where x_M, x'_M are the semi-axes of the ellipse itself.

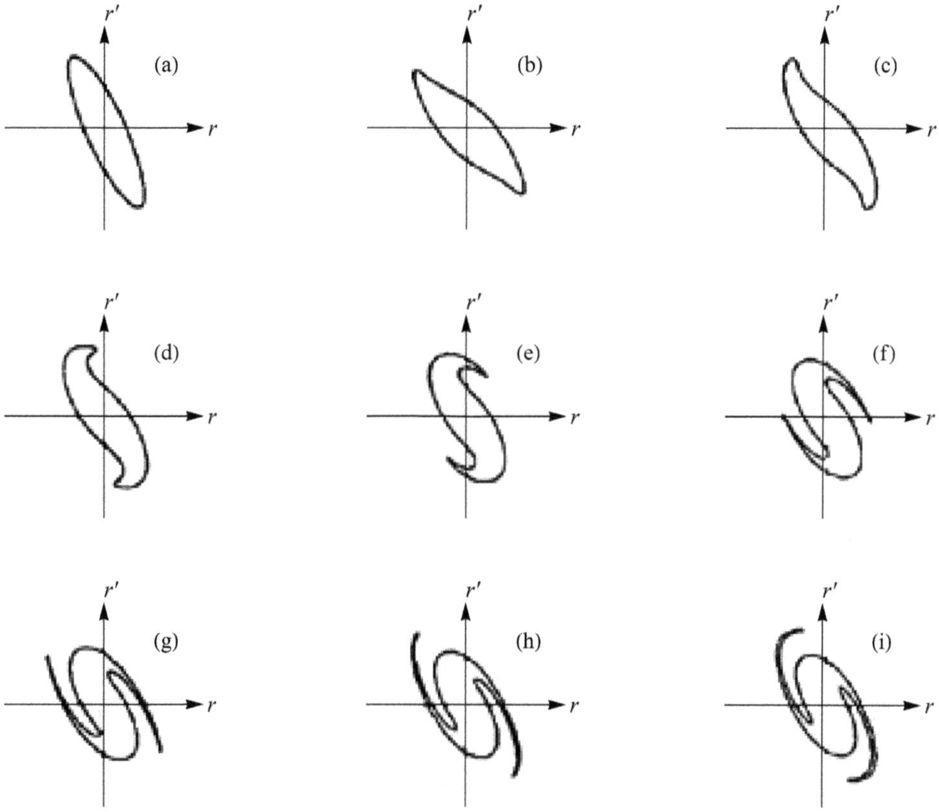

Figure 1.15. Courant–Snyder ellipse evolution through a channel whose elements exhibit non-linearities. The effect of the distortions eventually inducing filamentation are more evident on the tail of the distribution, where the non-linear contributions are more significant. The core of the bunch is essentially ruled by the linear part and therefore non-significant distortions are evident.

Now since, as we will see later, in a beam with uniform density and ruled by linear forces the following conditions hold

$$\sqrt{\langle x^2 \rangle} \cong \frac{x_M}{2}, \qquad \sqrt{\langle x'^2 \rangle} \cong \frac{x'_M}{2}, \tag{1.177}$$

in an ideal transport system the following relationship between rms emittance and trace space holds:

$$\tilde{\varepsilon}_x = \frac{A_x}{4\pi}. \tag{1.178}$$

Let us now provide an operative procedure to measure the beam emittance itself. Our measure set up can be idealized as follows (see figure 1.16). A slit with holes

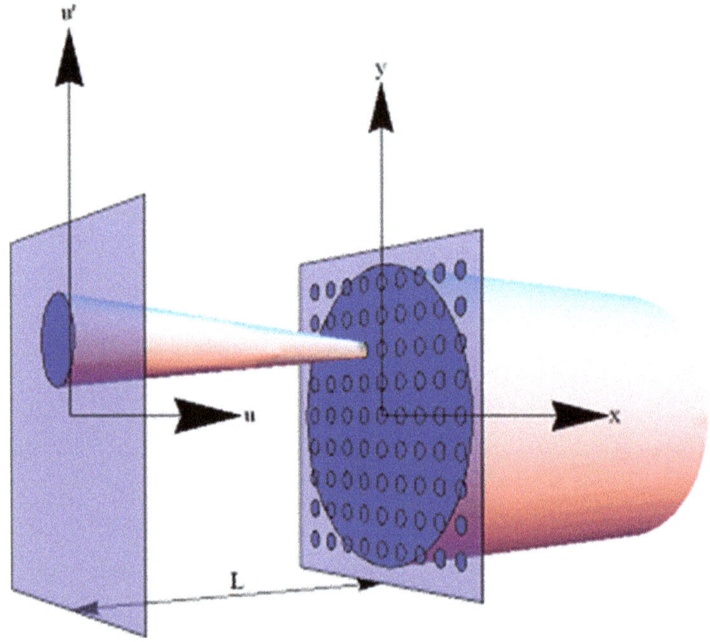

Figure 1.16. Hole diaphragm beamlet emittance measurement set-up.

intercepts the beam in one of the symmetry planes, thus producing a set of beamlets, collected on a screen at distance L from the slit.

The current values of each beamlet are recorded and the result is a series of Gaussian-like distributions around a peak value. Furthermore, two points on the screen are taken in such a way that the current at those point is (say) 10% of the maximum, the relevant distances from the axis are denoted by d_1, d_2.

The trace area space is accordingly straightforwardly defined: the hole defines the position coordinate x while the divergence is defined as

$$x' = \frac{d - x}{l}. \tag{1.179}$$

Collecting all the points according to the previous prescription one can draw the corresponding phase space area as indicated in figure 1.16. This is the operative procedure we were looking for, and it was exploited in one of the most beautiful measurements of beam emittance, performed in 1968 with the 10 MeV microtron at Frascati National Laboratories (Bordoni F, Letardi T and Placidi M 1968 *Nucl. Instrum. Methods* **65** 72).

Albeit clear and useful for our purposes, the definition of emittance we have given and which will be used in this book suffers from some drawbacks. For example the appearance of distributions with long tails (more akin to non-linear effects) may cause an apparent degradation of the emittance, while the effective core of the beam is well behaved in phase space. In this case appropriate cuts of the tail are necessary to guarantee a correct result and an appropriate definition of the beam quality.

The definitions of trace area we gave are not appropriate for comparison between beams of different energies. Those with higher energy may have smaller emittance, not because the quality of the beam is better, but because of relativistic effects due to the fact that the divergence decreases with increasing p_z. The emittance should be properly *normalized* to include this effect.

Accordingly we define the trace area as

$$\mathcal{A}_x = \frac{1}{P}\iint_{V_2} dx\, dp_x \cong \frac{1}{\beta\gamma m_0 c}\iint_{V_2} dx\, dp_x. \tag{1.180}$$

In the absence of acceleration $\beta\,\gamma$ is constant and therefore $\mathcal{A}_x$ is conserved but not otherwise.

It is accordingly convenient to introduce the *normalized* emittance

$$\tilde{\varepsilon}_n = \beta\gamma\tilde{\varepsilon}, \tag{1.181}$$

which is a relativistic invariant.

The definition given in (1.181) is an even more useful concept of its un-normalized counterpart. A beam undergoing an acceleration in a linear and decoupled transport system exhibits the conservation of the normalized emittance and its growth may be a measure of the deviation of the accelerating and of the transport system from the ideal conditions.

Before closing this section we will provide the other important figure of merit which often occurs in the jargon of accelerator physicists. The beam *brightness* $\mathcal{B}$ includes, in the phase-space considerations, the beam current itself, namely

$$\begin{aligned}\mathcal{B} &= \frac{J}{\Delta\Omega},\\ J &= \frac{I}{\Delta S}.\end{aligned} \tag{1.182}$$

The brightness is accordingly understood as the current per unit surface and unit solid angle,

$$\mathcal{B} = \frac{I}{\Delta S\Delta\Omega}. \tag{1.183}$$

Or, alternatively, as

$$\begin{aligned}\mathcal{B} &= \frac{I}{V_4},\\ V_4 &= \iint dSd\Omega.\end{aligned} \tag{1.184}$$

if we are interested in defining the brightness on a defined four-dimensional trace area. According to the discussion on the contour plot ellipses we can evaluate V_4 as

$$V_4 = \frac{1}{2}(\mathcal{A}_x)(\mathcal{A}_y) = \frac{1}{2}(4\pi)^2\tilde{\varepsilon}_x\tilde{\varepsilon}_y \tag{1.185}$$

and the associated normalized brightness is simply

$$\mathcal{B}_n = \frac{\mathcal{B}}{(\beta\gamma)^2}. \tag{1.186}$$

Other definitions are available in the literature; sometimes the brightness is written $\mathcal{B} = \frac{I}{\tilde{\varepsilon}_x \tilde{\varepsilon}_y}$, which yields values larger than those provided by the use of (1.186). Extreme care in the specific definitions are therefore recommended to avoid confusion in comparisons.

1.11 Beam transport elements, dipole, quadrupoles…

In the previous section we have stressed that charged particles may be transported, but we have not specified how. Since we know that charged particles are accelerated, deflected or, more in general, acted upon by static or electric magnetic fields, we can safely conclude that these are the most natural tools to accelerate, deflect and focus a charged beam.

Let us now suppose that we have an ensemble of particles, which are moving at relativistic energies, and we do not need to increase its energy, but only to bend its trajectory. To this aim we can exploit the device illustrated in figure 1.17 (the sector magnet), which the electrons are passed through, and in which they experience an (almost) constant (in space and time) field and execute a piecewise curved trajectory in the x–z-plane (sometimes, in particular for curved trajectories, the y-axis is called vertical, the x-axis horizontal and the z-axis the central orbit).

If, on the other side, we are interested in confining a beam in the transverse plane, we should use a device such as the one shown in figure 1.18, representing a quadrupole lens. In such a device the electrons emerging from the drawing are deflected toward the center of the axis (focused) in the x-direction and defocused in the other. A global focusing is therefore achieved after a judicious combination of focusing and defocusing quadrupole lenses, as shown in figure 1.19 and discussed later in this section.

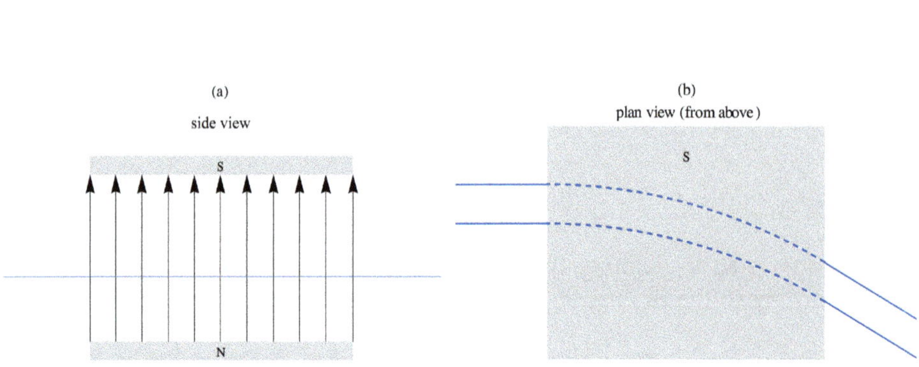

Figure 1.17. Schematic view of a sector magnet.

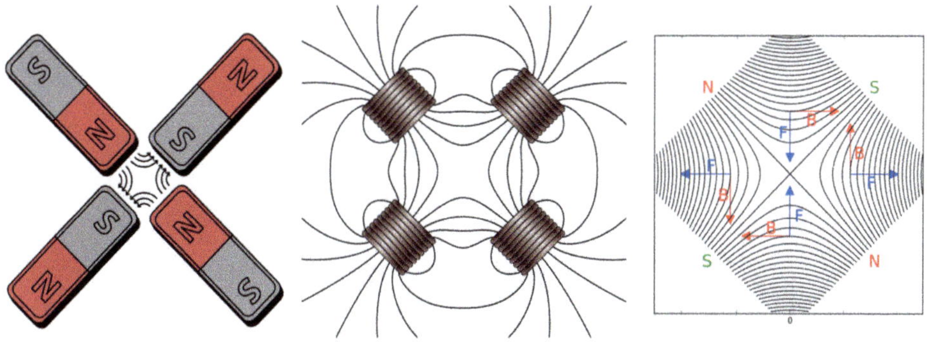

Figure 1.18. Schematic views of a quadrupole lens. The left image has been obtained by the authors from the Wikimedia website https://commons.wikimedia.org/wiki/File:Magnetic_quadrupole_moment.svg, where it is stated to have been released into the public domain. It is included within this article on that basis.

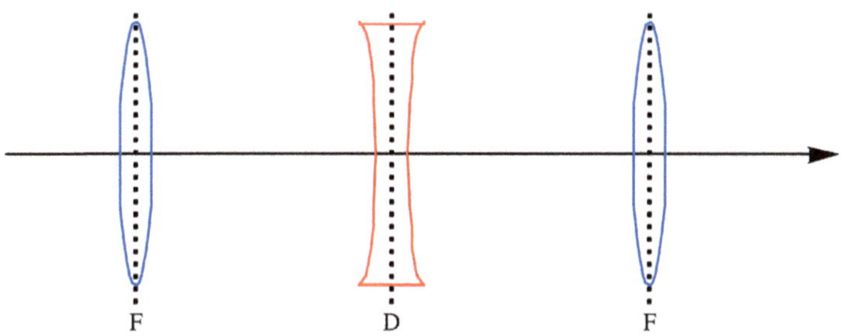

Figure 1.19. Arrangement of quadrupole lenses along a charged beam trajectory. We have used the symbols denoting the focusing (blue) and defocusing (red) lenses used in geometrical light ray optics. It must be stressed that a quadrupole, if focusing in the x–z-plane, is defocusing in the other plane. The focusing or defocusing shape usually refers to the quadrupole action in the x–z-plane.

In the description of the beam transport equation through actual magnetic devices, we will determine the ray equation of particles which deviate from that of the reference particles, namely the ideal particles following the reference trajectory, determining the axis. The derivation of the equations describing bending and focusing has been reported in rigorous terms in many authoritative books and will not be repeated here, where we will discuss the problem using a heuristic but effective procedure. In the following discussion we will reverse the presentation paradigm, in other words we will assume the existence of an 'item' producing a beam focusing, we will apply the relevant effect on the particle motion and then check if such a device is physically sound.

Let us consider the example of the focusing element, by assuming that a device can be inserted on the electron trajectory causing a kind of elastic restoring force in the x-direction (note the axis orientation reported in figure 1.20), namely

$$x'' = -kx. \tag{1.187}$$

We must specify that position x refers to a particle inside the beam shifted from its center, which is unaffected by the restoring force. Furthermore the apices do not

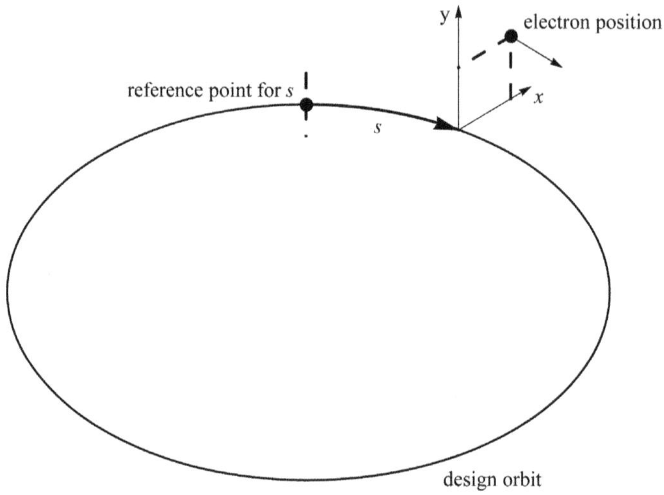

Figure 1.20. Axis orientation.

refer to time derivatives, but to the derivative with respect to the longitudinal coordinate, not necessarily following a linear path, and specified by

$$s = \beta_z ct. \tag{1.188}$$

We will further assume that the longitudinal and transverse motions are decoupled.

In the following we will treat the problem of beam focusing *ab initio* by considering the relevant derivation from the Lorentz equations of motion, which for a magnetic field with components B_x, B_y, B_z will explicitly be written as (the dot refers to the time derivative)

$$\begin{aligned} \ddot{y} &= -\frac{e}{m_e\gamma}(\dot{x}B_z - \dot{z}B_x), \\ \ddot{x} &= -\frac{e}{m_e\gamma}(\dot{z}B_y - \dot{y}B_z), \\ \ddot{z} &= -\frac{e}{m_e\gamma}(\dot{y}B_x - \dot{x}B_y). \end{aligned} \tag{1.189}$$

According to our assumptions we set

$$\begin{aligned} &\ddot{z} = 0,\ \dot{z} \cong \dot{s} \cong \beta c, \\ &\ddot{x} = -\frac{e}{m_e\gamma}\dot{z}B_y, \\ &\ddot{y} = \frac{e}{m_e\gamma}\dot{z}B_x. \end{aligned} \tag{1.190}$$

Therefore, we could realize a 'focusing' device if the following condition are satisfied

$$\begin{aligned} B_y &\cong gx, \\ B_x &\cong gy. \end{aligned} \tag{1.191}$$

If such a field exists it must satisfy the magneto-static equations

$$\vec{\nabla} \cdot \vec{B} = 0,$$
$$\vec{\nabla} \times \vec{B} = 0, \tag{1.192}$$

compatible with a field $\vec{B}$ having the following components:

$$\begin{aligned} &\vec{B} \equiv (gy, gx, 0), \\ &g = \frac{\partial}{\partial x} B_y \bigg|_{x,y=0} = \frac{\partial}{\partial y} B_x \bigg|_{x,y=0}, \\ &g \equiv \text{quadrupole gradient}. \end{aligned} \tag{1.193}$$

A static magnetic field with the previous characteristics is that provided by a quadrupole magnet, but before getting into specific aspects of the device we note that:

a. The equations we have derived describe the transverse motion of particles which are 'slightly' off those passing through the center of the magnet.
b. The particles explore a region of coordinates where the field exhibits a linear dependence on the transverse coordinates and therefore the resulting equations of motion in the transverse planes are linear and uncoupled.
c. The reference particles are unaffected by the field which is assumed to have a zero component along the coordinate of motion, and accordingly the curvature radius of the trajectory in both vertical planes is infinite.
d. The unwelcome feature emerging from equation (1.190) is that if the device focuses in the radial direction it defocuses in the vertical, and vice versa.

It is evident from the above points that the equations we have obtained are valid in a region around the reference particle where the non-linear effects do not provide significant contributions.

By keeping the derivatives with respect to the coordinate s, we find for the last two equations in (1.190)

$$\begin{aligned} &x'' = -kx, \\ &y'' = ky, \\ &k = \frac{e}{m_e \gamma \beta c} g \equiv \text{quadrupole strength}. \end{aligned} \tag{1.194}$$

In practical units the quadrupole strength is given by $k = \dfrac{0.2998\, g[\mathrm{T/m}]}{p[\mathrm{GeV}/c]}$. The units we have used are tesla/meter for the gradient (consistent with MKSA) while the momentum has been given in GeV/c (not MKSA) but more appropriate for the type of problems we are dealing with (see section 1.14 for further comments). The bending magnetic field has a more complicated structure than that reported in

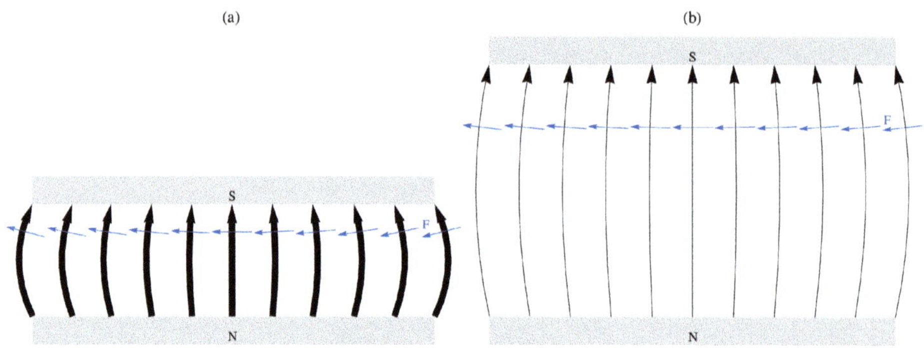

Figure 1.21. Schematic view of the bending magnet (x–y) focusing. The field fringe strength is more intense when the poles are closer (a). In case (b), vice versa, the field line forces become weaker.

figure 1.17. The component along the radial direction is in fact non-vanishing (see figure 1.18) and its presence also causes a focusing, as discussed below.

We have mostly discussed a kind of ideal situation in which non-linear corrections are absent and the transport elements have separate functions. In real-life accelerators this is only approximately true and bending magnets may have a focusing role as well. This can be easily understood by noting that the electrons leaving the magnet change slope (see figure 1.21), according to the identity

$$\Delta x' = -\frac{\Delta s}{\rho}, \tag{1.195}$$

where ρ is the curvature radius which, in terms of the field vertical component, is given by

$$\begin{gathered}\frac{e}{p}B_y = \frac{1}{\rho},\\ p = m_e\beta_z\gamma c.\end{gathered} \tag{1.196}$$

Accordingly we find (considering the quadrupole effect according to (1.191))

$$\Delta x' = -\Delta s\frac{eB_y}{p} \cong -\Delta s\frac{e\alpha x}{p}, \tag{1.197}$$

namely

$$\frac{\Delta x'}{\Delta s} \cong -\frac{e\alpha x}{p}. \tag{1.198}$$

It is furthermore evident that $\lim_{\Delta s\to 0}\frac{\Delta x'}{\Delta s} = x''$ and it is evident that (1.198) accounts for a focusing effect as in the case of the quadrupole. This is not the end of the story—particles moving in the radial plane trajectory moving away from the reference trajectory experience (see figure 1.21(b)) regions in which the field line force becomes weaker. This slightly reduces the focusing effect due to the quadrupole constant, which should be replaced by $k - \frac{1}{\rho^2}$.

Finally, in the case of dipole magnet the zeroth order of the y-component of the magnetic field is no longer vanishing at $x, y = 0$. Therefore, the second equation of (1.189) suggests the existence of a further contribution, associated with those particles exhibiting a non-zero linear momentum in the longitudinal direction. We find (see (1.196))

$$\begin{aligned} &\frac{e}{m_e\gamma}\dot{z}\, B_y \cong \\ &\cong \frac{e}{m_e\gamma c}\delta\beta_z B_y(0) \propto \frac{1}{\rho}\delta_p, \end{aligned} \tag{1.199}$$

where δ_p is the momentum spread associated with the velocity spread $\delta\beta_z$.

The guiding equations for the bending magnet can finally be written as

$$\begin{aligned} x'' + \left(k - \frac{1}{\rho^2}\right)x &= \frac{1}{\rho}\frac{\delta_p}{p}, \\ y'' &= ky. \end{aligned} \tag{1.200}$$

Further comments are provided in the section devoted to complements, where we give further details on the magnetic field transverse distribution.

Even though derived using a naïve procedure (amended in the concluding part of the chapter) we have obtained an effective idea of what tools we have to follow the beam dynamics, and our equations do describe the motion of particles around the reference orbit and along the path coordinate. The evolution of the beam distribution should therefore be accomplished by keeping this point of view in mind.

1.12 Beam transport and matrix formalism

In the previous sections we have described how the equation of motion of high-energy electrons in a transport channel should be treated. The equations we have described are similar to ray trace equations and the information which can be extracted from them can be obtained by the use of different formalisms involving either the use of the beam envelope equations, or the ray equation in a way similar to the geometrical optics. Let us therefore move on very quickly to the previous (complementary) procedures.

The quadrupole focusing equations can accordingly be derived from the Hamiltonian

$$\begin{aligned} H &= \frac{1}{2}x'^2 + V(x), \\ V(x) &= \frac{1}{2}kx^2, \end{aligned} \tag{1.201}$$

where k is the already defined quadrupole strength.

The associated Liouville equation (see (1.113)) is

$$\frac{\partial}{\partial s}F = -x'\frac{\partial}{\partial x}F + k\,x\frac{\partial}{\partial x'}F, \tag{1.202}$$

which can be solved exactly since it is just the Liouville equation for a harmonic oscillator.

A more pedagogical result is obtained from an approximate solution involving the so-called *leapfrog scheme*, which is based on the following approximation of the evolution operator

$$\hat{U} = e^{\delta s\left(-x'\frac{\partial}{\partial x} + V'(x)\right)} \cong e^{-\frac{\delta s}{2}x'\frac{\partial}{\partial x}} e^{\delta s\, V'(x)\frac{\partial}{\partial x'}} e^{-\frac{\delta s}{2}x'\frac{\partial}{\partial x}} + o(\delta s^3), \tag{1.203}$$

which yields

$$\begin{aligned} &e^{-\frac{\delta s}{2}x'\frac{\partial}{\partial x}} e^{\delta s\, V'(x)\frac{\partial}{\partial x'}} e^{-\frac{\delta s}{2}x'\frac{\partial}{\partial x}} F_0(x,\, x') \\ &= F_0\left(x - \delta s x' - \frac{\delta s^2}{2}V'\left(x - \frac{\delta s}{2}x'\right),\, x' + \delta s V'\left(x - \frac{\delta s}{2}x'\right)\right). \end{aligned} \tag{1.204}$$

The solution, based on a disentanglement of the exponential operator involving three distinct steps, can be interpreted as follows (referring in order to the three exponential terms on the rhs of equation (1.203)):

- Free propagation on a straight section of length *s*/2.
- Kick approximation of the force effect induced by the potential *V*.
- Free propagation on a straight section of length *s*/2.

The interpretation is the same as before for the solution of the Liouville equation in section 1.7, with the significant difference that the adopted exponential decoupling scheme (also called the *symmetric split* procedure) is more accurate.

As δs is a small step the solution going beyond the kick approximation is obtained by iterating the procedure n times, thus obtaining (see section 1.15 for further details)

$$\begin{aligned} &\hat{U}^n F_0(x,\, x') = F_0(x_n,\, x_n'), \\ &\quad x_n = x_{n-1} - \delta s\; x'_{n-1} - \frac{\delta s^2}{2}V'\left(x_{n-1} - \frac{\delta s}{2}x'_{n-1}\right), \\ &\quad x'_n = x'_{n-1} + \delta s V'\left(x_{n-1} - \frac{\delta s}{2}x'_{n-1}\right). \end{aligned} \tag{1.205}$$

The result of the iteration procedure is shown in figure 1.22 in which the step-by-step evolution of the phase-space distribution is shown.

It is also important to underline that each step of the iteration map represented by (1.205) provides a canonical transformation, as reported in the exercise.

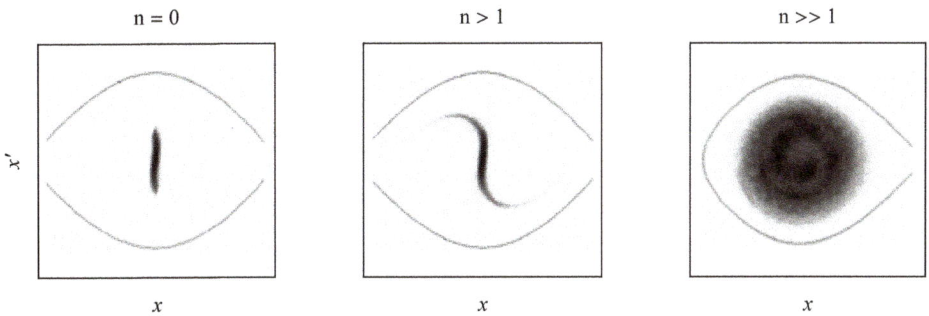

Figure 1.22. The different steps of the evolution are shown, at $n = 0$ the phase-space contour plots are represented by a canonical ellipse; with increasing n the effect of the non-linear terms in $V(x)$ becomes evident and for very large times (for strong non-linearity) a kind of vortex appears.

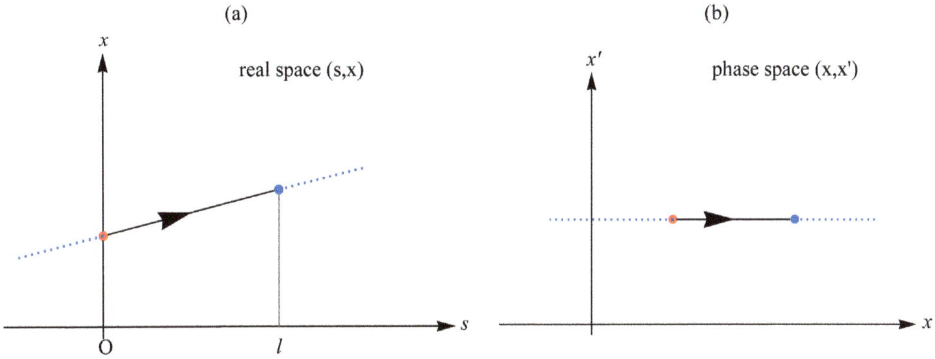

Figure 1.23. Real (a) and phase (b) space representation of free propagation (drift) (see (1.207)).

A complementary treatment of the transport is that of ray tracing. To appreciate this technique it is better to go back to the equations of motion for the quadrupole (1.194) and note that the relevant solution for a the focusing quadrupole with constant strength can be cast in the form

$$\begin{pmatrix} x \\ x' \end{pmatrix} = \begin{pmatrix} \cos(\sqrt{k}\ s) & \frac{1}{\sqrt{k}} \sin(\sqrt{k}\ s) \\ -\sqrt{k}\ \sin(\sqrt{k} s) & \cos(\sqrt{k} s) \end{pmatrix} \begin{pmatrix} x_0 \\ x_0' \end{pmatrix}. \tag{1.206}$$

The drift section is obtained as a particular case of (1.206) by setting $k = 0$ and indeed we find

$$\begin{pmatrix} x \\ x' \end{pmatrix} = \begin{pmatrix} 1 & s \\ 0 & 1 \end{pmatrix} \begin{pmatrix} x_0 \\ x_0' \end{pmatrix}, \tag{1.207}$$

which in the real and phase spaces can be represented as illustrated in figure 1.23(a) and (b), respectively.

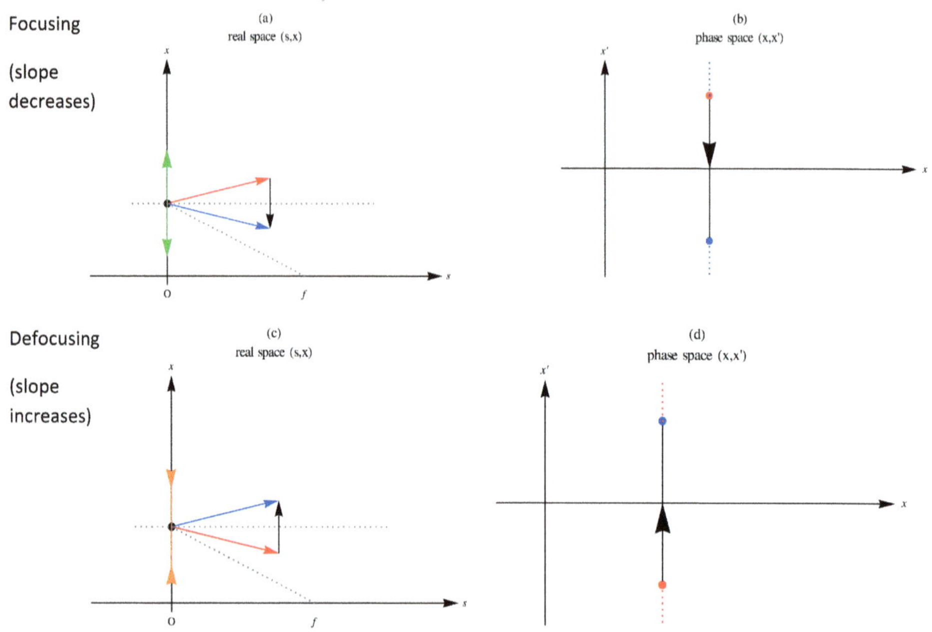

Figure 1.24. Real ((a) and (c)) and phase ((b) and (d)) space representations of the action of a quadrupole on the particle ray.

Let us now consider the case in which the quadrupole strength and thickness are such that $\sqrt{k}\ s \ll 1$, accordingly we can approximate the solution (1.206) as

$$\begin{pmatrix} x \\ x' \end{pmatrix} = \begin{pmatrix} 1 & 0 \\ -ks & 1 \end{pmatrix} \begin{pmatrix} x_0 \\ x_0' \end{pmatrix} \tag{1.208}$$

and note that the matrix resembles that of a *thin lens* with focal length

$$|f| = \frac{1}{ks}. \tag{1.209}$$

The action of a quadrupole on the *particle ray* in the thin lens approximation can accordingly be described in the respective space according to the picture reported in figure 1.24, describing the focusing and defocusing cases.

To reconcile the present picture with that of the evolution operator, the following correspondence should be noted

$$\begin{gathered} \begin{pmatrix} 1 & s \\ 0 & 1 \end{pmatrix} \Leftrightarrow e^{-sx'\frac{\partial}{\partial x}}, \\ \begin{pmatrix} 1 & 0 \\ -ks & 1 \end{pmatrix} \Leftrightarrow e^{ksx\frac{\partial}{\partial x'}}. \end{gathered} \tag{1.210}$$

We can now understand the spirit of the symmetric split approximation (SSA) summarized by (1.204), which can now be understood as

$$e^{-\frac{\delta s}{2}x'\frac{\partial}{\partial x}}e^{\delta s V'(x)\frac{\partial}{\partial x'}}e^{-\frac{\delta s}{2}x'\frac{\partial}{\partial x}} \Leftrightarrow \begin{pmatrix} 1 & \frac{s}{2} \\ 0 & 1 \end{pmatrix}\begin{pmatrix} 1 & 0 \\ -ks & 1 \end{pmatrix}\begin{pmatrix} 1 & \frac{s}{2} \\ 0 & 1 \end{pmatrix} = \begin{pmatrix} 1 - k\frac{s^2}{2} & s - \frac{ks^3}{4} \\ -ks & 1 - k\frac{s^2}{2} \end{pmatrix} \tag{1.211}$$

and viewed in the complementary spaces according to what has been sketched in figure 1.25.

Albeit the last matrix on the rhs of (1.211) is only an approximation of the exact solution, its determinant is 1. This is due to the nature of the approximation (in technical terms, called symplectic) which leaves the phase-space conservation law invariant.

Before going further let us note that the symmetric split disentanglement leading to (1.211), allowing the interpretation of the action of a quadrupole on a particle ray as half a drift section, a thin lens kick and half a drift section again, is not unique. It is indeed possible to use the alternative decomposition

$$e^{\frac{\delta s}{2}V'(x)\frac{\partial}{\partial x'}}e^{-\delta s x'\frac{\partial}{\partial x}}e^{\frac{\delta s}{2}V'(x)\frac{\partial}{\partial x'}} \Leftrightarrow \begin{pmatrix} 1 & 0 \\ -k\frac{s}{2} & 1 \end{pmatrix}\begin{pmatrix} 1 & s \\ 0 & 1 \end{pmatrix}\begin{pmatrix} 1 & 0 \\ -k\frac{s}{2} & 1 \end{pmatrix} \cong \begin{pmatrix} 1 - k\frac{s^2}{2} & s \\ -ks + \frac{k^2s^3}{4} & 1 - k\frac{s^2}{2} \end{pmatrix}, \tag{1.212}$$

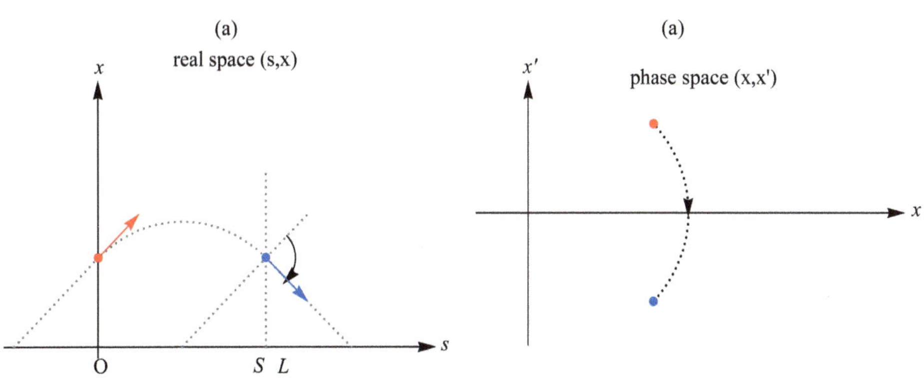

Figure 1.25. Real (a) and phase (b) space representations of the SSA (see (1.204) and (1.211)).

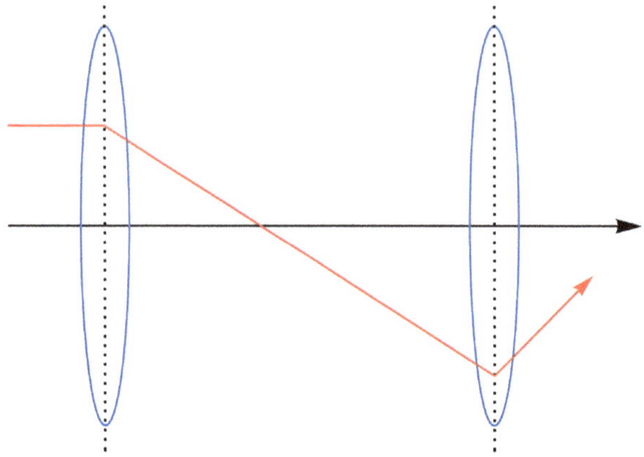

Figure 1.26. Interpretation of the action of a focusing quadrupole on a particle ray as a thin lens of thickness $s/2$, a drift of length s and again an $s/2$ thin lens.

understood as reported in figure 1.26, namely as a thin lens of thickness $s/2$, a drift of length s and again an $s/2$ thin lens (we invite the reader to check the unitarity of the last matrix). The matrices (1.211) and (1.212) are not completely equivalent, owing to the presence of second term $(\frac{k^2 s^3}{4})$ in the third entry of the matrix in (1.212). Such a term is, however, negligible within the limit of validity of the symmetric split technique which, as already stressed, holds at the order $o(s^3)$ (see section 1.15 for further comments).

Before proceeding further it is worth stating how the transport through a lens or a drift can be visualized in terms of a phase-space ellipse.

To this aim we recall that the argument of the two-variable distribution can be written as (see (1.119))

$$F_0(x, x') = \frac{1}{2\pi\varepsilon} e^{-\frac{1}{2\varepsilon}Q(x,x')}, \tag{1.213}$$

where

$$Q(x, x') = (x\ x') \begin{pmatrix} \gamma & \alpha \\ \alpha & \beta \end{pmatrix} \begin{pmatrix} x \\ x' \end{pmatrix}, \tag{1.214}$$

called, in accelerator physics, the Courant–Snyder (CS) ellipse. It is plotted in figure 1.27, where we have specified the associated Twiss parameters and the relevant physical interpretation.

As already mentioned, the orientation of the ellipse depends on the correlation parameter α and the beam is said to be in a waist when $\alpha = 0$ (see below for further comments). Any beam evolution through a magnetic element, which is not coupling the transverse planes and with negligible non-linearities, is characterized by area and shape invariant transformations of the CS ellipse. We expect therefore that during the transport the ellipse rotates around the origin, the axis may be either elongated or shrink but ε remains constant.

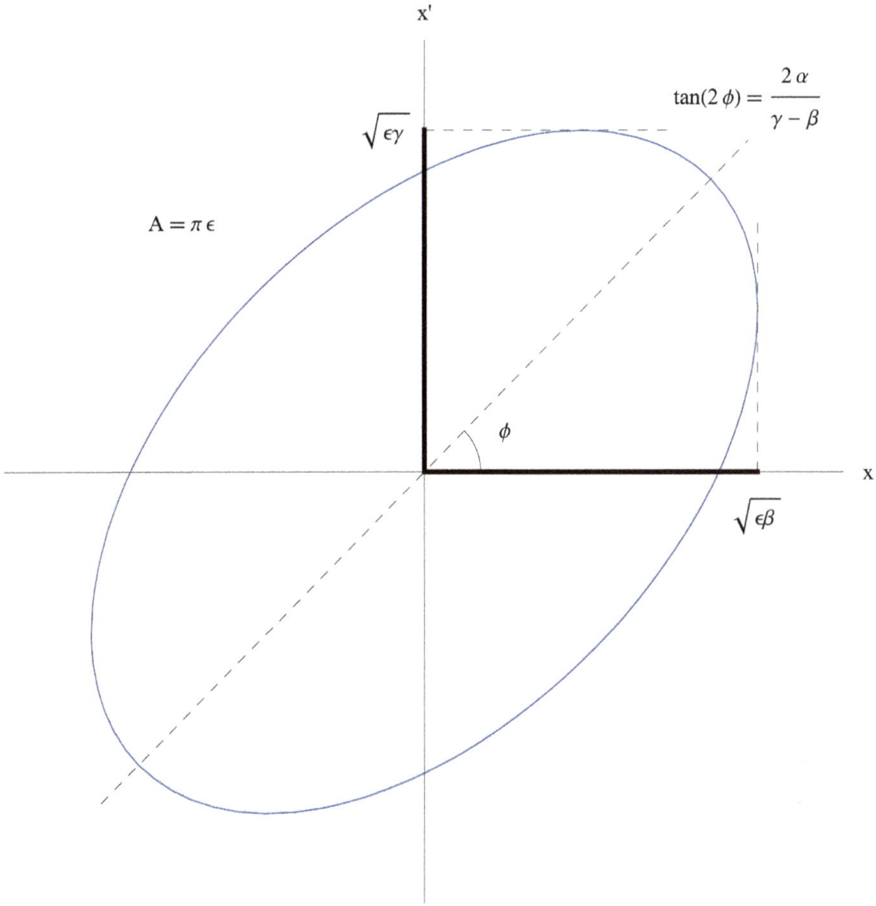

Figure 1.27. The CS ellipse.

Table 1.2. Physical units of the emittance and associated Twiss parameters.

	Unit
ε	m · rad
β	m
γ	m^{-1}
α	rad

The physical units are important and it is worth keeping in mind what is reported in table 1.2.

The evolution of the distribution along a drift section (or along any transport element discussed so far) can therefore be characterized by the ellipses transformations. This statement is particular transparent if we use the evolution operator method, which for a drifting beam yields

$$F(x, x'; s) = e^{-sx'\frac{\partial}{\partial x}}F_0(x, x') = \frac{1}{2\pi\varepsilon}e^{-\frac{1}{2\varepsilon}Q(x,x';s)}, \tag{1.215}$$

where

$$Q(x, x'; s) = (x\ x')\begin{pmatrix} 1 & 0 \\ -s & 1 \end{pmatrix}\begin{pmatrix} \gamma & \alpha \\ \alpha & \beta \end{pmatrix}\begin{pmatrix} 1 & -s \\ 0 & 1 \end{pmatrix}\begin{pmatrix} x \\ x' \end{pmatrix}. \tag{1.216}$$

By transferring the s dependence to the Twiss coefficients, we can rearrange (1.216) in the form

$$Q(x, x'; s) = (x\quad x')\begin{pmatrix} \gamma(s) & \alpha(s) \\ \alpha(s) & \beta(s) \end{pmatrix}\begin{pmatrix} x \\ x' \end{pmatrix},$$
$$\gamma(s) = \gamma,\ \alpha(s) = \alpha - \gamma s,\ \beta(s) = \beta - 2\alpha s + \gamma s^2. \tag{1.217}$$

In figure 1.28 we have reported the phase-space dynamics of a beam characterized by an initially correlated phase-space ellipse undergoing free propagation. The ellipse in figure 1.28(a), after a drift section, is turned into a rotated ellipse with the same β and the opposite correlation parameter (figure 1.28(b)), after a further step it acquires a more elongated form since β increases (figure 1.28(c)).

Regarding the behavior of the Twiss parameter versus s, we note that from the plot sketched in figure 1.29 it is evident that the section of the beam (proportional to $\sqrt{\beta}$) is minimized in correspondence of $\alpha = 0$, where the beam is said to be in a *waist*.

Regarding the thin lens quadrupole we can proceed as before, by considering that the Twiss coefficients of the quadratic form can be written as

$$Q(x, x'; s) = (x\quad x')\begin{pmatrix} 1 & ks \\ 0 & 1 \end{pmatrix}\begin{pmatrix} \gamma & \alpha \\ \alpha & \beta \end{pmatrix}\begin{pmatrix} 1 & 0 \\ ks & 1 \end{pmatrix}\begin{pmatrix} x \\ x' \end{pmatrix},$$
$$\gamma(s) = \gamma + 2\alpha ks + (ks)^2\beta,$$
$$\alpha(s) = \alpha + ks\beta, \tag{1.218}$$
$$\beta(s) = \beta.$$

In figure 1.30 we have reported an initially uncorrelated phase-space ellipse (figure 1.30(a)) undergoing the effect of a focusing (figure 1.30(b)) and defocusing (figure 1.30(c)) quadrupole. The sign of the tilt determining the correlation and the ellipse orientation depends on the nature of the quadrupole itself.

To further stress the CS ellipse's gymnastics during the transport, we report the examples in figure 1.31 and figure 1.32 for drift and quadrupole, respectively.

The phase-space ellipse dynamics induced by free propagation is easily understood if viewed as reported in figure 1.31. During the propagation the x-variable increases, while the ray divergence remains unaffected.

In figure 1.32 we have reported a slightly more complicated configuration, namely a quadrupole followed by a drift.

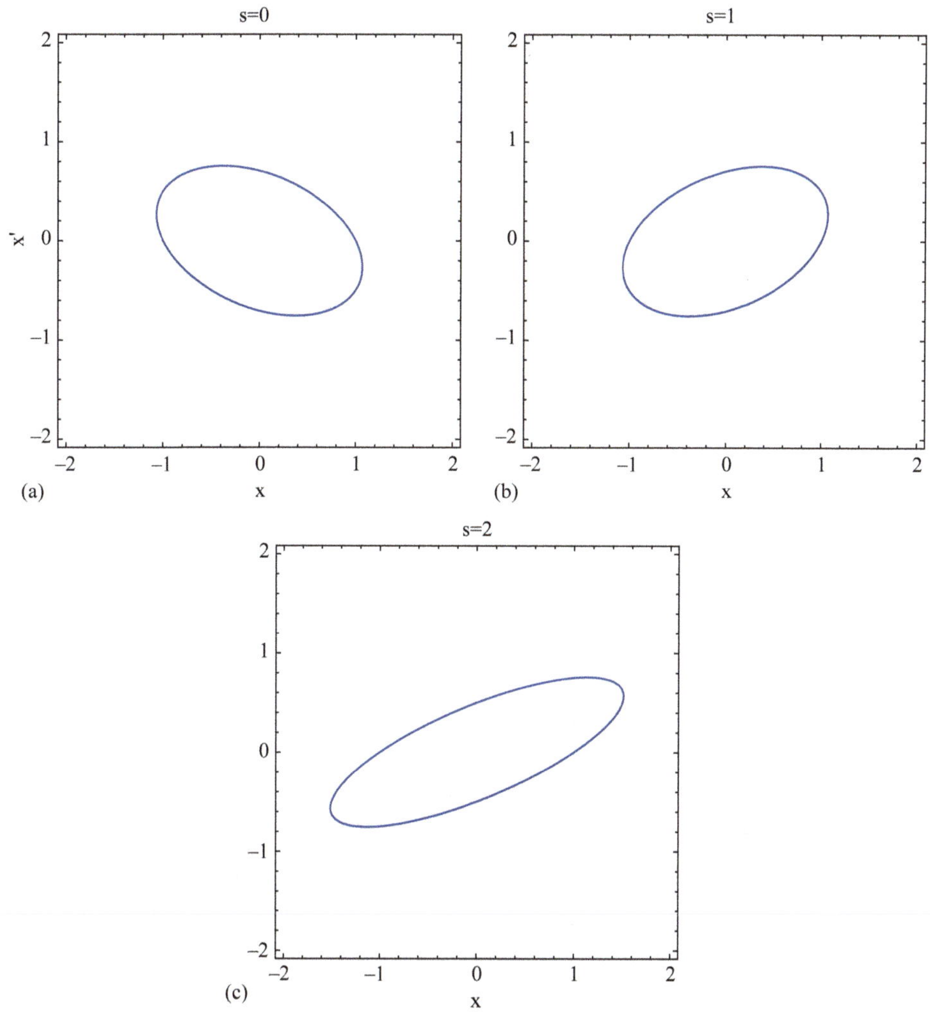

Figure 1.28. The phase-space dynamics of an initially correlated phase-space ellipse (a), after a drift section of 1 m it is simply rotated into an ellipse with the same Twiss parameters but with a negative correlation term (b), and after a further step β increases and the correlation term becomes larger in absolute value (c).

In this section we have provided the very first few elements to understand how charged beams and, in particular, electrons can be driven through a transport system and can be bent and focused as with ray beams in geometric optics. As we have mentioned the analogy with geometric optics, we want accordingly to stress that there is no 'wave optics' analogy between the charged beam and wave transport. Some speculation has been made in this direction in the past, but the output of such a point of view is not particularly encouraging.

In the forthcoming section we will provide further elements, allowing further confidence in the method.

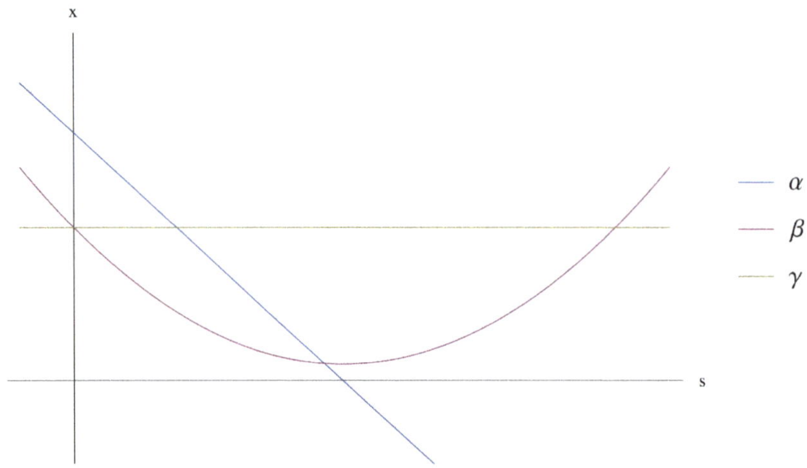

Figure 1.29. Sketch of the beam waist definition.

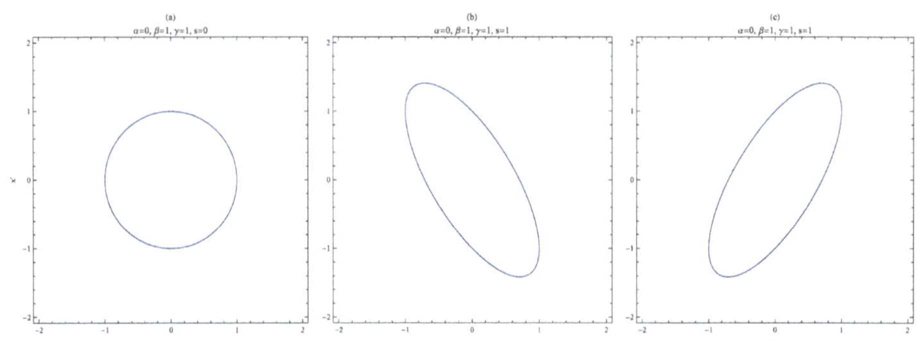

Figure 1.30. The phase-space dynamics of an initially uncorrelated phase-space ellipse (a) undergoing the effect of a focusing (b) and defocusing quadrupole (c).

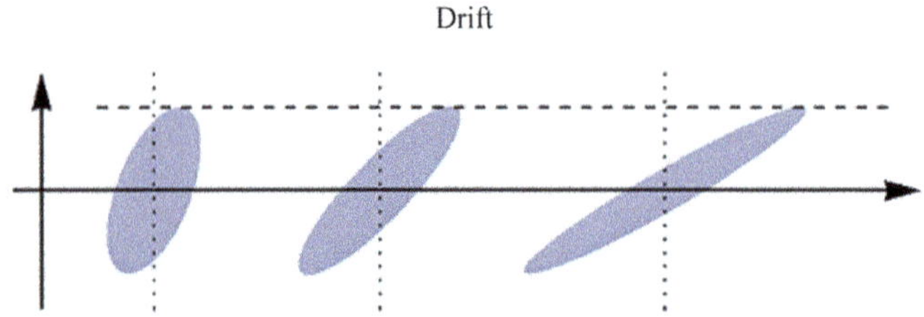

Figure 1.31. Ellipse gymnastic for drift.

1.13 Composed transport devices and the matrix formalism

Before entering the specific details of the discussion of this section, we want to underline that a given arrangement of transport elements—whose action on the beam is expressed either using an evolution operator or the corresponding matrix picture—requires the correct ordering of applications of the operators or of the

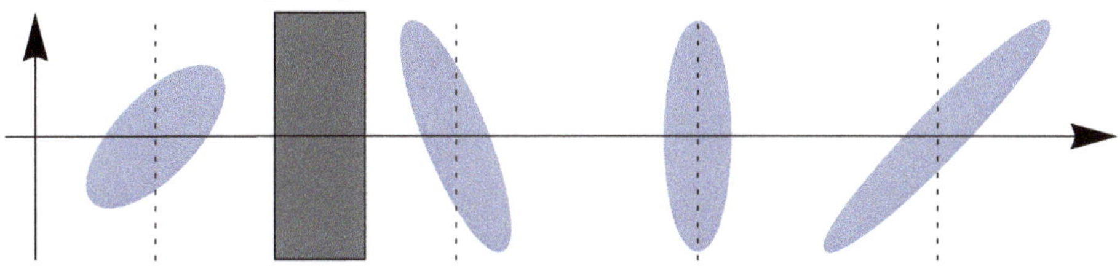

Figure 1.32. Ellipse gymnastics for the configuration of a quadrupole followed by a drift.

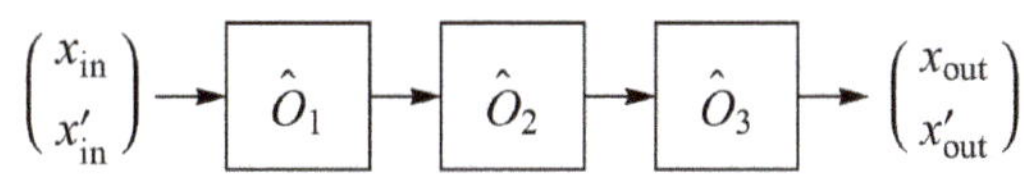

Figure 1.33. Effect of successive transport elements on an input ray beam, each element is represented by the operator $\hat{O}$.

corresponding matrix. In figure 1.33 we have reported such a sequence, in which the transport elements are denoted by $\hat{O}_n$, $n = 1, 2, 3$. The beam is assumed to cross the array from left to right and therefore we write

$$\begin{pmatrix} x_{out} \\ x'_{out} \end{pmatrix} = \hat{O}_3\left[\hat{O}_2\left[\hat{O}_1\begin{pmatrix} x_{in} \\ x'_{in} \end{pmatrix}\right]\right], \tag{1.219}$$

and the order is essential not only from the physical point of view, since, in general, the operators $\hat{O}_n$ are not commuting quantities. Albeit trivial, the above caveat should be considered very seriously to avoid catastrophic mistakes in the design of a transport line.

We will now consider a few specific examples regarding devices often exploited in charged beam transport devices, which suggest how such devices can be embedded to results in an actual transport line. To begin, we borrow examples from ray optics and describe a number of devices below. We invite the reader to 'translate' the device in terms of charged beam transport elements.

Consecutive thin lenses. According to the convention (in ordinary optics called the Cartesian convention) the ray travels from left to right and, assuming that the ray crosses the lenses in the sequence 1, 2, ..., we find

$$\begin{aligned} {}_L\hat{T} &= {}_{L_n}\hat{T}\,{}_{L_{n-1}}\hat{T}\,.....\,{}_{L_1}\hat{T} = \begin{pmatrix} 1 & 0 \\ -\frac{1}{f} & 1 \end{pmatrix}, \\ \frac{1}{f} &= \sum_{i=1}^{n}\frac{1}{f_i}. \end{aligned} \tag{1.220}$$

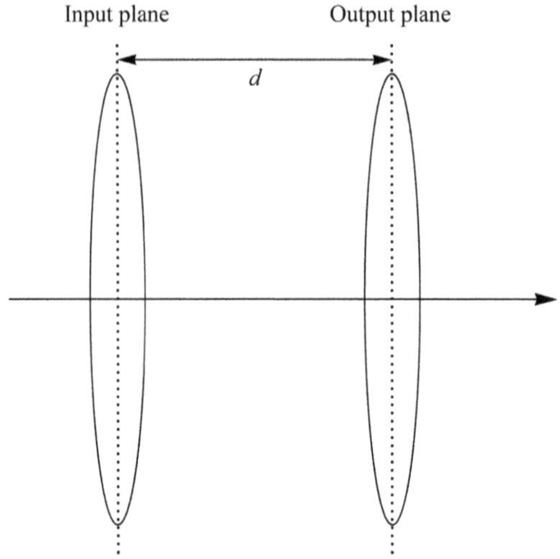

Figure 1.34. Two lenses separated by a drift section.

Lens–straight-section–lens. With reference to figure 1.34 we consider an optical device composed of a convex lens, a straight section and a second converging lens. By simply applying the rules stipulated so far, we obtain the following transfer matrix describing the ray transport system:

$$\hat{T} = \begin{pmatrix} 1 & 0 \\ -\dfrac{1}{f_2} & 1 \end{pmatrix} \begin{pmatrix} 1 & d \\ 0 & 1 \end{pmatrix} \begin{pmatrix} 1 & 0 \\ -\dfrac{1}{f_1} & 1 \end{pmatrix} = \begin{pmatrix} 1 - \dfrac{d}{f_1} & d \\ -\left(\dfrac{1}{f_2} + \dfrac{1}{f_1} - \dfrac{d}{f_1 f_2}\right) & 1 - \dfrac{d}{f_2} \end{pmatrix}. \tag{1.221}$$

We have dwelt on this point to put in evidence that the above equation corresponds to the already quoted representation of a quadrupole as the sequence thin lens–straight section–thin lens, addressed in (1.212). In optics this approximation is called the Gulstrand equation.

Afocal lens system. In figure 1.35 we show a particularly interesting application of the device discussed in the previous subsection, consisting of two lenses separated by a drift section $d = f_1 + f_2$. The lens system is said to be *afocal* in the sense that it transforms a parallel ray into another (inverted) parallel ray.

We find from (1.221) and an initially parallel ray

$$\begin{pmatrix} y_f \\ y'_f \end{pmatrix} = \begin{pmatrix} -\dfrac{f_2}{f_1} & f_1 + f_2 \\ 0 & -\dfrac{f_1}{f_2} \end{pmatrix} \begin{pmatrix} y_i \\ 0 \end{pmatrix} = \begin{pmatrix} -\dfrac{f_2}{f_1} y_i \\ 0 \end{pmatrix}. \tag{1.222}$$

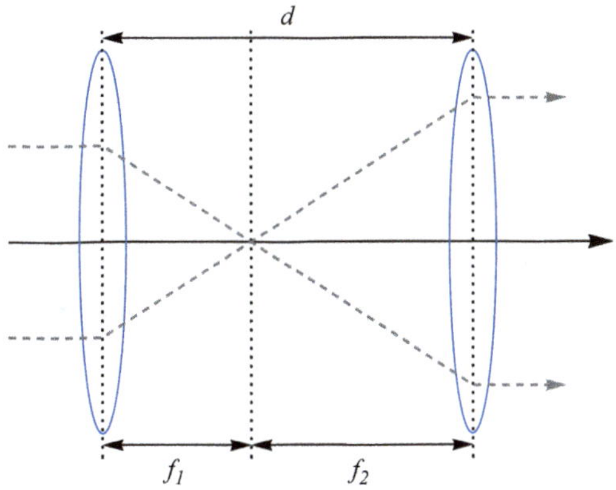

Figure 1.35. The geometry of an afocal optical device.

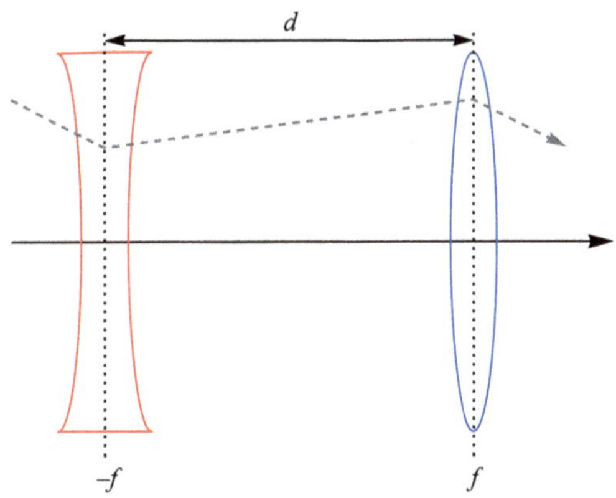

Figure 1.36. Optical device corresponding to an *FOD* configuration. With reference to this last point we consider the operator

It is worth stressing that the system acts as a telescope if $\frac{f_2}{f_1} > 1$. Further comments on telescopic systems can be found in the exercises at the end of the chapter.

The reader can now consider the optical device reported in figure 1.36 and show that the associated transport matrix can be written as

$$\hat{T} = \hat{F}\ \hat{O}\ \hat{D}, \tag{1.223}$$

with the convention $\hat{F}$ = focusing, $\hat{O}$ = straight section and $\hat{D}$ = defocusing,

$$\hat{T} = \begin{pmatrix} 1 + \frac{d}{f} & d \\ -\frac{d}{f^2} & 1 - \frac{d}{f} \end{pmatrix}. \tag{1.224}$$

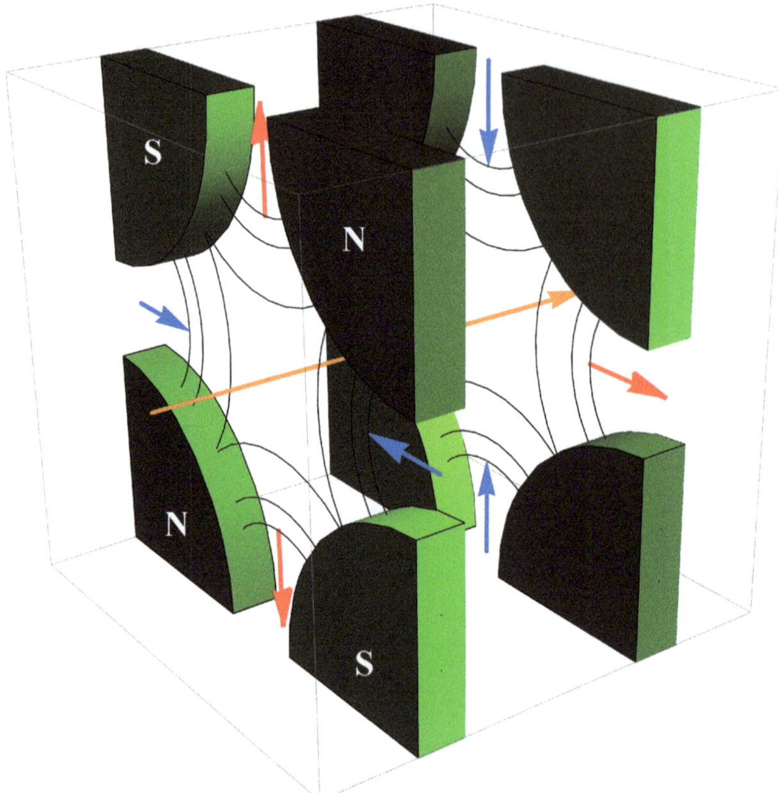

Figure 1.37. Three-dimensional picture of an $\hat{F}\ \hat{O}\ \hat{D}$ quadrupole arrangement.

We invite the reader to prove that the net action of the device is that of a focusing lens for a $0 < d \ll f$ result holding for the arrangement $\hat{D}\ \hat{O}\ \hat{F}$.

$$\hat{T}^{-1} = \hat{D}^{-1}\hat{O}^{-1}\hat{F}^{-1}, \tag{1.225}$$

equivalent to a kind of *'time reversal'* in which a converging lens is turned into its diverging counterpart, namely

$$\begin{aligned} \hat{D}^{-1} &\to \hat{F}, \\ \hat{F}^{-1} &\to \hat{D}, \end{aligned} \tag{1.226}$$

only because we have replaced $s \to -s$. In figure 1.37 we have provided a three-dimensional picture of an $\hat{F}\ \hat{O}\ \hat{D}$ quadrupole arrangement, including the force distribution, to give a more effective picture of the action of this device on the transport.

In figure 1.38 we have reported an *FODO* lens arrangement, namely a defocusing lens sandwiched between two focusing elements. This tool is of crucial importance in the design of beam transport lattices, whose elementary cell is realized as sketched in figure 1.37 and is provided by

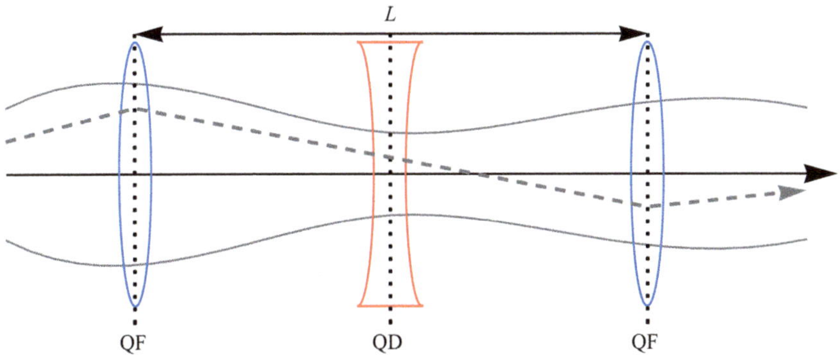

Figure 1.38. Effect of the *FODO* on the ray trajectory (dashed line) and on the beam envelope (continuous line).

$$\hat{T}_{FODO} = \left(\hat{F}_{\frac{d}{2}}\hat{O}_L\hat{D}_{\frac{d}{2}}\right) \times \left(\hat{D}_{\frac{d}{2}}\hat{O}_L\hat{F}_{\frac{d}{2}}\right). \tag{1.227}$$

If the focal lengths of the focusing and defocusing parts are $f_{1,2}$, respectively, we find

$$\hat{T}_{FODO} = \begin{pmatrix} 1 - 2\frac{L}{f^*} & 2L\left(1 - \frac{L}{2f_2}\right) \\ -\frac{2}{f^*}\left(1 - \frac{L}{2f_1}\right) & 1 - 2\frac{L}{f^*} \end{pmatrix}, \tag{1.228}$$
$$\frac{1}{f^*} = \frac{1}{2}\left(\frac{1}{f_1} + \frac{1}{f_2}\right) - \frac{L}{4f_1f_2}.$$

For $f_1 = -f_2 = f$ we find

$$\hat{T}_{FODO} = \begin{pmatrix} 1 - \frac{L^2}{2f^2} & 2L\left(1 + \frac{L}{2f}\right) \\ -\frac{L}{2f^2}\left(1 - \frac{L}{2f}\right) & 1 - \frac{L^2}{2f^2} \end{pmatrix}. \tag{1.229}$$

We have insisted on the above (tedious) exercises in matrix multiplication, which may sound useless in this epoch in which sophisticated computer codes calculate everything. It is indeed always mandatory to try a preliminary design of the transport elements with paper and pencil to obtain a qualitative understanding of how things may go and avoid serious mistakes. In the comments and exercises (section 1.15) we will refer to actual design configurations, on which the reader can practice and test his/her level of understanding.

We can now ask what happens to the phase-space ellipse (and thus to the beam distribution) in an *FODO* element. The relevant picture is shown in figure 1.39, where we show a beam initially at the waist ($\alpha = 0$), which, according to the

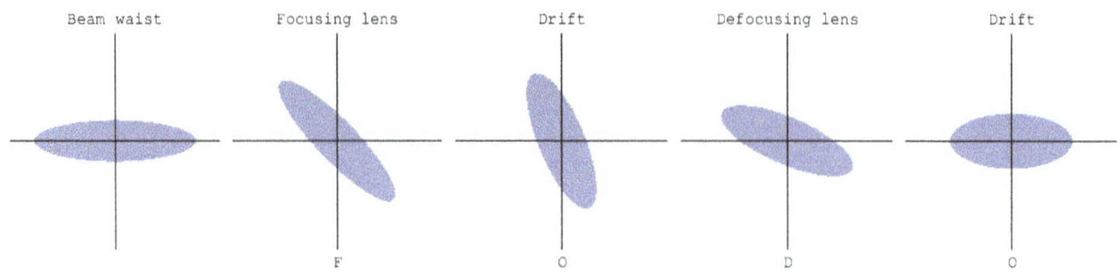

Figure 1.39. Phase-space ellipse evolution in an *FODO* element.

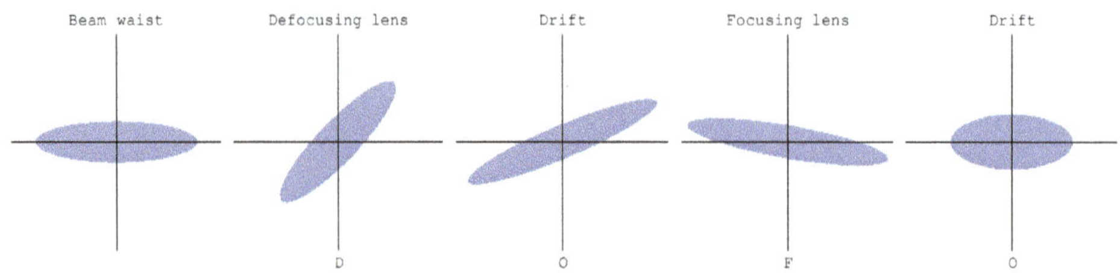

Figure 1.40. Phase-space ellipse evolution in a *DOFO* element.

discussion of the previous section, undergoes a tilt in the focusing quadrupole, acquiring an $x - x'$ correlation; in the drift section it experiences a reduction of the β parameter, leading to a reduction of the transverse size; finally a second tilt in the defocusing part brings the ellipse toward the waist condition, which is completely restored in the second drift. The effect of the device is that of providing a net (slight) focusing.

The specular configuration is shown in figure 1.40 along with the associate phase-space ellipse evolution, which does not need any further comment.

The considerations developed so far allow the understanding of the beam transport through a simple configuration of transport elements. In the forthcoming section we will discuss the role of non-linear transport elements.

What we have learned so far has provided the basic tools to understand the problems associated with charged particles in transport devices. We have considered particles brought to sufficiently high energy (via an accelerating mechanism which we will discuss in the following chapters) and we did not include any contribution from electric forces. We have, however, not addressed at least three important points:

a) Providing a sense of how the concepts we have developed work in a real-life transport design.
b) Providing an idea of how transverse plane coupling effects should be included in the scenario.
c) Discussing the role of the non-linearity of the transport elements which have been neglected thus far.

1.14 Beam transport: a few pragmatic remarks

In this section we will keep the discussion at a pragmatic level and we will avoid, whenever possible, mathematical details, which will be found in section 1.15.

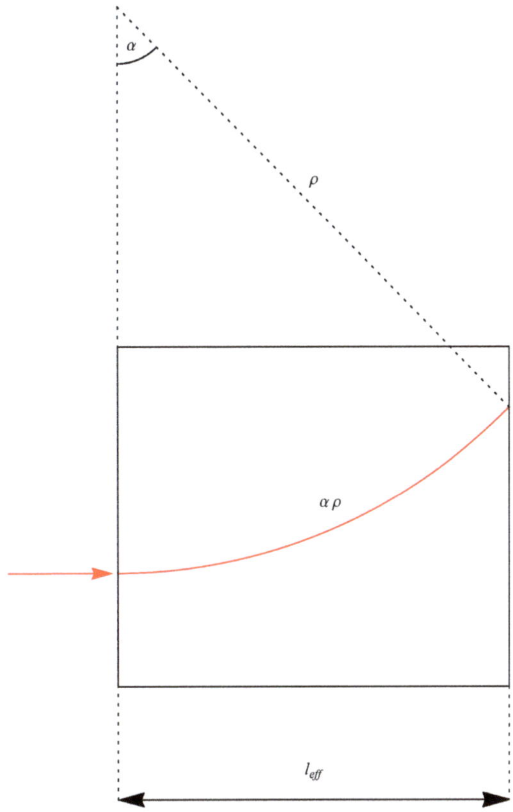

Figure 1.41. Definition of the magnet effective length l_{eff}.

We have mentioned the importance of the bending magnet as a transport element; here we will delve deeply into the relevant properties by discussing its associated uses as a measurement tool.

We will first show how the notions we have discussed can be combined to exploit a bending magnet to measure the particle momentum. According to the previous results and referring to figure 1.41, we can link particle momentum, magnetic field and the trajectory curvature radius of an electron moving in a bending magnet as

$$m_e \gamma v = eB\rho, \tag{1.230}$$

with

$$\rho = \frac{l_{\mathrm{eff}}}{\sin(\alpha)} \tag{1.231}$$

with l_{eff} denoting the magnet effective length, we find

$$m_e \gamma v = \frac{eBl_{\mathrm{eff}}}{\sin(\alpha)}. \tag{1.232}$$

Furthermore, by using table 1.1, we infer that the relativistic energy is linked to the bending parameters by

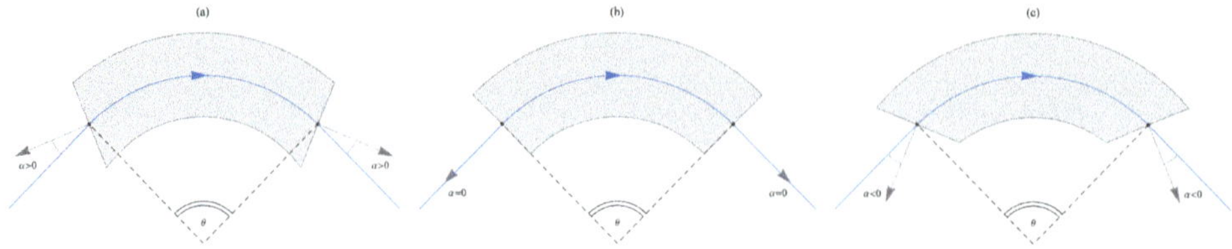

Figure 1.42. The trajectory of an electron entering a bending magnet at different angle with respect to the normal to the magnet pole face.

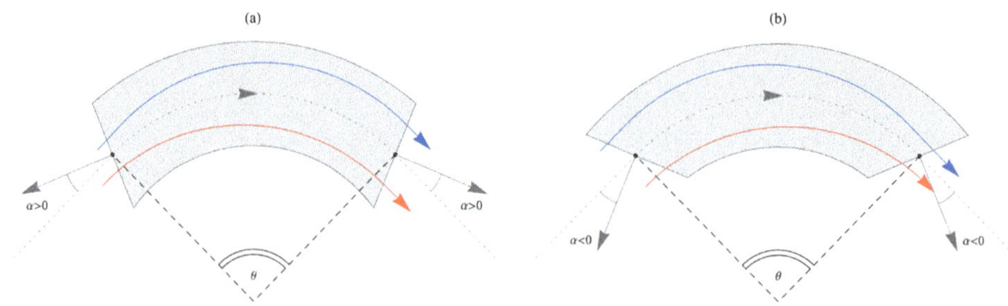

Figure 1.43. Defocusing for $\alpha > 0$, focusing for $\alpha < 0$.

$$\begin{aligned} \mathcal{E} &= m_e c^2 \sqrt{1 + \frac{K^2}{\sin(\alpha)^2}}, \\ K &= \frac{eBl_{\text{eff}}}{m_e c^2} \cong 5.87 \cdot 10^2 B[\text{T}] l_{\text{eff}}[\text{m}]. \end{aligned} \tag{1.233}$$

Therefore, once the magnetic field and its effective length are known, the value of the bending angle corresponds to that of the energy and thus to momentum, according to the summary of table 1.1.

The concept of magnet effective length is less trivial than might be thought; we invite the reader to consider this point carefully and examine the comments in section 1.15.

The transport in a bending magnet is more complicated than we have described. This is due either to geometrical reasons and to the fact that the bending magnetic fields is 'obliged' to fulfill Maxwell's equations. The geometrical reasons are summarized in figure 1.42: the trajectory of an electron entering a bending magnet is not always orthogonal to the magnet pole faces. If we denote by α the angle formed by the electron path and the normal to the pole face, we can draw significant conclusions on the nature of the focusing itself, according to the sign of α.

An idea of the relevant consequences is given in figure 1.43, where we have considered (panel (a))

$$\alpha > 0.$$

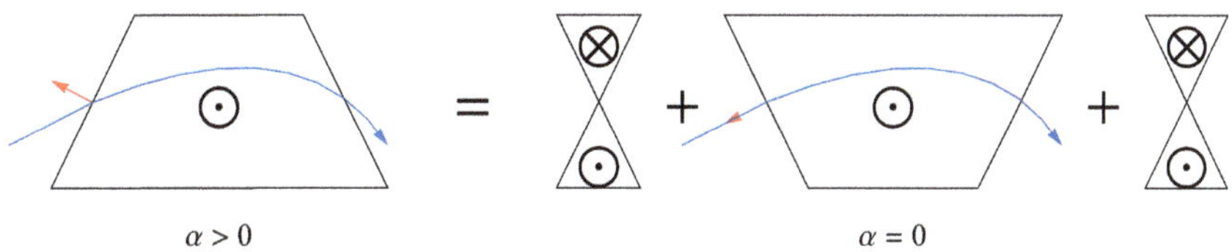

Figure 1.44. Bending magnet decomposed into a sector magnet sandwiched between two defocusing items.

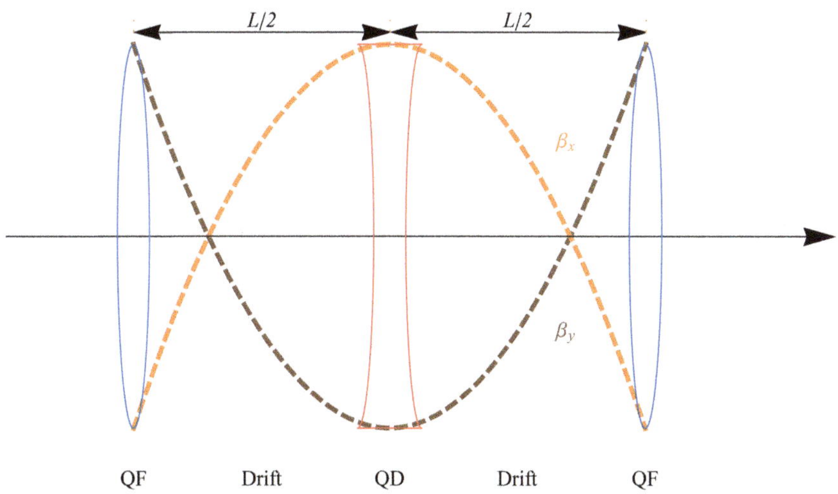

Figure 1.45. The behavior of the relevant β Twiss coefficients in an *FODO* cell.

The particles lying on the upper part experience less magnetic length (and thus less field) than those moving in the lower part, the net effect is therefore a defocusing (panel (b))

$$\alpha < 0.$$

The specular situation occurs with respect to the previous case and accordingly the net effect is a focusing.

To treat the problem of propagation in terms of transport matrices, we note that for the case $\alpha > 0$, the bending magnet can be decomposed as shown in figure 1.44(a), with a sector magnet sandwiched between two defocusing items. In these regions the ray trajectory behaves as reported in figure 1.44(b). In conclusion, as detailed in the next section of this chapter, the transport of the matrix can naïvely be written as the product of three matrices, each one corresponding to the magnetic split of figure 1.44.

We have so far described the building elements of a transport device which are then embedded to provide an actual guiding tool, which can be rather complicated, since it may be composed by different transport elements. In figure 1.45 we report an *FODO* cell to which we have superimposed the behavior of the $\beta_{x,y}$ Twiss coefficients, note the *DOFO*-like behavior of the vertical part. By a cell we mean the fundamental constituent of a sequence in which the Twiss parameters are required to have periodic behavior.

The question is now: however complicated such an arrangement, are we able to write the relevant transport matrix using what is in our hands, namely the focusing (defocusing) properties and the Twiss parameters?

From the mathematical point of view we should be able to find a convenient parameterization for the transport matrix

$$\begin{pmatrix} x \\ x' \end{pmatrix}_{\text{out}} = \hat{T}\begin{pmatrix} x \\ x' \end{pmatrix}_{\text{in}}, \tag{1.234}$$

which is valid however complicated the transport line is.

In the case in which the lattice is provided by n identical *FODO* cells, the total (start to end) lattice matrix is specified by

$$\hat{T} = [\hat{T}_{FODO}]^n. \tag{1.235}$$

Before entering the details of the discussion we recall that, given a 2 × 2matrix

$$\hat{A} = \begin{pmatrix} a & b \\ c & d \end{pmatrix} \tag{1.236}$$

with two distinct eigenvalues λ_+, λ_-, then, as a consequence of the Cayley–Hamilton theorem, its nth power can be written as

$$\hat{A}^n = \frac{\lambda_+^n\lambda_- - \lambda_-^n\lambda_+}{\lambda_- - \lambda_+}\hat{1} + \frac{\lambda_+^n - \lambda_-^n}{\lambda_+ - \lambda_-}\hat{A}. \tag{1.237}$$

In order to avoid instabilities we require that ($|\lambda_\pm| = 1$), and therefore the most natural choice is

$$\lambda_\pm = e^{\pm i\,\mu}. \tag{1.238}$$

To specify a convenient parameterization of the transport matrix, we note that

$$\begin{aligned} \lambda_\pm &= \frac{a + d \pm \sqrt{(a + d)^2 - 4|A|}}{2}, \\ a + d &= \lambda_+ + \lambda_-, \\ |A| &= \det(A). \end{aligned} \tag{1.239}$$

Therefore, according to our assumption (1.238), and to the fact that our transport matrix must be unit determinant, we have that

$$\begin{aligned} a + d &= 2\cos(\mu), \\ |A| &= 1, \end{aligned} \tag{1.240}$$

which is, e.g. satisfied by the rotation matrix

$$A = \begin{pmatrix} \cos(\mu) & \sin(\mu) \\ -\sin(\mu) & \cos(\mu) \end{pmatrix}, \tag{1.241}$$

which is not useful for our purposes, since the transport is not just a rotation of coordinates. Such a parameterization is, however, not unique and indeed we choose the alternative form

$$A = \begin{pmatrix} \cos(\mu) + K\sin(\mu) & L\sin(\mu) \\ -M\sin(\mu) & \cos(\mu) - K\sin(\mu) \end{pmatrix}, \tag{1.242}$$

which satisfies the conditions (1.240) provided that

$$K^2 - LM = -1. \tag{1.243}$$

This last identity suggests that the free quantities K, L, M can be identified with something playing the role of Twiss parameters. (These values should not be confused with the Twiss parameters of the beam, but those defined by the transport structure; the roles of the two sets of Twiss parameters is discussed at the end of this section.) Therefore in conclusion we find (l stands for lattice)

$$\hat{T}_{FODO} = \cos(\mu)\hat{1} + \sin(\mu)\begin{pmatrix} \alpha_l & \beta_l \\ -\gamma_l & -\alpha_l \end{pmatrix}. \tag{1.244}$$

The result is general and does not apply to *FODO* lattices only; it is called strong focusing parametrization.

Let us now describe how the previous parameterization can be exploited to treat the *FODO* lattice, consisting of a number of cells of the type reported in figure 1.45. In terms of drift space and quadrupole length we find

$$\hat{T}_{FODO} = \begin{pmatrix} 1 & \frac{l}{2} \\ 0 & 1 \end{pmatrix}\begin{pmatrix} 1 & 0 \\ \frac{1}{f} & 1 \end{pmatrix}\begin{pmatrix} 1 & \frac{l}{2} \\ 0 & 1 \end{pmatrix}\begin{pmatrix} 1 & 0 \\ -\frac{1}{f} & 1 \end{pmatrix} = \begin{pmatrix} 1 - \frac{l}{2f} - \left(\frac{l}{2f}\right)^2 & l + \frac{l^2}{4f} \\ -\frac{l}{2f^2} & 1 + \frac{l}{2f} \end{pmatrix}, \tag{1.245}$$

where we have assumed that the beam moves as indicated by the arrow in figure 1.45 and we have used the kick approximation. In comparison to the *FODO* matrix in (1.229), note that it has been derived with the assumption of 'thick lens' approximation and for $\frac{l}{2} = L$.

By comparing the entries of (1.245) and (1.244) keeping $\alpha = 0$, we end up with the following identifications,

$$\begin{aligned} 1 - \frac{1}{2}\left(\frac{l}{2f}\right)^2 &= \cos(\mu), \\ l + \frac{l^2}{4f} &= \beta\sin(\mu), \end{aligned} \tag{1.246}$$

which can be exploited to obtain the maximum and minimum values of the β coefficient inside the lattice.

The derivation proceeds as follows. From the first equation of (1.246) we find

$$\sin\left(\frac{\mu}{2}\right) = \pm\frac{l}{4f}, \tag{1.247}$$

while the second yields

$$\beta = \frac{l}{\sin(\mu)}\left(1 + \frac{l}{4f}\right) = \frac{l}{\sin(\mu)}\left(1 \pm \sin\left(\frac{\mu}{2}\right)\right). \tag{1.248}$$

Therefore, we end up with

$$\begin{aligned} \beta_{\mathrm{MAX}} &= \frac{l}{\sin(\mu)}\left(1 + \sin\left(\frac{\mu}{2}\right)\right), \\ \beta_{\mathrm{MIN}} &= \frac{l}{\sin(\mu)}\left(1 - \sin\left(\frac{\mu}{2}\right)\right). \end{aligned} \tag{1.249}$$

It is worth noting that (1.244) reduces to a rotation matrix for $\alpha = 0$, in such a way that

$$\hat{T}_{FODO} = \begin{pmatrix} \cos(\mu) & \beta\sin(\mu) \\ -\frac{1}{\beta}\sin(\mu) & \cos(\mu) \end{pmatrix}. \tag{1.250}$$

The previous results are however incomplete, because we have mixed up Twiss parameters (which are quantities describing the beam distributions) and transport parameters (which are associated with the transport elements themselves). We should therefore add a further element to the discussion by defining the concept of beam matching to the lattice. The idea is perhaps better expressed in figure 1.46, in which we have drawn the CS ellipse of the beam at the entrance of the lattice and that of the lattice itself. Matching is achieved when, at the entrance of the transport line, the beam ellipse 'coincides' with that of the lattice element.

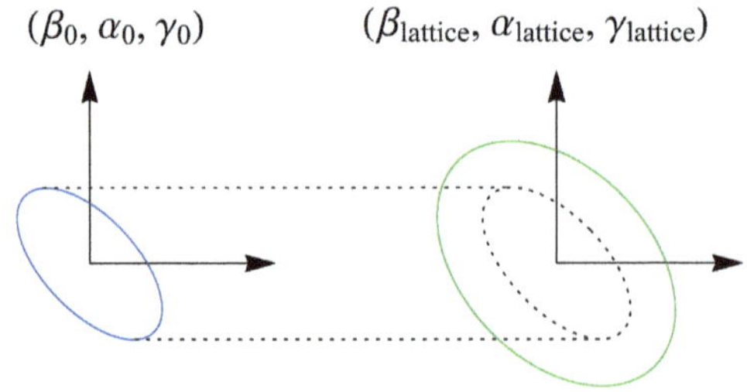

Figure 1.46. Matching is achieved when, at the entrance of the transport line, the beam ellipse 'coincides' with that of the lattice element.

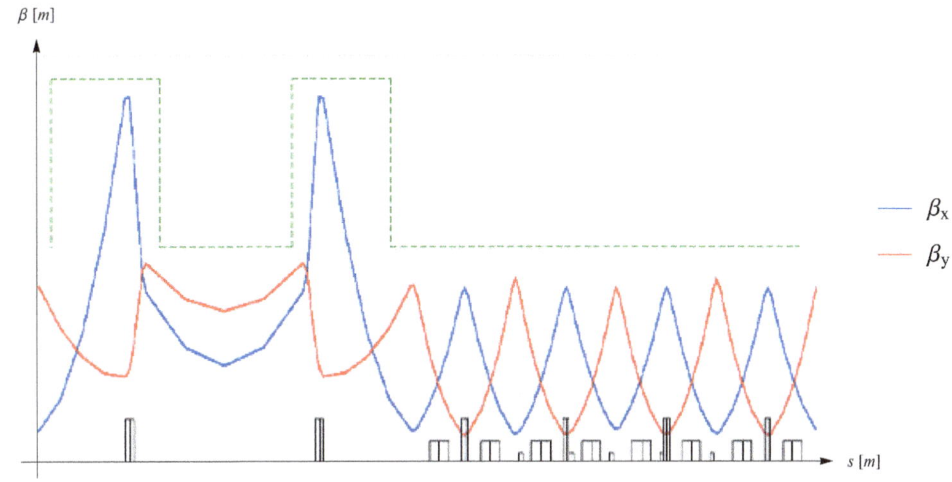

Figure 1.47. β parameter behavior inside the lattice. The graph shows the matching procedure at the entrance of the lattice, characterized by the strong variation of the Twiss parameters β. The matching is realized using ordinary quadrupoles. Inside the lattice β have the periodic behavior discussed previously.

The design procedure of a transport line should therefore be that reported in figure 1.47. The beam is transported to the lattice entrance in such a way that the beam ellipse is 'adapted' to that of the transport line and the β parameters behave as shown in the figure. Such a procedure allows the possibility of containing the beam inside the vacuum pipe (denoted by a dashed line in figure 1.47) of the accelerator without determining any losses.

The periodic behavior of the Twiss parameters inside the lattice and the fact that at the exit of any *FODO* the CS ellipse is the same, do not imply that the particles acquire the same position. This is a delicate and important concept, which albeit not particularly relevant for the topics treated in this book, will be touched on in the forthcoming part of the chapter devoted to comments and exercises.

1.15 Complements and exercises

1.15.1 Relativistic kinematics

Any event in the Minkowski space-time is specified by a quadrivector x^μ, accounting for its 'location' in time and space, namely

$$\begin{aligned} x^\mu &\equiv (\tau, \vec{x}), \\ \tau &= ct, \end{aligned} \tag{1.251}$$

with $\vec{x}$ being an ordinary vector in three-dimensional space. Along with x^μ we define its contravariant counterpart

$$x_\mu = \eta_{\mu,\nu} x^\nu, \tag{1.252}$$

where

$$\eta_{\mu,\nu} = \begin{pmatrix} 1 & 0 & 0 & 0 \\ 0 & -1 & 0 & 0 \\ 0 & 0 & -1 & 0 \\ 0 & 0 & 0 & -1 \end{pmatrix} \tag{1.253}$$

is the metric tensor, yielding its norm defined as (the convention of summing over the repeated indices is used)

$$\eta_{\mu,\nu} x^{\nu} x^{\mu} = x_{\mu} x^{\mu} = \tau^2 - |\vec{x}|^2. \tag{1.254}$$

The kinematic quantities of energy and momentum are contained in a quadrivector as well, specified by

$$p^{\mu} \equiv (E, \vec{p}\,c). \tag{1.255}$$

Exercise 1.1 Use the identity

$$p_{\mu} p^{\mu} = E^2 - c^2\, |\vec{p}\,|^2 = m^2 c^4 \tag{1.256}$$

to justify the content of table 1.1.

Exercise 1.2 Use the identity

$$\gamma^2 - (\gamma\beta)^2 = 1$$

to make the following identification

$$\begin{aligned} \gamma &= \cosh(\alpha), \\ \gamma\beta &= \sinh(\alpha). \end{aligned} \tag{1.257}$$

Restricting yourself to the case in which $\vec{x} \equiv (x, 0, 0)$, use figure 1.48 to provide the geometric interpretation of the Minkowski space time and prove that the Lorentz transformations can be written as

$$\begin{pmatrix} x' \\ \tau' \end{pmatrix} = \begin{pmatrix} \cosh(\alpha) & -\sinh(\alpha) \\ -\sinh(\alpha) & \cosh(\alpha) \end{pmatrix} \begin{pmatrix} x \\ \tau \end{pmatrix}. \tag{1.258}$$

Show that the above transformation preserves the 'norm'

$$x_{\mu} x^{\mu} = \tau^2 - x^2.$$

Show that two successive Lorentz transformations can be written as

$$\begin{pmatrix} x' \\ \tau' \end{pmatrix} = \begin{pmatrix} \cosh(\alpha_1 + \alpha_2) & -\sinh(\alpha_1 + \alpha_2) \\ -\sinh(\alpha_1 + \alpha_2) & \cosh(\alpha_1 + \alpha_2) \end{pmatrix} \begin{pmatrix} x \\ \tau \end{pmatrix}.$$

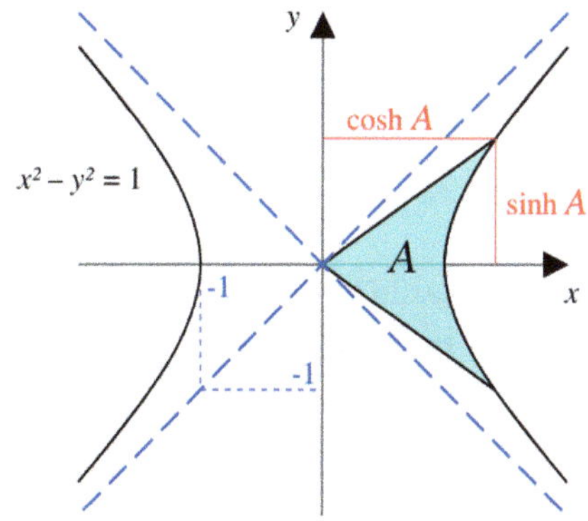

Figure 1.48. Definition of the hyperbolic sine and cosine. This image has been obtained by the author(s) from the Wikimedia website https://commons.wikimedia.org/wiki/File:Hyperbolic_functions.svg, where it is stated to have been released into the public domain. It is included within this article on that basis.

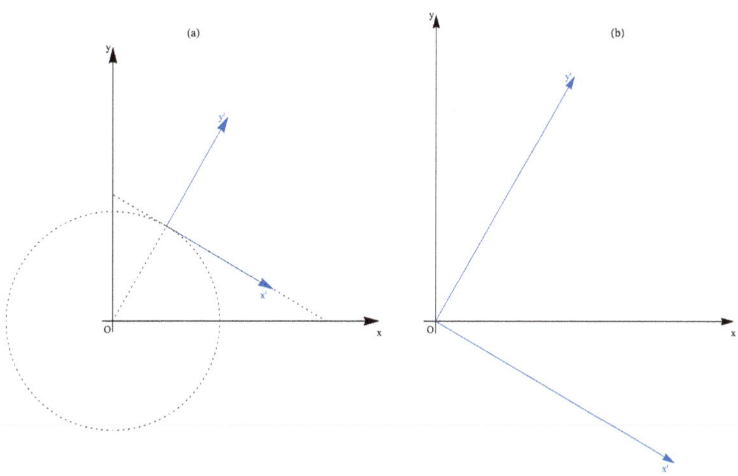

Figure 1.49. A circular rotation (a) follows from the prescription: (i) fix an angle and a corresponding point on the unitary circumference; (ii) draw two new axes one in the radial direction and the other along the tangent; (iii) after a shift to the original center of the axis one obtains the (circular) rotated system (b).

Show that the Lorentz transformation can be considered as a ‘hyperbolic rotation’, use figure 1.49 to figure out the difference between circular and hyperbolic rotations.

(Note that a circular rotation follows from the prescription

i. fix an angle and a corresponding point on the unitary circumference,
ii. draw two new axes, one in the radial direction and the other along the tangent,
iii. after a shift to the original center of the axis one obtains the (circular) rotated system.

The ‘hyperbolic rotation’ follows from the same procedure using the fundamental hyperbola as a reference.)

Exercise 1.3 Use (1.257) to obtain

$$\beta = \tanh(\alpha).$$

Show that the relativistic addition of velocities

$$\beta = \frac{\beta_1 + \beta_2}{1 + \beta_1\beta_2}$$

follows from

$$\beta = \tanh(\alpha_1 + \alpha_2),$$
$$\beta_1 = \tanh(\alpha_1),\ \beta_2 = \tanh(\alpha_2).$$

Exercise 1.4 Show that the hyperbolic rotation matrix in (1.258) can be written as

$$\begin{pmatrix} \cosh(\alpha) & -\sinh(\alpha) \\ -\sinh(\alpha) & \cosh(\alpha) \end{pmatrix} = e^{-\alpha\begin{pmatrix} 0 & 1 \\ 1 & 0 \end{pmatrix}}$$

and

$$\begin{pmatrix} \cos(\alpha) & -\sin(\alpha) \\ \sin(\alpha) & \cos(\alpha) \end{pmatrix} = e^{-\alpha\begin{pmatrix} 0 & 1 \\ -1 & 0 \end{pmatrix}}.$$

Hint: note that

$$e^{-\alpha\hat{A}} = \sum_{n=0}^{\infty} \frac{(-\alpha)^n}{n!}\hat{A}^n,$$

and

$$\begin{pmatrix} 0 & 1 \\ 1 & 0 \end{pmatrix}^{2n} = \begin{pmatrix} 1 & 0 \\ 0 & 1 \end{pmatrix}, \begin{pmatrix} 0 & 1 \\ 1 & 0 \end{pmatrix}^{2n+1} = \begin{pmatrix} 0 & 1 \\ 1 & 0 \end{pmatrix},$$
$$\begin{pmatrix} 0 & 1 \\ -1 & 0 \end{pmatrix}^{2n} = \begin{pmatrix} 1 & 0 \\ 0 & 1 \end{pmatrix}, \begin{pmatrix} 0 & 1 \\ -1 & 0 \end{pmatrix}^{2n+1} = \begin{pmatrix} 0 & 1 \\ -1 & 0 \end{pmatrix}.$$

Exercise 1.5 Use figure 1.50 to provide a geometric interpretation to the relativistic quantities (γ, β).

Hint: note that

$$\beta^2 + \frac{1}{\gamma^2} = 1 \Rightarrow \left\{ \begin{matrix} \beta = \sin(\vartheta) \\ \frac{1}{\gamma} = \cos(\vartheta) \end{matrix}, \gamma = \sec(\vartheta), \beta\,\gamma = \tan(\vartheta) \right\}.$$

Exercise 1.6 Use the above geometric pictures and figures 1.48–1.50 to find a corresponding justification for the relativistic contraction.

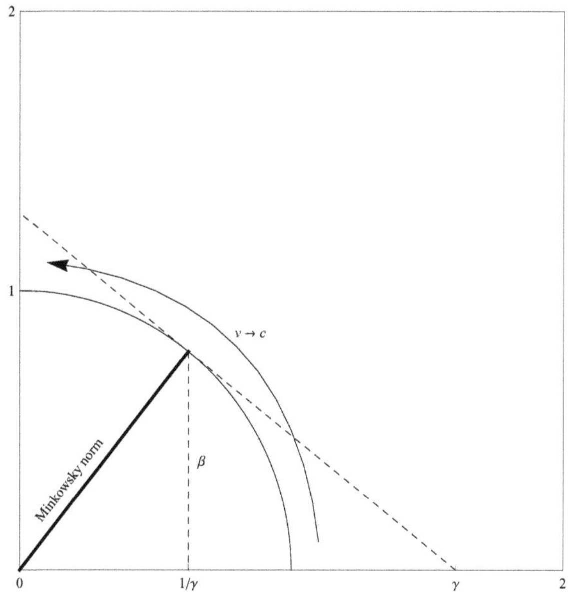

Figure 1.50. Definition of relativistic quantities in the Minkowski plane.

Exercise 1.7 Define the quantities

$$\Pi = \frac{pc}{E + m_0c^2},$$
$$M = \sqrt{\frac{2\,m_0c^2}{E + m_0c^2}},$$

and show that

$$\Pi^2 + M^2 = 1,$$
$$\Pi = \sqrt{\frac{T}{E + m_0c^2}} = \frac{\gamma\beta}{\gamma + 1},$$
$$M = \sqrt{\frac{2}{\gamma + 1}}.$$

Introduce the trigonometric parameterization

$$\cos(\varphi) = \Pi,$$
$$\sin(\varphi) = M,$$

and discuss the advantages (and disadvantages) of this picture. Identify the geometric meaning of the other kinematic quantities.

Show that the various relations in table 1.1 are compatible with elementary trigonometric identities.

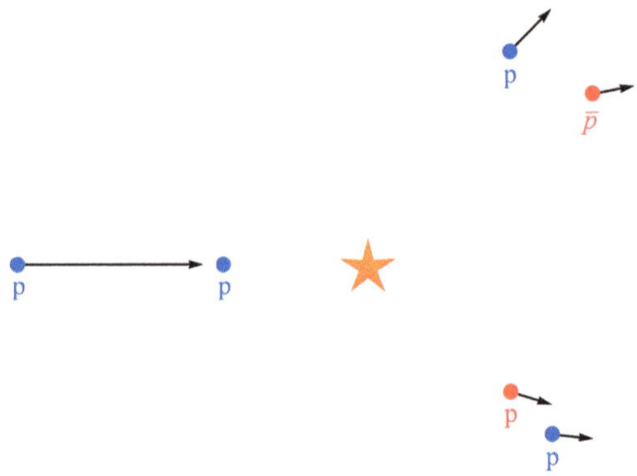

Figure 1.51. Kinematics of p–p scattering

Exercise 1.8 Consider the scattering process in figure 1.51 determining the production of antiprotons, namely

$$p + p \rightarrow p + \bar{p} + p + p.$$

Assuming that one of the protons is at rest, determine the threshold energy of the other allowing the process to occur.

Hint: labeling with 2, 1 the proton at rest and the moving proton, respectively, we have

$$\begin{aligned} &p_1^{\mu} \equiv \left(E_1,\, c\,\vec{p}_1\right),\, p_2^{\mu} \equiv \left(m_p c^2,\, 0\right) \\ &\rightarrow P^{\mu} \equiv \left(E_1 + m_p c^2,\, c\,\vec{p}_1\right) \rightarrow P^{\mu} P_{\mu} = (E_1 + m_p c^2)^2 - |\vec{p}_1 c|^2 = (4\, m_p c^2)^2 \\ &\rightarrow E_1 \equiv 7\, m_p c^2. \end{aligned}$$

Exercise 1.9 Consider a particle of mass M at rest in its reference frame, decaying into two daughter particles of masses $m_{1,2}$. Derive the energy and the momentum of the decay products, and the kinematics conditions allowing the decay.

Hint: use the energy momentum conservation

$$\begin{aligned} \vec{P} &= \vec{p}_1 + \vec{p}_2 = 0 \rightarrow |\vec{p}_1| = |\vec{p}_2| = p \\ Mc^2 &= E_1 + E_2 = c\left[\sqrt{p^2 + (m_1 c)^2} + \sqrt{p^2 + (m_2 c)^2}\right]. \end{aligned}$$

Exercise 1.10 Show that the velocity of a relativistic charged particle moving in a magnetic field is constant in modulus.

Hint: note that $\frac{d}{dt}(m_0 \gamma \vec{v}) = q\,\vec{v} \times \vec{B},\, \gamma^{-1} = \sqrt{1 - \frac{|\vec{v}|^2}{c^2}} \rightarrow \vec{v} \cdot \frac{d}{dt} \frac{\vec{v}}{\sqrt{1 - \frac{|\vec{v}|^2}{c^2}}} = 0.$

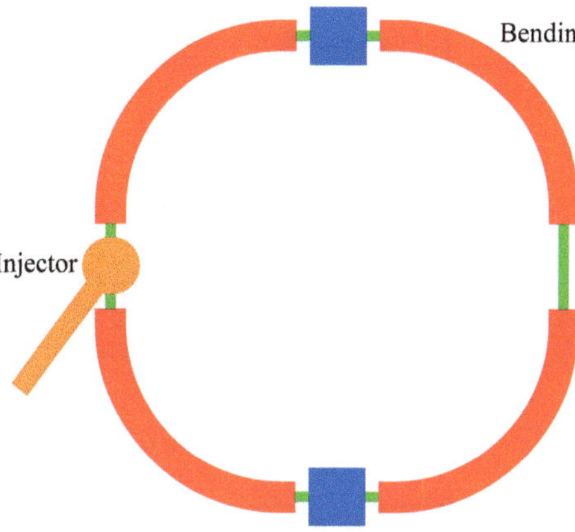

Figure 1.52. Schematic of the synchrotron.

Exercise 1.11 Apply the results of the exercise above to show that the particle in the field describes a circle, and determine the radius

$$\left(m\gamma\frac{v^2}{R} = qvB \rightarrow R = \frac{m\,\gamma v}{qB} \right).$$

Exercise 1.12 The synchrotron is a circular accelerator (figure 1.52) in which the charged particles (electrons, protons…) acquire energy from RF cavities placed along the orbits. The magnetic field is varied in such a way that it is compatible with the maximum radius of the orbit and with the maximum energy of the particle. Find the dimension of a protosynchrotron with a maximum energy of 30 GeV and a magnetic field of 3 T,

$$\left(R[\mathrm{m}] \cong 3.3\frac{p[\mathrm{GeV/c}]}{B[\mathrm{T}]} \rightarrow R \cong 33\ \mathrm{m} \right).$$

Exercise 1.13 In the Lawrence cyclotron (see figure 1.53) protons move in a constant magnetic field and acquire energy after having crossed the gap of the vacuum chamber, inside which an RF field oscillates. Assume that the radius of the accelerator is fixed and determines the maximum energy acquired and the maximum number of turns before the beam is extracted.

Hint:

$$E_f = m_p c^2\sqrt{1 + \left(\frac{e\,B\,R}{m_p c}\right)^2} \rightarrow E_K = E_f - m_p c^2 \rightarrow n = \frac{E_f - m_p c^2}{\Delta E}.$$

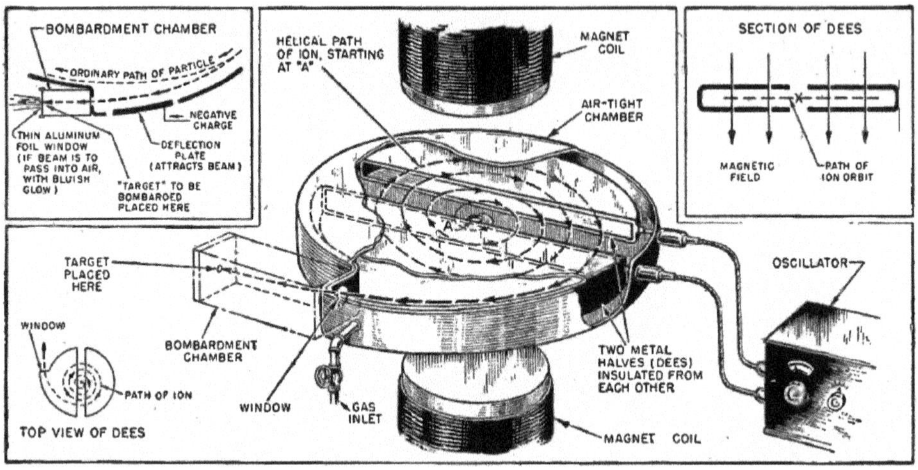

Figure 1.53. Schematic of the Lawrence cyclotron. This image has been obtained by the authors from the Wikimedia website https://commons.wikimedia.org/wiki/File:Cyclotron_diagram.png, where it is stated to have been released into the public domain. It is included within this article on that basis.

Exercise 1.14 Consider the Hamiltonian

$$\mathcal{H} = c\sqrt{p^2 + (mc)^2},$$

and derive the relevant equations of motion

Hint:

$$\dot{x} = \frac{\partial \mathcal{H}}{\partial p} = \frac{pc}{\sqrt{p^2 + (mc)^2}},$$

$$\dot{p} = -\frac{\partial H}{\partial x} = 0,$$

$$p = \frac{m\dot{x}}{\sqrt{1 - \left(\frac{\dot{x}}{c}\right)^2}}.$$

Exercise 1.15 The Hamiltonian

$$\mathcal{H} = c\sqrt{p^2 + (mc)^2} - Fx,$$

accounts for the motion of a relativistic particle ruled by a linear potential. Write and solve the relevant equations of motion.

Hint:

$$\dot{x} = \frac{pc}{\sqrt{p^2 + (mc)^2}}, \rightarrow p|_{t=0} = 0 \rightarrow \dot{x} = \frac{Fct}{\sqrt{(Ft)^2 + (mc)^2}}.$$
$$\dot{p} = F$$

Exercise 1.16 Show that the one-dimensional motion of a relativistic particle ruled by a linear potential is hyperbolic. Consider an electron initially at rest acted on by an electric field of 10 k V m^{-1} and determine when the relativistic effects become appreciable (>10%).

Exercise 1.17 Derive the equation of motion of a relativistic particle in a harmonic potential.

Hint:

$$\mathcal{H} = c\sqrt{p^2 + (mc)^2} + \frac{1}{2}kx^2 \rightarrow \quad \begin{aligned} \dot{x} &= \frac{pc}{\sqrt{p^2 + (mc)^2}}, \\ \dot{p} &= -kx \end{aligned}$$

and find that

$$\ddot{\Pi} = -\Omega^2 \frac{\Pi}{\sqrt{1 + \Pi^2}},$$
$$\Pi = \frac{p}{mc}, \; \Omega = \sqrt{\frac{k}{m}}.$$

Deduce that the motion is harmonic in the non-relativistic limit only $\Pi \ll 1$.

Exercise 1.18 Derive the motion invariant

$$I = \frac{1}{2}\dot{\Pi}^2 + \Omega^2\sqrt{1 + \Pi^2}$$

and explain the relevant physical meaning.

1.15.2 The Liouville and Fokker–Planck equations

As already discussed, we define the Liouville equation as a consequence of the Liouville theorem on phase-space conservation. The evolution of the phase-space distribution of a beam ruled by the Hamiltonian

$$\mathcal{H} = \frac{1}{2}p^2 + V(q), \tag{1.259}$$

is governed by the equation

$$\frac{\partial}{\partial t}\mathcal{F}(q, p; t) = \hat{\mathcal{L}}\mathcal{F}(q, p; t),$$
$$\hat{\mathcal{L}} = -p\frac{\partial}{\partial q} + V'(q)\frac{\partial}{\partial p}. \tag{1.260}$$

The relevant solution reads

$$\mathcal{F}(q, p; t) = e^{\hat{\mathcal{L}}t}\mathcal{F}_0(q, p),$$

where $e^{\hat{\mathcal{L}}t}$ is the Liouville evolution operator.

Exercise 1.19 Derive the Hamilton equations of motion and define an evolution operator for their solution.

The Hamilton equation of motion can be written in terms of Poisson brackets $\{.\, ,.\}$ as

$$\dot{q} = -\{\mathcal{H}, q\},$$
$$\dot{p} = -\{\mathcal{H}, p\},$$

where the brackets are explicitly defined as

$$\{\mathcal{G}(q, p), \mathcal{S}(q, p)\}$$
$$= \left(\frac{\partial}{\partial q}\mathcal{G}(q, p)\right)\left(\frac{\partial}{\partial p}\mathcal{S}(q, p)\right) - \left(\frac{\partial}{\partial p}\mathcal{G}(q, p)\right)\left(\frac{\partial}{\partial q}\mathcal{S}(q, p)\right).$$

If we define the two-component vector

$$\underline{\mathcal{Z}} = \begin{pmatrix} q \\ p \end{pmatrix},$$

we can cast the Hamilton equation as

$$\dot{\underline{\mathcal{Z}}} = -\hat{\mathcal{H}}\circ\underline{\mathcal{Z}},$$

where

$$\hat{\mathcal{H}}\circ\mathcal{G}(q, p) = \{\mathcal{H}, \mathcal{G}(q, p)\},$$
$$\hat{\mathcal{H}}^2\circ\mathcal{G}(q, p) = \hat{\mathcal{H}}\circ(\hat{\mathcal{H}}\circ\mathcal{G}(q, p)) = \{\mathcal{H}, \{\mathcal{H}, \mathcal{G}(q, p)\}\}.$$

The formal solution of the Hamilton equation above can therefore be written as

$$\underline{\mathcal{Z}} = e^{-t\,(\hat{\mathcal{H}}\circ)}\underline{\mathcal{Z}}_0.$$

Exercise 1.20 Define the Hamilton evolution operator $e^{-t\,(\hat{\mathcal{H}}\circ)}$ and discuss the relevant link with its Liouville counterpart.

Exercise 1.21 Use the evolution operator (either Hamilton or Liouville) for the Hamiltonian (1.259) and solve the evolution problem in the SSA. Show that at any step the evolution corresponds to a canonical transformation.

Hint: It has been shown that the solution of the Liouville equation in the SSA can be cast in the form

$$e^{\hat{\mathcal{L}}t}F(q, p) = F(q_n, p_n),$$
$$q_0 = q, \; p_0 = p,$$

$$\begin{aligned} q_n &= q_{n-1} - \delta\tau p_{n-1} - \frac{1}{2}(\delta\tau)^2 V'\left(q_{n-1} - \frac{\delta\tau}{2}p_{n-1}\right), \\ p_n &= p_{n-1}\delta\tau V'\left(q_{n-1} - \frac{\delta\tau}{2}p_{n-1}\right), \\ q_1 &= q - p\delta\tau - \frac{1}{2}(\delta\tau)^2 V'\left(q - \frac{\delta\tau}{2}p\right), \\ p_1 &= p + \delta\tau V'\left(q - \frac{\delta\tau}{2}p\right). \end{aligned} \tag{1.261}$$

The proof of the canonicity of the transformation is achieved by proving that the Jacobian determinant of the transformation

$$J = \begin{pmatrix} \dfrac{\partial q_n}{\partial q_{n-1}} & \dfrac{\partial p_n}{\partial q_{n-1}} \\ \dfrac{\partial q_n}{\partial p_{n-1}} & \dfrac{\partial p_n}{\partial p_{n-1}} \end{pmatrix}$$

is equal to unity. The SSA applied to the Hamilton equations yields the same recursive solution as in (1.261) provided that $\delta t \to -\delta t$.

Exercise 1.22 Consider the Hamiltonian

$$\mathcal{H} = \frac{1}{2}\nu^2 - \xi \cos(\zeta),$$

describing the free electron laser phase-space dynamics, with ν, ζ canonical variables (see chapter 3). Write the relevant Liouville equation and find a solution procedure

$$\left(\frac{\partial}{\partial\tau}\rho(\nu, \zeta, \tau) = \left(-\nu\frac{\partial}{\partial\zeta} + \xi\sin(\zeta)\frac{\partial}{\partial\nu}\right)\rho(\nu, \zeta, \tau)\right).$$

Exercise 1.23 Consider the two-dimensional Hamiltonian which can be cast in the form

$$H = \frac{1}{2}\left(p_x^2 + p_y^2\right) + V(x, y).$$

Derive the corresponding Liouville equation and obtain the relevant solution using the SSA

$$x_n = x_{n-1} - (p_x)_{n-1}\delta\tau - \frac{1}{2}(\delta\tau)^2\frac{\partial V}{\partial x}\left(x_{n-1} - \frac{\delta\tau}{2}(p_x)_{n-1}, y_{n-1} - \frac{\delta\tau}{2}(p_y)_{n-1}\right),$$

$$(p_x)_n = (p_x)_{n-1} + \delta\tau\frac{\partial V}{\partial x}\left(x_{n-1} - \frac{\delta\tau}{2}(p_x)_{n-1}, y_{n-1} - \frac{\delta\tau}{2}(p_y)_{n-1}\right),$$

$$y_n = y_{n-1} - (p_y)_{n-1}\delta\tau - \frac{1}{2}(\delta\tau)^2\frac{\partial V}{\partial y}\left(x_{n-1} - \frac{\delta\tau}{2}(p_x)_{n-1}, y_{n-1} - \frac{\delta\tau}{2}(p_y)_{n-1}\right),$$

$$(p_y)_n = (p_y)_{n-1} + \delta\tau\frac{\partial V}{\partial y}\left(x_{n-1} - \frac{\delta\tau}{2}(p_x)_{n-1}, y_{n-1} - \frac{\delta\tau}{2}(p_y)_{n-1}\right).$$

Write the corresponding Liouville equation, solve it using the SSA and show that at any iteration step the evolution corresponds to a canonical transformation.

Exercise 1.24 Find the solution of the diffusive evolution equation

$$\frac{\partial}{\partial t}F(x, t) = a\frac{\partial^2}{\partial x^2}F(x, t) + b\frac{\partial}{\partial x}F(x, t),$$
$$F(x, 0) = f(x).$$

Hint:

$$F(x, t) = e^{t\left(a\frac{\partial^2}{\partial x^2}+b\frac{\partial}{\partial x}\right)}f(x) = e^{bt\frac{\partial}{\partial x}}e^{at\frac{\partial^2}{\partial x^2}}f(x),$$

use the Gauss–Weierstrass transform

$$e^{k\frac{\partial^2}{\partial x^2}}f(x) = \frac{1}{2\sqrt{\pi k}}\int_{-\infty}^{+\infty} e^{-\frac{(x-\xi)^2}{4k}}f(\xi)d\xi$$

and finally obtain

$$F(x, t) = \frac{1}{2\sqrt{\pi at}}\int_{-\infty}^{+\infty} e^{-\frac{(x+at-\xi)^2}{4bt}}f(\xi)d\xi.$$

1.15.3 Uncoupled transport devices

The theory of beam transport has been covered in the main body of the chapter, here we will add a few remarks on some of the (many) topics touched upon or discussed qualitatively in the text. In the following the treatment of most of the analytical details will be necessarily schematic. We will omit excessive description, but, as far as possible, we will avoid shortcuts in the derivation of the formulae.

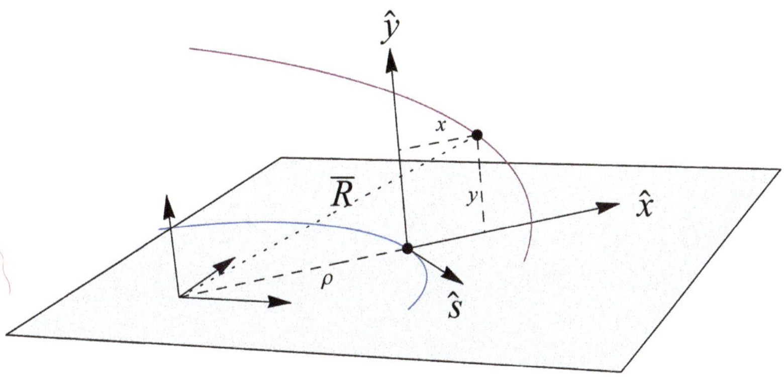

Figure 1.54. Reference frame of the motion.

The Lorentz force equations, accounting for the motion of charged relativistic particles, in the absence of an electric field (in the following we will consider only electrons), are written

$$\frac{d}{dt}\vec{p} = e\,\vec{v} \times \vec{B}, \qquad e = -\,|e|. \tag{1.262}$$

Assuming that the guiding element is a dipole magnet, we write the cross product (relevant to the frame in figure 1.54) as

$$\vec{v} \times \vec{B} = \begin{pmatrix} \hat{x} & \hat{y} & \hat{s} \\ v_x & v_y & v_s \\ B_x & B_y & 0 \end{pmatrix} = -v_s B_y \hat{x} + v_s B_x \hat{y} + \left(v_x B_y - v_y B_x\right)\hat{s}, \tag{1.263}$$

where the reference coordinates (figure 1.54) are specified by

$$\begin{aligned} &\vec{R} = \hat{x}\,r + \hat{y}\,y, \\ &r = \rho + x, \\ &\vartheta = \frac{s}{r}. \end{aligned} \tag{1.264}$$

To proceed further we note that for the time derivative (figure 1.55) we will use the identities

$$\begin{aligned} &\frac{d}{dt}\gamma \cong 0 \to \frac{d}{dt}(m_e\gamma\vec{R}) \cong m_e\gamma\dot{R}, \\ &\dot{\vartheta} = \frac{v_s}{r}, \\ &\begin{matrix} \dot{\hat{x}} = \dot{\vartheta}\hat{s} \\ \dot{\hat{s}} = -\dot{\vartheta}\hat{x} \end{matrix} \;\to \frac{d}{dt}\vec{R} = \dot{r}\hat{x} + r\dot{\vartheta}\hat{s} + \dot{y}\hat{y} \\ &\to \frac{d^2}{dt^2}\vec{R} = (\ddot{r} - r\dot{\vartheta}^2)\hat{x} + (2\dot{r}\dot{\vartheta} + r\ddot{\vartheta})\hat{s} + \ddot{y}\hat{y}. \end{aligned} \tag{1.265}$$

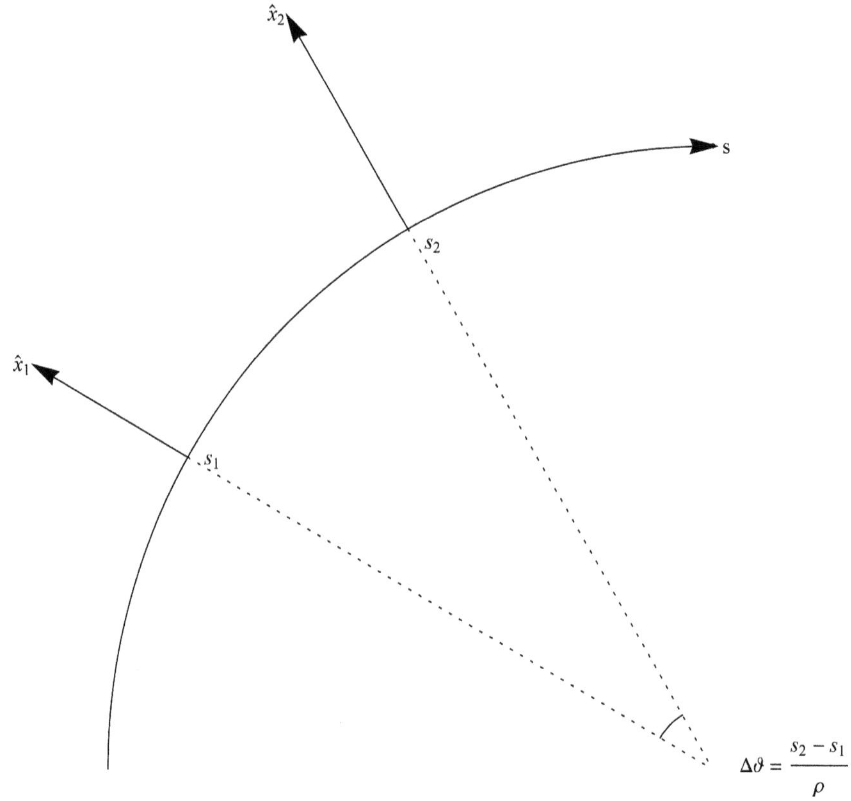

Figure 1.55. Curvilinear coordinates.

The Lorentz force components read

$$
\begin{aligned}
F_x &= m_e\gamma\left(\frac{d^2}{dt^2}\vec{R}\right)_x = -e\ v_s B_y, \\
\left(\frac{d^2}{dt^2}\vec{R}\right)_x &= (\ddot{r} - r\,\dot{\vartheta}^2) = \left(\ddot{r} - \frac{v_s^2}{r}\right), \\
r = \rho + x &\rightarrow \left(\frac{d^2}{dt^2}\vec{R}\right)_x = \ddot{x} - \frac{v_s^2}{r}, \\
F_x &= m_e\gamma\left(\ddot{x} - \frac{v_s^2}{r}\right) = -e\ v_s B_y, \\
F_y &= m_e\gamma\left(\frac{d^2}{dt^2}\vec{R}\right)_y = m_e\gamma\left(\frac{d^2}{dt^2}y\right) = e\ v_s B_x, \\
\vartheta = \frac{s}{r} = \frac{\beta ct}{r} &\rightarrow t = \frac{r\,\vartheta}{\beta c} \rightarrow
\begin{array}{l}
\rightarrow x'' + \left(\dfrac{e\ B_y}{\beta\ \gamma mc} r - 1\right) r = 0 \\
y'' - \dfrac{e\ B_x}{\beta\ \gamma mc} r^2 = 0
\end{array}.
\end{aligned}
\tag{1.266}
$$

Furthermore, by taking into account the approximations,

$$
\begin{aligned}
& r = \left(1 + \frac{x}{\rho}\right)\rho, \\
& \frac{x}{\rho}, \frac{y}{\rho} \ll 1 \rightarrow \text{paraxial approximation}, \\
& B_y \cong B_0 + \left.\frac{\partial B_y}{\partial x}\right|_{x=0} x + o\left(\left(\frac{x}{\rho}\right)^2\right), \\
& \vec{\nabla} \times \vec{B} = 0 \rightarrow \frac{\partial}{\partial x} B_y = \frac{\partial}{\partial y} B_x \rightarrow B_x \cong \left.\frac{\partial B_y}{\partial x}\right|_{x=0} y, \\
& \beta \cong \beta_0\, \rho = \frac{\beta \gamma m c}{e B_0}, \\
& x'' + \left(\frac{e\, B_y}{\beta \gamma m c} r - 1\right) r = 0 \rightarrow x''\left(\left(1 + \frac{1}{B_0} \left.\frac{\partial B_y}{\partial x}\right|_{x=0} x\right)\left(1 + \frac{x}{\rho}\right) - 1\right) \\
& \qquad \times \rho\left(1 + \frac{x}{\rho}\right) = 0 \\
& \rightarrow x''\left(\frac{\rho}{B_0} \left.\frac{\partial B_y}{\partial x}\right|_{x=0} + 1\right) x = 0 + o\left(\left(\frac{\rho}{x}\right)^2\right), \\
& y'' - \frac{e\, B_x}{\beta \gamma m c} r^2 = 0 \rightarrow y'' - \left(1 + \frac{2x}{\rho}\right) \frac{\rho}{B_0} \left.\frac{\partial B_y}{\partial x}\right|_{x=0} y = 0,
\end{aligned}
\tag{1.267}
$$

introducing the *field index*

$$
n = -\frac{\rho}{B_0} \left.\frac{\partial B_y}{\partial x}\right|_{x=0}
\tag{1.268}
$$

and by using the paraxial approximation, we end up with the equation

$$
\begin{aligned}
x'' + (1 - n)x &= 0, \\
y'' + n\, y &= 0,
\end{aligned}
\tag{1.269}
$$

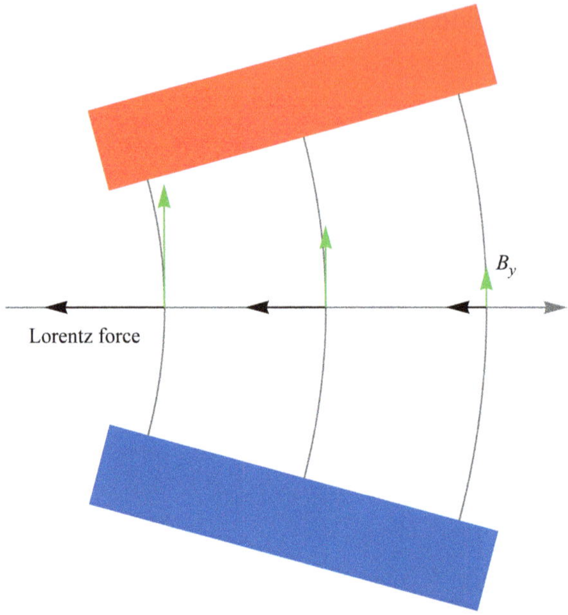

Figure 1.56. Weak focusing mechanism.

Figure 1.57. Weak focusing magnet consisting of the focusing term of length $\frac{\pi\rho}{2}$ and drift section l_0.

which account for the *weak focusing* in bending magnets. The physical mechanisms underlying the weak focusing effect are shown in figure 1.56, where we have reported two magnets with tilted poles (ensuring a non-zero gradient $\frac{\partial B_y}{\partial x}|_{x=0} \neq 0$). According to the field line distribution we can conclude that:

a. The electrons experience less focusing on the external edge and more focusing inside; the net effect is a dilution of the focusing.
b. The field lines are more intense outward than inward, thus providing more focusing.

Exercise 1.25 In figure 1.57 we show a cell provided by a weak focusing magnet of effective length $\frac{\pi\rho}{2}$; a beam of electrons is moving as indicated below. Derive the relevant transport matrix (for the x-component).

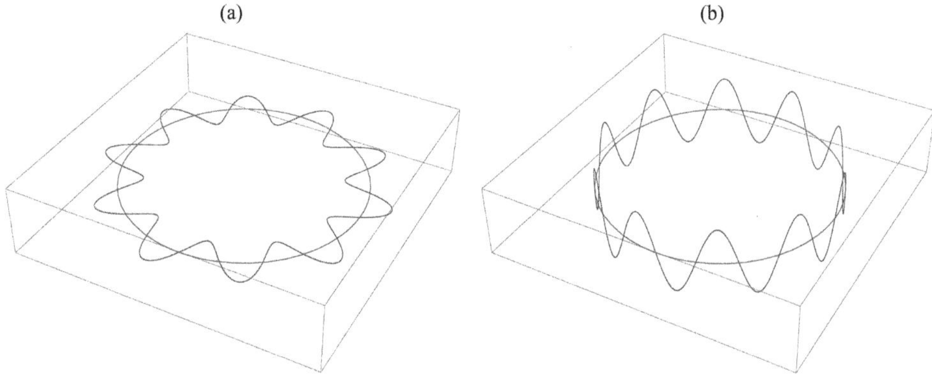

Figure 1.58. Horizontal (a) and vertical (b) betatron oscillations.

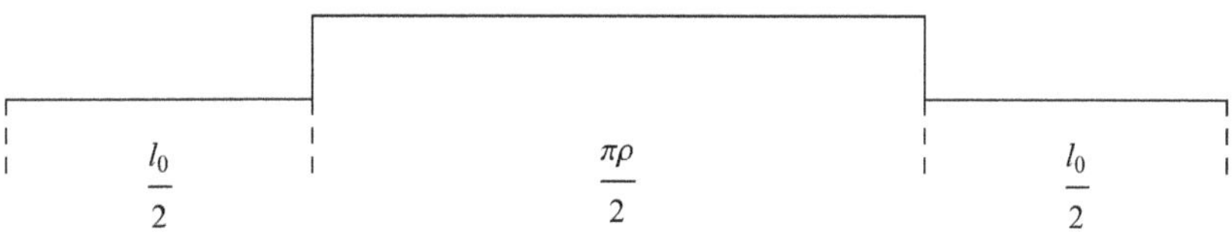

Figure 1.59. The same as figure 1.57, except the magnet has been divided in three parts according to the SSA.

Hint: note first that

$$\hat{T} = \hat{T}_d \hat{T}_{\frac{\pi}{2}} = \begin{pmatrix} 1 & l_0 \\ 0 & 1 \end{pmatrix} \begin{pmatrix} \cos\left(\frac{\sqrt{1-n}}{2}\pi\right) & \frac{\rho}{\sqrt{1-n}} \sin\left(\frac{\sqrt{1-n}}{2}\pi\right) \\ -\frac{\sqrt{1-n}}{\rho} \sin\left(\frac{\sqrt{1-n}}{2}\pi\right) & \cos\left(\frac{\sqrt{1-n}}{2}\pi\right) \end{pmatrix}, \ n < 1.$$

Comment: we have denoted the magnet transport matrix with $\hat{T}_{\frac{\pi}{2}}$ because the magnet is supposed to transport the beam on a quarter of the arc of circumference.

Exercise 1.26 Derive the transport matrix for horizontal and vertical positions, in the case of a generic bending arc ϑ

$$_{x,y}\hat{T}_\vartheta = \begin{pmatrix} \cos\left(Q_{x,y}\vartheta\right) & \frac{\rho}{Q_{x,y}} \sin\left(Q_{x,y}\vartheta\right) \\ -\frac{Q_{x,y}}{\rho} \sin\left(Q_{x,y}\vartheta\right) & \cos\left(Q_{x,y}\vartheta\right) \end{pmatrix}.$$

$$Q_x = \sqrt{1-n}, \ Q_y = \sqrt{n}.$$

Explain why this type of magnet is focusing in both planes.

Where $Q_{x,y}$ are the *betatron tunes*, explain the relevant physical meaning.

Hint: use figure 1.58 as a reference.

Exercise 1.27 Consider the SSA applied to the problem in exercise 1.25 (see figure 1.59), and show that the relevant transport matrix is written

$$
{}_x\hat{T}_{\frac{\pi}{2}} = \begin{pmatrix} \cos\left(Q_x\frac{\pi}{2}\right) - \frac{l_0}{2\rho}\sin\left(Q_x\frac{\pi}{2}\right) & \frac{\rho}{Q_x}(1-\phi)\sin\left(Q_x\frac{\pi}{2}\right) + l_0\cos\left(Q_x\frac{\pi}{2}\right) \\ -\frac{Q_x}{\rho}\sin\left(Q_x\frac{\pi}{2}\right) & \cos\left(Q_x\frac{\pi}{2}\right) - \frac{l_0 Q_x}{2\rho}\sin\left(Q_x\frac{\pi}{2}\right) \end{pmatrix},
$$

$$
\phi = 1 - \left(\frac{l_0}{2\rho}\right)^2 Q_x.
$$

Show that for $\frac{l_0}{2\rho} \ll \frac{\pi}{2}$ the above matrix can be parameterized as

$$
\hat{T} = \begin{pmatrix} \cos(\mu_x) & \beta_x \sin(\mu_x) \\ -\frac{1}{\beta_x}\sin(\mu_x) & \cos(\mu_x) \end{pmatrix}, \quad \mu_x = \left(Q_x + \frac{l_0}{\rho\,\pi}\right)\frac{\pi}{2}
$$

$$
\beta_x = \frac{2\rho}{\mu_x \pi}\left(1 + \frac{2\,l_0}{\pi\,\rho}\right).
$$

Derive the analogous expressions for the vertical plane.

Exercise 1.28 Show that a further parameterization of the problem in exercise 1.25 can be obtained by the use of the 'split' scheme shown in figure 1.60, write the corresponding matrix and compare with the previous results.

Exercise 1.29 Include in (1.269) the effect of beam momentum spread, namely

$$
p = \beta\,\gamma m_e c \to p\left(1 + \frac{\delta_p}{p}\right),
$$

$$
\frac{\delta_p}{p} \ll 1.
$$

Hint: note that

$$
x'' + \left(\frac{1}{1+\frac{\delta_p}{p}}\left(1 + \frac{1}{B_0}\frac{\partial B_y}{\partial x}\bigg|_{x=0} x\right)\left(1 + \frac{x}{\rho}\right) - 1\right)\rho\left(1 + \frac{x}{\rho}\right) = 0
$$

$$
\to x'' + \left(\left(1 - \frac{\delta_p}{p}\right)\left(1 + \frac{1}{B_0}\frac{\partial B_y}{\partial x}\bigg|_{x=0} x\right)\left(1 + \frac{x}{\rho}\right) - 1\right)\rho\left(1 + \frac{x}{\rho}\right) = 0,
$$

and use the paraxial condition to prove that

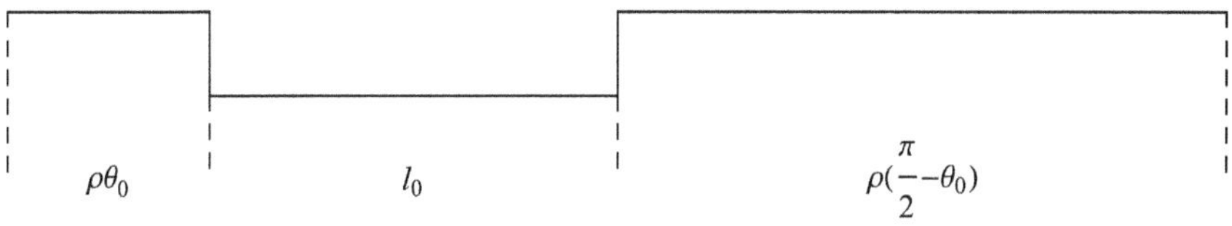

Figure 1.60. Same as figure 1.59 with an SSA.

$$x'' + (1 - n)x = \rho\frac{\delta_p}{p}. \tag{1.270}$$

Explain why the equation describing the vertical motion is left unchanged.

Exercise 1.30 Find the solution of (1.270).
Hint: use the following transformation of variable

$$X = x - \frac{\rho}{1 - n}\frac{\delta_p}{p}.$$

Assume that $\frac{\delta_p}{p}$ is independent of the integration variable and write

$$X'' - (1 - n)\,X = 0,$$

$$X(0) = x_0 - \frac{\rho}{1 - n}\frac{\delta_p}{p},$$

$$X'(0) = x_0'.$$

Obtain the solution in the form

$$\begin{pmatrix} X \\ X' \end{pmatrix} = \begin{pmatrix} \cos(Q_x\vartheta) & \frac{\rho}{Q_x}\sin(Q_x\vartheta) \\ -\frac{Q_x}{\rho}\sin(Q_x\vartheta) & \cos(Q_x\vartheta) \end{pmatrix}\begin{pmatrix} X(0) \\ X'(0) \end{pmatrix}$$

and finally end up with the matrix identity

$$\begin{pmatrix} x \\ x' \\ \frac{\delta_p}{p} \end{pmatrix} = \begin{pmatrix} \cos(Q_x\vartheta) & \frac{\rho}{Q_x}\sin(Q_x\vartheta) & \frac{\rho}{Q_x}(1 - \cos(Q_x\vartheta)) \\ -\frac{Q_x}{\rho}\sin(Q_x\vartheta) & \cos(Q_x\vartheta) & \frac{\sin(Q_x\vartheta)}{Q_x} \\ 0 & 0 & 1 \end{pmatrix}\begin{pmatrix} x_0 \\ x_0' \\ \frac{\delta_p}{p} \end{pmatrix}.$$

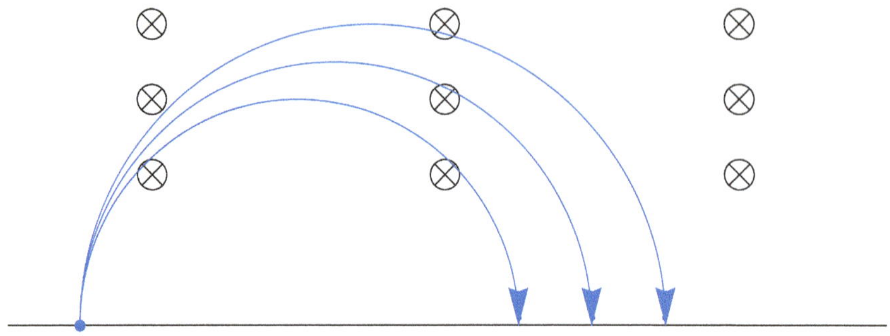

Figure 1.61. Beam broadening in the radial direction.

Exercise 1.31 With reference to figure 1.61 use the previous equation to show that one of the consequences of the momentum spread is a beam broadening in the radial direction.

Hint: note that the particles make half a turn inside a uniform magnet ($n = 1$), therefore

$$\begin{pmatrix} x \\ x' \\ \dfrac{\delta_p}{p} \end{pmatrix} = \begin{pmatrix} -1 & 0 & 2\rho \\ 0 & -1 & \pi \\ 0 & 0 & 1 \end{pmatrix} \begin{pmatrix} x_0 \\ x_0' \\ \dfrac{\delta_p}{p} \end{pmatrix}.$$

Furthermore since the spread may have either sign, we find

$$x(\pi) = -x_0 \pm 2\,\rho\,\frac{\delta_p}{p},$$

as illustrated in figure 1.61.

Exercise 1.32 Show that the Hamiltonian corresponding to the equation of motion (1.270) is

$$\mathcal{H} = \frac{1}{2}\left(x'^2 + Q_x x^2\right) - \frac{\rho}{Q_x}\frac{\delta_p}{p}x.$$

Write the associated Liouville equation and discuss the evolution problem within such a context.

Exercise 1.33 Find a way to exploit the mechanism illustrated in figure 1.61 to reduce the momentum spread of a beam.

Exercise 1.34 The synchrocyclotron is an accelerating device with constant radius. The particles are injected inside the accelerator, as illustrated in figure 1.52, and are accelerated during each turn by an RF cavity. The radius is kept constant by varying the bending magnet field with increasing energy.

Use the results of the problem (1.27) to prove that the synchrotron cell can be written as the product of four identical cells with

$$
{}_x M = \begin{pmatrix} \cos(\mu_x) & \beta_x \sin(\mu_x) \\ -\dfrac{1}{\beta_x}\sin(\mu_x) & \cos(\mu_x) \end{pmatrix},
$$

where for $\dfrac{l_0}{2\,\rho} \ll 1$

$$
\begin{aligned}
&\mu_x = \left(1 + \frac{l_0}{\rho\,\pi}\right)\frac{Q_x \pi}{2},\ \beta_x = \frac{\rho\left(1 + \dfrac{2\,l_0}{\pi\,\rho}\right)}{Q_x}, \\
&\tilde{Q}_x = \frac{\mu_x}{(\pi/2)} \equiv \text{betatron tune per cell.}
\end{aligned}
\tag{1.271}
$$

Exercise 1.35 The conclusion of the previous exercise holds if one neglects the effect of the momentum spread in the horizontal plane. What do we expect from the introduction of this further contribution?

Qualitatively speaking and according to exercise 1.32, particles with non-identical momentum will diffuse while propagating around the magnet, and the circumferences described by the different particles are not the same.

We need therefore a quantity quantifying the dependence of the circumference variations on the beam momentum spread. We note that the length of the circumference of a synchrocyclotron is

$$
\begin{aligned}
&L = 2\,\pi\,\rho + 4\,l_0, \\
&\rho = \frac{p}{eB},
\end{aligned}
$$

and introduce the identity

$$
\frac{\delta L}{L} = \alpha_c \frac{\delta_p}{p},
$$

where α_c($\equiv$ momentum compaction) is the constant specifying the proportionality between fractional deviation of the path length and fractional deviation of momentum. We can now combine the previously introduced quantities to obtain an explicit definition of the momentum compaction.

In the drift regions l_0 we do not expect any variation of the circumference length, therefore

$$\delta L = 2\,\pi\,\delta\rho \rightarrow \frac{\delta L}{L} = \left(1 + 2\,\frac{l_0}{\pi\,\rho}\right)^{-1}\frac{\delta\rho}{\rho}$$

$$\frac{\delta_p}{p} = \frac{\delta\rho}{\rho} + \frac{\delta B}{B} \cong \left(1 + \frac{\rho\,\delta B}{B_0\,\delta\rho}\right)\frac{\delta\rho}{\rho}$$

$$\rightarrow \frac{\delta L}{L} = \left[\left(1 + 2\,\frac{l_0}{\pi\,\rho}\right)\left(1 + \frac{\rho\,\delta B}{B_0\,\delta\rho}\right)\right]^{-1}\frac{\delta_p}{p}.$$

The use of the gradient index $n = -\frac{\rho\,\delta B}{B\,\delta\rho}$ allows one to cast the momentum compaction in the form

$$\alpha_c = \left[\left(1 + 2\,\frac{l_0}{\pi\,\rho}\right)(1 - n)\right]^{-1} \cong \frac{1 - 2\,\frac{l_0}{\pi\,\rho}}{1 - n}.$$

According to the last equations of (1.271)

$$\tilde{Q}_x \cong \left(1 + \frac{l_0}{\rho\,\pi}\right)\sqrt{1 - n}.$$

Finally, since $\frac{l_0}{\rho\,\pi} \ll 1$, we end up with the following identity between the vertical term and momentum compaction

$$\alpha_c \cong Q_x^{-2} \cong \frac{\left(1 - 2\frac{l_0}{\rho\,\pi}\right)}{1 - n}.$$

What are the consequences of this relation for weak focusing devices?

Exercise 1.36 Consider the slanting dipole magnet in figures 1.43 and 1.44, use the notions discussed in this chapter and in these complements to prove that

$$_y\hat{T}_{Sl} \cong {}_y\hat{T}_{W\,y}\hat{T}_{S\,y}\hat{T}_W,$$

$$_y\hat{T}_W = \begin{pmatrix} 1 & 0 \\ \frac{\tan(\alpha)}{\rho} & 1 \end{pmatrix},$$

$$_y\hat{T}_S \cong \begin{pmatrix} 1 & \rho\vartheta \\ 0 & 1 \end{pmatrix},$$

$$
{}_x\hat{T}_{SI} \cong {}_x\hat{T}_{W\,x}\hat{T}_{S\,x}\hat{T}_W,
$$

$$
{}_x\hat{T}_W = \begin{pmatrix} 1 & 0 & 0 \\ \dfrac{\tan(\alpha)}{\rho} & 1 & 0 \\ 1 & 0 & 1 \end{pmatrix},
$$

$$
{}_x\hat{T}_S = \begin{pmatrix} 1 & \rho\vartheta \\ 0 & 1 \end{pmatrix}.
$$

Hint: approximate the magnet as a *DOD*.

1.15.4 Magnet multipole expansion and coupled transport devices

The properties of static magnetic fields are summarized by the Maxwell equations

$$
\begin{aligned}
\vec{\nabla} \times \vec{B} &= 0, \\
\vec{\nabla} \cdot \vec{B} &= 0.
\end{aligned} \tag{1.272}
$$

The above two equations are compatible with a definition of $\vec{B}$ involving a scalar field Φ_M, namely

$$
\vec{B} = -\vec{\nabla}\Phi_M. \tag{1.273}
$$

The magnetic field is therefore perpendicular to the equipotential lines. This is illustrated in figures 1.17 and 1.18 for the case of a bending and of a quadrupole magnet, in both cases the equipotentials are parallel to the pole faces of the magnet (see below for further comments).

Equation (1.273) combined with the second equation of (1.272) yields, for the scalar potential, the equation

$$
\vec{\nabla}^2\Phi_M = 0. \tag{1.274}
$$

The three-dimensional Laplace operator $\vec{\nabla}^2$ can be reduced to a two-dimensional form, since the field dependence on the longitudinal coordinate can usually be neglected. In the following we will therefore deal with a two variable operator such that

$$
\vec{\nabla}^2 = \frac{\partial^2}{\partial x^2} + \frac{\partial^2}{\partial y^2}. \tag{1.275}
$$

Exercise 1.37 Check the validity, for the problem under study, of neglecting the derivative with respect to the longitudinal coordinate:

a. Set

$$
\left(\frac{\partial^2}{\partial x^2} + \frac{\partial^2}{\partial y^2}\right)\Phi_M = -\frac{\partial^2}{\partial z^2}\Phi_M.
$$

b. Define

$$\overline{\Phi}_M(x, y) = \frac{1}{Z_2 - Z_1}\int_{Z_1}^{Z_2} \Phi_M(x, y, z)dz,$$

and obtain

$$\left(\frac{\partial^2}{\partial x^2} + \frac{\partial^2}{\partial y^2}\right)\overline{\Phi}_M = -\frac{1}{Z_2 - Z_1}\int_{Z_1}^{Z_2} \frac{\partial^2}{\partial z^2}\Phi_M dz.$$

c. Use (1.273) to establish the identity

$$\left(\frac{\partial^2}{\partial x^2} + \frac{\partial^2}{\partial y^2}\right)\overline{\Phi}_M = \frac{1}{Z_2 - Z_1}(B_z(x, y, Z_2) - B_z(x, y, Z_1)).$$

The boundaries $Z_{1,2}$ can be chosen where the field longitudinal components are vanishing so the rhs vanishes and the two-dimensional Laplace equation is satisfied. If, e.g., the magnet is relatively short the integration end points can be chosen well outside the region where the longitudinal magnetic field is absent; or, for longer magnets, Z_1 can be chosen well inside and Z_2 well outside, where the field is two-dimensional.

It is evident that the possibility of neglecting the second derivative with respect to the longitudinal coordinate is less trivial than might be thought. It should be understood to be the consequence of several facts, including the measurement criteria and the shortness of the magnet compared to the beam betatron oscillations.

Exercise 1.38 Show that in polar coordinates the two-dimensional Laplacian (1.275) is

$$\vec{\nabla}^2 = \frac{1}{r}\partial_r r\partial_r + \frac{1}{r^2}\partial_\vartheta^2.$$

Hint: recall that

$$x = r\cos(\vartheta)$$
$$y = r\sin(\vartheta)$$

and that

$$\frac{\partial}{\partial x} = \frac{\partial r}{\partial x}\partial_r + \frac{\partial\vartheta}{\partial x}\partial_\vartheta.$$

Exercise 1.39 Find a solution for the equation

$$\frac{1}{r}\partial_r r\partial_r\Phi_M + \frac{1}{r^2}\partial_\vartheta^2\Phi_M = 0$$

using a Trigonometric expansion.

Hint: try with

$$\Phi_M = \frac{B_0}{\mu_0}\,\mathrm{Re}\left[\sum_{n=0}^{\infty}\varphi_n\right],$$

$$\varphi_n = \frac{A_n}{n+1}R_0\left(\frac{r}{R_0}\right)^{n+1} e^{i\left[(n+1)\,\vartheta+\frac{\pi}{2}\right]},\; A_n = \gamma_n + i\,\delta_n = \kappa_n e^{-i\,\phi_n}.$$

In the above expressions $[a]$ is the integer part of a (e.g. $[3.1] \equiv 3$).

Exercise 1.40 Write the above solution in Cartesian coordinates.
Hint: note that $r^n e^{in\vartheta} = (x + i\,y)^n$ and obtain

$$\varphi_n = \frac{\kappa_n e^{-i\left(\phi_n-\frac{\pi}{2}\right)}}{n+1}R_0\left(\frac{x+i\,y}{R_0}\right)^{n+1} = \frac{\kappa_n}{R_0^n(n+1)}e^{-i\left(\phi_n-\frac{\pi}{2}\right)}(\tau_n(x, y) + i\,\pi_n(x, y)),$$

$$\pi_n(x, y) = (n+1)!\sum_{s=0}^{\left[\frac{n}{2}\right]}\frac{(-1)^s y^{2s+1}x^{n-2s}}{(n-2s)!(2s+1)!},$$

$$\tau_n(x, y) = (n+1)!\sum_{s=0}^{\left[\frac{n+1}{2}\right]}\frac{(-1)^s y^{2s}x^{n+1-2s}}{(n+1-2s)!(2s)!}.$$

Exercise 1.41 Write the explicit polynomial form of the scalar potential,

$$\Phi_M = \frac{B_0}{\mu_0}\,\mathrm{Re}\left[\sum_{n=0}^{\infty}\varphi_n\right] = -\frac{B_0}{\mu_0}\sum_{n=0}^{\infty}\frac{1}{R_0^n}(\bar{\gamma}_n\pi_n(x, y) + \bar{\delta}_n\tau_n(x, y)),$$

$$\bar{\delta}_n = \frac{\delta_n}{n+1},\; \bar{\gamma}_n = \frac{\gamma_n}{n+1}.$$

Exercise 1.42 Show that

$$\sum_{n=0}^{\infty}\frac{t^n}{n!}\pi_{n-1}(x, y) = e^{xt}\cos(yt),$$

$$\sum_{n=0}^{\infty}\frac{t^n}{n!}\tau_{n-1}(x, y) = e^{xt}\sin(yt),$$

and that

$$
\begin{aligned}
\partial_x \pi_n(x, y) &= -\partial_y \tau_n(x, y), \\
\partial_x \pi_{n-1}(x, y) &= n \pi_{n-2}(x, y), \\
\pi_n &= x\, \pi_{n-1} + y\, \tau_{n-1}, \\
\tau_n &= x\, \tau_{n-1} + y\, \pi_{n+1}.
\end{aligned}
$$

Exercise 1.43 Derive the first few orders and speculate on the relevant meaning:

n	$\pi_n(x, y)$	$\tau_n(x, y)$
0	y	x
1	$-2xy$	$y^2 - x^2$
2	$y^3 - 3x^2y$	$3xy^2 - x^3$
3	$-4xy(x^2 - y^2)$	$6x^2y^2 - x^4 - y^4$
...		

It is evident that the expressions above represent the potential lines of each multipole component.

In the case of $n = 1$ we have two different equipotential curves:

$$
\begin{aligned}
xy &= k, \\
x^2 - y^2 &= k,
\end{aligned}
$$

which are geometrically represented by hyperbolas (orthogonal and equilateral, respectively), while, for $n = 2$ we have

$$
\begin{aligned}
x^3 - 3xy^2 &= k, \\
y^3 - 3x^2y &= k.
\end{aligned}
$$

It is not difficult to end up with the conclusion that the potential lines, associated with the previous polynomial expansions are, from the geometrical point of view, rotated by an angle $\dfrac{\pi}{2\,(n+1)}$ with respect to each other. From now on the π_n-components will be called *skew* and the τ_n *normal.* Examples of normal and skew magnets are given in figure 1.62 (see also table 1.3).

Exercise 1.44 Interpret the previous mathematical formulae in terms of actual magnetic devices.

We have described the fields and the associated transport items, but we did not say anything about their actual realization. The forthcoming exercise yields an idea

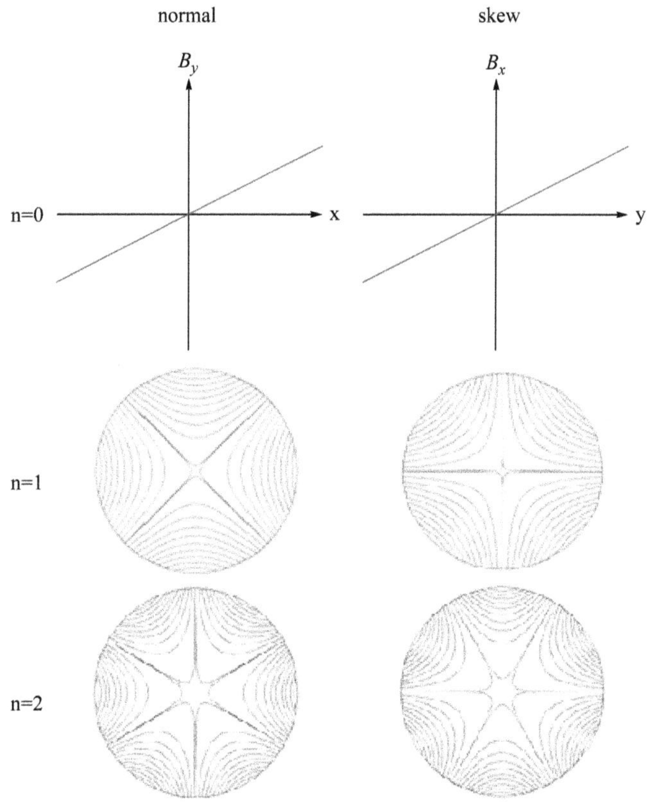

Figure 1.62. Equipotential lines for normal and skew magnets for different values of *n*.

Exercise 1.45 Write the radial and tangential components of the magnetic field.

Hint: recall that $\vec{B} = -\mu_0 \vec{\nabla} \Phi_M$ and find therefore

$$\begin{aligned} B_r(r, \vartheta) &= -\partial_r \Phi_M = B_0 \sum_{n=0}^{\infty} |\kappa_n| \left(\frac{r}{R_0} \right)^n \sin\left[(n+1)\vartheta - \phi_n\right], \\ B_\vartheta(r, \vartheta) &= -\frac{1}{r} \partial_\vartheta \Phi_M = B_0 \sum_{n=1}^{\infty} |\kappa_n| \left(\frac{r}{R_0} \right)^n \cos\left[(n+1)\vartheta - \phi_n\right]. \end{aligned} \tag{1.276}$$

Exercise 1.46 Use the previous results to write the Cartesian components of the magnetic field.

Hint: use figure 1.63 to establish that

$$\begin{pmatrix} B_x \\ B_y \end{pmatrix} = \begin{pmatrix} \cos(\vartheta) & -\sin(\vartheta) \\ \sin(\vartheta) & \cos(\vartheta) \end{pmatrix} \begin{pmatrix} B_r \\ B_\vartheta \end{pmatrix},$$

Table 1.3. Multipole nomenclature and corresponding items.

n	2(n+1)		angle between poles
0	Dipole	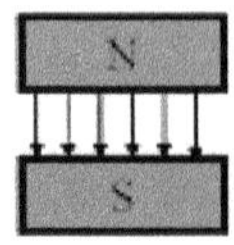	180
1	Quadrupole	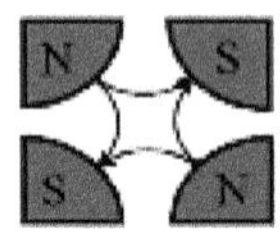	90
2	Sextupole		60
3	Octupole		45

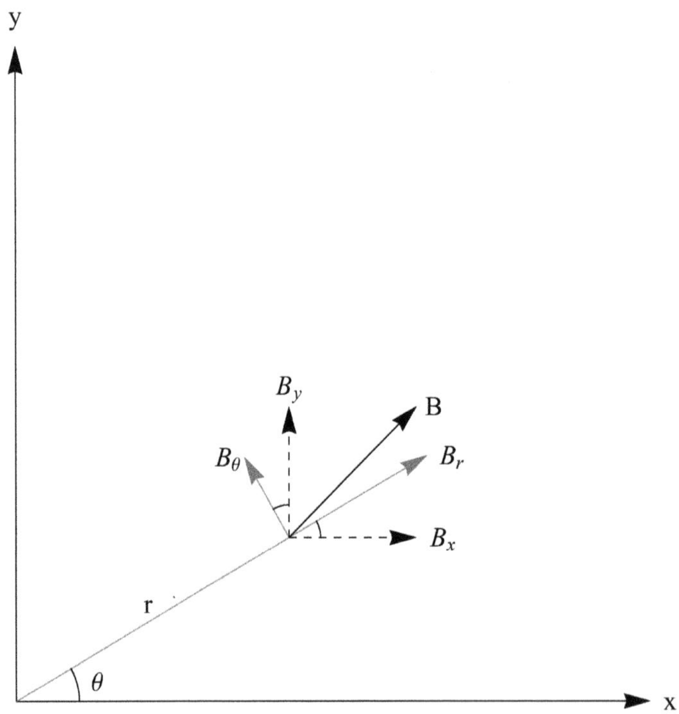

Figure 1.63. Reference frame for exercise 1.47.

and obtain

$$
\begin{aligned}
B_x(r, \vartheta) &= B_0 \sum_{n=0}^{\infty} |\kappa_n| \left(\frac{r}{R_0}\right)^n \sin\left[n\,\vartheta - \phi_n\right] \\
&= B_0 \sum_{n=0}^{\infty} \left(\frac{r}{R_0}\right)^n [\gamma_n \sin(n\,\vartheta) + \delta_n \cos(n\vartheta)], \\
B_y(r, \vartheta) &= B_0 \sum_{n=0}^{\infty} |\kappa_n| \left(\frac{r}{R_0}\right)^n \cos\left[n\,\vartheta - \phi_n\right] \\
&= B_0 \sum_{n=0}^{\infty} \left(\frac{r}{R_0}\right)^n [-\delta_n \sin(n\,\vartheta) + \gamma_n \cos(n\vartheta)].
\end{aligned}
\tag{1.277}
$$

Write

$$
\begin{pmatrix} B_x^{(n)}(r, \vartheta) \\ B_y^{(n)}(r, \vartheta) \end{pmatrix} = B_0 \left(\frac{r}{R_0}\right)^n \begin{pmatrix} \cos(n\,\vartheta) & \sin(n\,\vartheta) \\ -\sin(n\,\vartheta) & \cos(n\,\vartheta) \end{pmatrix} \begin{pmatrix} \delta_n \\ \gamma_n \end{pmatrix}.
$$

The terms containing the cosine are called *normal multipole contributions,* those containing the sine are called *skew multipole contributions* (see exercise 1.43).

Exercise 1.47 Introduce the complex field

$$B(z) = B_y(x, y) + iB_x(x, y)$$

and derive the relevant expression in terms of Cartesian coordinates.

Hint: use (1.276) and (1.277) to write

$$
\begin{aligned}
B(z) &= B_0 \sum_{n=0}^{\infty} b_n(z), \\
b_n(z) &= \left[|\kappa_n| e^{-i\,\phi_n} \left(\frac{z}{R_0}\right)^n \right], \\
z &= x + i\,y = re^{i\vartheta}.
\end{aligned}
\tag{1.278}
$$

Exercise 1.48 Explain why $B(z)$ is an analytic function of z.

Hint: recall that a given function

$$w(z) = u(x, y) + iv(x, y)$$

is said to be analytic if it satisfies the Cauchy–Riemann conditions, which impose that

$$
\begin{aligned}
\partial_x u &= \partial_y v, \\
\partial_y u &= -\,\partial_x v.
\end{aligned}
$$

Use the Maxwell equations ($\vec{\nabla} \cdot \vec{B} = 0,\ \vec{\nabla} \times \vec{B} = 0$) to prove the analyticity of $B(z)$.

Exercise 1.49 Use the operator $\partial_x + i\partial_y$ to prove that

$$(\partial_x + i\partial_y)\,(B_y + iB_x) = (\vec{\nabla} \times \vec{B})_z + \vec{\nabla} \cdot \vec{B}.$$

Exercise 1.50 Prove that

$$(\partial_x + i\,\partial_y)\,(x + i\,y)^{n-1} = 0.$$

Exercise 1.51 Use the previous identity to prove that

$$F_n(x,\, y) = C_n(x + i\,y)^{n-1}$$

can be considered a solution of the Maxwell equations for a magnetostatic field.

Exercise 1.52 Write the $B_{x,y}(x,\, y)$ components in terms of the π_n, τ_n polynomials:

$$b_{n,x}(x,\, y) = -\frac{n}{R_0^n}\,(\bar{\gamma}_n\pi_{n-1} - \bar{\delta}_n\tau_{n-1}),$$
$$b_{n,y}(x,\, y) = -\frac{n}{R_0^n}\,(\bar{\delta}_n\pi_{n-1} + \bar{\gamma}_n\tau_{n-1}).$$

We therefore obtain the results in table 1.4.

Exercise 1.53 In the derivation of the previous quantities we have left unexplained the physical meaning of R_0. Find an argument to specify its role.

Hint: note that (1.278) can be understood as a series expansion in the x, y variables.

Therefore,

$$\gamma_n = \frac{R_0^n}{n!}\partial_x^n B_y\bigg|_{x=0,y=0},$$
$$\delta_n = \frac{R_0^n}{n!}\partial_y^n B_y\bigg|_{x=0,y=0}.$$

Table 1.4. Multipole expansion of magnetic field in terms of π_n, τ_n polynomials.

n	$b_{n,x}(x,\, y)$	$b_{n,y}(x,\, y)$
1	$-\frac{1}{R_0}(\gamma_1 x - \delta_1 y)$	$-\frac{1}{R_0}(\delta_1 x + \gamma_1 y)$
2	$-\frac{1}{R_0^2}\left(\gamma_2 x\, y - \delta_2\frac{x^2 - y^2}{2}\right)$	$-\frac{1}{R_0^2}\left(\gamma_2 x\, y - \delta_2\frac{x^2 - y^2}{2}\right)$
3	$-\frac{1}{R_0^3}\left(\gamma_3\frac{y^3 - 3x^2y}{3} - \delta_3\frac{x^3 - 3xy^2}{3}\right)$	$-\frac{1}{R_0^3}\left(\gamma_3\frac{y^3 - 3x^2y}{3} + \delta_3\frac{x^3 - 3xy^2}{3}\right)$

From which it follows that R_0 is a reference radius fixing the radius of convergence of the expansion. It is an arbitrary value depending on the specific problem under study, usually chosen as between 50%–70% of the magnet aperture.

Exercise 1.54 Use figure 1.64 to express the field rotation.

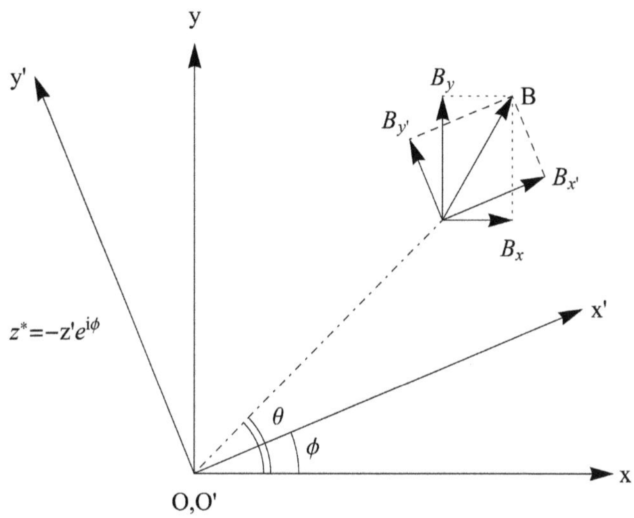

Figure 1.64. Reference frames for exercise 1.54.

Exercise 1.55 Study the field properties under coordinate reflections.
Hint: use figure 1.65 to state that

$$z^* = x - i\,y = -x' - iy' = -z',$$

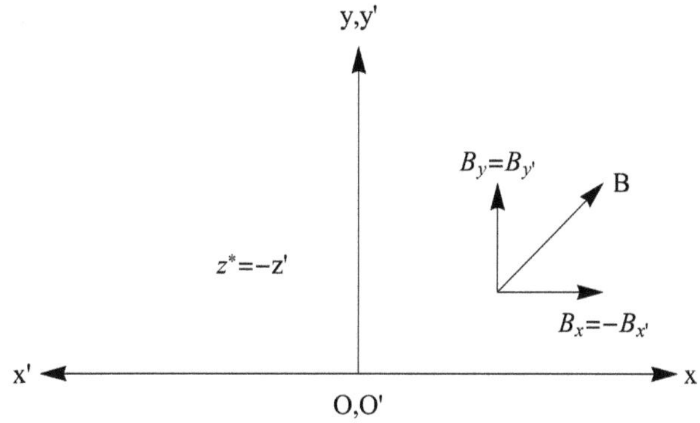

Figure 1.65. Reference frames for exercise 1.55.

$$B_{y'} + i\,B_{x'} = B_y - iB_x = \left(B_y + iB_x\right)^*,$$
$$\sum_{n=1}^{\infty} \chi_n^*\left(\frac{z^*}{R_0}\right)^n = \sum_{n=1}^{\infty} \chi_n^*(-1)^n\left(\frac{z'}{R_0}\right)^n,$$
$$\chi_n = |\kappa_n|e^{-in\phi_n} = \gamma_n + i\,\delta_n,$$

and finally obtain

$$\sum_{n=1}^{\infty} \chi_n'\left(\frac{z'}{R_0}\right)^n = \sum_{n=0}^{\infty} \chi_n^*(-1)^n\left(\frac{z'}{R_0}\right)^n,$$
$$\gamma_n' = -(-1)^n\gamma_n,\ \delta_n' = (-1)^n\delta_n,\ n \geqslant 1.$$

In the case of the dipole only the zeroth term survives which provides a constant field. For $n = 1$ we have the quadrupole which, according to (1.277), consists of two contributions with cos and sin terms. The sin parts can be viewed as rotated by $\frac{\pi}{2\,n}$ and are accordingly called skew.

Exercise 1.56 Consider the magnets in figure 1.66. Explain the type of magnets and the relevant differences.

Hint: use the notion of normal and skew magnets to state that (a) is a normal quadrupole and (b) is skew, discuss the relevant differences in terms of equipotential lines and fields.

We have described the fields and the associated transport items, we did not say anything about their actual realization. The following exercise yields an idea of how they can be designed. Before proceeding further we recall that the Ampère theorem

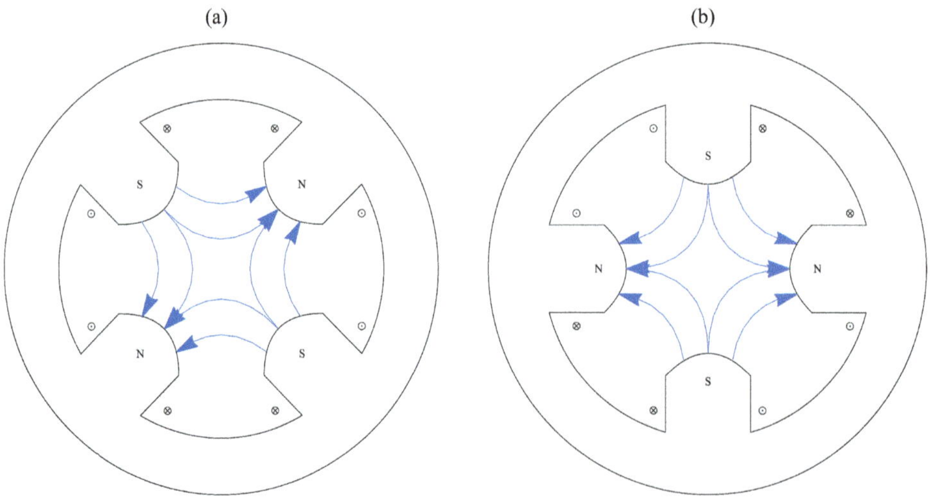

Figure 1.66. Normal quadrupole (a) and skew quadrupole (b).

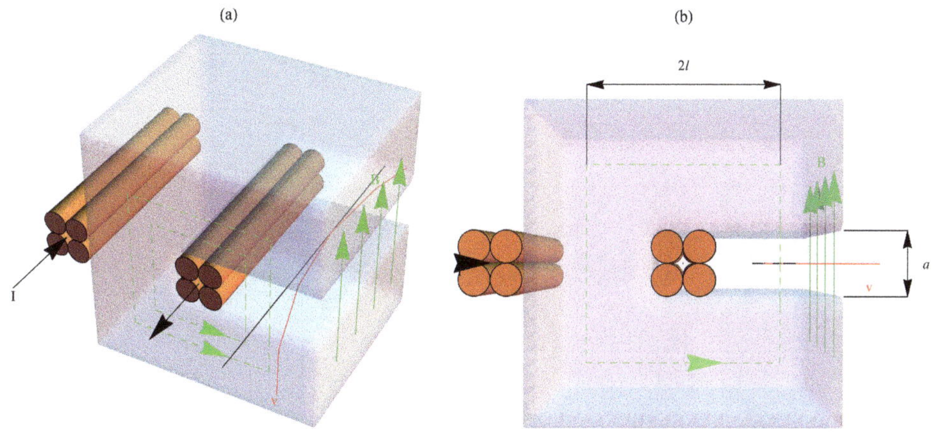

Figure 1.67. A dipole magnet section along with the field force line and the trajectory bending of a negatively charged particle (a) and integration path specifications (b).

states that the magnetic field along a closed line encircling a wire carrying a current I satisfies the well-known integral identity

$$\oint \vec{B} \cdot d\vec{l} = \mu_0 I.$$

The extension of the theorem to a more complicated geometry involves many turns around a conductor (or the vacuum).

Exercise 1.57 Use the Ampère law to evaluate the field inside the dipole magnet gap (make reference to figure 1.67).

Hint: in figure 1.67 we have reported the integration path, the use of the Ampère theorem yields $\oint B\, dl = \int_{C_{\mathrm{Fe}}} B\, dl + \int_{C_{\mathrm{Vac}}} B\, dl$, where we have broken the integration contour into two parts (inside the iron and inside the vacuum). Note that inside the iron $B_{\mathrm{Fe}} = N \dfrac{\mu_0}{\mu_{\mathrm{Fe}}} \dfrac{I}{4l - a}$, assuming that the iron permeability μ_{Fe} is infinity we find inside the gap $B = N\mu_0 \dfrac{I}{a}$.

Exercise 1.58 Show that in practical units we have

$$NI \cong \frac{1.5}{\pi} 10^6 B[\mathrm{T}] a[\mathrm{m}].$$

Exercise 1.59 Use the same arguments as exercise 1.58 to derive the relation between the currents in the coils and the quadrupole field.

The solution of the problem requires more care. The integration contour is that reported in figure 1.68. We again break the integral in two parts and note that the

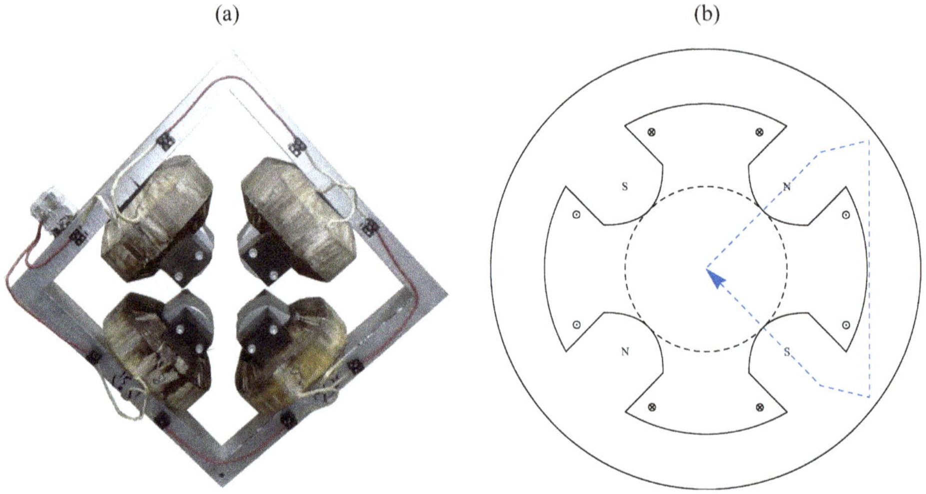

Figure 1.68. Real quadruple (a) and transverse section (b) including the quadrupole radius and the direction of the enveloping coils, the arrows denote the region where the integration contour is calculated. The left image has been obtained and adapted by the authors from the Wikimedia website https://commons.wikimedia.org/wiki/File:Quadrupole_Magnet.jpg, where it is stated to have been released into the public domain. It is included within this article on that basis.

field contribution along the path with $\vartheta = 0$ is vanishing because the field is orthogonal to the integration line, we are therefore left with the radial component of the field provided by

$$\int_0^a B_r dr = N\,\mu_0 I.$$

The evaluation of B_r follows from (1.276), which yields

$$B_r = B_0 \frac{r}{a},$$

thus finding

$$B_0 = \frac{2\,\mu_0 NI}{a}.$$

Exercise 1.60 Derive the equations of motion for a skew quadrupole.

Hint: note that

$$B_y + iB_x = \frac{B_0}{R_0}(\gamma_1 + i\delta_1)(x + iy)$$

$$\Rightarrow \begin{pmatrix} B_y \\ B_x \end{pmatrix} = \frac{B_0}{R_0}\begin{pmatrix} \gamma_1 & -\delta_1 \\ \delta_1 & \gamma_1 \end{pmatrix}\begin{pmatrix} x \\ y \end{pmatrix}.$$

The skew terms are those that are off-diagonal, accordingly the field components are given by

$$B_y = -\frac{B_0}{R_0}\delta_1 y,$$
$$B_x = \frac{B_0}{R_0}\delta_1 x.$$

The relevant equations of motion should accordingly be written as

$$\begin{pmatrix} x'' \\ y'' \end{pmatrix} = -\begin{pmatrix} 0 & k \\ k & 0 \end{pmatrix}\begin{pmatrix} x \\ y \end{pmatrix}.$$

Vertical planes are therefore coupled.

Exercise 1.61 Write the transfer matrix of a skew quadrupole.
Hint: note that defining

$$\underline{Z} = \begin{pmatrix} x \\ y \end{pmatrix}$$

we find

$$\underline{Z}'' = -\hat{A}\underline{Z},$$
$$\hat{A} = k\,{}_h\hat{1},$$
$${}_h\hat{1} = \begin{pmatrix} 0 & 1 \\ 1 & 0 \end{pmatrix}.$$

The solution of the previous matrix equation can be obtained using the analogy with second order ordinary differential equations, thus finding

$$\underline{Z} = e^{i\sqrt{\hat{A}}s}\underline{\alpha} + e^{-i\sqrt{\hat{A}}s}\underline{\beta},$$

where the vectors $\underline{\alpha}$ and $\underline{\beta}$ are defined through the initial conditions

$$\underline{Z}_0 = \underline{\alpha} + \underline{\beta},$$
$$Z_0' = i\sqrt{\hat{A}}\,(\underline{\alpha} - \underline{\beta}).$$

Furthermore, according to (1.237), we end up with

$$\sqrt{\hat{A}} = \frac{\sqrt{k}}{2}[(\hat{1} + {}_h\hat{1}) + i\,(\hat{1} + {}_h\hat{1})].$$

Therefore, we finally end up with the transport matrix

$$
{}_{SQ}\hat{T} = \begin{pmatrix} C_+ & \frac{S_+}{\sqrt{k}} & C_- & \frac{S_-}{\sqrt{k}} \\ -\sqrt{k}\,S_- & C_+ & -\sqrt{k}\,S_+ & C_- \\ C_- & \frac{S_-}{\sqrt{k}} & C_+ & \frac{S_+}{\sqrt{k}} \\ -\sqrt{k}\,S_+ & C_- & -\sqrt{k}\,S_- & C_+ \end{pmatrix},
$$

$$
C_+ = \frac{\cos(\vartheta) + \cosh(\vartheta)}{2}, \quad C_- = \frac{\cos(\vartheta) - \cosh(\vartheta)}{2},
$$

$$
S_+ = \frac{\sin(\vartheta) + \sinh(\vartheta)}{2}, \quad S_- = \frac{\sin(\vartheta) - \sinh(\vartheta)}{2},
$$

$$
\vartheta = \sqrt{k}\,s.
$$

Prove that the determinant of the matrix above is 1.

Exercise 1.62 Study the properties of the functions $C_\pm$, $S_\pm$.

Hint: derive the relevant series expansion, differential equations, addition formulae...

(This exercise may be skipped at a first reading of the complements, however, when a good deal of understanding is reached, the understanding of these trigonometric-like functions will be extremely important.)

Exercise 1.63 An ideal solenoid is a magnet device providing a magnetic field along the beam propagation axis, depending on the longitudinal coordinate only. It is evident that such a configuration requires an almost infinite magnet. Neglecting therefore the end coil effects, shows that the particle equations of motion in such a device are

$$
x'' + 2gx' + g'y = 0,
$$
$$
y'' - 2\,gy' - g'x = 0,
$$
$$
g = \frac{e\;B_{||}(s)}{2\,p}.
$$

Hint: use the Lorentz force equation.

Exercise 1.64 Derive the solenoid transfer matrix.

Hint: introduce the complex variable

$$
Z = x + iy
$$

and show that

$$
Z'' - 2\,i\,g\,Z' - i\,g'Z = 0.
$$

Use the transformation

$$Z = \tilde{Z}e^{i\vartheta},$$
$$\vartheta = \int_0^s g\,ds,$$

and explain the relevant physical meaning. Find the following equation for $\tilde{Z}$,

$$\tilde{Z}'' + g^2\tilde{Z} = 0,$$

and obtain the transport matrix for

$$\begin{pmatrix} \tilde{Z} \\ \tilde{Z}' \end{pmatrix} = \begin{pmatrix} \tilde{x} \\ \tilde{x}' \\ \tilde{y} \\ \tilde{y}' \end{pmatrix},$$

$$_s\hat{\tilde{T}} = \begin{pmatrix} \cos(\vartheta) & \frac{1}{g}\sin(\vartheta) & 0 & 0 \\ -g\sin(\vartheta) & \cos(\vartheta) & 0 & 0 \\ 0 & 0 & \cos(\vartheta) & \frac{1}{g}\sin(\vartheta) \\ 0 & 0 & -g\sin(\vartheta) & \cos(\vartheta) \end{pmatrix}.$$

Discuss the physical meaning of the previous result.

Go back to the original variables

$$\begin{pmatrix} x \\ y \end{pmatrix} = \begin{pmatrix} \cos(\vartheta) & \sin(\vartheta) \\ -\sin(\vartheta) & \cos(\vartheta) \end{pmatrix} \begin{pmatrix} \tilde{x} \\ \tilde{y} \end{pmatrix}.$$

Define

$$\begin{pmatrix} x \\ x' \\ y \\ y' \end{pmatrix} = \hat{M} \begin{pmatrix} \tilde{x} \\ \tilde{x}' \\ \tilde{y} \\ \tilde{y}' \end{pmatrix},$$

$$\hat{M} = \begin{pmatrix} \cos(\vartheta) & \sin(\vartheta) & 0 & 0 \\ -g\sin(\vartheta) & g\cos(\vartheta) & 0 & 0 \\ 0 & 0 & \cos(\vartheta) & \sin(\vartheta) \\ 0 & 0 & -g\sin(\vartheta) & g\cos(\vartheta) \end{pmatrix},$$

and obtain

$$_s\hat{T} = \hat{M}^T\, {_s\hat{\tilde{T}}}\, \hat{M}.$$

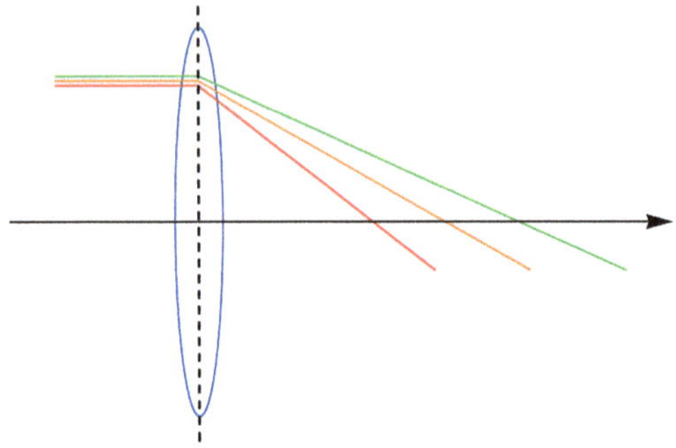

Figure 1.69. Chromatic effect in focusing quadrupoles.

Exercise 1.65 Explain why in a quadrupole the focal length is not the same for all the particles in the beam but depends on the particle energy. Find an analogy with the chromatic aberrations in ordinary optics.

Hint: Prove first that a quadrupole focal length is

$$f_Q = \frac{1}{k_Q d} \propto \gamma_0 \left(1 + \frac{\delta_p}{p}\right).$$

In the kick approximation we find therefore

$$\begin{pmatrix} x_o \\ x_o' \end{pmatrix} = \begin{pmatrix} 1 & 0 \\ -\frac{1}{f_Q^{(0)}}\left(1 - \frac{\delta_p}{p}\right) & 1 \end{pmatrix} \begin{pmatrix} x_i \\ x_i' \end{pmatrix}$$

and

$$x_o' - x_i' = -\frac{1}{f_Q^{(0)}} x_i + \frac{x_i}{f_Q^{(0)}} \frac{\delta_p}{p},$$

where $f_Q^{(0)}$ is the focal length corresponding to the 'nominal' energy (corresponding to the average energy value), and the term after the + sign on the rhs is the angle offset due to the chromatic aberration. Accordingly, particles with larger momentum have longer focal length and vice versa (see figure 1.69).

Exercise 1.66 Derive the equation of motion for a sextupole and analyze the relevant focusing properties

a. Use the results of exercise 1.50 and 1.51 to show that

$$B_x = 2g_S\, xy,$$
$$B_y = g_S(x^2 - y^2),$$
$$g_S = \frac{3\mu_0 NI}{a^3}.$$

b. Use the Lorentz force equation to write

$$x'' = -k_S(x^2 - y^2),$$
$$y'' = 2k_S xy,$$
$$k_S = \frac{e g_S}{m_e \gamma c},$$

and note that they are non-linear and coupled x- and y-components.

c. Introduce the complex variable

$$Z = x + iy$$

and obtain

$$Z'' = -k_S Z^2.$$

Exercise 1.67 Explain why sextupoles may correct the quadrupole chromatic aberration.

The field of a sextupole can be viewed in a region around the minimum as a quadrupole which adds or subtracts focusing according to the values of x. We therefore can use the arrangement reported in figure 1.70, where

a. A bending magnet is used to select particles with different momenta at different positions in x.
b. The sextupole, provides focusing or defocusing according to the previous prescription.
c. The focal point of the quadrupole is finally corrected.

According to exercise 1.32 the bending magnet provides a dispersion (in x) proportional to $D\,\frac{\delta_p}{p}$, where D is a quantity called magnet dispersion length; we have therefore, at the entrance of the sextupole, that the electron ray is located at

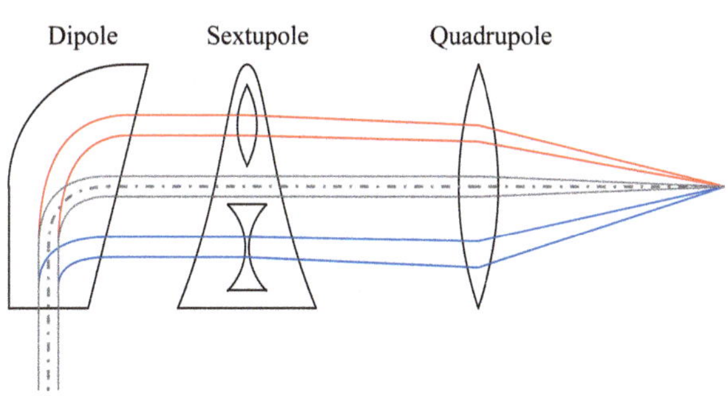

Figure 1.70. Corrections with sextupoles of the chromaticity induced by a bending magnet.

Table 1.5. Synopsis of non-linear terms in the Hamiltonian of exercise 1.68.

$-b\, x\, \frac{\delta_p}{p} + \frac{b^2}{2} x^2$	Dipole (dispersive and focusing)
$\frac{k_Q}{2}(x^2 - y^2)$	Quadrupole (focusing and defocusing)
$\frac{k_S}{3}(x^3 - 3\, x\, y^2)$	Sextupole

$$x = x_0 + D\, \frac{\delta_p}{p},$$

where x_0 is the position without the effect of the dispersion. Inside the sextupole the angular ray deflection is (we keep terms to the first order in $\frac{\delta_p}{p}$)

$$\Delta\, x'_S = -k_S d(x^2 + y^2) \cong -k_S d\left(x_0^2 + y_0^2\right) - 2k_S d\; x_0 D\, \frac{\delta_p}{p},$$

which once added to that induced by the quadrupole yields

$$\Delta\, x'_S + \Delta\, x'_Q \cong -k_S d\left(x_0^2 + y_0^2\right) - 2k_S d\; x_0 D\, \frac{\delta_p}{p} + -\frac{1}{f_Q^{(0)}} x_0 + \frac{x_0}{f_Q^{(0)}} \frac{\delta_p}{p}.$$

The chromatic contributions are cancelled whenever

$$2k_S d\; D \;=\; \frac{1}{f_Q^{(0)}}.$$

We invite the reader to consider the role of the extra terms.

Exercise 1.68 Write the Hamiltonian including the magnetic non-linearities up to sextupoles, see table 1.5 for a synopsis of the non-linear terms.

$$\mathcal{H} = \frac{1}{2\,(1+\delta)}(x'^2 + y'^2) + \left(-b\, \frac{\delta_p}{p}\, x + \frac{b^2}{2} x^2\right) + \frac{k_Q}{2}(x^2 - y^2) + \frac{k_S}{3}(x^3 - 3\, xy^2).$$

Exercise 1.69 Write the Liouville equation corresponding to the above Hamiltonian. Use the symmetric split procedure to obtain the relevant (approximate) solution.

Further reading for chapter 1

Special relativity and relativistic kinematics

Harris P *Special Relativity* http://web.stanford.edu/~oas/SI/SRGR/notes/srHarris.pdf

Artioli M, Babusci D and Dattoli G 2012 *Appunti di relatività ristretta e Cinematica Relativistica* https://www.openaccessrepository.it/record/21050/files/INFN-12-18-LNF.pdf (in Italian)

Morin D 2014 *Problems and Solutions in Introductory Mechanics* (Scotts Valley, CA: CreateSpace) chapter 11

Classical mechanics and classical electromagnetism

Goldstein H 1965 *Classical Mechanics* (Menlo Park, CA: Addison Wesley)

Jackson J D 1975 *Classical Electrodynamics* (New York: Wiley)

Child–Langmuir and klystron

Goldston R J and Rutherford P H 2000 *Introduction to Plasma Physics* (Philadelphia, PA: IOP Publishing) chapter 1 pp 3–7

Umstattd R J, Carr C G, Frenzen C L, Luginsland J W and Lau Y Y 2005 A simple physical derivation of Child–Langmuir space-charge-limited emission using vacuum capacitance *Am. J. Phys* **73** 160–3

For a complete account of klystron physics, concepts and technology see the lecture series available from www-group.slac.stanford.edu/kly/Lecture_Series/slac_klystron_lecture_series.htm

Tsimring S E 2007 *Electron Beams and Microwave Vacuum Electronics* (Hoboken, NJ: Wiley)

Particle motion in a combined electric and magnetic field

Freidberg J P *Plasma Physics and Fusion Energy* (Cambridge: Cambridge University Press) chapter 8 pp 139–182

Johnson B R, Hirschfelder J O and Yang K 1983 *Rev. Mod. Phys.* **55** 109

Babusci D, Dattoli G and Sabia E 2010 Operational methods and Lorentz-type equations of motion, arXiv: 1009.3818v2 [math-ph]

Beam transport and phase space

Reiser M 2004 *Theory and Design of Charged Particle Beams* (Weinheim: Wiley)

Wiedemann H 2007 *Particle Accelerator Physics* (Berlin: Springer)

Steffen K 1990 Fundamentals of accelerator optics *Proc. CERN-Accelerator School (CAS) on Synchrotron Radiation and Free Electron Lasers* ed S Turner CERN 90–03

Barletta W A *Lecture 4, Basic Properties of Particle Beams (Introduction to Accelerators)* US Particle Accelerator School http://uspas.fnal.gov/materials/09UNM/Unit_2_Lecture_4_Beam_Properties.pdf

Evolution operator method

Dragt A J, Neri F, Rangarajan G, Douglas D R, Healy L M and Ryan R D 1988 Lie algebraic treatment of linear and non-linear beam dynamics *Ann. Rev. Nucl. Part. Sci* **38** 455–96

Dattoli G, Galli M and Ottaviani P L 1993 Anharmonic betatron motion in linearly polarized undulators *J. Appl. Phys.* **73** 7046–53

IOP Publishing

Giuseppe Dattoli, Andrea Doria, Elio Sabia and Marcello Artioli

Chapter 2

Linear accelerators

2.1 Introduction

In the previous chapter we have considered a beam of charged particles (electrons) moving at relativistic energies and we did not specify how this beam is produced. We merely noticed that some accelerating mechanism exists, which brings the electrons to kinetic energies, exceeding their rest mass by orders of magnitude. We have seen that the most efficient way to accelerate and/or guide charged particles is through electric and magnetic fields.

The use of an electromagnetic field in a vacuum to accelerate charged particles cannot be recommended. Indeed, the Lawson–Woodward theorem states that, under specific conditions, the net exchange of energy between charged particles and electromagnetic radiation in a vacuum is zero (see section 2.8 for further comments). In the previous chapter we considered static fields, while here we will see how a radio-frequency (RF) field, feeding ad hoc designed accelerating cavities, is the most suitable tool to bring electrons (and protons and ions) to high energies. In the first part of this chapter we deal with electron linear accelerators (linac) and we touch upon other types of devices in section 2.8.

An idea of how these accelerators work is given by figure 2.1. A particle beam is injected inside a chain of coupled cavities, powered by an amplified and suitably split RF signal. The electric field inside the cavities acts on the particles and brings them to higher and higher energies. The cavities are adjusted in such a way that the electrons find the fields with the appropriate accelerating phase.

In the figure we have shown an idealized structure, in which the accelerating field is nicely placed within the cavity and the particles experience the acceleration only outside the drift tube. Let us simplify the problem even more, by considering the example in figure 2.2(a), showing a charged particle entering two infinitely extended

doi:10.1088/978-0-7503-1239-4ch2

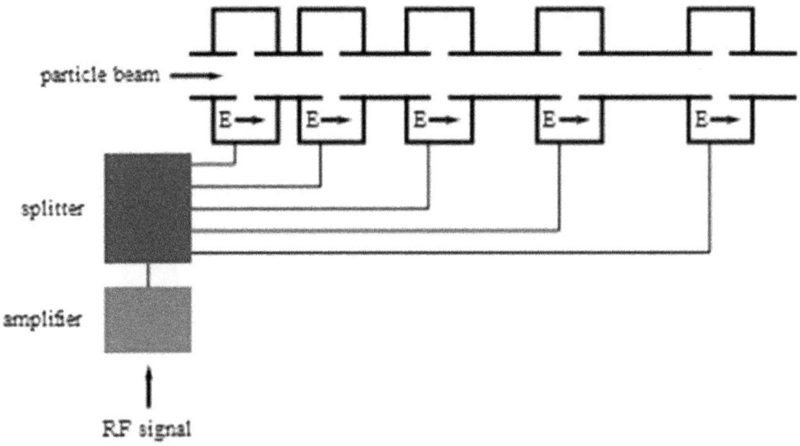

Figure 2.1. Chain of coupled accelerating cavities.

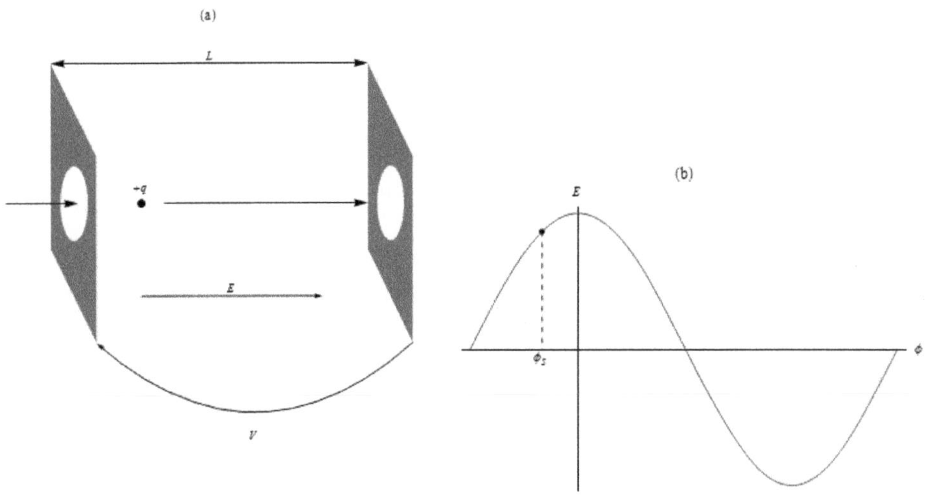

Figure 2.2. (a) Schematic view of a charged particle in the electric field of two infinite plate and (b) particle field phase mismatch.

parallel plates separated by a distance L powered by a sinusoidal voltage, in such a way that inside the plates the electric field is

$$E(t) = E_0 \cos(\omega t + \phi). \tag{2.1}$$

The reference geometry is that reported on the left side of the figure, furthermore the particle at $t = 0$ is at the center of the gap and ϕ measures the phase of the field relative to $t = 0$ (see figure 2.2(b)).

The energy gain of a particle moving inside the plates is just provided by

$$\Delta \mathcal{E} = q \int_{-\frac{L}{2}}^{\frac{L}{2}} E_z dz = qE_0 \int_{-\frac{L}{2}}^{\frac{L}{2}} \cos(\omega t(z) + \phi) dz. \tag{2.2}$$

The dependence of the time on z can be obtained by solving exactly the equation of motion of the particle inside the field. However, by assuming that the particle velocity remains constant during its transit across the plates, we can make the following approximation,

$$\omega t(z) \cong \frac{\omega z}{v} = \frac{2\pi c}{\lambda}\frac{z}{v} = \frac{2\pi z}{\beta\lambda}, \tag{2.3}$$

thus ending up with

$$\begin{aligned} &\Delta\mathcal{E} = qV_0\,\mathcal{T}\cos(\phi), \\ &V_0 = E_0 L, \\ &\mathcal{T} = \frac{\sin(\Lambda)}{\Lambda}, \qquad \Lambda = \frac{\pi L}{\beta\lambda}, \end{aligned} \tag{2.4}$$

The previous equation, even though it is extremely simple, implies at least two important points:

a) The energy gain depends on the phase ϕ.
b) The transit factor $\mathcal{T}$ should be such that it provides the largest energy increase for the crossing electron (see figure 2.3).

If we choose to operate at $\Lambda = \frac{\pi}{2}$ the corresponding gap length should be adjusted at $L = \frac{\beta\lambda}{2}$, thus finding for the transit factor $\mathcal{T} \cong 0.637$. Regarding the values of λ we note that the typical accelerating frequencies of these devices are in the range of GHz, the RF wavelength λ is therefore tens of centimeters. Accordingly, we obtain a first rough idea on the length of the accelerating cavities.

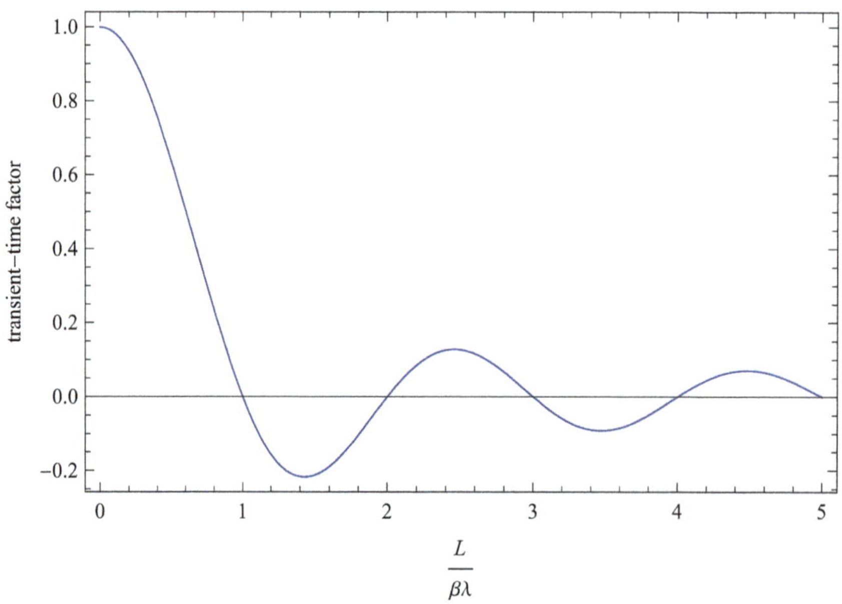

Figure 2.3. Transit time factor versus accelerating length.

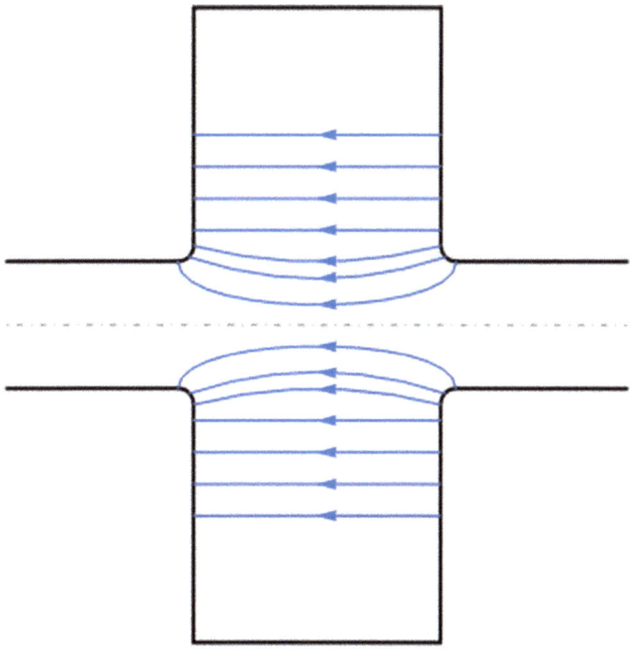

Figure 2.4. Field line distortion near the cavity walls.

The physical meaning of the transit time factor is that of a figure of merit defining the energy efficiency transfer from the field to the particle. This quantity is therefore of crucial importance and will be further commented on in section 2.8. It should, however, be stressed that the derivation of equation (2.4), albeit substantially correct, is valid only if the assumption of field homogeneity inside the cavity holds. An idea of the field line distortion near the cavity walls is given in figure 2.4, the factor $\mathcal{T}$ should therefore be calculated accordingly.

Postponing the problem of a more appropriate quantitative definition to section 2.8, here we stress that a particle crossing all the cavities at the same phase ϕ_s (the subscript s stands for synchronous) will experience an energy gain per accelerating gap summarized by

$$\gamma_{n+1}\,\beta_{n+1} - \gamma_n\,\beta_n = \frac{qV_0\,\mathcal{T}}{m_q c^2}\cos(\phi_s), \tag{2.5}$$

where γ_n and β_n are the relativistic factor and the reduced velocity at the nth cavity. The velocity gain per gap is, accordingly, provided by

$$\beta_{n+1} = \frac{\gamma_n\beta_n + \dfrac{qV_0\,\mathcal{T}}{m_q c^2}\cos(\phi_s)}{\sqrt{1 + \left[\gamma_n\beta_n + \dfrac{qV_0\,\mathcal{T}}{m_q c^2}\cos(\phi_s)\right]^2}}. \tag{2.6}$$

We can therefore understand when the particle acquires a sufficient amount of energy to become relativistic. In figure 2.5 we have reported a picture comparing the velocity and kinetic energy gain per gap, using equations (2.5) and (2.6). Assuming that the particle is an electron, the peak accelerating field 5 MV m^{-1}, $L = 10$ cm, $\mathcal{T} = 0.8$ and $\phi_s \cong \frac{\pi}{9}$ we obtain $\frac{qV_0\mathcal{T}}{m_q c^2}\cos(\phi_s) = 0.75$, we infer that the particle reaches the relativistic regime in the first 2–3 cavities and a final energy of a few MeV in 1 m of accelerating cavities.

In the previous discussion we have provided a description of the energy gain by a single particle, intercepting all the cavities at a specific value of the phase, which has been denoted by ϕ_s and is usually defined as the 'synchronous phase'. The conditions ensuring such a synchronous crossing can be easily stated, but we first mention that, according to figure 2.1, the cavities are fed at different phases, namely

$$\begin{aligned} V_n &= V_0\cos(\omega t), \\ V_{n+1} &= V_0\cos(\omega t + \Delta\phi_{n+1}), \\ V_{n+2} &= V_0\cos(\omega t + \Delta\phi_{n+1} + \Delta\phi_{n+2}). \\ &\dots \end{aligned} \tag{2.7}$$

By taking into account that, in each cavity, the particle acquires a velocity v_n and then moves along a drift section of length d, the phase it meets when entering the $(n + 1)$th cavity, is

$$\Phi_{n+1} = \frac{\omega d}{v_n} + \phi_s + \Delta\phi_{n+1}, \tag{2.8}$$

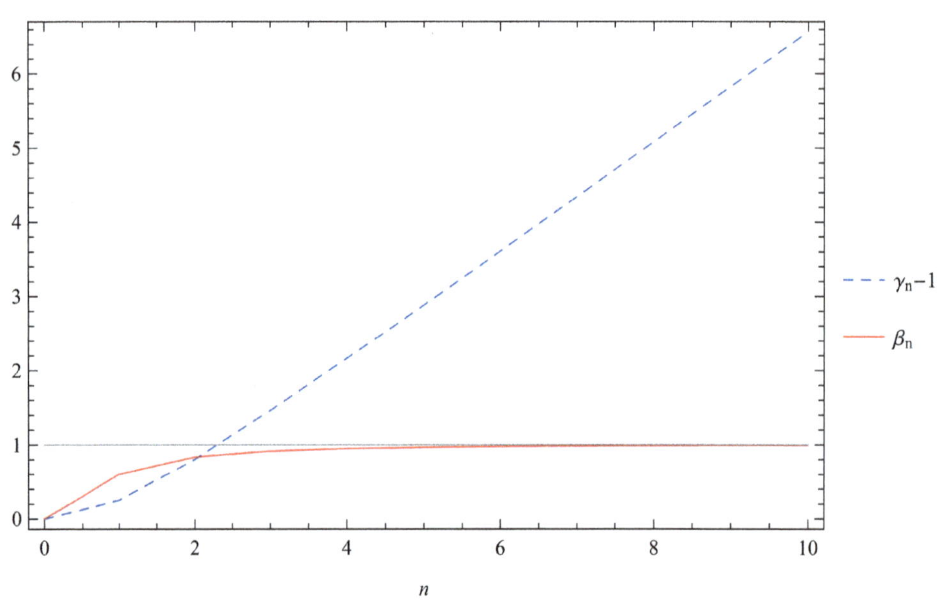

Figure 2.5. Kinetic energy gain (dot line) and reduced velocity gain per cavity (continuous line) for $L = 10$ cm, $E_0 = 5$ MV m^{-1}, $L = 0.1$ m, $\mathcal{T} = 0.8$ and $\phi_s \cong \frac{\pi}{9}$.

where $\Delta\phi_{n+1}$ is the phase of the $(n+1)$th cavity. The condition ensuring synchronism is therefore

$$\Phi_{n+1} = \phi_s, \tag{2.9}$$

which is granted if the phase of each cavity be adjusted in such a way that

$$\Delta\phi_{n+1} = -\frac{\omega d}{v_n}. \tag{2.10}$$

When the particle approaches the relativistic regime, the velocity is frozen at ($v_n \cong c$) and therefore all $\Delta\phi_{n+1}$ are identical.

To have a very primitive idea of the effect of the phase mismatch, we compare the energy gain dynamics for two particles, injected at different phases, with the cavities adjusted to compensate the phase retardation of one of the two. The relevant dynamics is that reported in figure 2.6, where we show the kinetic energy and velocity gain per cavity for the same parameters as figure 2.5, including a second particle with a different injection phase ($\phi \cong \frac{\pi}{3.65}$). The physical meaning is clear; the synchronous phase correction (2.10) does not prevent the de-phasing due to the different velocities. After each accelerating section the following phase mismatch will be accumulated,

$$\delta\phi_{n+1} = -\omega d\frac{v_n' - v_n}{v_n v'_n}, \tag{2.11}$$

where v_n' is the velocity acquired by the non-synchronous particle. It is evident from the figure that the out-of-phase particles are accelerated less efficiently, both because the accelerating potential is lower ($\cos(\phi) < \cos(\phi_s)$) and because the accumulated phase mismatch during the transition from the non-relativistic to relativistic regime

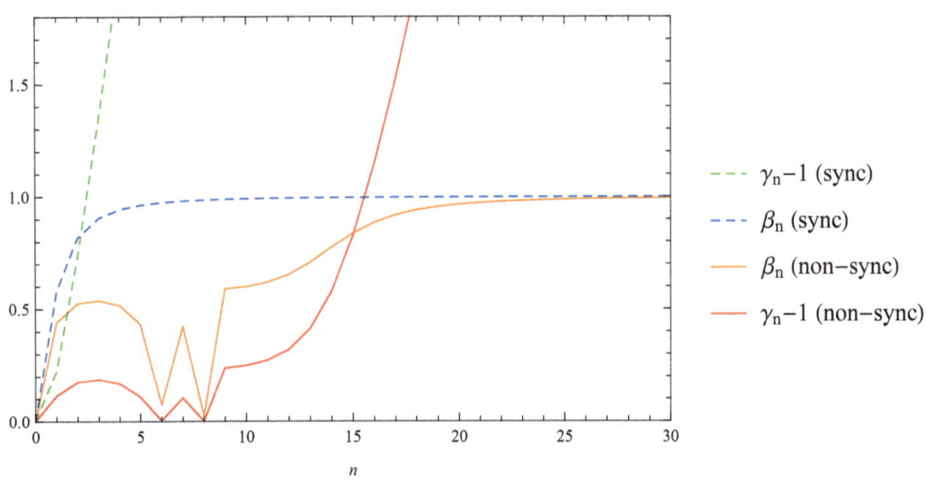

Figure 2.6. The same as figure 2.5. The continuous and dashed curves (kinetic and velocity gain per cavity) refer to the non-synchronous particle ($\phi \cong \frac{\pi}{3.65}$).

creates a further reduction of the acceleration efficiency (the term efficiency has been used improperly, a more appropriate discussion of this important concept will be provided in following sections).

In the following we will discuss two linac design conceptions, based on the different ways of feeding the accelerating structures. In the *standing wave* (SW) configuration a stationary RF field in the cavities accelerates electrons, while in *travelling wave* (TW) configurations, the electrons ride the RF field propagating inside the accelerating tube. The previous remarks, albeit more appropriate for the first case, are general enough to provide a feeling of the problems we will meet later in this chapter, when we will develop a less elementary treatment.

We have just touched upon the surface of the role of the interplay in acceleration between synchronous and non-synchronous particles. These mechanisms play a central role in the understanding of linac devices and of the relevant performance characteristics. An appropriate treatment of the associated phenomenology requires a more careful description of its various components.

2.2 Accelerating cavities, shunt impedance and quality factor

The accelerating cavities, discussed so far in abstract terms, are devices where the electromagnetic power to be delivered to the electrons is stored.

Actual cavities are characterized by unavoidable losses and not all the power feeding the resonator, $\mathcal{P}_{\mathcal{RF}}$, is transferred to the electron current. The power losses, $\mathcal{P}_d$, can be calculated from the following balance equation (we will provide a more complete analysis of the power balance in a linac in forthcoming parts of the chapter),

$$P_E = \mathcal{P}_{\mathcal{RF}} - \mathcal{P}_d, \tag{2.12}$$

where P_E is the electron-beam power.

The dissipated power is evidently a limiting factor of the performances of the cavity itself, and can be exploited to introduce an appropriate figure of merit, usually referred as *shunt impedance* (SI) R_s. If we view the accelerating structure as an electrical circuit element, we can just take advantage of Ohm's law, to introduce the lost power as

$$\mathcal{P}_d = \frac{V_0^2}{R_s}, \tag{2.13}$$

where V_0 is the peak accelerating voltage. It is evident that the SI is a useful quantity summarizing different effects leading to power dissipation (e.g., oscillating currents induced by the RF on the cavity surfaces and thus dissipating power). It depends therefore on the geometrical characteristics of the structure itself, on the composing materials, on the operating RF and so on. The relevant computation requires a significant effort, with ad hoc developed codes, as will be mentioned later in this chapter. More often, along with R_s, the SI per meter (r_s) is given. Just to quote reference numbers, we note that typical values range around hundreds of $M\Omega$.

Accordingly the power dissipated in a structure of length $\mathcal{L}$ is

$$\mathcal{L} = \frac{V_0^2}{r_s \mathcal{P}_d}, \tag{2.14}$$

thus obtaining from equation (2.12)

$$\mathcal{L} = \frac{1}{\mathcal{P}_{\mathcal{RF}} - P_E} \frac{V_0^2}{r_s}, \tag{2.15}$$

where P_E denotes the electron-beam power, whose 'Ohm-like' version has been discussed in the previous chapter and is linked to the product of current and energy.

Equation (2.15) can be exploited to obtain an idea of how long a linac should take to reach a given energy with a given (average) current, once the characteristic SI per unit length is known. Writing equation (2.15) in practical units as

$$\mathcal{L}[\mathrm{m}] = \frac{1}{\mathcal{P}_{\mathcal{RF}}[\mathrm{MW}] - I[\mathrm{A}]\, E[\mathrm{MeV}]} \frac{E[\mathrm{MeV}]^2}{r_s[M\Omega\ \mathrm{m}^{-1}]}. \tag{2.16}$$

Keeping, e.g., $E = 500$ MeV, $\mathcal{P}_{\mathcal{RF}} = 250$ MW, $I = 0.06$ A, $r_s = 50\ \mathrm{M}\Omega\ \mathrm{m}^{-1}$, we find $\mathcal{L} \cong 23$ m. This is nothing but a rough estimation, but it yields an idea of how the various elements should be merged to end up with conclusions of practical interest.

Along with SI, a figure of merit in linac design is the cavity quality factor, defined as the ratio of the stored energy per cycle to the dissipated power, namely

$$Q = \frac{\omega \mathcal{U}_{st}}{\mathcal{P}_d}. \tag{2.17}$$

The link to SI is accordingly evident; by the relevant definitions we indeed obtain

$$Q^{-1} R_s = \frac{V_0^2}{\omega \mathcal{U}_{st}}, \tag{2.18}$$

which can be exploited to fix, e.g., the accelerating voltage and frequency in terms of the maximum power which can be stored in the cavity itself, compatible with the SI and Q factors.

To provide a sense of the engineering problems involved in the design of linac cavities, we note (see forthcoming sections) that

$$R_s \propto \omega^{\frac{1}{2}}. \tag{2.19}$$

An accelerating field with higher frequencies should therefore be considered more efficient for acceleration. Higher frequencies impose less severe limitations due to breakdown problems associated with high electric field on the cavity surface, but at the same time require smaller aperture gaps, which might be not suitable to deal with intense electron beams, as required for the operation of fourth-generation synchrotron radiation sources.

After these remarks we will start a deeper discussion on the field distribution inside accelerating cavities.

2.3 Electromagnetic fields and accelerating cavities

According to the previous discussion the electric field accelerating the electrons inside the cavity should be directed along the longitudinal direction. Electromagnetic waves in a vacuum are therefore not suitable since they are transverse electric and magnetic (TEM) waves, namely the electric and magnetic field they carry are orthogonal to the wave propagation direction.

The electromagnetic fields in wave guides or cavities, with perfectly conducting walls, have different distributions of electric and magnetic vectors. The relevant propagation equation is specified by boundary conditions (such as the vanishing of tangential component of the electric field and the normal component of the magnetic field on the surface boundaries, see section 2.8) allowing the presence of transverse electric (TE) waves (with magnetic field along the longitudinal direction) or of transverse magnetic (TM) waves (with electric field along z).

Postponing the details of the mathematical analysis, we note that the conditions of the electric and magnetic field with null tangent and normal component, respectively, ($E_t = 0,\ B_n = 0$) is ensured in the electromagnetic wave propagation in a cavity in which the radiation enters at an incidence angle α and is reflected inside the perfectly conducting plates of the resonator (see figure 2.7).

According to the reflection-bouncing picture of figure 2.7 the wavelength of the propagating mode in the optical cavity is no longer that of the incident radiation, but that of the mode propagating in the longitudinal direction which, with reference to figure 2.8, is simply

$$\lambda_p = \frac{\lambda}{\cos(\alpha)}. \tag{2.20}$$

The corresponding wave vector is

$$k_p = k\cos(\alpha) \tag{2.21}$$

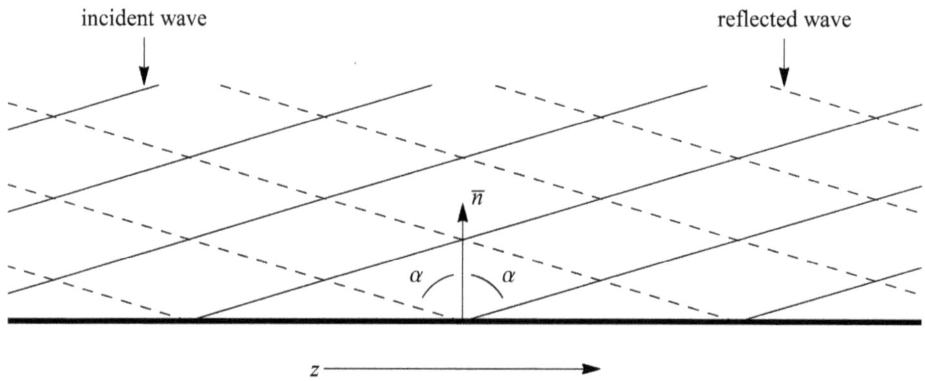

Figure 2.7. Electromagnetic wave propagation between perfectly conducting planes.

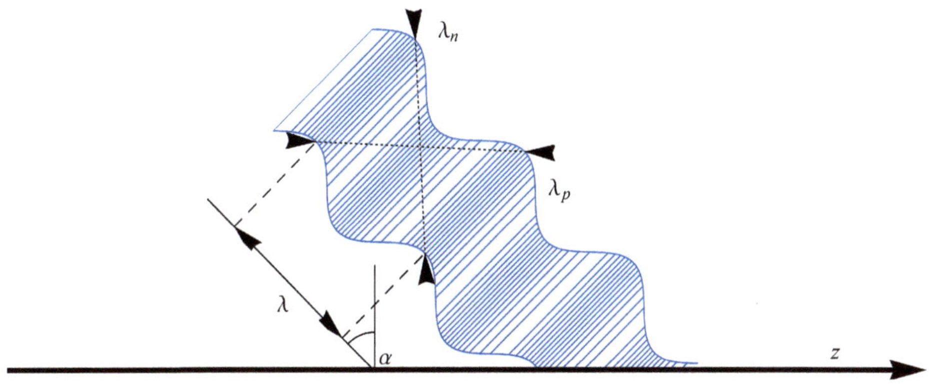

Figure 2.8. Free wave propagation wavelength λ and wavelength of the propagating mode λ_p.

and the phase velocity is accordingly given by

$$v_p = \frac{\omega}{k_p} = \frac{c}{\cos(\alpha)} > c. \tag{2.22}$$

Finally, the group velocity, namely the velocity of propagation of the power carried by the wave is

$$v_g = c\cos(\alpha). \tag{2.23}$$

The incidence angle is not subject to any constraint, in the case of only one conducting plane. When the radiation is bounced back, if a second plane is included to guide the field, the situation is different. With reference to figure 2.9(a), we note that boundary conditions on normal λ_n and parallel λ_p wavelengths must be imposed. By requiring, e.g., that an integer number (m) of half-wavelengths can be accommodated in the normal direction we find $\frac{\lambda_n}{2\sin(\alpha)} = m$.

In figure 2.9(b) we have reported a rectangular metallic wave guide where TE modes (with a longitudinal magnetic field component) and TM (with a longitudinal electric field component) can propagate. The modes are usually labelled by two indices, $TE_{m,n}$, $TM_{m,n}$, being the number of half wavelengths in the two transverse directions (x, y). In the case of bounded (resonant) cavities a third index labelling the modes in the longitudinal direction is also added.

From the mathematical point of view the orthogonal modes of a rectangular wave guide are expressed in terms of trigonometric functions (see section 2.8). The mode indices are determined by the boundary conditions that fix the confinement and thus the vanishing of the electric and magnetic fields on the cavity walls.

The same recipe applies to other cavity configurations. In the case of a cylindrical cavity (see figure 2.10), the search for the proper orthogonal modes is made slightly more complicated by the bounding surface geometry. A more elaborate set of functions, naturally emerging from the symmetry of the problem under study, is indeed required.

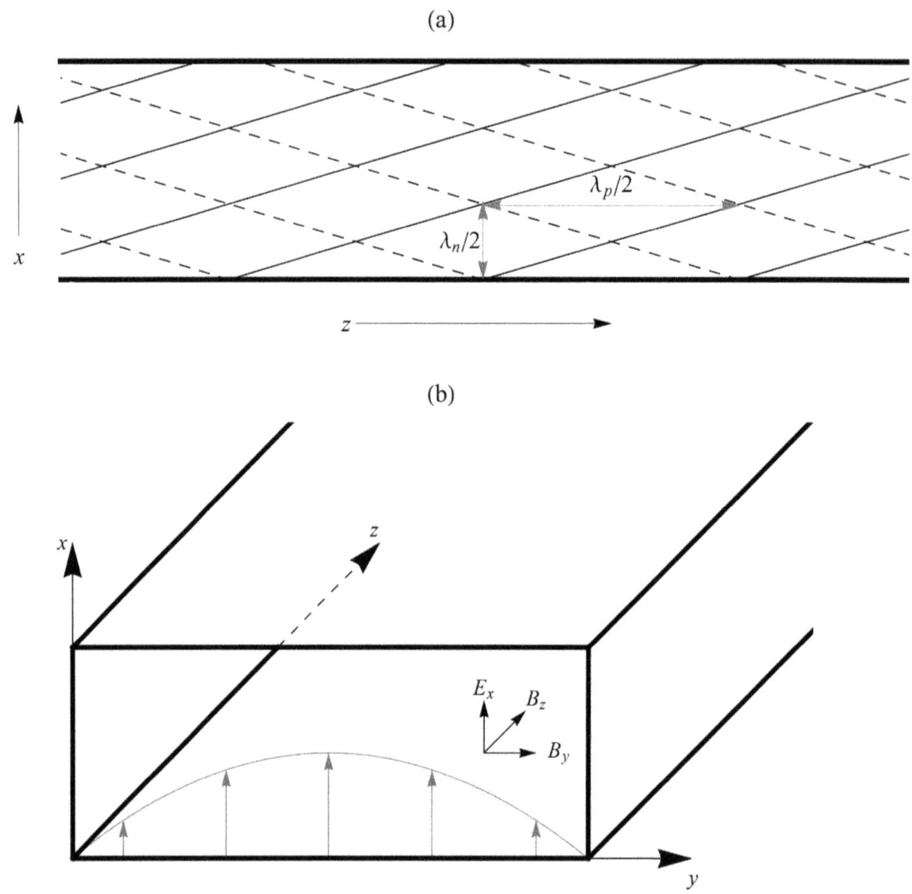

Figure 2.9. (a) Boundary conditions for parallel and normal wavelengths. (b) Field distribution of TE_{01} mode.

The discussion that follows is very schematic; further and more detailed comments can be found in section 2.8.

a. The electric and magnetic fields satisfy the D'Alembert equation, namely

$$\left(\frac{1}{c^2}\frac{\partial^2}{\partial t^2} - \nabla^2\right)F(x, y, z, t) = 0. \tag{2.24}$$

b. In cylindrical coordinates the Laplacian is written

$$\begin{aligned}\nabla^2 &= \left(\frac{\partial}{\partial x}\right)^2 + \left(\frac{\partial}{\partial y}\right)^2 + \left(\frac{\partial}{\partial z}\right)^2 \\ &= \frac{1}{\rho}\frac{\partial}{\partial \rho}\rho\frac{\partial}{\partial \rho} + \frac{1}{\rho^2}\frac{\partial^2}{\partial \phi^2} + \frac{\partial^2}{\partial z^2}.\end{aligned} \tag{2.25}$$

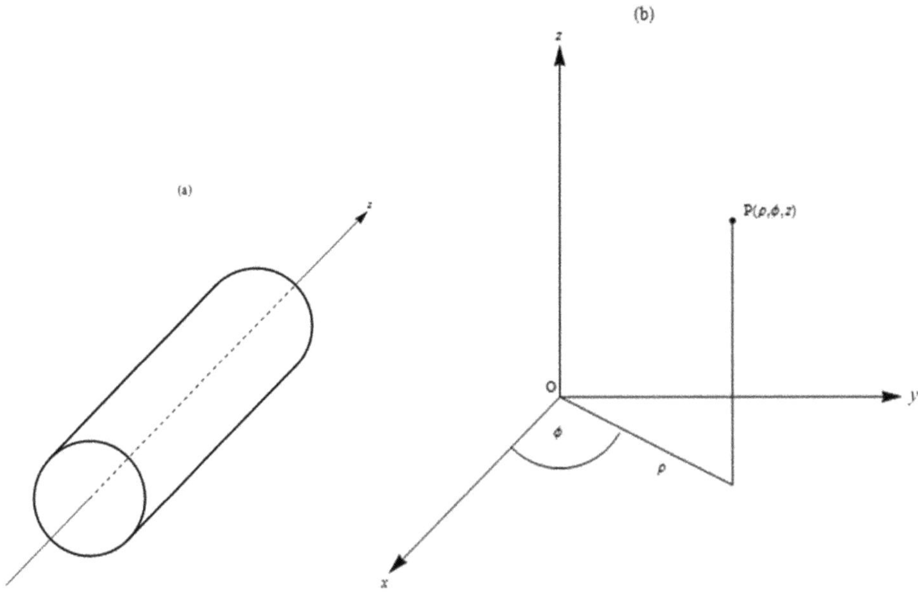

Figure 2.10. Schematic of cylindrical waveguide and relevant coordinates.

Thus obtaining for (2.25)

$$\left(\frac{1}{c^2}\frac{\partial^2}{\partial t^2} - \left[\frac{1}{\rho}\frac{\partial}{\partial \rho}\rho\frac{\partial}{\partial \rho} + \frac{1}{\rho^2}\frac{\partial^2}{\partial \phi^2}\right] + \frac{\partial^2}{\partial z^2}\right)\Phi(\rho, z, \phi, t) = 0. \tag{2.26}$$

c. The method of separation of variables suggests that

$$\Phi(\rho, \phi, z, t) = R(\rho)\,\chi(\phi)f(z)\,\Omega(t). \tag{2.27}$$

d. The wave structure of equations (2.24)–(2.26) allows one to combine all the functions appearing in equation (2.27), except for $R(\rho)$, in the following oscillatory form:

$$\chi(\phi)f(z)\,\Omega(t) = e^{i\,n\,\phi - i\,\omega t + i\,k\,z}, \quad k \equiv \text{wave vector along the longitudinal direction.} \tag{2.28}$$

Inserting equations (2.27) and (2.28) in equation (2.26) we end up with the following equation ruling the dependence of the solution on the radial coordinate,

$$\left[\frac{1}{\rho}\frac{\partial}{\partial \rho}\rho\frac{\partial}{\partial \rho} + \left(\frac{\omega^2}{c^2} - k^2 - \frac{n^2}{\rho^2}\right)\right]R(\rho) = 0, \tag{2.29}$$

which is easily recognized to be the Bessel equation and the relevant solutions can be written as

$$R(\rho) = a_1 J_n\left[\sqrt{\left(\frac{\omega}{c}\right)^2 - k^2}\rho\right] + a_2 Y_n\left[\sqrt{\left(\frac{\omega}{c}\right)^2 - k^2}\rho\right], \tag{2.30}$$

where $J_n(-)$, $Y_n(-)$ are the cylindrical Bessel functions of first and second kind, respectively, discussed in some detail in section 2.8. For completeness we note that they are the two independent solutions of the Bessel equation (we recall that the Bessel equation is $x^2y'' + x\,y' + (x^2 - n^2)y = 0,\ y'' = \frac{d^2}{dx^2}y,\ y' = \frac{d}{dx}y$). They play the role of the sine and cosine functions for the harmonic oscillator equation, and the relevant damped oscillating behavior is shown in figure 2.11.

As functions of the second kind are infinite at $\rho = 0$, we will consider in the following cylindrical functions the first kind only and recall that they are defined by the infinite series

$$J_n(x) = \sum_{r=0}^{\infty} \frac{(-1)^r\left(\frac{x}{2}\right)^{n+2r}}{(n+r)!\,r!}. \tag{2.31}$$

As already stressed, the TM modes have a vanishing on-axis magnetic field component and, according to the Maxwell equations, we have (see section 2.8) the following field components in cylindrical coordinates (omitting the time dependence)

$$\begin{aligned}
&E_z \propto J_n\left[K_{n,m}\frac{\rho}{\rho_0}\right]\cos(n\phi)\cos(kz),\\
&E_\rho \propto -J_n'\left[K_{n,m}\frac{\rho}{\rho_0}\right]\cos(n\phi)\sin(kz),\\
&E_\phi \propto J_n\left[K_{n,m}\frac{\rho}{\rho_0}\right]\cos(n\phi)\sin(kz),\\
&K_{n,m} = \sqrt{\left(\frac{\omega\rho_0}{c}\right)^2 - (k\rho_0)^2}, \qquad \rho_0 \equiv \text{pill Box radius},
\end{aligned} \tag{2.32}$$

and

$$\begin{aligned}
&B_z = 0,\\
&B_\phi \propto -J_n'\left[K_{n,m}\frac{\rho}{\rho_0}\right]\cos(n\phi)\sin(kz),\\
&B_\rho \propto -iJ_n'\left[K_{n,m}\frac{\rho}{\rho_0}\right]\sin(n\phi)\sin(kz), \qquad i = \text{imaginary unit}.
\end{aligned} \tag{2.33}$$

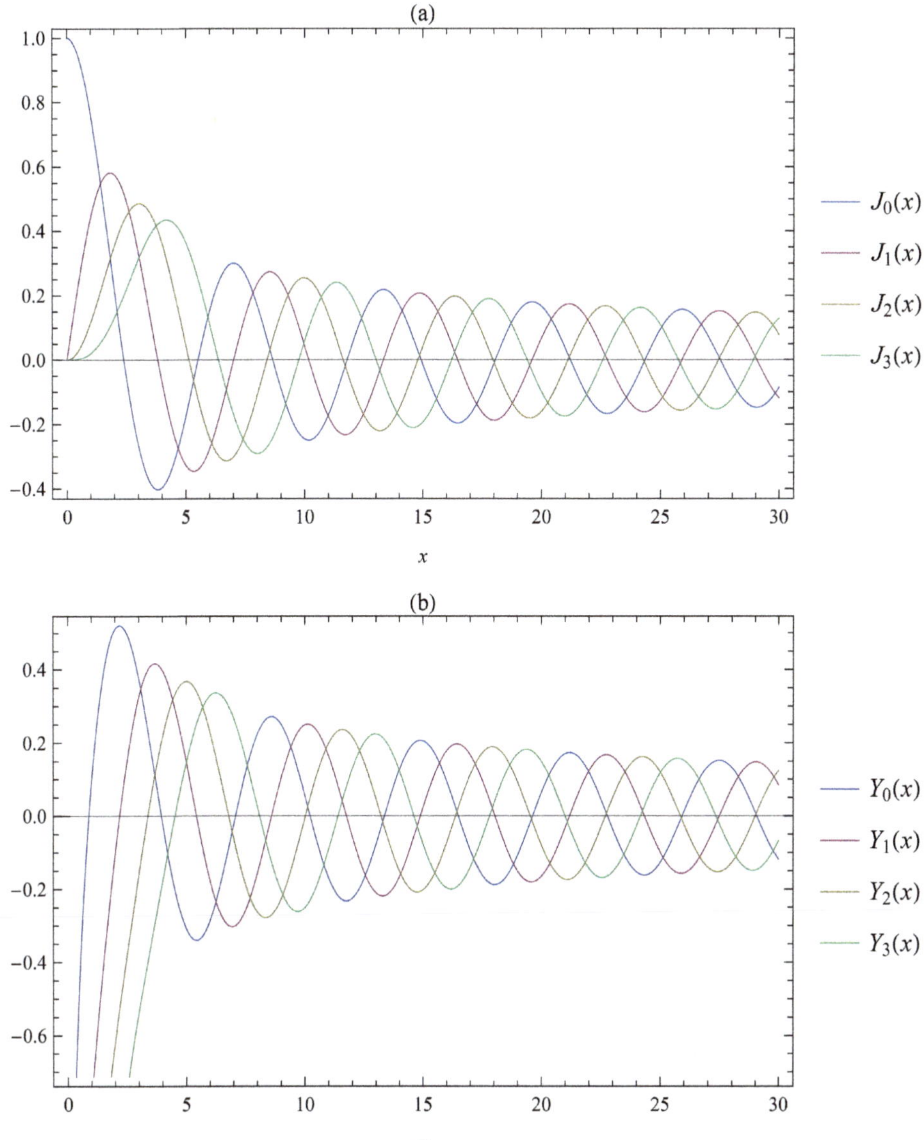

Figure 2.11. Cylindrical Bessel functions of (a) the first kind and (b) the second kind.

There are three indices labeling the mode:

- n associated with the nth order of the Bessel function.
- p associated with the longitudinal confinement in a box of length L.
- m associated with the mth zero of the nth Bessel function, namely

$$J_n\left[K_{n,m}\frac{\rho}{\rho_0}\right]\bigg|_{\rho=\rho_0} = 0.$$

In equations (2.32) and (2.33) we have denoted by $J_n'(x)$ the first order derivative of the Bessel functions, which satisfy the following recurrences:

$$\begin{aligned} J_n'(x) &= \frac{1}{2}[J_{n-1}(x) - J_{n+1}(x)], \\ nJ_n(x) &= \frac{x}{2}[J_{n-1}(x) + J_{n+1}(x)], \end{aligned} \tag{2.34}$$

and reflection property

$$J_{-n}(x) = (-1)^n J_n(x). \tag{2.35}$$

Postponing to the forthcoming parts of the chapter the discussion of the properties of the $TM_{n,m,(p)}$ modes, we consider here the mode $TM_{0,1,(0)}$ specified by the field components

$$\begin{aligned} E_z &= E_0 J_0(k_c\rho)\cos(kz), \\ E_\rho &= \frac{k}{k_c}E_0 J_1(k_c\rho)\sin(kz), \\ B_\phi &= \frac{\kappa}{ck_c}E_0 J_1(k_c\rho)\sin(kz), \\ J_0(K_{0,1}) &= 0 \to K_{0,1} \cong 2.404\,825, \\ \kappa &= \frac{2\pi}{\lambda}, \qquad k = p\,\frac{\pi}{\mathcal{L}}, \qquad k_c = \frac{K_{0,1}}{\rho_0}, \end{aligned} \tag{2.36}$$

where $\mathcal{L}$ is the length of the waveguide.

We have kept the mode p between parentheses because we will essentially refer to the case $p = 0$, which is called the accelerating mode, characterized by $E_z \neq 0,\ E_\rho = 0,\ B_\phi = 0$. An example of transverse and longitudinal mode distribution is given in figure 2.12.

Let us now understand the physical meaning of the previous relations. The dispersion relation we obtain from the boundary condition is

$$\begin{aligned} \left(\frac{\omega}{c}\right)^2 &= \left(\frac{\omega_c}{c}\right)^2 + k^2, \\ \omega_c &= ck_c, \\ k_c &= \frac{K_{0,1}}{\rho_0}, \qquad \lambda_c = \frac{2\pi}{K_{0,1}}\rho_0 \cong 2.613\,\rho_0. \end{aligned} \tag{2.37}$$

By ω_c, k_c and λ_c we have denoted the cut-off frequency, cut-off wave vector and cut-off wavelength, respectively. The relevant Brillouin diagram is shown in figure 2.13, the phase velocity is defined as

$$v_p = \frac{\omega}{k} = \frac{c}{\sqrt{1 - \left(\frac{\omega_c}{\omega}\right)^2}} \tag{2.38}$$

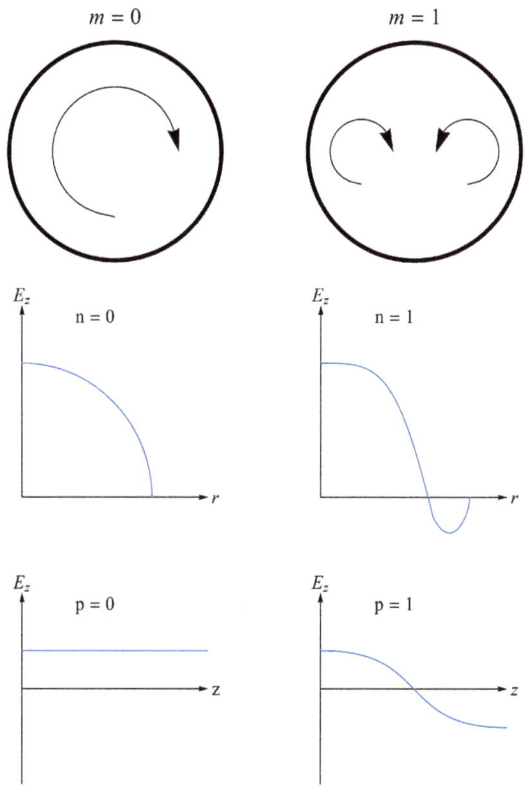

Figure 2.12. Behavior of the longitudinal electric field of modes $TM_{0,0,(0)}$ and $TM_{1,1,(1)}$ versus radial and longitudinal coordinates.

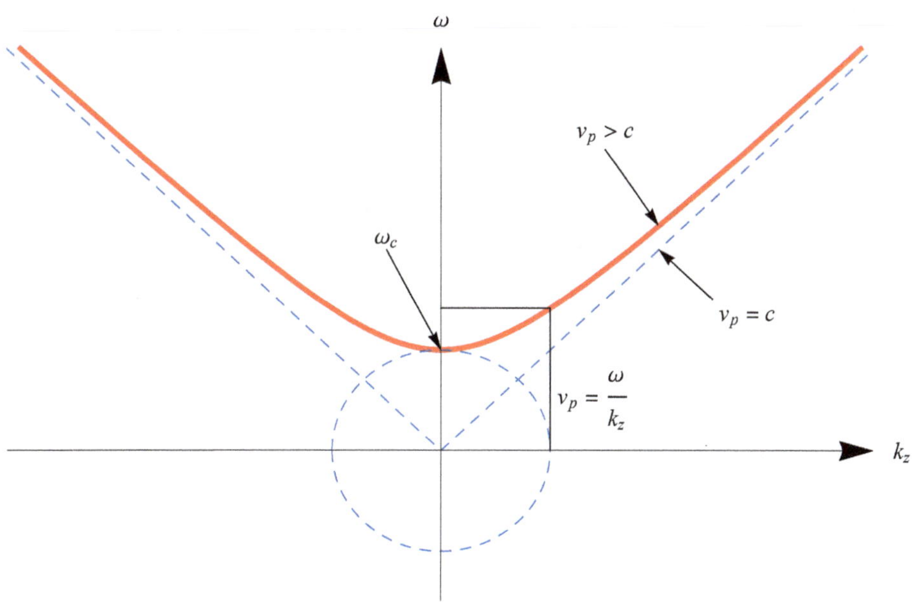

Figure 2.13. Frequency-wave-vector Brillouin diagram.

and the group velocity is

$$v_g = \frac{\partial \omega}{\partial k} = c\sqrt{1 - \left(\frac{\omega_c}{\omega}\right)^2}. \tag{2.39}$$

It is also worth noting that v_g can be expressed in terms of energy flow as (see section 2.8.)

$$v_g = \frac{P}{u_0}, \tag{2.40}$$

where P is the power of the propagating wave and u_0 is the energy per unit length.

It is evident that the phase velocity is larger than the light velocity. A 'naïve' pillbox cavity cannot be exploited to accelerate electrons, since the wave phase velocity and the electron velocity do not match. The way to overcome such a problem will be discussed in a forthcoming section.

Before closing this section, to reconcile the discussion with the introductory comments, we note that the wave incidence angle of equation (2.20) can be defined as

$$\begin{aligned} \cos(\alpha(\omega)) &= \sqrt{1 - \left(\frac{\omega_c}{\omega}\right)^2}, \\ \sin(\alpha(\omega)) &= \frac{\omega_c}{\omega}. \end{aligned} \tag{2.41}$$

It is evident that according to this picture the incidence angle depends on the frequency of the propagating wave (or vice versa). In particular we note that (see figure 2.13) for the cutoff frequency $\alpha(\omega_c) = \frac{\pi}{2}$, while, in correspondence to $v_p = c$, $\alpha(\infty) = 0$. For a further geometrical view see figure 2.13.

2.4 Disk loaded cavities and accelerating conditions

As already mentioned, current linacs are based on travelling wave (TW) or standing wave (SW) accelerating cavities. There are no definite reasons to prefer one or the other paradigm, but to provide a distinction we can say that TWs are exploited to accelerate light particles while SWs are preferred for protons or ions.

In a TW (figure 2.14), an RF field is injected inside the structure at the input port and extracted at the end. Charged particles are captured by the RF field and are accelerated and travel with the wave. In an SW the injected RF field fills the coupled cavities and establish a standing wave pattern.

In a TW device the radiation should be slowed down to bring the phase velocity below (or close to) that of light, to ensure synchronism with electrons and thus allowing energy exchange. The solution is therefore that of placing inside the

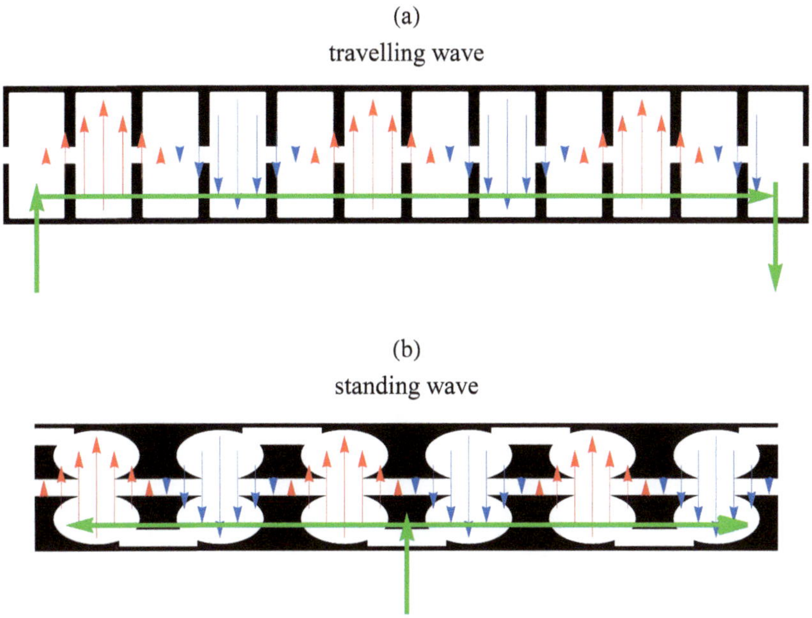

Figure 2.14. Field distributions in (a) TW and (b) SW structures.

wave-guide a suitable number of 'obstacles' providing the desired effect. In figure 2.15 we have sketched a disk charged cavity, which is the most commonly adopted configuration in TW structures.

The loaded structure is a periodic array of disks with distance $\mathcal{L}$ and an iris aperture. The insertion of these obstacles has a number of consequences (see section 2.8):

a. The superimposed periodic structure along z modifies the cavity modes in such a way that the function $\Phi(\rho, \phi, z, t)$ in equation (2.27) should be written as

$$\Phi(\rho, \phi, z, t) = R(\rho, z)\,\chi(\phi)\,\Omega(t), \tag{2.42}$$

where we take into account the superimposed periodic structure by expanding the function $R(\rho, z)$ in Fourier series,

$$R(\rho, z) = \sum_{r=-\infty}^{\infty} a_r \Phi_r(\rho)\, e^{-i\left(\frac{2\pi\, r}{\mathcal{L}} + k_0\right)z}, \tag{2.43}$$

which automatically ensures the periodicity condition $R(\rho, z + \mathcal{L}) = R(\rho, z)$, (to be more rigorous we should invoke the Floquet theorem, according to which the solution of a second order differential equation with a periodic potential has the same periodicity as the potential itself). The mode amplitudes a_r decrease with increasing $|r|$.

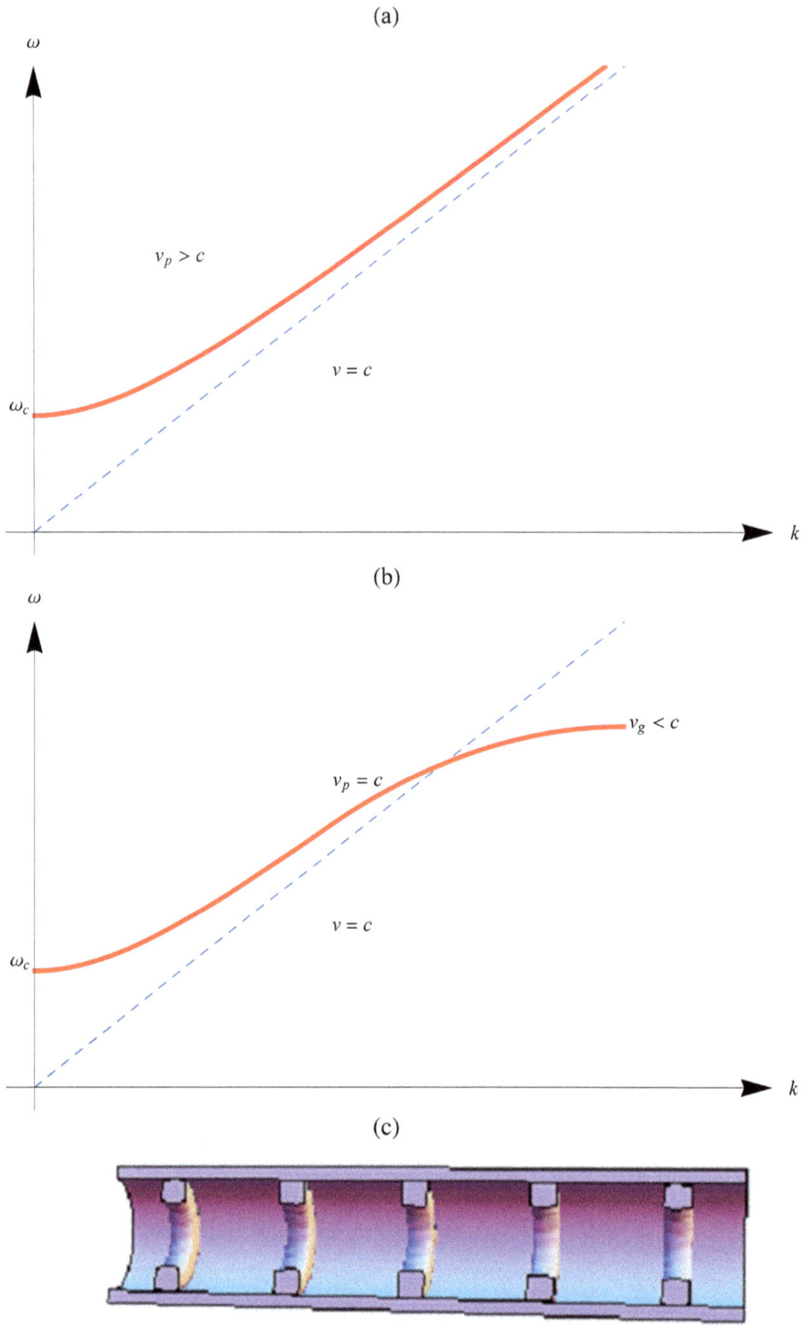

Figure 2.15. A Brillouin diagram for (a) empty and (b) disk charged cavities. The obstacles (disks sketched in (c)) set inside the accelerating tube (wave-guide) change the Brillouin diagram of the loaded wave guide in such a way that the phase velocity is reduced and the dispersion curve intercepts the light line (and therefore also ultra-relativistic beam lines). The group velocity is always less than c and is represented by the slope of the tangent to the curve and vanishes at the cutoff frequency.

b. The modes we have labelled m are now projected on the modes labelled with the index r, and will be specified by the dispersion relation

$$\left(\frac{\omega}{c}\right)^2 = \left(\frac{\omega_c}{c}\right)^2 + \left(k_0 + \frac{2r\pi}{\mathcal{L}}\right)^2, \tag{2.44}$$

where k_0 is the wave vector associated with the zeroth order component. The mode $TM_{0,1}$ is accordingly projected onto the previously defined Fourier modes and therefore we write the relevant field components as

$$\begin{aligned} E_z &= E_0 \sum_{r=-\infty}^{\infty} a_r J_0(k_c^{(r)}\rho)\cos(k_r z - \omega t), \\ E_\rho &= E_0 \sum_{r=-\infty}^{\infty} a_r \frac{k_r}{k_c^{(r)}} J_1(k_c^{(r)}\rho)\sin(k_r z - \omega t), \\ B_\phi &= \frac{E_0}{c} \sum_{r=-\infty}^{\infty} a_r \frac{\kappa}{k_c^{(r)}} J_1(k_c^{(r)}\rho)\sin(k_r z - \omega t), \\ \kappa &= \frac{2\pi}{\lambda}. \end{aligned} \tag{2.45}$$

The phase velocity is

$$\begin{aligned} v_p^{(r)} &= \frac{\omega}{k_r}, \\ k_r &= \left(k_0 + \frac{2r\pi}{\mathcal{L}}\right). \end{aligned} \tag{2.46}$$

Requiring $v_p < c$ implies

$$v_p^{(r)} = \frac{\omega}{\left(k_0 + \frac{2r\pi}{\mathcal{L}}\right)} < c \rightarrow \left(\frac{\omega}{c}\right)^2 - \left(k_0 + \frac{2r\pi}{\mathcal{L}}\right)^2 < 0. \tag{2.47}$$

It is evident that the geometrical picture summarized by the Brillouin diagram should be reconsidered by including the periodicity of the superimposed structure.

The effect of the loading is that of modifying the waveguide dispersion, as shown in figure 2.16, which reflects the periodical behavior implicit in the Fourier expansion (2.43). In the figure we have drawn the Brillouin diagram for the loaded wave guide, and the relevant dispersion curve is shown for the different modes r, versus the associated wave vector. We have reported $r = -1$, 0, 1 and for $r = 0$ we have also repeated, for comparison, the unloaded dispersion diagram. The interaction between radiation and charged particles is allowed in the region for which $v_p \leqslant c$, where the intersection between the electron dispersion, indicated by the straight line ($\frac{v_e}{c} = \beta$), and the wave guide dispersion curves.

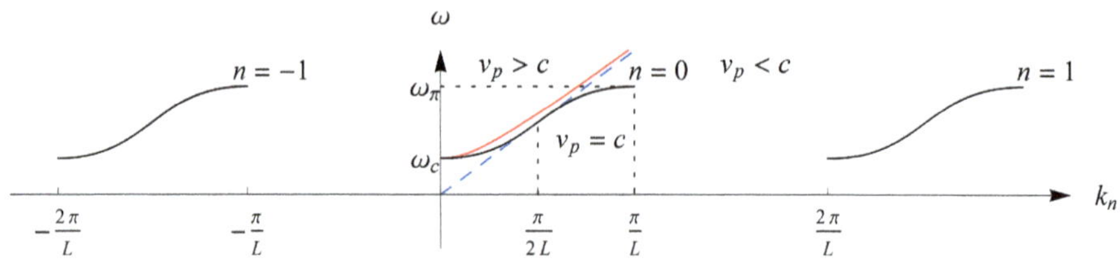

Figure 2.16. Brillouin diagram for mode harmonics propagating in the positive z-direction with different values of the phase velocity.

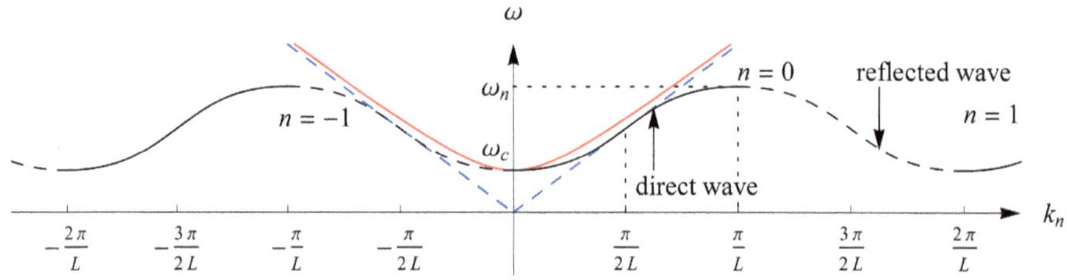

Figure 2.17. Forward and backward wave Brillouin diagram.

The discussion of the physical meaning of figure 2.16 is particularly instructive and implements the highlights reported in the caption of figure 2.15. It is therefore worth noting that

- For a given mode $\boldsymbol{r}$ there is a frequency gap allowing the transmission of the mode itself. The group velocity ($v_g = \frac{\partial \omega}{\partial k_r} = \frac{\partial \omega}{\partial k}$) indeed vanishes when the tangent to the dispersion curve vanishes. These frequencies, denoted by $\omega_{c,\pi}$ in the figure, are located at $k\mathcal{L} = m\,\pi$ and in correspondence to these points the velocity of propagation of the TW energy flow is zero.
- In correspondence to a given frequency (inside the pass band gap) there is an infinite number of harmonics from $n = -\infty$ to $n = +\infty$, propagating with the same group velocity but different phase velocities (when both group and phase velocities are in the same direction the wave is said to be forward, when they are in opposite directions it is said backward).
- The wave representative of the Brillouin diagram of figure 2.16 is forward and propagating in the $+z$-direction.

In figure 2.17 we complete our description by including the wave vector region with negative slopes, corresponding to a condition supporting a backward wave.

What is the moral to be drawn from the above discussion? The radiation propagating in a linac, can either be dissipated at the end of the line or can be reflected back by shortening the cavity. In the first case we have a TW structure and in the second an SW structure.

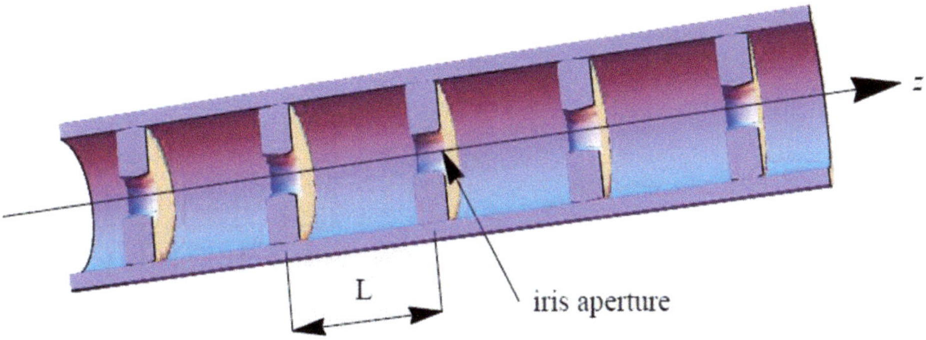

Figure 2.18. Sketch of a disk loaded cavity.

When the particles (electrons) have been brought to modest energies, say $\gamma \cong 10$, the corresponding reduced velocity is already very close to unity, $\beta \cong 0.995$. We can therefore choose as the operating mode $n = 0$ (which has the largest amplitude a_0 and $v_p = c$), the curve crosses the beam line with $\beta = 1$, between the allowed band (see figure 2.16) in correspondence to $\frac{\pi}{2\,\mathcal{L}} \leqslant k_0 \leqslant \frac{2}{3}\frac{\pi}{\mathcal{L}}$, corresponding to a phase advance of $\frac{\pi}{2}, \frac{2}{3}\pi$. (In the case of heavier particles, such as protons, the reduced velocity is much lower, at 10 MeV kinetic energy it is only 0.145, and therefore modes with $n > 0$ should be used).

There are different operating modes for the cavities, depending on the number of sub-cavities (cells) realizing an accelerating section. In figure 2.18 we have reported a disk loaded cavity in which each cell has length $\mathcal{L}$.

The cells are coupled and, as we will see in section 2.8, such a coupling can be understood by exploiting the analogy between the cavity and an LC electrical circuit, which is in turn analogous to an array of coupled harmonic oscillators (see figure 2.19). The associated oscillation modes are summarized in figure 2.19 and are indicated by $(0, \frac{\pi}{2}, \frac{2}{3}\pi, \pi)$, specifying the phase difference between one cell and the successive cell (for further comments see section 2.8; we invite the reader to consider both sides of figure 2.19 carefully and explain the relevant physical meaning).

We have provided a fairly general description of the linac architecture, which far from being exhaustive, allows a very preliminary understanding of the problems one faces in accelerating charged particles with linear structures. In the following section we will mention a few practical details yielding a first hint on how 'numbers' should be collected to obtain a preliminary design of a linac.

2.5 The linac design key quantities

The concepts we have introduced so far are too general to be helpful in recognizing the parameters to be selected to start thinking about the design of an actual linac accelerating device. Let us therefore try to aggregate a few elements of practical interest.

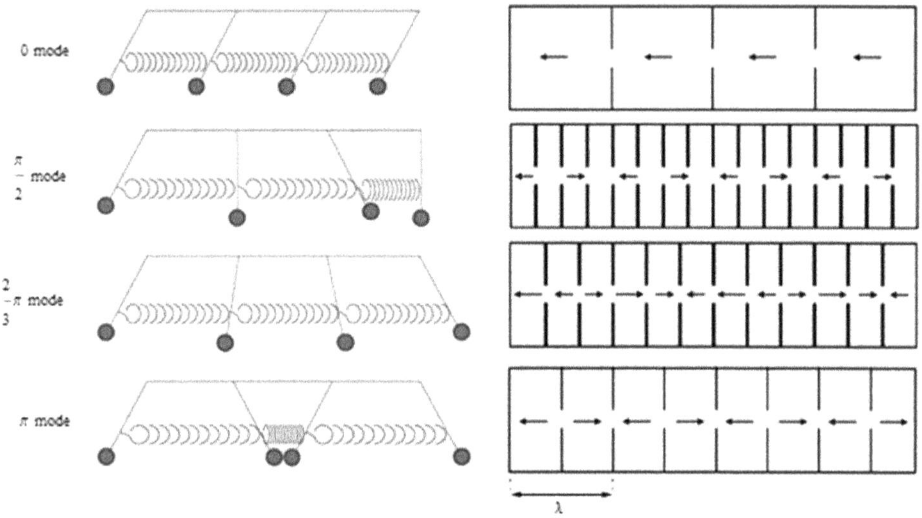

Figure 2.19. Cavity operating modes for different phase difference of the field from one cell to the other, on the left we have provided an analogy with string coupled harmonic oscillators.

Table 2.1. Linac figures of merit.

$V = \int E_z(z, t)dz$	$E_{\mathrm{acc}} = V/L$
$R_s \equiv \frac{V^2}{P_d}$	$R_s/L = \frac{E_{\mathrm{acc}}^2}{(P_c/L)}$
$R_s/Q \equiv \frac{V^2}{2\omega U}$	$(R_s/L)/Q = \frac{E_{\mathrm{acc}}^2}{(P_c/L)}$
$Q \equiv \frac{\omega U}{P_d}$	$Q = \frac{\omega(U/L)}{(P_c/L)}$

Quantities of crucial importance to frame a linac accelerator as an actual device are the accelerating voltage, the accelerating electric field, the SI and the cavity quality factor. These are introduced and discussed in section 2.2, and are summarized for later convenience in table 2.1, along with their mutual relationships.

We have made a distinction between accelerating parameters per cavity (left) and per unit length (right). We have denoted by E the accelerating field, which is in turn linked to the SI and to the quality factor.

Let us to put things together by considering an accelerating structure with *constant impedance*, such as, e.g., the disk loaded cavity with all identical cells.

In this configuration the radiation flows inside the structure and its power is decreased by the dissipative losses. The accelerating gradient is therefore decreased and the energy gain should be corrected by including this effect. A slightly more elaborated version of the energy gain equation is therefore needed, which accounts for the reduction of the peak electric field along the linac section.

In the following we will sketch the various steps for dimensioning a constant impedance structure.

Step 1: Cavity dimensions. Assuming an RF operating at 3 GHz, we can fix (see equation (2.37)) the cavity radius $\rho_0 \cong \dfrac{\lambda}{2.613} \cong 3.83$ cm.

Step 2: Power propagation and attenuation inside the accelerating structure. The RF power will fill the structure in a characteristic time

$$t_f = \frac{L}{v_g}. \tag{2.48}$$

Writing the power losses per unit length as $\dfrac{dP}{dz}$, we find from the identities in table 2.1

$$\begin{aligned} \frac{dP}{dz} &= -\frac{\omega u_0}{Q},\ u_0 = \frac{P}{v_g}, \\ \frac{dP}{dz} &= -\frac{\omega}{Q v_g} P, \end{aligned} \tag{2.49}$$

which once integrated yields

$$P(z) = P_0 e^{-\frac{\omega}{Q v_g} z}. \tag{2.50}$$

Step 3: Evaluation of the electric RF field. The electric field along z is just provided by

$$E(z) = E_0 e^{-\frac{1}{2}\frac{\omega}{Q v_g} z}. \tag{2.51}$$

Integrating on the whole length we find the accelerating potential

$$\begin{aligned} V &= \int_0^L E(z)\, dz = V_0\left[1 - e^{-\frac{1}{2}\frac{\omega}{Q v_g} L}\right], \\ V_0 &= 2\, E_0 v_g \frac{Q}{\omega}. \end{aligned} \tag{2.52}$$

Since, as is easily checked, the electric field intensity can be written in terms of the power as

$$E_0^2 = \frac{\omega}{v_g}\frac{r_s}{Q} P_0, \tag{2.53}$$

we end up with the following identity yielding the total accelerating voltage:

$$\begin{aligned} V(\tau) &= v(\tau)\sqrt{2 P_0 r_s L}, \\ v(\tau) &= \left[\frac{1 - e^{-\tau}}{\sqrt{\tau}}\right], \\ \tau &= \frac{1}{2}\frac{\omega L}{Q v_g} = \frac{1}{2}\frac{\omega}{Q} t_f. \end{aligned} \tag{2.54}$$

The interplay between field decay and filling time is shown in figure 2.20(a) and figure 2.20(b) reports the plot of the function $v(\tau)$.

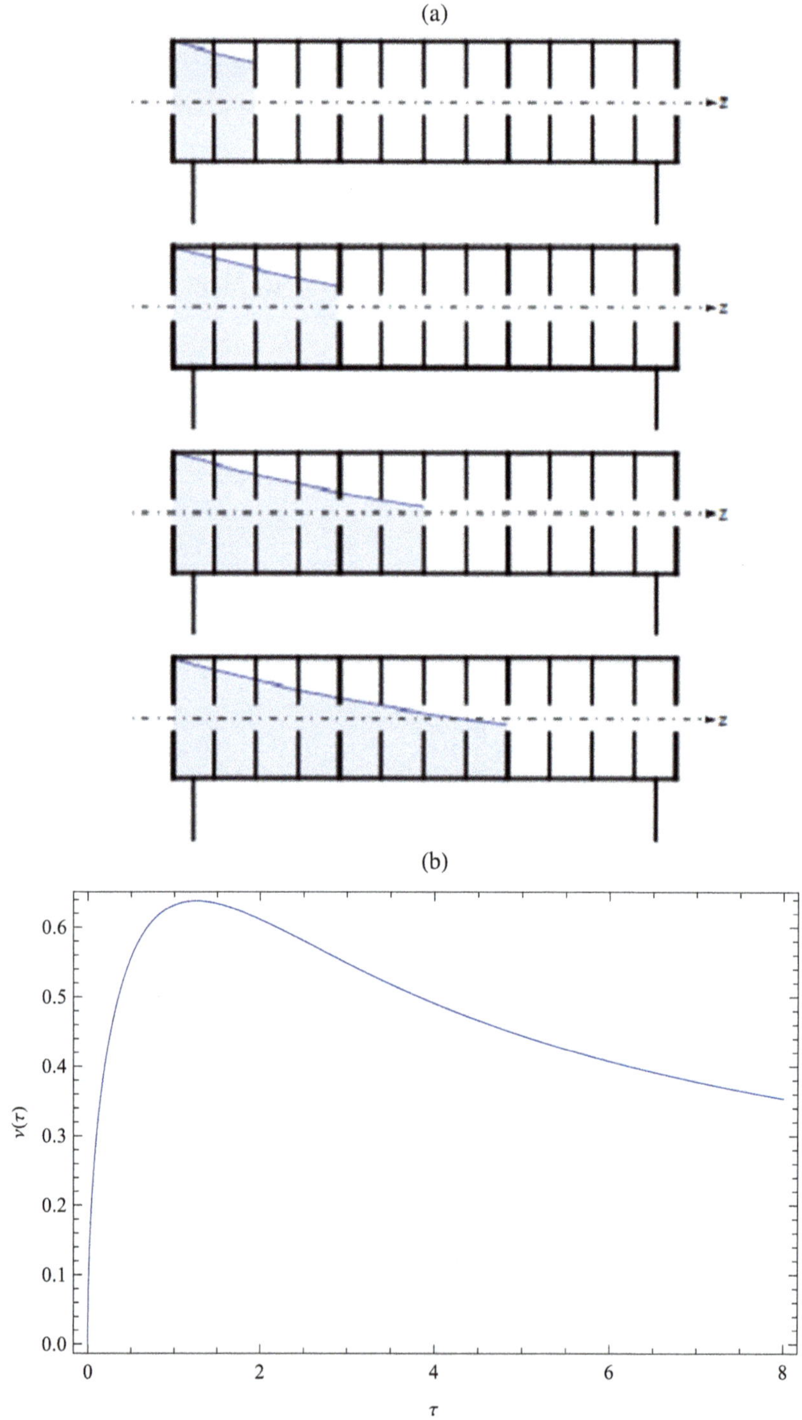

Figure 2.20. (a) Filling and decay of a linac accelerating cavity and (b) $\nu(\tau)$ versus τ.

Step 4*: Wrap Up.* A first criterion to be chosen is that of designing the accelerating structure in order to maximize the function $\nu(\tau)$, thus obtaining

$$t_f = \frac{2Q}{\omega}\tau^* \to L = \frac{2\,Q}{\omega}v_g\tau^*, \qquad \tau^* \cong 1.3. \tag{2.55}$$

Equation (2.55) yields the 'optimum' accelerating length, once we have fixed the cavity quality factor, the RF frequency and the group velocity.

This last quantity depends on the operating mode and on the iris aperture ($2a \equiv$ iris diameter). If we consider, e.g., a loaded cavity set at the $\frac{2}{3}\pi$ operating mode, the group velocity can be approximately expressed by (see section 2.8) $v_g[\mathrm{ms}^{-1}] \cong \frac{(2a[\mathrm{cm}])^{3.23}}{8.91}3\times10^6$ we obtain

$$L[\mathrm{m}] \cong 1.31\times10^6\times(a[\mathrm{cm}])^{3.23}\times\frac{Q}{\nu[\mathrm{Hz}]}. \tag{2.56}$$

At $\tau^* \cong 1.3$ we have $\nu(\tau^*) \cong 0.638$, thus finding

$$\begin{aligned} P_0[\mathrm{MW}] &\cong \left(\frac{V^*[\mathrm{MV}]}{0.638}\right)^2\frac{1}{2\,r_s[M\Omega\,\mathrm{m}^{-1}]L[\mathrm{m}]} \\ &\cong 9.38\times10^{-7}\frac{\nu[\mathrm{Hz}]\times(V^*[\mathrm{MV}])^2}{r_s[M\Omega\,\mathrm{m}^{-1}]\times Q\times(a[\mathrm{cm}])^{3.23}}. \\ V^* &= V(\tau^*) \end{aligned} \tag{2.57}$$

In conclusion with equation (2.56) we evaluate the length of the accelerating structure, with equation (2.57) the required power if we want to have an accelerating voltage $V(\tau^*)$.

The previous example has covered the case of a structure with constant impedance. In the following we develop a few design remarks concerning a device with *constant gradient structure.* In a constant gradient structure the power losses are compensated by slightly tapering the iris radius of the loaded disks, as shown in figure 2.21, in such a way that the peak field increases.

When this solution is adopted, the field attenuation is no longer constant and should be designed in such a way that the losses are compensated. The impedance and the group velocity are no longer constant. The relevant feedback on the energy gain should therefore be properly accounted for to get the correct energy gain along the structure.

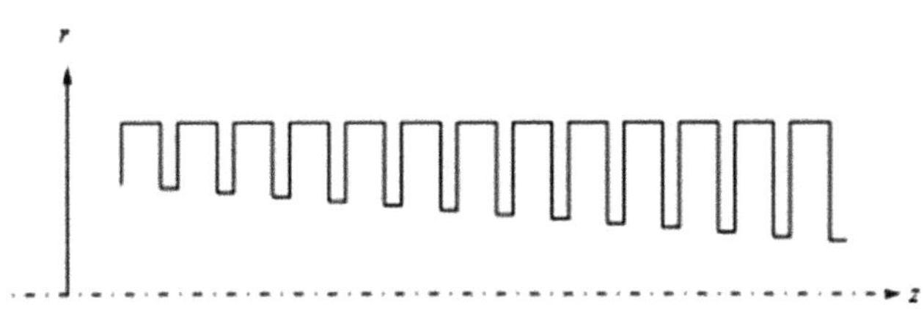

Figure 2.21. Constant gradient structure with iris tapering.

The starting point of the discussion of the constant impedance structure has been the equation

$$\frac{dP}{dz} = -2\alpha P, \qquad \alpha = \frac{\omega}{2Qv_g}, \tag{2.58}$$

in which α is independent of the longitudinal coordinate. Such an assumption is no longer valid if we are interested in a structure with a constant accelerating gradient for which $\frac{dE}{dz} = 0$.

In order to obtain an identity such as equation (2.54), linking the accelerating potential to the SI, quality factor etc, we write the electric field intensity in terms of r_s and the power lost per unit length as

$$r_s = \frac{E^2}{-\left(\frac{dP}{dz}\right)} \rightarrow E^2 = -r_s\left(\frac{dP}{dz}\right). \tag{2.59}$$

The assumption of constant accelerating field implies that $\left(\frac{dP}{dz}\right) =$ constant and therefore that P is a linear function of z. It is therefore natural to set

$$P(z) = P_0 + \frac{P(L) - P_0}{L} z. \tag{2.60}$$

By keeping $\alpha \equiv \alpha(z)$ in equation (2.58) we find

$$P(z) = P_0 e^{-2\chi(z)}, \qquad \chi(z) = \int_0^z \alpha(z')\, dz', \tag{2.61}$$

$$\frac{P_0 - P(L)}{P_0} = 1 - e^{-2\chi(L)}. \tag{2.62}$$

The dependence of the power on the longitudinal coordinate, inside the accelerating structure, can accordingly be written as

$$P(z) = P_0\left(1 - \frac{1 - e^{-2\chi(L)}}{L} z\right). \tag{2.63}$$

This last expression can be exploited to obtain the functional form of $\alpha(z)$, which from equations (2.61) and (2.63) we find to be

$$2\alpha(z)P(z) = \frac{1 - e^{-2\chi(L)}}{L} P_0 \rightarrow \alpha(z) = \frac{1}{2L} \frac{1 - e^{-2\chi(L)}}{1 - \frac{z}{L}(1 - e^{-2\chi(L)})}. \tag{2.64}$$

According to the second equation of (2.58) the quantity which physically varies along the longitudinal direction is the group velocity, which can be written as

$$v_g(z) = \frac{\omega L}{Q}\left[\frac{1 - e^{-2\chi(L)}}{1 - \frac{z}{L}(1 - e^{-2\chi(L)})}\right]^{-1}. \tag{2.65}$$

The most significant result of the present computation is that since the electron energy gain in a constant gradient structure is

$$\Delta U = eEL \tag{2.66}$$

and

$$E = \sqrt{-r_s\left(\frac{dP}{dz}\right)}, \tag{2.67}$$

we end up with

$$\begin{aligned} \Delta U &= e\sqrt{r_s L P_0 A(L)}, \\ A(L) &= (1 - e^{-2\chi(L)}). \end{aligned} \tag{2.68}$$

It is evident that either for *constant impedance* or *constant gradient* structures, the energy gain depends on $\sqrt{r_s L P_0}$, which puts in direct evidence the pivotal role of the SI. The importance of the quality factor is implicitly contained in the 'attenuation' factor $A(L)$.

In this section we have provided a number of practical issues yielding the essential tools for a linac design. In the forthcoming chapters we will see how these concepts can be exploited to design accelerating devices dedicated, e.g., to free electron laser (FEL) operation. In section 2.8 we have collected further 'design recipes' which we hope will provide useful suggestions for more complete studies in this direction.

2.6 Phase stability in linacs

After the previous 'crude' remarks let us go back to a less naïve environment, by recalling that

a. The electrons should be injected inside the linac at the synchronous phase, which is conveniently chosen in a region where the accelerating field is growing (see figure 2.22). The electrons, accelerated in a region around this synchronous phase value, are clustered in short bunches characterized by a specific length and a definite energy spread.
b. The oscillations around the synchronous phase are lost for ultra-relativistic particles and therefore it is convenient to allow the particles to ride the rest of the wave.

Accordingly the problem of phase stability should have relevance during the first accelerating stages or for heavy particles (protons or ions).

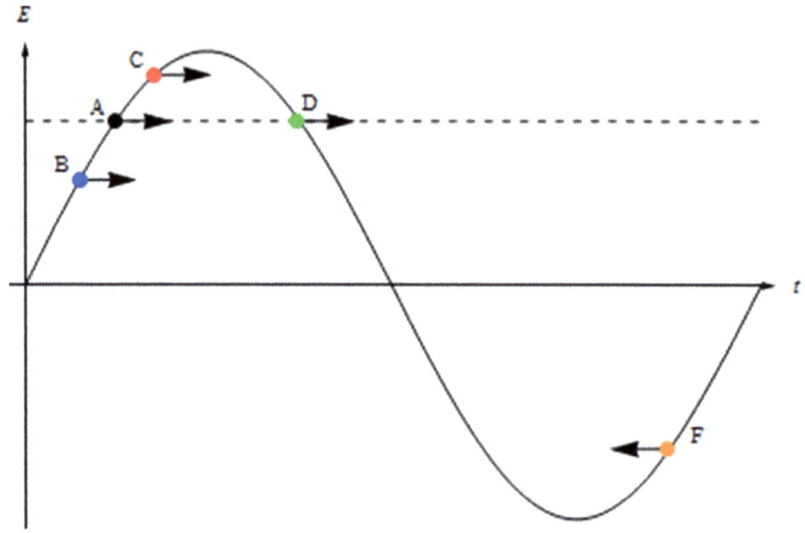

Figure 2.22. The accelerating gradient experienced by different particles in an accelerating cavity as a function of the time (t) at which they arrive at the cavity. The synchronous particle A is between particle B, at higher energy, and C at lower energy. Particle D is on the boundary of the stability region. Finally, F is out of phase and is decelerated.

It should, however, be noted that riding the maximum becomes problematic for short and high-charge bunches, in this case (as we will see better in the following) additional energy spread may be induced and it is suggested that the center of the beam be placed out of the crest.

Regarding the energy spread, we note that not all the particles in a bunch have the same energy and that a perfectly mono-energetic beam is unattainable. A small spread is one of the crucial parameters for many applications including FEL, therefore dedicated machines are carefully designed to minimize effects contributing to the increase of such a quantity.

The mechanisms leading to a stable motion in the longitudinal phase space will be discussed in the forthcoming part of this section.

The mathematics is rather elementary, we will therefore privilege in our description the associated physical content.

2.6.1 Longitudinal equation of motion

Adapting equation (2.5) to the highly relativistic case, we write the electron energy increment per cell as

$$\begin{aligned} \delta\gamma_s &= \gamma_{n+1} - \gamma_n \cong \frac{eE_0\,\mathcal{T}}{m_e c^2} L_c \cos(\phi_s), \\ \delta\gamma &\cong \frac{eE_0\,\mathcal{T}}{m_e c^2} L_c \cos(\phi), \end{aligned} \tag{2.69}$$

where the sub-index s denotes the particles ruled by the stable phase.

The energy gain difference between non-synchronous and synchronous particles is

$$\Delta\gamma = \delta\gamma - \delta\gamma_s \cong \frac{eE_0\,\mathcal{T}}{m_e c^2} L_c\big(\cos(\phi) - \cos(\phi_s)\big). \tag{2.70}$$

The use of the identity $\Delta\gamma = \Delta(1/\sqrt{1-(\frac{v}{c})^2}) \cong \gamma^3(\frac{v}{c})(\frac{\Delta v}{c})$, with the synchronous particle as reference, yields

$$\delta\gamma - \delta\gamma_s \cong \gamma^3\beta_s\Delta\beta. \tag{2.71}$$

We obtain the following identity yielding the particle velocity difference in terms of the phase mismatch

$$\frac{\Delta\beta}{L_c} \cong \frac{eE_0\,\mathcal{T}}{m_e c^2\gamma^3\beta_s}(\cos(\phi) - \cos(\phi_s)). \tag{2.72}$$

The phase difference is in turn linked to the longitudinal coordinate by

$$\phi - \phi_s \cong -\frac{z - z_s}{\beta_s}\frac{2\pi}{\lambda}. \tag{2.73a}$$

By setting $\Delta\beta = \beta - \beta_s$ we end up with

$$\Delta\beta \cong -\frac{\beta_s^2\lambda}{2\pi}\frac{d\Delta\phi}{dz}. \tag{2.73b}$$

By replacing $\frac{\Delta\beta}{L_c}$ with $\frac{d}{dz}\Delta\beta$ in equation (2.72) and by combining with equation (2.73) we find

$$\frac{d}{dz}\Delta\beta \cong \frac{eE_0\,\mathcal{T}}{m_e c^2\gamma^3\beta_s}(\cos(\phi) - \cos(\phi_s)), \tag{2.74}$$

eventually leading to (without any loss of generality we have set $\frac{d\Delta\phi}{dz} \to \frac{d\phi}{dz}$)

$$-\frac{d}{dz}\frac{\beta_s^2\lambda}{2\pi}\frac{d\phi}{dz} = \frac{eE_0\,\mathcal{T}}{m_e c^2\gamma^3\beta_s}(\cos(\phi) - \cos(\phi_s)). \tag{2.75}$$

The above equation is not the end of the game, we consider its lhs and expand it as

$$\frac{d}{dz}\left(\frac{\beta_s^2\lambda}{2\pi}\frac{d\phi}{dz}\right) = \lambda\frac{\beta_s}{\pi}\left(\frac{d\beta_s}{dz}\right)\frac{d\phi}{dz} + \frac{\beta_s^2\lambda}{2\pi}\frac{d^2\phi}{dz^2}. \tag{2.76}$$

The phase dynamics described by equation (2.75) should therefore be understood as that of a pendulum motion with a damping. If we assume that the velocity variation per unit length be negligible, we can reduce our problem to a pendulum like equation, namely

$$\phi'' = -\Omega_s^2(\cos(\phi) - \cos(\phi_s)),$$
$$\phi' = \frac{d\phi}{dz}, \qquad \phi'' = \frac{d^2\phi}{dz^2}, \tag{2.77}$$
$$\Omega_s^2 = \frac{2\pi}{\lambda}\frac{eE_0\,\mathcal{T}}{m_ec^2\gamma^3\beta_s^3}.$$

To better exploit the analogy with pendulum dynamics, we consider the previous equation from a Hamiltonian point of view. Let us therefore note that

$$\phi'\phi'' = \frac{1}{2}\frac{d}{dz}[(\phi')^2 - (\phi_0')^2],$$
$$\phi'(\cos(\phi) - \cos(\phi_s)) = \frac{d}{dz}(\sin(\phi) - \phi\cos(\phi_s) - (\sin(\phi_0) - \phi_0\cos(\phi_s))), \tag{2.78}$$

where ϕ_0', ϕ_0 are the initial constants of the problem. If we multiply both sides of equation (2.77) by ϕ' and then integrate over z, we end up with

$$\mathcal{H} = \frac{1}{2}(\phi')^2 + \Omega_s^2 V(\phi),$$
$$V(\phi) = \left(\sin(\phi) - \phi\cos(\phi_s)\right), \tag{2.79}$$

where $\frac{1}{2}(\phi')^2$ represents the kinetic energy of the system and the second term the potential (note that ϕ' is proportional to the velocity variation, see equation (2.73)). The previous dynamical system is conservative and indeed $\mathcal{H} = \frac{1}{2}(\phi_0')^2 + \Omega_s^2(\sin(\phi_0) - \phi_0\cos(\phi_s))$.

All the dynamics can now be understood in a fairly straightforward way. According to equation (2.79) the problem is not a simple pendulum, but also includes a linear term. The associated Liouville equation is written

$$\partial_t\mathcal{L}(\phi, \phi'; t) = \left[-\phi'\partial_\phi + \Omega_s^2\left(\cos(\phi) - \cos(\phi_s)\right)\partial_\phi\right]\mathcal{L}(\phi, \phi'; t). \tag{2.80}$$

The physical meaning of the canonical variables ϕ and ϕ' is obvious. The Liouville distribution $\mathcal{L}(\phi, \phi'; t)$ indeed characterizes a packet of electrons and $\langle\phi^2\rangle$, $\langle\phi'^2\rangle$ are, respectively, linked to the bunch length and to the *absolute energy spread.* The evolution of an ensemble of particles ruled by this potential is discussed in section 2.7, the underlying physics can, however, be easily understood with the help of figure 2.23, where we show the phase–space plot $(\phi'/\sqrt{\mathcal{H}}, \phi)$, for different values of $\Omega_s^2/\mathcal{H}$ and ϕ_s.

The previous plots include open and closed curves. The latter are associated with stable 'buckets', whose height depends on the strength of the potential $\Omega_s^2/\mathcal{H}$.

The accelerated particles inside the structure are those around the synchronous particle (see section 2.8 for further comments). The stable and unstable points are obtained by plotting the phase–space curves along with the Hamiltonian. It is evident from figure 2.24 that the stable and unstable points are located in correspondence of the minimum ($\phi = -\phi_s$) (see below to clarify the origin of the sign) and the maximum ($\phi = \phi_s$).The contour line denoting the transition between

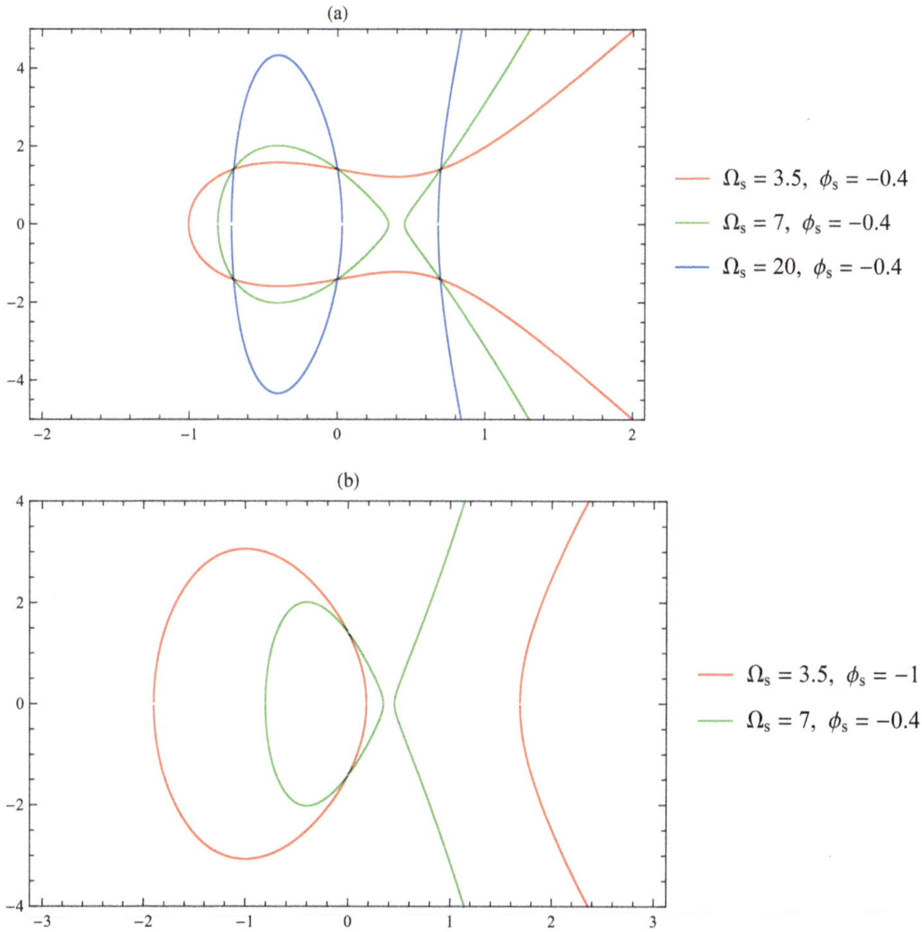

Figure 2.23. Longitudinal phase–space plots (a) $\phi_s = -0.4$, $\Omega_s^2/\mathcal{H} = 3.5$ (red), $\phi_s = -0.4$, $\Omega_s^2/\mathcal{H} = 7$ (green), $\phi_s = -0.4$, $\Omega_s^2/\mathcal{H} = 20$ (blue). (b) $\phi_s = -1$, $\Omega_s^2/\mathcal{H} = 3.5$ (red), $\phi_s = -0.4$, $\Omega_s^2/\mathcal{H} = 7$ (green).

the stable and unstable regions is called the separatrix and is located between the boundaries $\phi^* < \phi < \phi_s$, where $V(\phi^*) = V(-\phi_s)$. A more comprehensive portrait of the stability and instability region is reported in figure 2.25 and further comments can be found in section 2.8.

In closing this section we mention the case of small oscillations, which occur for small values of $\delta = \phi - \phi_s$. In these conditions the equation of motion describing the relevant dynamics is that of a harmonic oscillator and reads (see section 2.8)

$$\frac{d^2}{dz^2}\delta = \left[\Omega_s^2 \sin(\phi_s)\right]\delta, \tag{2.81}$$

which requires $\phi_s < 0$ to ensure stable oscillations.

It is evident that within the present picture the phase oscillations are ruled by both the synchronous phase and by the coupling constant Ω_s, in such a way that the oscillation period is provided by

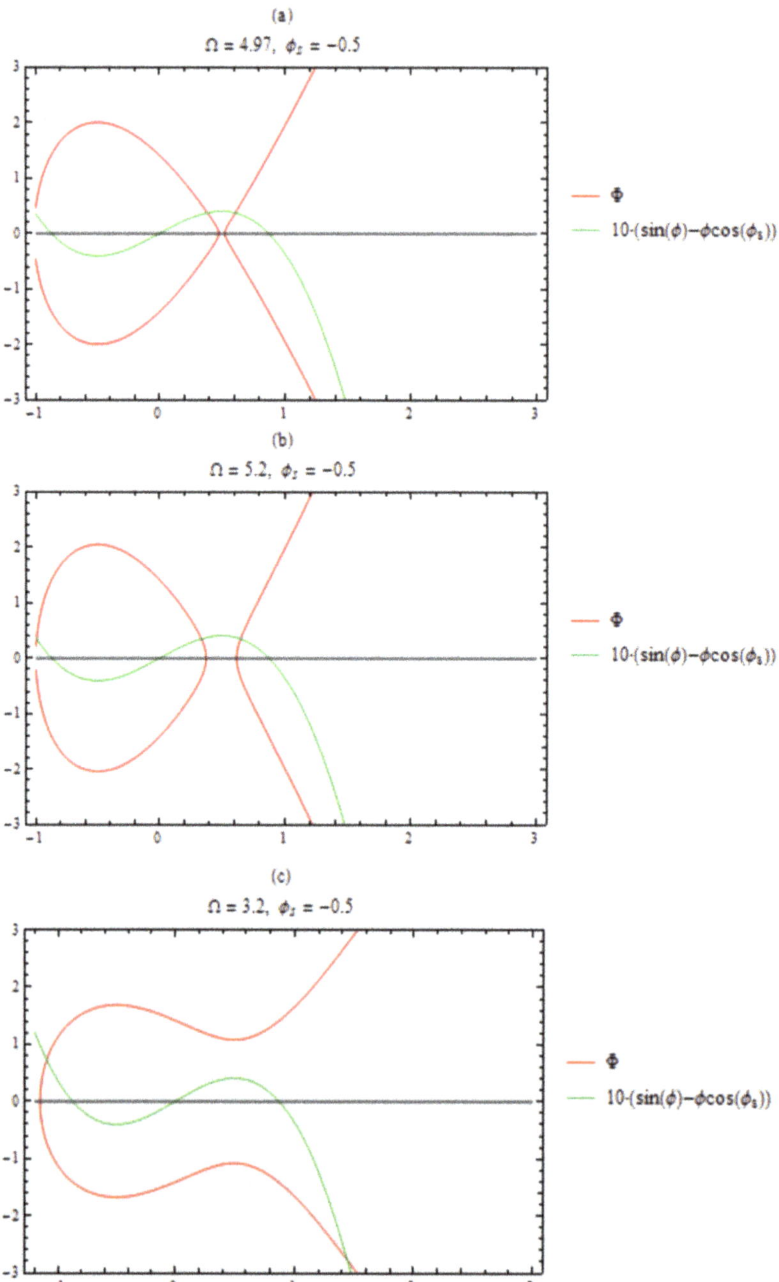

Figure 2.24. Longitudinal phase–space plots (red) and potential (see equation (2.79)) (green): (a) $\phi_s = -0.5, \quad \Omega_s^2/\mathcal{H} = 4.97$, (b) $\phi_s = -0.5, \quad \Omega_s^2/\mathcal{H} = 5.2$, (c) $\phi_s = -0.5, \quad \Omega_s^2/\mathcal{H} = 3.2$.

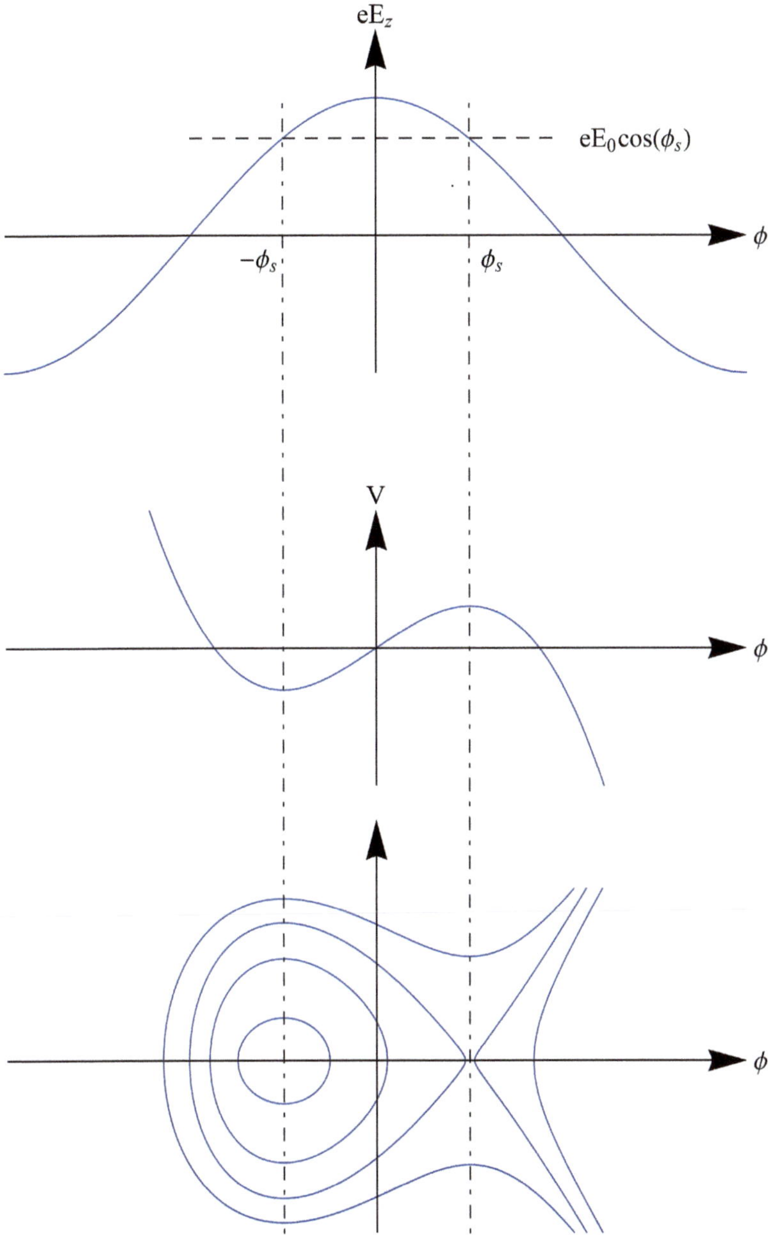

Figure 2.25. Stability and instability region in the longitudinal linac bucket.

$$T = \sqrt{\frac{2\pi}{\sin(\phi_s)}}\sqrt{\frac{\lambda m_e c^2 \gamma_s^3 \beta_s^3}{eE_0 \mathcal{T}}}. \tag{2.82}$$

We infer therefore that, with increasing energy and a small stable phase the oscillation period becomes longer.

According to the discussion in the previous chapter, we can take advantage of the simple equation (2.81) by considering the accelerating cavity as a kind of focusing element. Such a point of view will be further developed in section 2.8, where we apply what we have learned on beam transverse guiding and on the Twiss parameters to infer further results of practical interest.

If we introduce a longitudinal emittance as ε_L (we will later specify the relevant meaning) we easily end up with the following identities for the second moments of the distribution in the longitudinal phase space $(\delta,\ \delta')$

$$\sigma_\delta = \sqrt{\varepsilon_L\left(\left|\sin(\phi_s)\right|\right)^{-1}(\Omega_s)^{-1}},$$
$$\sigma_{\delta'} = \sqrt{\varepsilon_L\left|\sin(\phi_s)\right|\ \Omega_s}. \tag{2.83}$$

As already stressed the bunch length and the energy spread are, respectively, associated with σ_δ and $\sigma_{\delta'}$ (see section 2.8), it is worth noting that:

a. small values of the synchronous phase may lead to a long bunch and small energy spread;
b. as $\Omega_s \propto (\beta_s\gamma_s)^{-\frac{3}{2}}$, we find that with increasing energy the bunch length increases and the *absolute* energy spread decreases.

The above results are naïve but important; further consequences on the beam qualities due to the accelerating mechanism will be discussed in forthcoming sections.

2.7 Beam qualities and accelerating field

In the previous sections we have slowly moved toward a picture of the linac as an actual device and started to clarify how the various elements can be merged to outline a primitive, but effective, design of the accelerator.

Apart from technical and technological details, we have left open many points and we will try to fill the gap by discussing:

1. The capture conditions.
2. The effect of the intra-cavity accelerating transverse field distribution on the beam emittance.
3. The wake fields and their consequences on the longitudinal dynamics.

The discussion of point 1 will be aimed at clarifying the conditions underlying the capture of non-relativistic electrons from the cathode output and bringing them to the desired energy at the end of the accelerating tube.

We recall that relativistic electrons met the synchronism conditions in a fairly natural way. The electrons indeed move at the speed of light and intercept all the cavities with the correct synchronism. The dynamics of the system is therefore 'frozen'; the phase spread remains constant and, if it is not too large, the absolute energy spread also remains constant, while the relative part decreases with increasing energy.

The dynamics is less straightforward for non-relativistic particles, which are moving at a speed not close to that of light, as happens for the electrons at the cathode output. The charges experience the effect of the RF field inside the accelerating structure, but since the radiation moves faster it will overcome the particles by a distance

$$dl = (c - v)dt. \tag{2.84}$$

Such a slippage measured in terms of the field phase is

$$dl = \frac{\lambda}{2\pi}d\phi, \tag{2.85}$$

where λ is the RF wavelength. The above phase mismatch may be responsible for a loss of efficiency of capture. In other words the electrons may be unable to remain in phase with the accelerating field and, if it is not sufficiently intense, to provide enough acceleration to restore the matching, and some of them are lost. Our strategy will be that of deriving a condition (involving electron initial energy and peak electric field) ensuring optimal capture. On the basis of a naïve physical argument we may also conclude that such a condition is achieved whenever the work done on the electrons by the electric field in one RF wavelength is at least equal to the electron rest mass. This is just a rough estimate and a more quantitative result is achieved by putting together phase slippage and electron dynamics. To this aim we couple the equation of motion of electrons with the phase mismatch, namely by using the identities

$$\begin{gathered}\frac{d}{dt}\left(\frac{\beta}{\sqrt{1-\beta^2}}\right) = \frac{eE_0}{m_e c}\sin(\phi),\\ \frac{d\phi}{dt} = \omega(1-\beta),\end{gathered} \tag{2.86}$$

the second of which has been derived by merging equations (2.84) and (2.85).

The use of the results of exercise (1.5) suggests that the trigonometric representation of the relativistic variables (note that with respect to exercise (1.5) $\vartheta \to \vartheta - \frac{\pi}{2}$, therefore zero velocity corresponds to a 'relativistic' phase of $\frac{\pi}{2}$),

$$\beta = \cos(\vartheta), \ \gamma = \frac{1}{\sin(\vartheta)}, \tag{2.87}$$

is a useful trick to introduce a kind of kinematic phase, to be combined with that of the field phase, and obtain physically transparent results. Accordingly we write the first equation of (2.86) as

$$\frac{d\vartheta}{dt} = -\frac{eE_0}{m_e c}\sin(\phi)(\sin(\vartheta))^2, \tag{2.88}$$

and by noting that

$$\frac{d\phi}{dt} = \frac{d\phi}{d\vartheta}\frac{d\vartheta}{dt}, \tag{2.89}$$

we end up with the identity

$$\frac{2\pi}{\lambda}\frac{m_e c^2}{E_0}\frac{(1 - \cos(\vartheta))}{(\sin(\vartheta))^2}d\vartheta = -\sin(\phi)\, d\phi. \tag{2.90}$$

After integrating both sides we find

$$\frac{2\pi}{\lambda}\frac{m_e c^2}{eE_0}\left[\tan\left(\frac{\vartheta_1}{2}\right) - \tan\left(\frac{\vartheta_0}{2}\right)\right] = \cos(\phi_1) - \cos(\phi_0). \tag{2.91}$$

If we use the elementary trigonometry identity to express the tangent in terms of the reduced velocity

$$\tan\left(\frac{\vartheta}{2}\right) = \sqrt{\frac{1-\beta}{1+\beta}}. \tag{2.92}$$

Noting that at relativistic velocities ($\beta_1 \cong 1$) the tangent almost vanishes, we find from equation (2.91)

$$\frac{2\pi}{\lambda}\frac{m_e c^2}{eE_0}\sqrt{\frac{1-\beta_0}{1+\beta_0}} = \cos(\phi_0) - \cos(\phi_1). \tag{2.93}$$

As $|\cos(\phi_0) - \cos(\phi_1)| < 2$, we obtain the following condition on the peak electric field

$$E_0 > \frac{\pi}{\lambda}\frac{m_e c^2}{e}\sqrt{\frac{1-\beta_0}{1+\beta_0}}, \tag{2.94}$$

which yields the strength of the electric field necessary to bring an electron with initial velocity β_0 to the relativistic regime.

The physical content of equation (2.94) is clear; $eE_0\lambda$ is indeed the work done by the peak electric field on the electron, therefore our original statement that such a quantity must exceed the electron rest mass is substantively correct.

The following exercise may be useful:

Express the rhs of equation (2.94) *in terms of the electron kinetic energy and fix the value of the peak electric field for different values of the RF wavelength.*

Setting $\delta = \frac{T}{m_e c^2}$ we write the rhs in the practical form

$$E_{th}[\mathrm{MV}/m] \cong \frac{50\pi}{\lambda[\mathrm{cm}]}\frac{\sqrt{\delta + 2} - \sqrt{\delta}}{\sqrt{\delta + 2} + \sqrt{\delta}}, \tag{2.95}$$

where the subscript *th* stands for threshold. In figure 2.26 we have reported this quantity versus δ for kinetic energies not exceeding that of the electron rest mass energy. The capture condition could be hardly satisfied for short wavelength RF fields. It is therefore convenient to use a pre-buncher cavity with variable phase velocity, which brings the electron energy to a few MeV, thus improving the capture efficiency (for further comments see section 2.8).

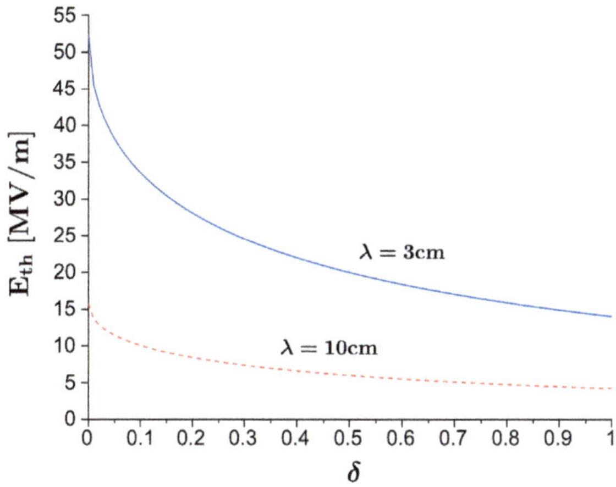

Figure 2.26. Threshold peak accelerating field versus $\frac{T}{m_e c^2}$ for two different RF field wavelengths.

In our study of the longitudinal dynamics of section 2.6, we did not include the contributions derived from the transverse structure of the accelerating field. We will amend this deficiency and we will show the importance of the associated effects on the beam quality itself.

The strategy we follow is fairly simple, we derive indeed the dependence of the accelerating field on the radial coordinate around the propagation axis, by adopting an uncomplicated mathematical procedure.

We recall that in cylindrical coordinates the longitudinal field associated with the $TM_{0,1}$ mode is ruled by the Bessel equation

$$\left[\frac{1}{\rho}\frac{\partial}{\partial\rho}\rho\frac{\partial}{\partial\rho} + \frac{\partial^2}{\partial z^2} + k^2\right]E_z(\rho,\ z) = 0, \tag{2.96}$$

and forget that we have already found the relevant solution by the use of a rigorous procedure and make recourse to a kind of 'blasphemy' to treat it in a way more useful for our purposes. We write therefore equation (2.96) in the form

$$\frac{\partial}{\partial\rho}E_z(\rho,\ z) = -\frac{1}{\rho}\int\rho\left[\frac{\partial^2}{\partial z^2} + k^2\right]E_z(\rho,\ z)\,d\rho \tag{2.97}$$

and expand the field at the second order in the radial coordinate, namely

$$E_z(\rho,\ z) \cong E_z(0,\ z) + \frac{\rho^2}{2}\left[\frac{\partial^2}{\partial\rho^2}E_z(\rho,\ z)\right]_{\rho=0} + o(\rho^4). \tag{2.98}$$

The series expansion is justified by the fact that the beam has a finite extension around the axis and experiences only this part of the accelerating field (see below for a more appropriate quantification).

If we insert equation (2.97) into equation (2.96) we obtain (at the same order of accuracy)

$$\frac{\partial}{\partial \rho} E_z(\rho,\ z) \cong -\frac{\rho}{2}\left[\left(\frac{\partial^2}{\partial z^2} E_z(0,\ z)\right) + k^2 E_z(0,\ z)\right], \tag{2.99}$$

which once 'integrated' yields

$$E_z(\rho,\ z) \cong E_z(0,\ z) - \frac{\rho^2}{4}\left[\left(\frac{\partial^2}{\partial z^2} E_z(0,\ z)\right) + k^2 E_z(0,\ z)\right]. \tag{2.100}$$

It is evident from the previous identities that the validity of our expansion is limited to a region around the axis limited by $k\rho \ll 1$ yielding $\rho \ll \frac{\lambda}{2\pi}$, which, for RF wavelength of 3 cm, amounts to less than 0.1 cm around the axis. The beam transverse dimensions are largely contained within these limits.

This is, however, not the end of the story, because equation (2.100) does not provide a clear indication of how the dependence of the accelerating field component affects the longitudinal beam qualities. To this aim we include the time dependence and write explicitly the dependence on the longitudinal coordinate according to

$$E_z(0,\ z,\ t) \cong E_0 \cos\left(\omega t - \frac{2\pi}{L_c} z + \phi\right). \tag{2.101}$$

According to the synchronism condition, namely that the phase and the electron velocities are equal, we obtain

$$\frac{L_c}{2\pi}\omega = v, \tag{2.102}$$

which imposes the following constraint on the cavity length

$$L_c = \beta c T, \tag{2.103}$$

namely that the cavity length be proportional to the RF period. Putting everything together we find

$$E_z(\rho,\ z,\ t) \cong \left\{1 - \frac{\rho^2}{4}\left[-\left(\frac{2\pi}{L_c}\right)^2 + k^2\right]\right\} E(0,\ z,\ t), \tag{2.104}$$

noting that

$$\begin{aligned} -\left(\frac{2\pi}{L_c}\right)^2 + k^2 &= k^2\left(1 - \left(\frac{2\pi}{kL_c}\right)^2\right) \\ &= k^2\left(1 - \left(\frac{2\pi}{k\beta c T}\right)^2\right) = -k^2\left(1 - \left(\frac{1}{\beta}\right)^2\right) = -\left(\frac{2\pi}{\beta\gamma\lambda}\right)^2. \end{aligned} \tag{2.105}$$

We eventually end up with

$$E_z(\rho,\ z,\ t) \cong E(0,\ z,\ t)\left(1 - \left(\frac{\pi\rho}{\beta\gamma\lambda}\right)^2\right). \tag{2.106}$$

It is thus clear that the dependence of the field on the transverse coordinate will more substantively affect the longitudinal beam quality (namely the energy spread) for larger values of $\frac{\pi\rho}{\beta\gamma\lambda}$. Accordingly, for larger energy and smaller extension of the beam transverse section σ_ρ with respect to the RF wavelength, the impact of the accelerating field inhomogeneity is less significant (for further comments see section 2.8).

On the other side (see equation (2.36)) the electric field transverse component can be approximated as (note that at lowest the cylindrical Bessel function can be approximated as $J_1(x) \cong \frac{x}{2}$ and that from the Maxwell equations $\nabla \cdot \vec{E} = 0$)

$$E_\rho(\rho,\ z,\ t) \cong \rho\,\frac{\pi}{\beta\lambda} E_z(0,\ z,\ t), \tag{2.107}$$

whose effects on the longitudinal dynamics can be counteracted by operating with a beam whose transverse section satisfies the condition $\sigma_\rho \ll \lambda$. However, regarding the magnetic field, we find equation (2.36)

$$B_\phi \cong \rho\frac{\pi}{c\lambda}\ E_z(0,\ z,\ t). \tag{2.108}$$

The forces induced on the electrons by the field (2.107)–(2.108) act in the same direction but are opposite in sign, we obtain therefore the following expression

$$F_\rho = e\ (E_\rho - v_z B_\phi) \cong -(1 - \beta^2)\ |e|\ E_\rho(\rho,\ z,\ t). \tag{2.109}$$

It is evident that for ultra-relativistic particles the transverse force induced by the accelerating field inhomogeneity vanishes, while for $\beta^2 < 1$ the effect is that of defocusing.

A similar effect arises due to space charge effects in the low-energy phase. Very crudely speaking the forces acting on the electrons inside the bunches are due to the intra-beam Coulomb force

$$F_C = \frac{e^2}{\varepsilon_0\rho}\int_0^\rho n(r)\ dr \tag{2.110}$$

and to the Lorentz force induced by the magnetic field generated by electron-beam current (see figure 2.18)

$$F_L = -e\ |\vec{v} \times \vec{B}| = -\frac{e^2\beta^2}{\varepsilon_0\rho}\int_0^\rho n(r)\ dr, \tag{2.111}$$

which sum up to give

$$F_{SC} = (1 - \beta^2)\, F_C = \frac{F_C}{\gamma^2}. \tag{2.112}$$

Also, for space charge forces the relevant effect on the beam becomes less significant with increasing energy.

Let us finally touch on the beam quality dilution effects due to the wake field. The notion of wake field is fairly straightforward. A charged beam interacts with the accelerating environment, which is not only provided by the accelerating field, but by the cavity walls, valves, the transport tube with its discontinuities and so on. When the beam crosses any variation in the chamber profile (or other perturbing element) it loses energy which sustains an electromagnetic field (the wake), which does not vanish after the passage of the beam and may affect the next bunches and even the trailing part of the beam itself. According to this statement, the tail of a bunch crossing the cavity experiences the field left behind by its head and the result is an increase of the relevant energy spread. Very roughly speaking we can use a two-macro-particle model of the bunch itself and compute the different effects of the wake potential on the head and tail.

Denoting by $W_{||}(z)$ the longitudinal component of the wake potential, and by assuming that the two macro-charges, separated by a distance d, constituting the bunch are $\frac{Ne}{2}$, we find that the head and tail energy variations are

$$\begin{aligned} \frac{d}{dz}\mathcal{E}_h &= -\frac{Ne^2}{4} W_{||}(0), \\ \frac{d}{dz}\mathcal{E}_{ht} &= -\frac{Ne^2}{4} W_{||}(0) - \frac{Ne^2}{2} W_{||}(d). \end{aligned} \tag{2.113}$$

The wake potential is usually given in V $\text{pC}^{-1}\,\text{m}^{-1}$ and, as we will see in section 2.8, depends on the bunch length, the accelerating frequency, the cavity shapes and materials, etc. In the case of the SLAC linac cavities (2.856 GHz) with a 3 mm bunch length, $W_{||}(0) = 225$ V $\text{pC}^{-1}\,\text{m}^{-1}$, $W(3\text{ mm}) = 57$ V $\text{pC}^{-1}\,\text{m}^{-1}$.

We have provided an idea of the interplay between accelerating field and beam qualities during the acceleration process. We will discuss further details in section 2.8, dedicated to complements and exercises.

The consequences of electron-beam quality degradation on FEL performance are carefully described in chapters 4 and 5.

2.8 Complements and exercises

2.8.1 Electromagnetic fields inside cylindrical metallic structures

In this complement we report a detailed derivation of the electromagnetic field in cylindrical guiding structures. The lecture in this section is not compulsory for the understanding of most of the aspects of the topics treated in this book. It is, however, strongly recommended because it provides a rigorous analysis of the subject which

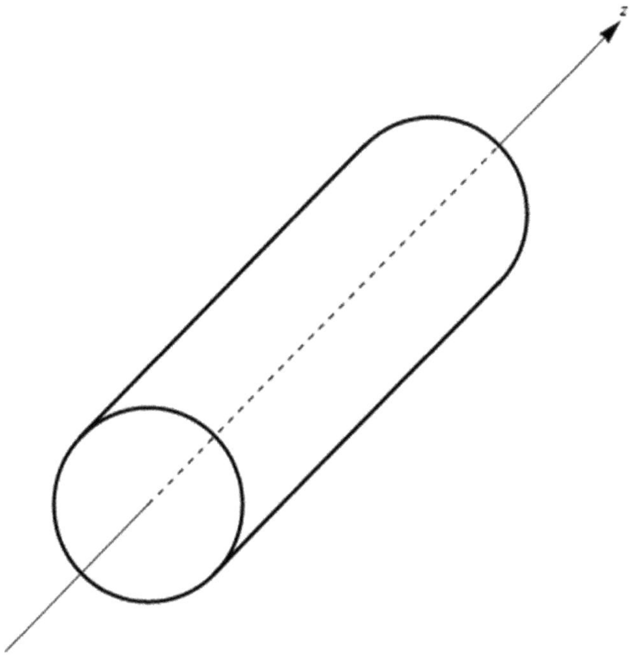

Figure 2.27. Cylindrical waveguide.

completes the heuristic background acquired in the main body of the chapter. It can however be skipped in an introductory lecture.

Cylindrical guiding structures, regardless of their cross-section geometry, sustain electromagnetic fields and transport the associated energy in a very convenient way and thus are widely used in many applied scientific and industrial applications.

In the following we will use the magnetic field H instead of the magnetic induction B. We recall that (in a vacuum) $B = \mu_0 H$, and since we use Gaussian units the conversion between the equations derived here and those in the main body of the chapter are obtained by the use of the conversion table available in the foreword.

The equation ruling the dynamics of these fields is directly derived by the Maxwell equations applied to the geometry described in figure 2.27.

The cylindrical metallic surface has a general cross-section contour that infolds a dielectric medium with a dielectric constant ε and a magnetic permeability μ. Assuming a sinusoidal time dependence $\exp(i\omega t)$ for the fields, we can write the Maxwell equations as follows:

$$\begin{cases} \vec{\nabla} \times \vec{E} = i\omega\mu\vec{H} \quad \vec{\nabla} \cdot \vec{B} = 0, \\ \vec{\nabla} \times \vec{H} = -i\varepsilon\omega\vec{E} \quad \vec{\nabla} \cdot \vec{D} = 0. \end{cases} \tag{2.114}$$

We can now apply the curl operator to the first equation of (2.114), making use of the other equations and exploiting the well-known expression of differential vector calculus ($\vec{\nabla} \times (\vec{\nabla} \times \vec{A}) = \vec{\nabla}(\vec{\nabla} \cdot \vec{A}) - (\vec{\nabla} \cdot \vec{\nabla})\vec{A}$), we obtain

$$\begin{aligned}\vec{\nabla} \times (\vec{\nabla} \times \vec{E}) = i\omega\mu(\vec{\nabla} \times \vec{H}) \Rightarrow \vec{\nabla} \cdot (\vec{\nabla} \cdot \vec{E}) - \nabla^2\vec{E} = i\omega\mu(\vec{\nabla} \times \vec{H}) \\ \Rightarrow \nabla^2\vec{E} = -\mu\varepsilon\omega^2\vec{E} \Rightarrow \nabla^2\vec{E} + \mu\varepsilon\omega^2\vec{E} = 0.\end{aligned} \tag{2.115}$$

The constitutive relations states that $\mu = \mu_r\mu_0$ for the magnetic permeability, and $\varepsilon = \varepsilon_r\varepsilon_0$ for the dielectric constant; the subscript r indicates the properties of the medium 'relative' to the vacuum that is described by μ_0 and ε_0. Considering the vacuum as dielectric medium we obtain $\mu_r = \varepsilon_r = 1$, and using the relation that links the medium electromagnetic properties with the speed of the electromagnetic wave ($\mu_0\varepsilon_0 = \frac{1}{c^2}$) we obtain the expression for the Helmholtz equation in vacuum:

$$\begin{aligned}\nabla^2\vec{E} + \left(\frac{\omega}{c}\right)^2\vec{E} = 0, \\ \nabla^2\vec{H} + \left(\frac{\omega}{c}\right)^2\vec{H} = 0.\end{aligned} \tag{2.116}$$

The cylindrical symmetry of our geometry allows us to separate the vector fields in two separate components, the longitudinal and the transverse: $\vec{E} = \vec{E}_T + \vec{E}_z$. It is easy to verify that, defining as $\hat{z}$ the longitudinal unit vector in the z-direction, we have

$$\begin{cases}\vec{E}_z = (\hat{z} \cdot \vec{E}) \cdot \hat{z} \\ \vec{E}_T = (\hat{z} \times \vec{E}) \times \hat{z}.\end{cases} \tag{2.117}$$

In the same framework, the 'Nabla' operator can be separated as well; assuming, furthermore, a longitudinal sinusoidal dependence of the field ($\exp(-ikz)$), we obtain $\vec{\nabla} \Rightarrow \vec{\nabla}_T - ik\hat{z}$. The rotational Maxwell equation for the electric field can thus be rewritten, by means of equation (2.117), as

$$\begin{aligned}&\vec{\nabla} \times \vec{E} = (\vec{\nabla}_T - ik\hat{z}) \times (\vec{E}_T + \vec{E}_z)e^{-ikz} = -i\omega\mu(\vec{H}_T + \vec{H}_z)e^{-ikz} \\ &\Rightarrow \vec{\nabla}_T \times \vec{E}_T - ik\hat{z} \times \vec{E}_T - ik\hat{z} \times \vec{E}_z - \vec{\nabla}_T \times \vec{E}_z = -i\omega\mu\vec{H}_T - i\omega\mu\vec{H}_z \\ &\Rightarrow \vec{\nabla}_T \times \vec{E}_T - ik\hat{z} \times \vec{E}_T - \hat{z} \times (\vec{\nabla}_T E_z) = -i\omega\mu\vec{H}_T - i\omega\mu\vec{H}_z,\end{aligned} \tag{2.118}$$

with $\hat{z} \times \vec{E}_z = 0$ and $\vec{\nabla}_T \times \vec{E}_z = \vec{\nabla}_T \times \hat{z}E_z = -\hat{z} \times (\vec{\nabla}_T E_z)$. Equation (2.118) can be more clearly divided into two equations putting in comparison, separately, the longitudinal and transverse components

$$\begin{cases}\vec{\nabla}_T \times \vec{E}_T = -i\omega\mu\vec{H}_z \\ \hat{z} \times (\vec{\nabla}_T E_z) + ik\hat{z} \times \vec{E}_T = i\omega\mu\vec{H}_T.\end{cases} \tag{2.119}$$

An analogous analysis on the rotational equation for the magnetic field leads to:

$$\begin{cases} \vec{\nabla}_T \times \vec{H}_T = i\omega\varepsilon\vec{E}_z \\ \hat{z} \times (\vec{\nabla}_T H_z) + ik\hat{z} \times \vec{H}_T = i\omega\varepsilon\vec{E}_T. \end{cases} \tag{2.120}$$

Analogous considerations hold for the field divergence equations (for the magnetic field, for instance)

$$\begin{aligned} \vec{\nabla} \cdot \vec{H} &= (\vec{\nabla}_T - ik\hat{z})(\vec{H}_T + \vec{H}_z)e^{-ikz} \\ &= [\vec{\nabla}_T \cdot \vec{H}_T + \vec{\nabla}_T \cdot \vec{H}_z - ik\hat{z} \cdot \vec{H}_T - ik\hat{z} \cdot \vec{H}_z]e^{-ikz} = 0, \end{aligned}$$

giving another set of important equations for the magnetic and electric field (with $\vec{\nabla}_T \cdot \vec{H}_z = 0$ and $\hat{z} \times \vec{H}_T = 0$)

$$\begin{cases} \vec{\nabla}_T \cdot \vec{H}_T = ik\hat{z} \cdot \vec{H}_z \\ \vec{\nabla}_T \cdot \vec{E}_T = ik\hat{z} \cdot \vec{E}_z. \end{cases} \tag{2.121}$$

Equations (2.119), (2.120) and (2.121) are extremely important because they state a relation between the transverse and longitudinal components of the same field (magnetic or electric), as for equation (2.121), or between different fields. The first equation of (2.119), for instance, establishes a link between the longitudinal magnetic field component and the TE field component. This will be particularly important because it allows us to reduce the number of equations to be solved for calculating the electromagnetic field in cylindrical metallic guiding structures.

In the following we will consider the propagating modes associated with the solutions of the previous set of equations, called

a. *Transverse electric* (TE), with vanishing electric longitudinal field component ($E_z = 0$).
b. *Transverse magnetic* (TM), with vanishing magnetic longitudinal field component ($H_z = 0$).

2.8.1.1 TE modes

In the TE modes the field component H_z plays the role of the pivotal function, because it is the only longitudinal component present. It can be derived by solving equation (2.116), which can be separated in transverse and axial parts as previously done for the derivation of equation (2.118):

$$\begin{cases} \nabla_T^2 H_z(\vec{r}) + \gamma^2 H_z(\vec{r}) = 0 \\ \nabla_T^2 \vec{H}_T(\vec{r}) + \gamma^2 \vec{H}_T(\vec{r}) = 0 \end{cases} \quad \text{where } \gamma^2 = \left(\frac{\omega}{c}\right)^2 - k^2, \tag{2.122}$$

where a longitudinal $\exp(-ikz)$ dependence has been assumed again and $\vec{r}$ represents the transverse coordinate vector. The quantity γ (not to be confused with the electron relativistic factor) represents the so-called cut-off momentum and can be determined by imposing the boundary conditions, it is therefore strongly dependent on the geometry of the guide environment.

Once we have solved equation (2.122) for $H_z(\vec{r})$, the other field components can be found by solving equations (2.119), (2.120) and (2.121), imposing the condition $E_z(\vec{r}) = 0$ for the TE modes. From the second equation of (2.119) we obtain

$$k[\hat{z} \times (\hat{z} \times \vec{E}_T)] = k[(\hat{z} \cdot \vec{E}_T)\hat{z} - (\hat{z} \cdot \hat{z})\vec{E}_T] = -k\vec{E}_T = \omega\mu(\hat{z} \times \vec{H}_T)$$

from the first equation of (2.121), and using equation (2.122) we end up with

$$\nabla_T(\nabla_T \vec{H}_T) = ik\nabla_T \vec{H}_z \Rightarrow \vec{H}_T = -\frac{ik}{\gamma^2}\nabla_T \vec{H}_z.$$

Finally, after introducing the wave-guide impedance

$$Z = \frac{\omega/v}{k} Z_M, \text{ where } Z_M = \sqrt{\frac{\mu}{\varepsilon}},$$

and denoting by v the light velocity in the medium

$$v = 1/\sqrt{\varepsilon\mu} = 1/\left(\sqrt{\varepsilon_0\mu_0}\sqrt{\varepsilon_r\mu_r}\right) = c/\sqrt{\varepsilon_r\mu_r} = c/n,$$

we can write the equations defining the TE modes of a metallic cylindrical waveguides as follows

$$\begin{cases} \nabla_T^2 H_z(\vec{r}) + \gamma^2 H_z(\vec{r}) = 0 \\ \vec{H}_T = -\dfrac{ik}{\gamma^2}\vec{\nabla}_T H_z(\vec{r}) \qquad \text{where } \gamma^2 = \left(\dfrac{\omega}{c}\right)^2 - k^2 \\ \vec{E}_T = -Z(\hat{z} \times \vec{H}_T) \end{cases} \tag{2.123}$$

in the vacuum, and for a rectangular geometry, we obtain

$$H_x = -\frac{ik}{\gamma_{mn}^2}\frac{\partial H_z}{\partial x}; \quad H_y = -\frac{ik}{\gamma_{mn}^2}\frac{\partial H_z}{\partial y};$$

$$E_x = ZH_y = -i\frac{\omega}{c\gamma_{mn}^2}\sqrt{\frac{\mu}{\varepsilon}}\frac{\partial H_z}{\partial y}; \quad E_y = -ZH_x = i\frac{\omega}{c\gamma_{mn}^2}\sqrt{\frac{\mu}{\varepsilon}}\frac{\partial H_z}{\partial x}.$$

2.8.1.2 TM modes

For the TM modes the situation is reversed with respect to that of the previous discussion. The E_z field component plays the role of the pivot, because it is the only longitudinal component present. These modes can be therefore considered as the dual of the TE modes and the derivation follows the same algebraic steps as before, thus yielding (the waveguide impedance is now $Z = \frac{k}{\omega/v} Z_M$, where again $Z_M = \sqrt{\frac{\mu}{\varepsilon}}$).

$$\begin{cases} \nabla_T^2 E_z(\vec{r}) + \gamma^2 E_z(\vec{r}) = 0 \\ \vec{E}_T = -\dfrac{ik}{\gamma^2}\vec{\nabla}_T E_z(\vec{r}) \qquad \text{where } \gamma^2 = \left(\dfrac{\omega}{c}\right)^2 - k^2 \\ \vec{H}_T = -\dfrac{1}{Z}(\hat{z} \times \vec{E}_T) \end{cases} \tag{2.124}$$

in the vacuum and for a rectangular geometry

$$E_x = -\frac{ik}{\gamma_{mn}^2}\frac{\partial E_z}{\partial x};\ E_y = -\frac{ik}{\gamma_{mn}^2}\frac{\partial E_z}{\partial y};$$
$$H_x = -\frac{1}{Z}E_y = i\frac{\omega}{c\gamma_{mn}^2}\sqrt{\frac{\varepsilon}{\mu}}\frac{\partial E_z}{\partial y};\ H_y = \frac{1}{Z}E_x = -i\frac{\omega}{c\gamma_{mn}^2}\sqrt{\frac{\varepsilon}{\mu}}\frac{\partial E_z}{\partial x}.$$

It is evident that the procedure we have envisaged requires the complete knowledge of E_z, which can be determined from the first equation of (2.124), which is then used to specify the transverse components of the magnetic and electric fields, calculated from the other two equations. The boundary conditions are related to the geometry of the guiding structure itself and fix the cut-off eigenvalues for the momentum γ^2.

2.8.1.3 Circular section waveguide

The propagation of TE modes in a circular wave guide is a paradigmatic example because it is a very common tool in applications (particle accelerators, electromagnetic radiation generators or, simply, electromagnetic field transport systems). According to the natural symmetry of the problem we study equation (2.123) in cylindrical coordinates; the axial magnetic field thus becomes

$$H_z(\rho, \phi, z) = h_z(\rho, \phi)\exp(-ikz).$$

The first equation of (2.123), the Helmholtz equation, therefore becomes

$$\left(\frac{\partial^2}{\partial\rho^2} + \frac{1}{\rho}\frac{\partial}{\partial\rho} + \frac{1}{\rho^2}\frac{\partial^2}{\partial\phi^2} + \gamma^2\right)h_z(\rho, \phi) = 0. \tag{2.125}$$

We factorize the equation by separating the variables and set $h_z(\rho, \phi) = R(\rho)P(\phi)$, thus obtaining

$$\frac{1}{R}\frac{d^2R}{d\rho^2} + \frac{1}{\rho R}\frac{dR}{d\rho} + \frac{1}{\rho^2 P}\frac{d^2P}{d\phi^2} + \gamma^2 = 0$$
$$\Rightarrow \frac{\rho^2}{R}\frac{d^2R}{d\rho^2} + \frac{\rho}{R}\frac{dR}{d\rho} + \rho^2\gamma^2 = -\frac{1}{P}\frac{d^2P}{d\phi^2}.$$

The two members of this last expression only depend on a single coordinate variable, hence they are separately equal to the same constant, which we define as n^2 and therefore we obtain

$$\begin{cases} \rho^2\dfrac{d^2R}{d\rho^2} + \rho\dfrac{dR}{d\rho} + (\rho^2\gamma^2 - n^2)R = 0 \\ \dfrac{d^2P}{d\phi^2} + n^2P = 0. \end{cases} \tag{2.126}$$

The second equation of the (2.126) is a harmonic equation that tells us that the electromagnetic field has a $2\pi n$ periodicity in the angular variable ϕ inside the waveguide, where n is an integer index. Its general solution can be therefore written as

$$P(\phi) = A\sin(n\phi) + B\cos(n\phi),$$

where A and B are arbitrary constant. The first equation of (2.126) is a differential Bessel's equation with two independent solutions ($J_n(\gamma\rho)$ and $Y_n(\gamma\rho)$) called Bessel functions of the first and second kind, respectively, and of order n. The general solution can then be written as

$$R(\rho) = CJ_n(\gamma\rho) + DY_n(\gamma\rho).$$

It can easily be accepted that the Bessel function $Y_n(\gamma\rho)$ can be neglected because it becomes infinite at the origin $\rho = 0$ and thus it has no physical meaning for our problem. The general solution for the axial magnetic field is thus

$$h_z(\rho, \phi) = [A\sin(n\phi) + B\cos(n\phi)]J_n(\gamma\rho). \tag{2.127}$$

The cut-off momentum γ can now be evaluated by imposing the boundary conditions. As is well known the tangential component of the electric field, as well as the normal component of the magnetic field, on a metallic surface must be equal to zero. For the TE modes in a circular waveguide we have $E_\phi(\rho, \phi) = 0\big|_{\rho=a}$. But due to the general rules reported in equation (2.123) we obtain

$$E_\phi(\rho, \phi) = i\frac{\omega}{c\gamma^2}\frac{\partial H_z(\rho, \phi)}{\partial\rho}\sqrt{\frac{\mu}{\varepsilon}} = i\frac{\omega}{c\gamma}[A\sin(n\phi) + B\cos(n\phi)]J_n'(\gamma\rho)e^{-ikz}.$$

To satisfy the boundary condition at $\rho = a$ and for any value of the angle ϕ we must require that $J_n'(\gamma a) = 0$. The root of the Bessel $J_n'(x)$ function can be defined as P_{mn}' in such a way that $J_n'(P_{mn}') = 0$. The final results for the cut-off momentum and for the transverse momentum of the modes as expressed in equation (2.123) are

$$\gamma_{mn} = \frac{P_{mn}'}{a} \Rightarrow k_{mn} = \sqrt{\left(\frac{\omega}{c}\right)^2 - \gamma_{mn}^2} = \sqrt{\left(\frac{\omega}{c}\right)^2 - \left(\frac{P_{mn}'}{a}\right)^2}. \tag{2.128}$$

The *TM modes* can be treated in a completely analogous way. The axial electric field is the reference function, it is factorized as $E_z(\rho, \phi, z) = e_z(\rho, \phi)\exp(-ikz)$, thus finding from the Helmholtz equation

$$\left(\frac{\partial^2}{\partial\rho^2} + \frac{1}{\rho}\frac{\partial}{\partial\rho} + \frac{1}{\rho^2}\frac{\partial^2}{\partial\phi^2} + \gamma^2\right)e_z(\rho, \phi) = 0. \tag{2.129}$$

Following the previous procedure and by making use of the separation of variable method, we obtain the following general solution

$$e_z(\rho, \phi) = [A\sin(n\phi) + \cos(n\phi)]J_n(\gamma\rho). \tag{2.130}$$

The cut-off momentum γ can be calculated again by imposing the boundary conditions for the tangential electric field, which for the TM modes becomes $E_z(\rho, \phi) = 0\big|_{\rho=a}$. This condition requires that $J_n(\gamma a) = 0$. The root of the Bessel $J_n(x)$ function can be defined as P_{mn} in such a way that $J_n(P_{mn}) = 0$. The final results

Table 2.2. Electric and magnetic fields for the TE and TM modes.

TE modes

$$\begin{cases} H_z(\rho, \phi) = [A\sin(n\phi) + B\cos(n\phi)]J_n(\gamma_{mn}\rho)e^{-ikz} \\ H_\rho(\rho, \phi) = -i\dfrac{k}{\gamma_{mn}^2}\dfrac{\partial H_z}{\partial \rho} = -i\dfrac{k}{\gamma_{mn}}[A\sin(n\phi) + B\cos(n\phi)]J_n'(\gamma_{mn}\rho)e^{-ikz} \\ H_\phi(\rho, \phi) = -i\dfrac{k}{\gamma_{mn}^2}\dfrac{1}{\rho}\dfrac{\partial H_z}{\partial \phi} = -i\dfrac{kn}{\gamma_{mn}^2\rho}[A\cos(n\phi) - B\sin(n\phi)]J_n(\gamma_{mn}\rho)e^{-ikz} \\ E_\rho(\rho, \phi) = -i\dfrac{\omega}{c\gamma_{mn}^2}\dfrac{1}{\rho}Z_M\dfrac{\partial H_z}{\partial \phi} = -i\dfrac{\omega}{c\gamma_{mn}^2}\dfrac{n}{\rho}Z_M[A\cos(n\phi) - B\sin(n\phi)]J_n(\gamma_{mn}\rho)e^{-ikz} \\ E_\phi(\rho, \phi) = i\dfrac{\omega}{c\gamma_{mn}^2}Z_M\dfrac{\partial H_z}{\partial \rho} = i\dfrac{\omega}{c\gamma_{mn}}Z_M[A\sin(n\phi) + B\cos(n\phi)]J_n'(\gamma_{mn}\rho)e^{-ikz} \end{cases}$$

where $Z_M = \sqrt{\dfrac{\mu}{\varepsilon}}$ and $\gamma_{mn} = \dfrac{P'_{mn}}{a}$

TM modes

$$\begin{cases} E_z(\rho, \phi) = [A\sin(n\phi) + B\cos(n\phi)]J_n(\gamma_{mn}\rho)e^{-ikz} \\ E_\rho(\rho, \phi) = -i\dfrac{k}{\gamma_{mn}^2}\dfrac{\partial E_z}{\partial \rho} = -i\dfrac{k}{\gamma_{mn}}[A\sin(n\phi) + B\cos(n\phi)]J_n'(\gamma_{mn}\rho)e^{-ikz} \\ E_\phi(\rho, \phi) = -i\dfrac{k}{\gamma_{mn}^2}\dfrac{1}{\rho}\dfrac{\partial E_z}{\partial \phi} = -i\dfrac{kn}{\gamma_{mn}^2\rho}[A\cos(n\phi) - B\sin(n\phi)]J_n(\gamma_{mn}\rho)e^{-ikz} \\ H_\rho(\rho, \phi) = i\dfrac{\omega}{c\gamma_{mn}^2}\dfrac{1}{\rho}\dfrac{1}{Z_M}\dfrac{\partial E_z}{\partial \phi} = i\dfrac{\omega}{c\gamma_{mn}^2}\dfrac{n}{\rho}\dfrac{1}{Z_M}[A\cos(n\phi) - B\sin(n\phi)]J_n(\gamma_{mn}\rho)e^{-ikz} \\ H_\phi(\rho, \phi) = -i\dfrac{\omega}{c\gamma_{mn}^2}\dfrac{1}{Z_M}\dfrac{\partial E_z}{\partial \rho} = -i\dfrac{\omega}{c\gamma_{mn}}\dfrac{1}{Z_M}[A\sin(n\phi) + B\cos(n\phi)]J_n'(\gamma_{mn}\rho)e^{-ikz} \end{cases}$$

where $Z_M = \sqrt{\dfrac{\mu}{\varepsilon}}$ and $\gamma_{mn} = \dfrac{P_{mn}}{a}$

for the cut-off momentum and for the transverse momentum of the modes as expressed in the equation set (2.124) are

$$\gamma_{mn} = \frac{P_{mn}}{a} \Rightarrow k_{mn} = \sqrt{\left(\frac{\omega}{c}\right)^2 - \gamma_{mn}^2} = \sqrt{\left(\frac{\omega}{c}\right)^2 - \left(\frac{P_{mn}}{a}\right)^2}. \tag{2.131}$$

Table 2.2 summarises the results, yielding the expression for the TE and TM modes using the computing procedure discussed so far

2.8.2 The longitudinal dynamics in a linac

In the previous sections of this chapter we have studied, in some detail, the problems associated with phase stability during the linac acceleration process. The analysis we have developed applies to TW and SW structures, but also to the dynamics in

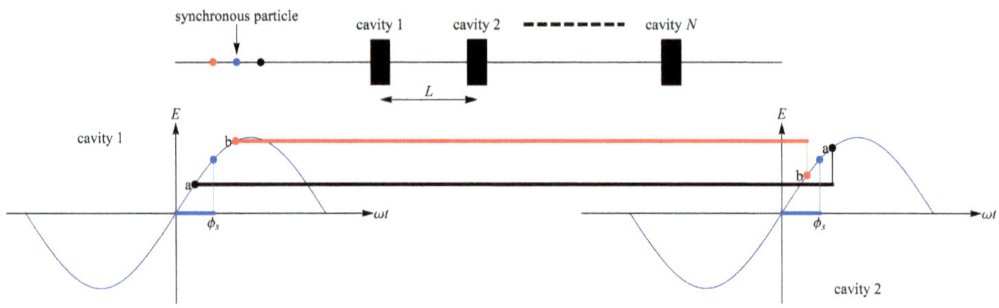

Figure 2.28. Cavity array for a linac acceleration: L denotes the 'synchronous distance' ensuring the phase stability condition from the first to the second cavity.

circular accelerators. According to the principle of phase stability it is possible to accelerate an ensemble of particles (a 'bunch' in the jargon of accelerator physics) characterized by an energy and a temporal spread. This last quantity defines the length of the bunch itself.

In figure 2.28 we show a cavity array, with the relevant separation provided by $L = h\beta_s\lambda$ (h is the harmonic number, namely the number of RF cycles inside the cavities) to ensure synchronism. The figure shows a very rough picture of the phase stable region, along with the fact that individual particles oscillate around the synchronous phase according to whether they gain more or less energy during the acceleration process inside the cavity.

The exercises that follow are aimed at clarifying some points left open, or loosely treated, in the main body of the chapter. We will exploit a slightly modified notation with respect to our previous treatment and characterize the bunch ensemble in terms of time and energy instead of adopting dimensionless quantities and, in view of the fact that $\phi = \omega t$, we will rewrite the equations in section 2.6 using the following substitutions

$$\phi_n \to t_n,$$
$$\gamma_n \to E_n,$$

where n refers to the nth cavity. Furthermore, instead of the longitudinal coordinate we will use the discrete variable n to denote the 'time' at which the particle crosses the nth cavity

Exercise 2.1 Use the previous prescription to recast the equations of section 2.6, accounting for longitudinal phase–space dynamics, in the form

$$\begin{aligned}
\frac{dt_n}{dn} &= \frac{\lambda}{E_s\beta_s^2 c}\eta_c(\Delta E_n),\\
\frac{d(\Delta E_n)}{dn} &= eV[\sin(\omega t_n) - \sin(\omega t_s)],\\
\Delta E_n &= E_n - E_{s,n}, \quad \eta_c = -\frac{1}{\gamma^2},
\end{aligned} \tag{2.132}$$

and discuss the role of η_c.

Hint: denoting by T_n the intra-cavity transit time and with $T_{s,n}$ the same quantity for the synchronous particle, we find

$$t_{n+1} = t_n + T_n - T_{s,n}$$
$$\rightarrow t_{n+1} - t_n \cong \frac{dt_n}{dn} \cong T_n(E_n) - T_{s,n}(E_{s,n}),$$
$$T_n(E_n) \cong T_{s,n}(E_{s,n}) + \left.\frac{dT}{dE}\right|_{E=E_{s,n}} \Delta E.$$

Note then that

$$\frac{dt}{t} \cong \eta_c \frac{dp}{p},$$
$$\frac{dp}{p} \cong \frac{1}{\beta^2}\frac{dE}{E} \rightarrow \frac{dt}{t} \cong \frac{\eta_c}{\beta^2}\frac{dE}{E} \rightarrow \frac{dt}{dE} \cong \frac{\eta_c}{\beta^2}\frac{t}{E}.$$

Exercise 2.2 Show that, by assuming that both the synchronous energy and synchronous phase vary slowly with n, we can derive from equation (2.132) the following second order differential equation

$$\frac{d^2 t_n}{dn^2} = \frac{eV\lambda\eta_c}{E_s\beta_s^2 c}[\sin(\omega t_n) - \sin(\omega t_s)]. \tag{2.133}$$

Hint:

$$\frac{d^2 t_n}{dn^2} = \frac{d}{dn}\left[\frac{\lambda}{E_s\beta_s^2 c}\eta_c(\Delta E_n)\right] = \frac{\lambda}{E_s\beta_s^2 c}\eta_c\left(\frac{d}{dn}(\Delta E_n)\right) + (\Delta E_n)\frac{1}{c}\frac{d}{dn}\left[\frac{\lambda\eta_c}{E_s\beta_s^2}\right],$$

neglect the last term by assuming that $\frac{\lambda\eta_c}{E_s\beta_s^2} \cong$ constant. Such an assumption also allows us to keep $\eta_c = -\frac{1}{\gamma_s^2}$.

Exercise 2.3 Use the approximation $\omega\Delta t_n \ll 1$ to reduce equation (2.133) to

$$\begin{aligned} \frac{d^2\Delta t_n}{dn^2} &\cong (2\pi Q_s)^2 \Delta t_n, \\ \Delta t_n &= t_n - t_s, \\ Q_s^2 &= \frac{eVh\eta_c\cos(\phi_s)}{2\pi E_s\beta_s^2}. \end{aligned} \tag{2.134}$$

Hint: $\sin(\omega(\Delta t_n + t_s)) - \sin(\omega t_s) \cong \omega\Delta t_n \cos(\phi_s)$.

Exercise 2.4 Use the matrix formalism to prove that

$$\begin{pmatrix} \Delta t \\ \Delta E \end{pmatrix}_n = \begin{pmatrix} \cos(\Phi) & \beta_L \sin(\Phi) \\ -\dfrac{1}{\beta_L}\sin(\Phi) & \cos(\Phi) \end{pmatrix} \begin{pmatrix} \Delta t \\ \Delta E \end{pmatrix}_0,$$
$$\Phi = 2\pi Q_s n, \qquad \beta_L = \frac{1}{Q_s \mathcal{P}},$$
$$\mathcal{P} = \frac{\eta_c \omega E_s}{2\pi h}. \tag{2.135}$$

Comment on the physical meaning of $\mathcal{P}$ by noting that it has the dimension of power.

Hint: note that in analogy with the transverse case, β_L is the longitudinal Twiss parameter and by noting that

$$\beta_L = \frac{\lambda}{c\beta_s}\sqrt{-\frac{\eta_c h}{2\pi e V E_s \cos(\phi_s)}}, \tag{2.136}$$

when $0 < \phi_s < \frac{\pi}{2}$ the effect of the cavity can be viewed as that of a focusing quadrupole.

Exercise 2.5 Comment on the meaning of the 'invariant'

$$\frac{1}{\beta_L}(\Delta t_n)^2 + \beta_L(\Delta E_n)^2 = \varepsilon_L. \tag{2.137}$$

Hint: think in terms of longitudinal emittance.

Exercise 2.6 Show that within such a framework the bunch length and energy spread can be written as

$$\sigma_t \cong \sqrt{\beta_L \varepsilon_L},$$
$$\sigma_E \cong \sqrt{\frac{\varepsilon_L}{\beta_L}}, \tag{2.138}$$

and compare to the results of section 2.6, equation (2.83).

Hint: make reference to figure 2.29 to comment on the physical meaning of the longitudinal phase–space ellipse and note that

$$\frac{\sigma_E}{\sigma_t} = \mathcal{P}. \tag{2.139}$$

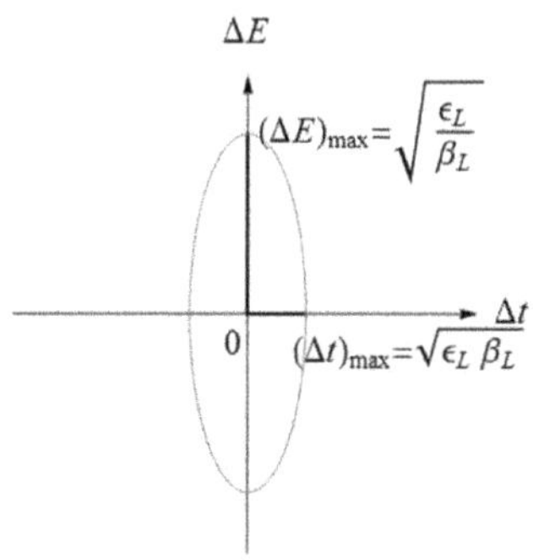

Figure 2.29. Longitudinal phase–space ellipse.

Note finally that the bunch length and relative energy spread are

$$\sigma_\gamma = \frac{\sigma_E}{\gamma_s},$$
$$\sigma_z = \beta_s c \sigma_t. \tag{2.140}$$

Exercise 2.7 Use the previous results to design a proton linac with an energy spread of 0.1%, a bunch length of 3 mm, a current (per bunch) of 1 A, with a 3 GHz RF ($h = 1/2$) and a maximum energy of 20 MeV.

Exercise 2.8 Relax the assumption of small oscillation amplitudes and prove that

$$\frac{1}{2}\left[(\Delta E)^2 - (\Delta E)_0^2\right] = \frac{\left[\cos(\phi) + \phi\,\sin(\phi_s)\right] - \left[\cos(\phi_0) + \phi_0\,\sin(\phi_s)\right]}{\cos(\phi_s)\,\omega^2\beta_L^2}. \tag{2.141}$$

Hint: use the result of section 2.6 and explain the differences with respect to equation (2.83).

Exercise 2.9 Use the phase ϕ_0 for which $(\Delta E)_0 = 0$ and draw the corresponding phase–space curve.

Exercise 2.10 Derive the values at which the curve crosses the line $\Delta E = 0$ and define the trajectory bounds.

Hint: solve the equation $\cos(\phi_1) + \phi_1\,\sin(\phi_s) = \cos(\phi_0) + \phi_0\,\sin(\phi_s)$.

Exercise 2.11 Determine the maximum extension in ϕ of the bound phase–space trajectories and show that

$$\phi_0^* = \pi - \phi_s.$$

Hint: plot the function $F(\phi) = [\cos(\phi) + \phi \sin(\phi_s)] - [\cos(\phi_0) + \phi_0 \sin(\phi_s)]$ as shown in figure 2.30.

Exercise 2.12 Derive the height (namely the maximum ΔE) of the bucket.

Hint: such a quantity corresponds to the height of the separatrix provided by

$$(\omega\beta_L)^2(\Delta E^*)^2 = \frac{2}{\cos(\phi_s)}\left[\cos(\phi_s) - (\pi - \phi_s)\sin(\phi_s)\right]$$
$$\to \Delta E^* = \frac{1}{\omega\beta_L}\sqrt{2\left[1 - (\pi - \phi_s)\tan(\phi_s)\right]}. \tag{2.142}$$

Exercise 2.13 Discuss the physical meaning of ΔE^*.

Hint: think in terms of RF acceptance.

Exercise 2.14 Verify that when ϕ_s decreases the area in the phase–space of the bucket decreases. Clarify the meaning of this result and evaluate the bucket area in phase–space and plot it versus ϕ_s.

Hint: make reference to the figure 2.31.

Exercise 2.15 Write equation (2.93) in the form

$$\frac{2\pi m_e c^2}{eE_0\lambda}\left[\sqrt{1+\Pi^2} - \Pi\right] = \cos(\phi_0) - \cos(\phi_1),$$
$$\Pi = \frac{p}{m_e c}. \tag{2.143}$$

Draw the corresponding phase–space plot and show that the separatrices are specified by $\cos(\phi_1) = \pm 1$.

Hint: check figure 2.32; note that the separatrix corresponds to $\cos(\phi_1) = -1$.

Exercise 2.16 Show that the minimum injection momentum is

$$p_m = \frac{1-\Delta^2}{2\Delta} m_e c,$$
$$\Delta = \frac{eE_0\lambda}{\pi m_e c^2}. \tag{2.144}$$

Evaluate the minimum momentum and kinetic energy for $E_0 = 11$ MV m^{-1} and $\lambda = 10.5$ cm.

Hint: evaluate Π_m for $\phi_0 = 0$, $\phi_1 = \pi$. Note that $T_m = \left[\frac{(1-\Delta)^2}{2\Delta}\right] m_0 c^2$.

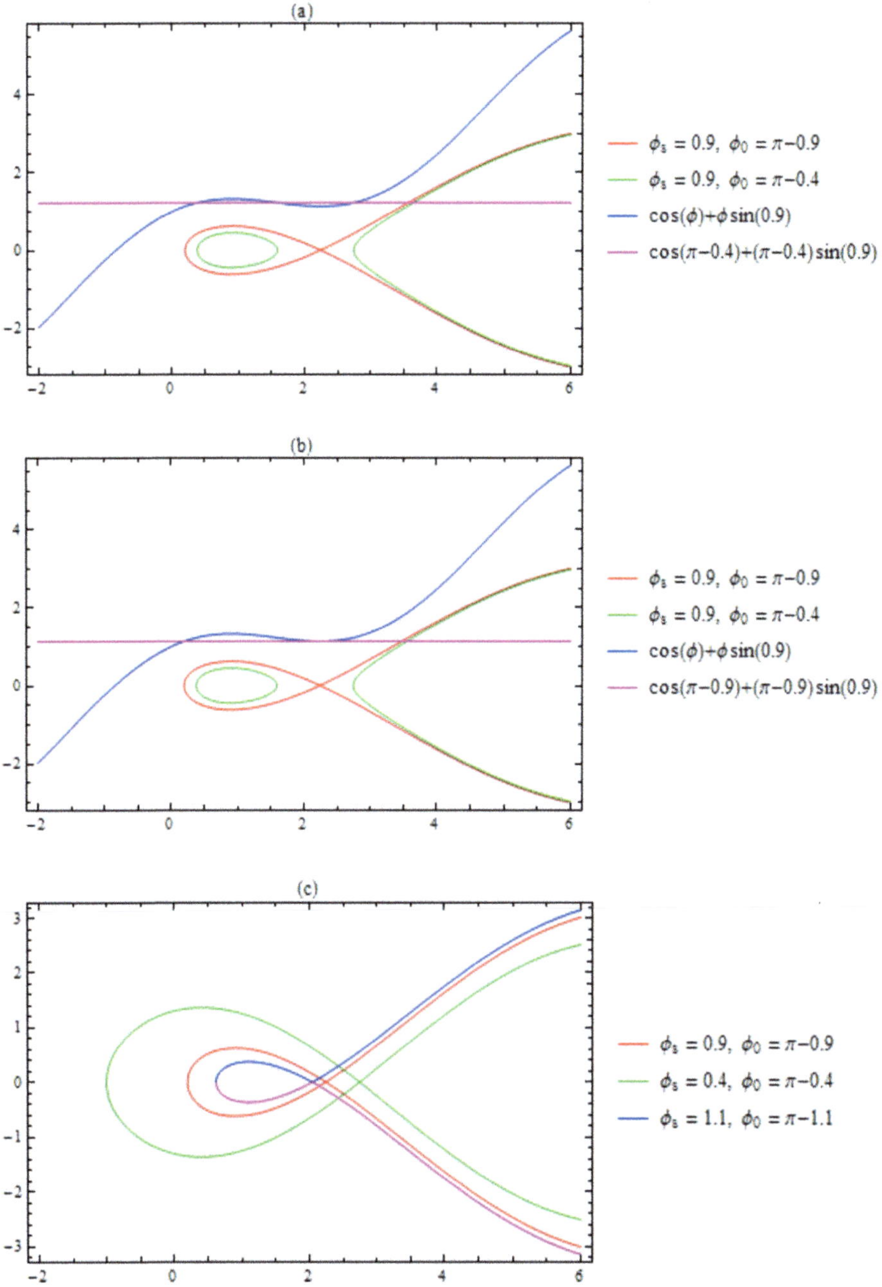

Figure 2.30. Longitudinal phase–space plots versus ϕ open and bound curves for fixed ϕ_s and different values of ϕ_0: (a) $\phi_s = 0.9,\ \ \phi_0 = \pi - 0.4$; (b) $\phi_s = 0.9,\ \ \phi_0 = \phi_0^* = \pi - 0.9$; (c) separatrices for different values of ϕ_s.

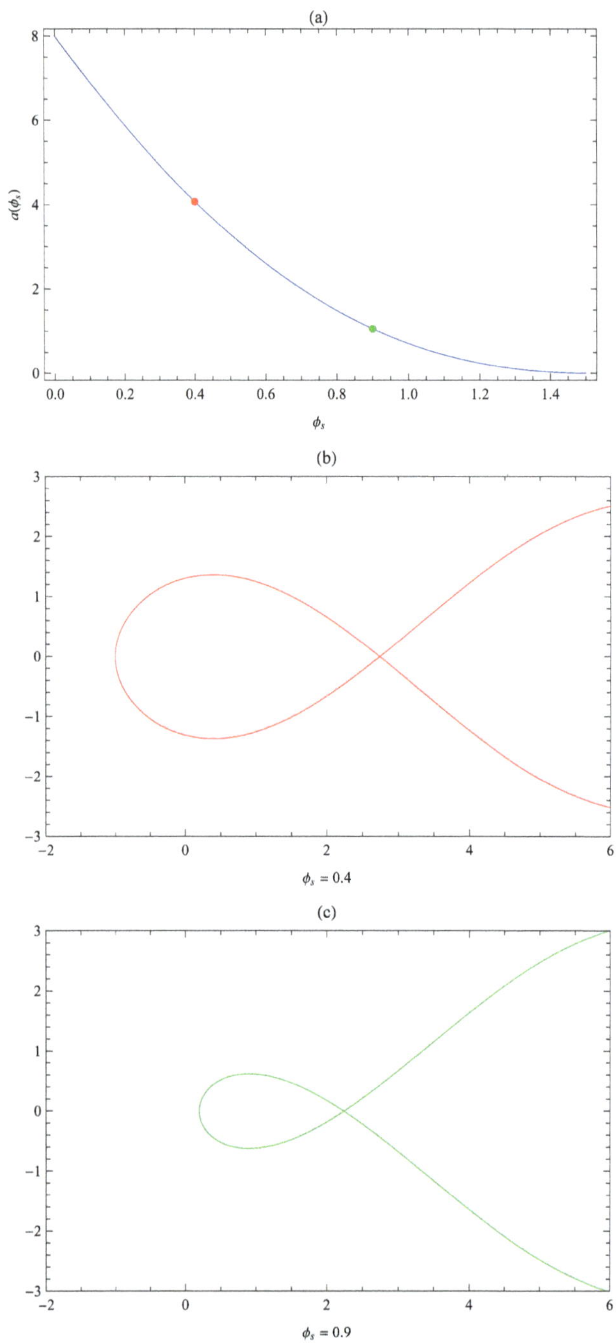

Figure 2.31. (a) Bucket area within the separatrix $a(\phi_s) = \frac{1}{A}\int_{\phi_s}^{\phi*} \sqrt{2 \times [f(\phi, \phi_s) - f(\phi*, \phi_s]} d\phi$, $A = 16$, $S_B = \dfrac{A\, a(\phi_s)}{\omega^2 \beta_L}$. The dots on the curve correspond to the area inside the separatrix for ϕ_s =0.4 (red) and for $\phi_s = 0.9$ (green). The separatrix for $\phi_s = 0.4$ and for $\phi_s = 0.9$ are indicated in (b) and (c), respectively.

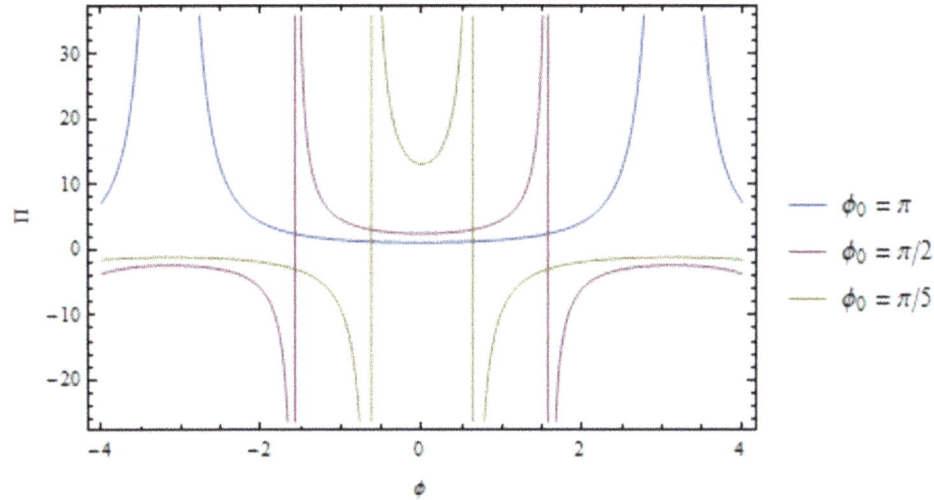

Figure 2.32. Phase space (Π, ϕ) for $\frac{2\pi m_e c^2}{eE_0\lambda} = 0.2$ and different values of ϕ_0.

In the previous remarks we have stressed some of the most significant elements characterizing the linac accelerating process. We have left open the problems associated with the injections from the cathode. In particular, electrons extracted from DC thermionic guns with pulsed grids are provided in pulses of nanosecond duration, which should be shrunk to avoid problems with the capture efficiency and with the energy spread at the end of the acceleration process.

In figure 2.33 we show an example of how the prebuncher works. The electrons are extracted from the cathode and then accelerated by the anode voltage, a successive modulated potential (chopper) at a given frequency is then applied to produce a train of bunches. The bunches (of the order of nanosecond length) are then sent into a prebuncher cavity, at 'zero crossing' the front part of the bunch is accelerated, and the tail decelerated, as the regime is non-relativistic; after a dispersive section the bunch is compressed.

Exercise 2.17 Evaluate the drift section necessary to compress the initial bunch by a factor of 5.

Hint: use non-relativistic mechanics and find $L_d \cong \frac{\lambda_{\mathrm{RF}}}{10 \times \pi} \frac{mv^2}{eE_0 l}$.

2.8.3 Accelerating cavities and electric circuits

We will now make a short detour from our journey by considering the concepts of phase and group velocities and evanescent waves. We have seen that the first two are pivotal elements to understand the mechanisms involving charge–radiation interaction, notions associated with evanescent waves will be touched upon here

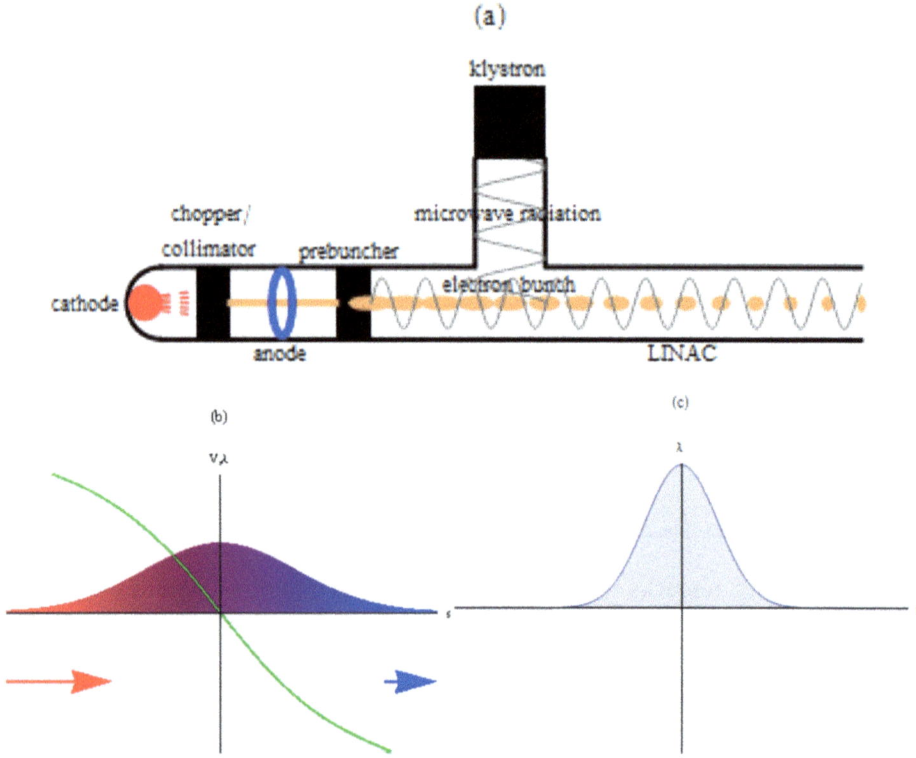

Figure 2.33. Bunching at zero crossing: the electrons are accelerated by the positive part of the wave, the right part is decelerated. The final result is a reduction of the bunch length.

because they will be exploited later in this chapter and in the forthcoming part of the book.

The previous discussion has been kept on rather general grounds and we did not explain the role of the various operating cavities and how they may be coupled. We will try to treat this topic using elementary tools. It is therefore worth clarifying the meaning of the often quoted statement that a cavity is equivalent to an electrical circuit. We have insisted on notions such as SI, cavity quality factor, dissipated power, etc, all concepts which can be transposed to a circuit, such as that shown in figure 2.34.

The origin of the analogy is fairly evident by looking at figure 2.34 and at its captions. We underscore that in a cavity, where the electromagnetic field is confined, it interacts with the cavity environment by exciting, e.g., currents on the surface walls, which causes ohmic losses. In figure 2.35 we have further summarized the genesis of such an interplay:

a. In figure 2.35(a) we show the electric field inside the cavity, whose conducting planes confine the field, thus playing a role analogous to *capacitors*.

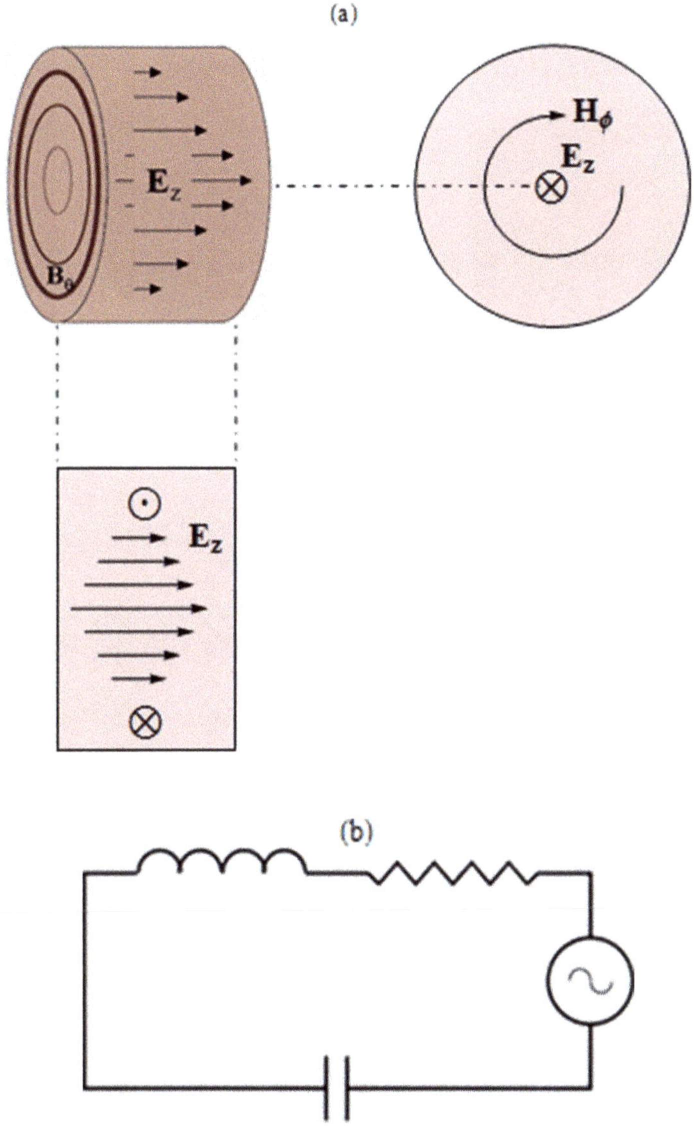

Figure 2.34. (a) Pillbox cavity with the electric and magnetic field confined, and (b) circuit analogy.

b. The charges are accumulated on the plates and current will flow along the edges (figure 2.35(b) and (c)).
c. The current travels around and gives rise to a magnetic field, therefore the cavity has some *inductance* (figure 2.35(d)).
d. The cavity has a finite conductivity, the surface current will flow inside the skin depth and produce cavity heating and thus power loss. In figure 2.35(e)

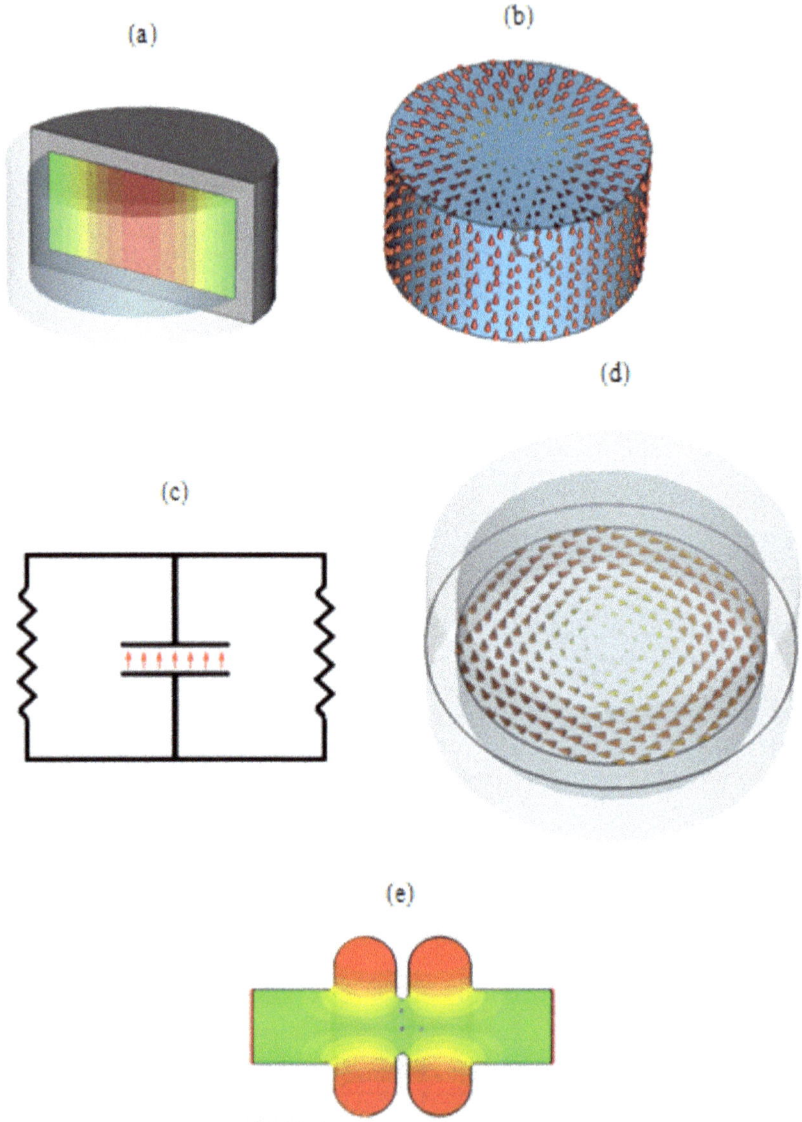

Figure 2.35. Genesis of the cavity circuit analogy.

we have indicated the hot spot of the dissipated power in an accelerating cavity, which is more significantly lost along the surfaces.

e. It is almost natural to associate the cavity surface *resistance* to the SI and close the circuit.

The previously outlined elements justify the analogy and most of the formulae exploited in the theory of the circuit can be used to specify the cavity parameters.

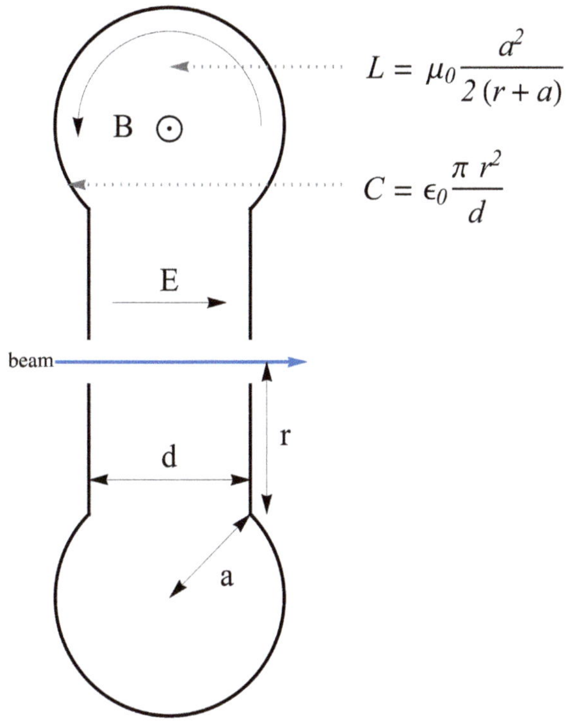

Figure 2.36. A section of a cylindrical cavity with a superimposed re-entrance.

Exercise 2.18 With reference to figure 2.36, which shows a section of a cylindrical cavity with a superimposed re-entrance, recognize the origin of the circuit elements and find the relevant expression.

Exercise 2.19 With reference to the previous exercise evaluate the equivalent circuit oscillating frequency

$$\omega_0 = \frac{1}{\sqrt{LC}} = c\,\frac{\sqrt{2\,d(r+a)}}{\pi r a}.$$

Exercise 2.20 Use the formulae derived in the last two exercises to design a cavity operating at 3 GHz.

Exercise 2.21 Check the correctness of the following statement relevant to the simple pillbox geometry:

Increasing the radius implies a reduction of the accelerating frequency ($\omega_0 \propto r^{-1}$) and a reduction of the stored energy.

Hint: recall that

$$U = \frac{CV_{\text{acc}}^2}{2}, \; V_{\text{acc}} = TV_0, \quad T = \frac{\sin\left(\dfrac{L}{\beta\lambda}\right)}{\left(\dfrac{L}{\beta\lambda}\right)}.$$

Exercise 2.22 Use the previous definitions to show that

$$Q = \frac{\sqrt{\dfrac{L}{C}}}{R_s} = \frac{\omega_0 L}{R_s}.$$

Hint: recall that $Q = \dfrac{\omega_0 \text{ energy stored}}{\text{power lost}}, \; U = \dfrac{1}{2}LI^2, \; P_d = \dfrac{1}{2}R_s I^2.$

The linac performances are accordingly improved if we decrease the SI, thus reducing the losses and optimizing the efficiency of RF–electron-beam power transfer. The power losses inside the cavity occur due to the finite conductivity of the cavity walls and the associated surface currents, which penetrate inside the cavity walls at the so-called skin depth

$$\delta = \sqrt{\frac{2}{\mu_0 \mu_w \sigma_w \omega}},$$

where we have denoted by μ_w, σ_w the wall permittivity and conductivity.

Exercise 2.23 Show that the SI of a pillbox cavity is

$$r_s = \frac{1}{2\pi b}\sqrt{\frac{\mu_0 \mu_w \omega}{2\sigma_w}}.$$

Hint: recall that the resistance per unit length of a conductor of volume Ad is $\dfrac{R_c}{d} = \dfrac{1}{\sigma_c A}$, in our case $A = 2\pi b\delta$.

The SI for copper is, e.g., given by $r_s[\text{M}\Omega\,\text{m}^{-1}] \cong 1.28\sqrt{f[\text{MHz}]}$.

Exercise 2.24 Use the cavity-circuit analogy to find a link between the quality factor and the width of the circuit resonance curve.

Solving the problem is almost straightforward by recalling a few elementary notions. In an RLC circuit the equation describing the evolution of the current can be written as

$$L\frac{dI}{dt} + \frac{1}{C}\int I dt + RI = V_0 e^{i\omega t}.$$

Making the substitution $I = I_0 e^{i\phi} e^{i\omega t}$ we can turn the previous equation in the following algebraic equation

$$ZI_0 e^{i\phi} = V_0,$$

$$Z = R + i\omega L\left(1 - \frac{\omega_0^2}{\omega^2}\right),$$

which is recognized as the Ohm law written in complex form. The dissipated power in an RLC circuit is

$$P_d = \mathrm{Re}\left(\frac{|V|^2}{Z}\right),$$

$$P_d = \frac{V_0^2 R}{R^2 + \left(\frac{\omega L}{R}\right)^2\left(1 - \frac{\omega_0^2}{\omega^2}\right)^2}.$$

With the quality factor provided by $Q = \frac{\omega_0 L}{R}$ and with the full width half maximum of the resonance curve $\Delta\omega = \frac{L}{R}$ (see figure 2.37) we find

$$Q = \frac{\omega_0}{\Delta\omega}.$$

Accelerating lines may be composed by structures which repeat after one, two, three or four cavities. Each constituting sub-cavity is called a cell. These arrays of cavities constituting the accelerating line cannot be considered decoupled. Within our circuital point of view, such a coupling can be considered as a kind of mutual inductance (see figure 2.38). It happens therefore that the oscillation frequency is different from that provided by the uncoupled operating configuration. Furthermore, different modes with different phases may be oscillating, as happens in circuit theory or in coupled harmonic oscillators. The relevant physical meaning and consequences will be clarified by the following exercises.

Exercise 2.25 Show that two coupled RLC cavities with $M = k\ L$ and $Q \gg 1$ have two distinct oscillating modes with frequencies:

$$\omega_+ = \frac{\omega_0}{\sqrt{1+k}}, \text{ 0 mode};$$

$$\omega_- = \frac{\omega_0}{\sqrt{1-k}}, \ \pi \text{ mode}.$$

Corresponding to modes oscillating with the same phase (0 mode) or out of phase π mode.

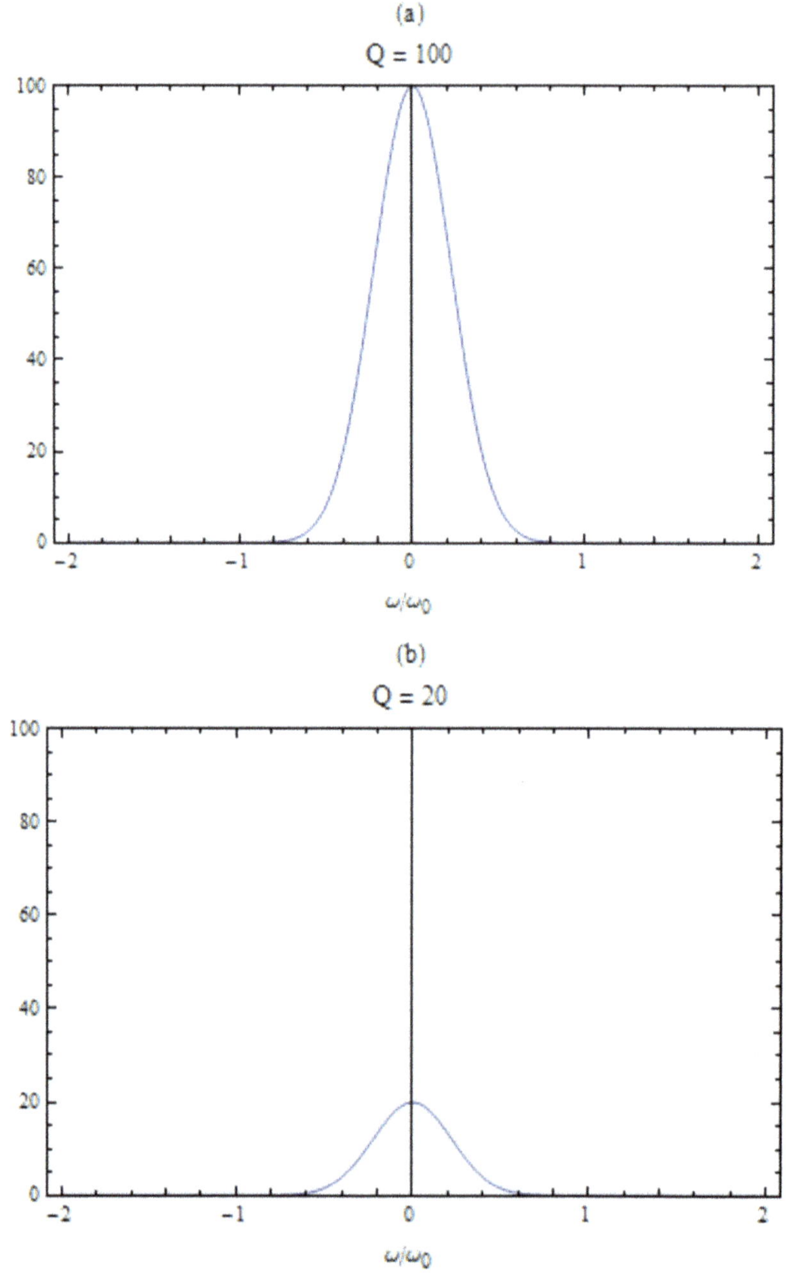

Figure 2.37. RLC resonance curve for different quality factor values.

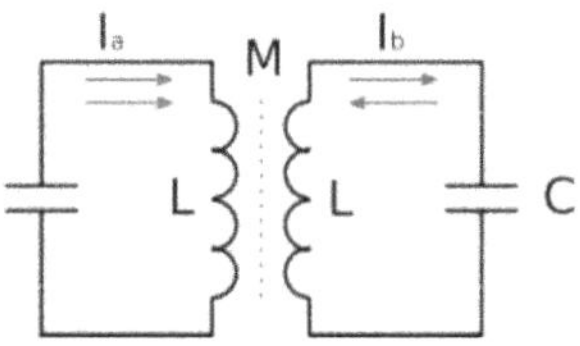

Figure 2.38. Two coupled LC circuits, 0 mode (blue) and π mode (red).

Hint: note that from the Kirchoff law, written using the complex impedance representation

$$
\begin{aligned}
V_0 &= Z\mathcal{I}_0 + M\mathcal{I}_1,\\
V_1 &= Z\mathcal{I}_1 + M\mathcal{I}_0,\\
\mathcal{I}_\alpha &= \mathcal{I}_\alpha e^{i\,\phi_\alpha}
\end{aligned}
$$

by keeping $V_{0,1} = 0$ we can cast the previous algebraic system in the form

$$
\begin{aligned}
X_0\left(1 - \left(\frac{\omega_0}{\omega}\right)^2 - i\frac{\omega_0}{\omega Q}\right) + kX_1 = 0,\\
X_1\left(1 - \left(\frac{\omega_0}{\omega}\right)^2 - i\frac{\omega_0}{\omega Q}\right) + kX_0 = 0,\\
X = \sqrt{L}\mathcal{I}.
\end{aligned}
$$

The problem has accordingly been reduced to an eigenvalue exercise, namely

$$
\begin{pmatrix} X_0 \\ X_1 \end{pmatrix} = \begin{pmatrix} \left(\frac{\omega_0}{\omega}\right)^2 + \frac{i\omega_0}{\omega Q} & -k \\ -k & \left(\frac{\omega_0}{\omega}\right)^2 + \frac{i\omega_0}{\omega Q} \end{pmatrix} \begin{pmatrix} X_0 \\ X_1 \end{pmatrix},
$$

which, assuming $Q \gg 1$, yields the condition

$$
\left[\left(\frac{\omega_0}{\omega}\right)^4 - 1\right] = k^2.
$$

Hence for the two eigenvalues $\omega_\pm$ the corresponding eigenvectors are easily recognized to be

$$
\underline{X}_+ \propto \begin{pmatrix} 1 \\ +1 \end{pmatrix}, \quad \underline{X}_- \propto \begin{pmatrix} 1 \\ -1 \end{pmatrix},
$$

corresponding to the current circulating in the two circuits with the same or opposite phases $\phi = 0$ or $\phi = \pi$, respectively. The operating modes are accordingly said to be the 0- and ‘pi’ modes.

Exercise 2.26 Extend the previous result to three coupled cavities, according to the scheme reported in figure 2.39, and comment on the meaning of the oscillating modes.

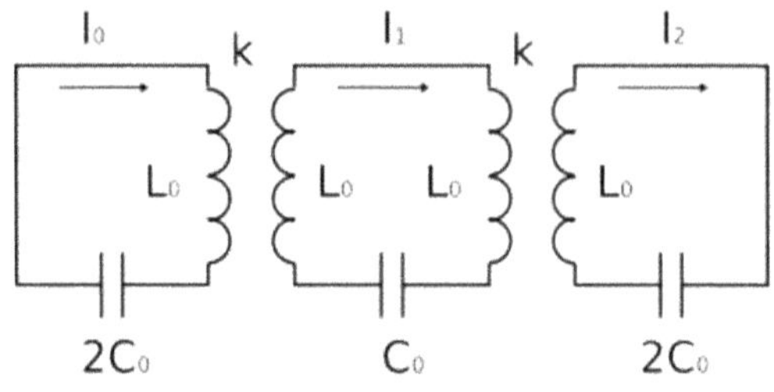

Figure 2.39. Three coupled LC circuits.

Hint: note that according to the previously outlined procedure the eigenvalue equation for the above problem is as below:

$$\hat{T}\underline{X} = \frac{1}{\omega^2}\underline{X},$$

$$\hat{T} = \begin{pmatrix} \omega_0^{-2} & k \cdot \omega_0^{-2} & 0 \\ \frac{k}{2} \cdot \omega_0^{-2} & \omega_0^{-2} & \frac{k}{2} \cdot \omega_0^{-2} \\ 0 & k \cdot \omega_0^{-2} & 0 \end{pmatrix}.$$

The relevant eigenvalues and eigenvectors are

$$\omega^{(q)} = \frac{\omega_0}{\sqrt{1 + k \cdot \cos\left(q\frac{\pi}{2}\right)}},$$

$$\underline{X}_q = \begin{pmatrix} 1 \\ \cos\left(q\frac{\pi}{2}\right) \\ \cos\left(2\,q\frac{\pi}{2}\right) \end{pmatrix}.$$

The order of the mode is $q = 0, 1, 2$, therefore we have

$$\underline{X}_0 = \begin{pmatrix} 1 \\ 1 \\ 1 \end{pmatrix}, \quad \underline{X}_1 = \begin{pmatrix} 1 \\ 0 \\ -1 \end{pmatrix}, \quad \underline{X}_2 = \begin{pmatrix} 1 \\ -1 \\ 1 \end{pmatrix},$$

which are, respectively, the 0, $\frac{\pi}{2}$ and π modes (the phase difference between the adjacent cells is indeed 0, $\frac{\pi}{2}$, π).

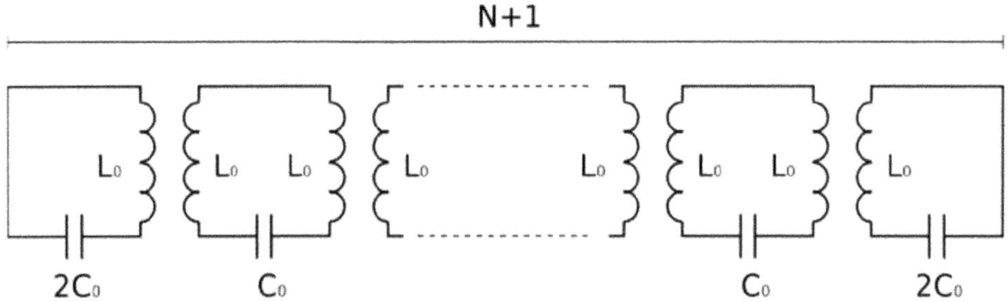

Figure 2.40. N coupled LC circuits corresponding to $N + 2X(1/2)$ coupled cavities.

The previous result can be extended to $N + 2X(1/2)$ coupled cavities (see figure 2.40), and prove that the eigenfrequencies and eigenvectors are

$$\omega^{(q)} = \frac{\omega_0}{\sqrt{1 + k\cos\left(\frac{q\pi}{N}\right)}}, \quad X_q \propto \begin{pmatrix} 1 \\ \cos\left(\frac{\pi q}{N}\right) \\ \cos\left(\frac{\pi q}{N}2\right) \\ \dots \\ \cos\left(\frac{\pi q}{N}n\right) \\ \dots \\ \cos\left(\frac{\pi q}{N}\right) \end{pmatrix}.$$

Hint: use the same procedure as before and write the eigenvalue problem as below:

$$X_0\left(1 - \left(\frac{\omega_0}{\omega}\right)^2\right) + kX_1 = 0,$$
$$\dots$$
$$X_n\left(1 - \left(\frac{\omega_0}{\omega}\right)^2\right) + \frac{k}{2}(X_{n-1} + X_{n+1}) = 0,$$
$$\dots$$
$$X_N\left(1 - \left(\frac{\omega_0}{\omega}\right)^2\right) + kX_{N-1} = 0.$$

The phase difference between adjacent cell is $\frac{\pi q}{N}$. In the case of $N = 3$ we have four oscillating modes, whose distribution is sketched in figure 2.41, which shows the field distribution calculated with the code CST.

Exercise 2.27 Consider the frequency dispersion curve for 7 and 25 coupled cavities with $k = 0.1$ and comment on the relevant physical meaning.

In figure 2.42 we have reported ω_q versus q for $N = 7$, 25 and $k = 0.1$ for a cavity operating at 3 GHz. It is evident that the modes are more dense for large numbers of

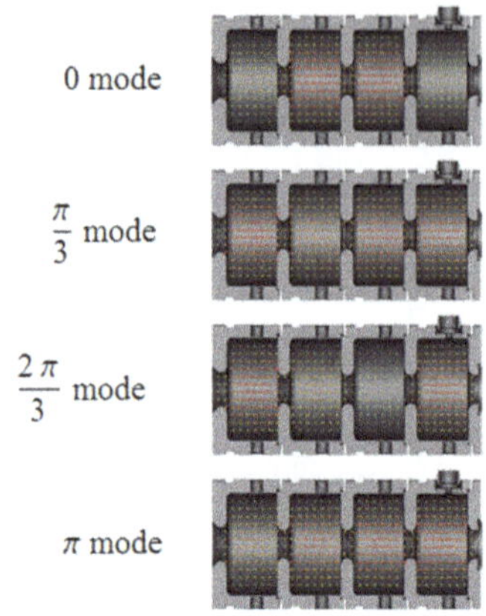

Figure 2.41. Four coupled cavities (cells) and the associated oscillating modes.

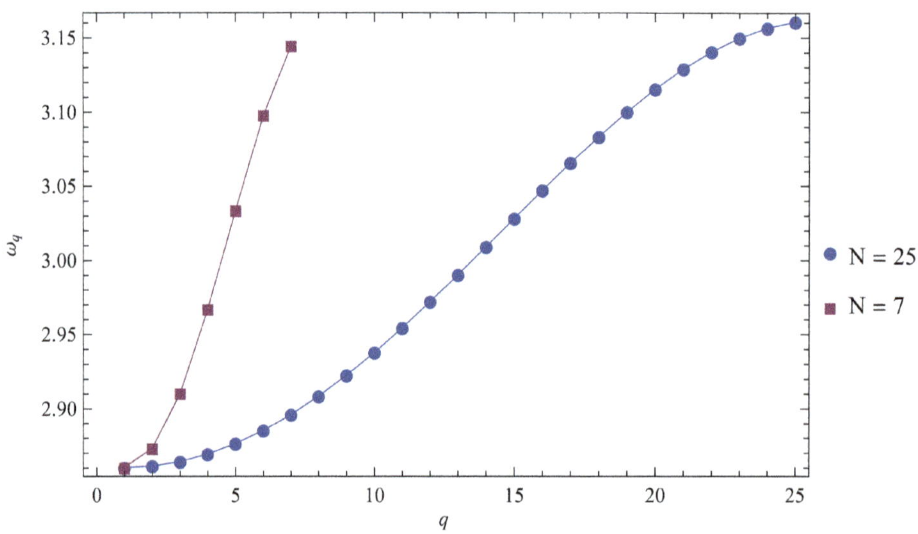

Figure 2.42. ω_q versus q for $N = 7$ (red) and $N = 25$ (blue) coupled cavities.

coupled cavities. Regarding this point it is worth stressing that, for a small coupling parameter k, $\delta\omega = \omega^{(N)} - \omega^{(0)} \cong k\ \omega_0$, when losses are included, each normal mode has a width $\Delta\omega \cong \frac{\omega_0}{Q}$. When N increases, to avoid mode overlapping it is required that $kQ \gg 1$, otherwise simultaneous excitation of two modes (and thus problems with the synchronism condition) may arise.

Exercise 2.28 Find a difference between the operation with $\frac{\pi}{2}$ and π modes.

Hint: note that for the $\frac{\pi}{2}$ operating mode the separation between adjacent modes is

$$\frac{\delta\omega}{\omega_0} = \frac{\omega^{\left(\frac{N+1}{2}\right)} - \omega^{\left(\frac{N-1}{2}\right)}}{\omega_0} \cong k\ \frac{\pi}{2N},$$

while for the π mode

$$\frac{\delta\omega}{\omega_0} = \frac{\omega^{(N+1)} - \omega^{(N-1)}}{\omega_0} \cong k\left(\frac{\pi}{2N}\right)^2.$$

The relative mode separation is therefore larger for the $\frac{\pi}{2}$ case.

Exercise 2.29 Evaluate the group velocity of π, 0 and $\frac{\pi}{2}$, and draw the relevant conclusions.

Hint: recall that $\omega_g = \frac{\partial\omega}{\partial k}$, therefore (see the previous exercise) $\omega_g^{(0,\pi)} \cong 0$, $\omega_g^{(\frac{\pi}{2})} \cong k\omega_0\frac{L_c}{2}$, the $\frac{\pi}{2}$ mode is therefore that with the largest group velocity and thus the smallest filling time, see section 2.5 for further details.

2.8.4 Linac design formulae

In the forthcoming parts of the book we will often refer to this chapter regarding the design of linac structures for FEL operation. We will therefore report in these complements a kind of pocket compendium of the often exploited formulae to obtain a quick understanding of the accelerator performance. Most of the subject has already been covered in section 2.5, and we simply recall here the notation to specify how these relations should be exploited.

2.8.4.1 The energy gain in a constant impedance structure

According to section 2.5 in a *constant impedance structure* we find an energy gain

$$U(\tau) = |e|v(\tau)\sqrt{2P_0r_sL},$$
$$v(\tau) = \left[\frac{1 - e^{-\tau}}{\sqrt{\tau}}\right],$$
$$\tau = \frac{1}{2}\frac{\omega L}{Qv_g} = \frac{1}{2}\frac{\omega}{Q}t_f.$$

The parameters in the previous equations summarize the design details of the structure itself. In figure 2.43 we have reported a DLS with an iris aperture

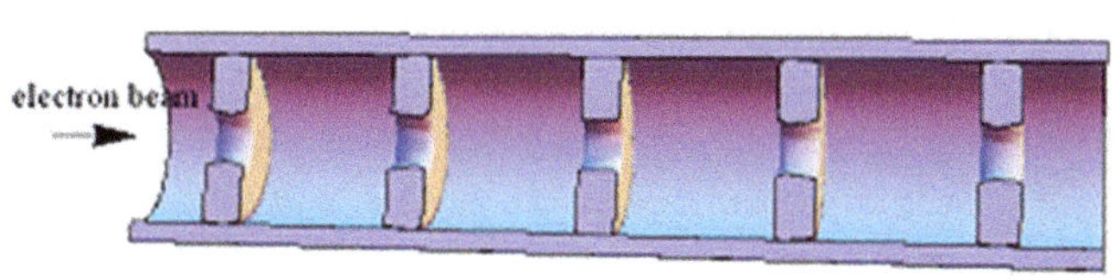

Figure 2.43. DLS with iris.

of $2a$. In the case of a DLS fed by a 3 GHz RF, the group velocity can be computed as

$$\beta_g \cong \frac{[2a[\mathrm{cm}]]^{3.23}}{891},$$

$$\beta_g = \frac{v_g}{c}.$$

The SI also has a dependence on the iris diameter and is indeed provided by

$$r_s \cong r_s^0 - \kappa\,(2a[\mathrm{cm}])^2,$$

where r_s^0 and κ depend on the material, frequency, etc.

The previous remarks allow the solution of the following exercise.

Exercise 2.30 Evaluate the energy gain of an electron beam in a 1 m long copper DLS with an iris diameter of 1.97 cm, operating at 3 GHz, 2/3 π.

Hint: use table 2.3 and obtain an accelerating potential linked to the cavity power by

$$V[\mathrm{MV}] \cong 4.93\sqrt{P[\mathrm{MW}]}.$$

If we are interested in an energy gain of 30 MeV m^{-1} we should require an RF power of about 37 MW.

Usually the RF power from the klystron (see the next chapter) is split into different sections; in the case of SLAC a 120 MW klystron is split four ways; in our present example at the end of the four sections the electrons would have achieved an energy larger than 100 MeV.

2.8.4.2 The energy gain in constant gradient structures

In section 2.5 it has been proven that for a *constant gradient structure* we have

$$\Delta U = |e|\,\Lambda(L)\sqrt{2r_s L\,P_0},$$

$$\Lambda(L) = \sqrt{\frac{1 - e^{-2\chi(L)}}{2}}.$$

Table 2.3. Parameters used in exercise 2.30.

v_g	$0.01c$	
$r_s[\mathrm{M\Omega/m}]$	72	$r_s^0 \cong 86$, $\kappa \cong 3.6$
$t_f[\mu\mathrm{s}]$	0.33	
τ	0.207	$\nu(\tau) \cong 0.411$

The function $\Lambda(L)$ is linked to the fractional attenuation of the RF power along the structure through the factor $\chi(L)$ which is defined by means of equation (2.63) as

$$P(z) = P_0\left(1 - \frac{1 - e^{-2\chi(L)}}{L}z\right).$$

Introducing the quantity

$$\delta = \frac{P(L) - P_0}{P_0},$$

we find

$$\chi(L) = \frac{|\ln(\delta)|}{2}.$$

Assuming an attenuation of 10% and using the same parameters of the exercise 2.30 we obtain

$$V[\mathrm{MV}] \cong 4.67\sqrt{P[\mathrm{MW}]}$$

and therefore a corresponding energy gain.

Exercise 2.31 Derive the filling time of the structure.

Hint: note that according to equation (2.65) the following identity holds

$$v_g(z) = \frac{\omega L}{Q}\left[\frac{1 - \delta}{1 - \frac{z}{L}(1 - \delta)}\right]^{-1}.$$

Then

$$dt = \frac{Q}{\omega L}\frac{1 - \delta}{1 - \frac{z}{L}(1 - \delta)}.... \to t_f \cong \frac{Q}{\omega}|\ln(\delta)|.$$

2.8.4.3 Beam loading effects

In figure 2.44 we have shown a TW linac operating in the 2/3 π mode. The RF source (klystron) provides a set of macro-pulses, pulses lasting for tens of µs, separated by the klystron repetition rate. The RF pulse grows and a series of micro-bunches, with a separation given by $h\frac{\lambda_{\mathrm{RF}}}{c}$, are accelerated in the cavity (see figure 2.45).

When successive bunches are accelerated and extracted within the same RF pulse, the energy stored in the cavity is lowered. The effect goes through a transient regime until it reaches an equilibrium at lower energies.

The inclusion of the beam loading in the previous equation is almost straightforward. We study this effect, starting from the fact that the field E_b excited in an empty

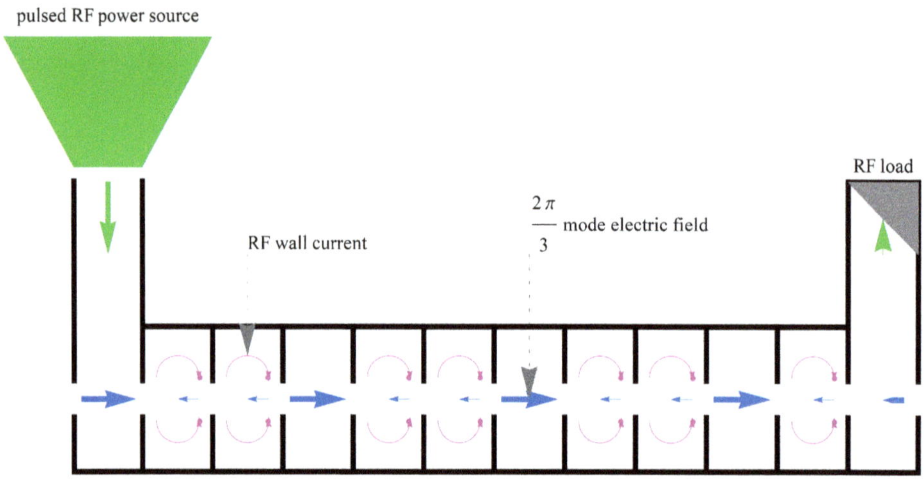

Figure 2.44. TW linac operating in the 2/3 π mode.

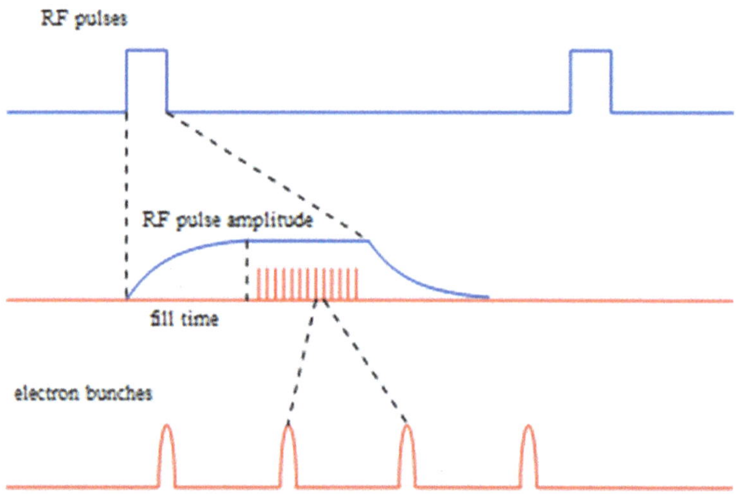

Figure 2.45. Bunch separation = n/f, where n is the number of RF buckets between electron bunches.

cavity by a current i is due to the beam deceleration. Without specifying the deceleration mechanism, we write that the power per unit length given to the cavity can be derived from the balance equation

$$\frac{dP}{dz} = iE_b - 2\alpha P,$$

where i is the current over the RF pulse, E_b is the field due to beam deceleration and the second term on the right is the local power lost in the structure. Going back to equation (2.53) we note that

$$E_b^2 = 2\alpha P,$$

Which, once differentiated with respect to z, yields

$$E_b\frac{dE_b}{dz} = P\frac{d\alpha}{dz} + \alpha\frac{dP}{dz}$$

$$\rightarrow \frac{dE_b}{dz} = \frac{P}{E_b}\frac{d\alpha}{dz} + \alpha\left(i - 2\alpha\frac{P}{E_b}\right)$$

$$\rightarrow \frac{dE_b}{dz} = \frac{P}{E_b}\left(\frac{d\alpha}{dz} - 2\alpha^2\right) + \alpha i.$$

Thus finally ending up with

$$\frac{dE_b}{dz} = -\alpha E_b\left(1 - \frac{1}{2\alpha^2}\frac{d\alpha}{dz}\right) + \alpha i.$$

We can now distinguish between constant impedance and constant gradient structure cases so that

$$\frac{dE_b}{dz} = \begin{matrix} -\alpha E_b + \alpha\, i & \text{constant impedance} \\ \alpha\, i & \text{constant gradient structure} \end{matrix}.$$

Exercise 2.32 Show that for constant impedance structures the beam loading electric field and potential are

$$E_b(z) = r_s i\,(1 - e^{-\alpha z}),$$

$$V_b(z) = r_s i\; L\left(1 - \frac{1 - e^{-\tau}}{\tau}\right).$$

Hint: solve the differential equation defining E_b for constant impedance with the initial condition $E_b(0) = 0$, recall that

$$V_b(z) = \int_0^L E_b(\zeta)\, d\zeta.$$

Exercise 2.33 Show that for a constant gradient structure we have

$$E_b(z) = -\frac{r_s i}{2}\ln\left(1 - \frac{1 - e^{-2\tau}}{L}z\right).$$

Then evaluate the associated voltage along the whole structure

$$V_b = I_0 r L\left[\frac{1}{2} - \frac{\tau\, e^{-2\tau}}{1 - e^{-2\tau}}\right].$$

Hint: integrate the equation for the constant gradient structure case and use the dependence of the attenuation coefficient α derived from equations (2.60), (2.61) and (2.62).

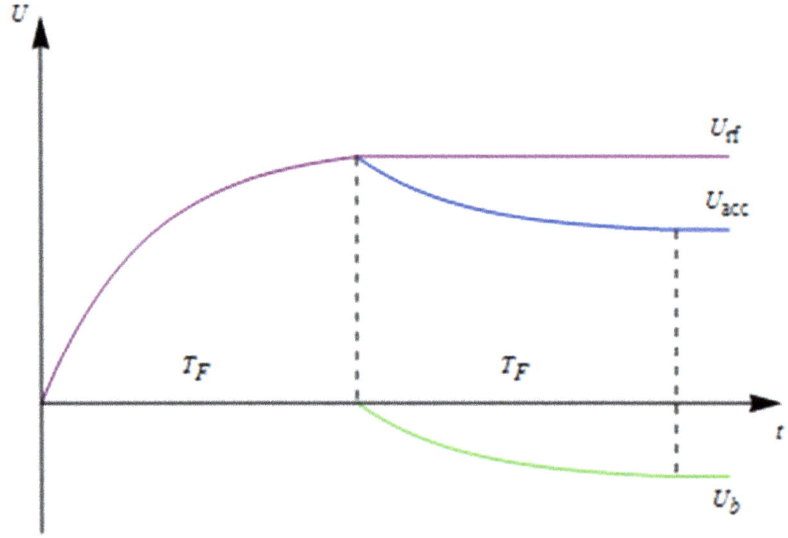

Figure 2.46. Accelerating potential U_{acc}, RF potential U_{rf} and beam decelerating potential U_{b} versus time.

Exercise 2.34 Explain why the accelerating voltage can be written as

$$V = V_{\text{RF}} - V_b.$$

Consider figure 2.46 and explain why the results of exercises 2.32 and 2.33 apply to the steady state regime.

Further reading for chapter 2

Linac theory and design

Turner S 1987 Second general accelerator physics course *CERN Accelerator School (Scanticon Conference Centre, Aarhus, Denmark, 1986)* CERN 87-10

Wangler T P 1993 Introduction to linear accelerator theory *Los Alamos National Laboratory* LA-UR-93-805

Nguyen D, Russell S and Moody N 2009 Theory and practice of free electron lasers *US Particle Accelerator School*

Nguyen D, Lewellen J and Duffy L 2014 RF linacs for high gain FELs *US Particle Accelerator School (16–20 June)*

Lapostolle P and Weiss M Formulae and procedures useful for the design of linear accelerators *CERN* CERN-PS-2000-001 (DR)

Reiser M 1994 *Theory and Design of Charged Particle Beams* (New York: Wiley)

RF transport and handling linacs

Slater J C 1951 *Microwave Electronics* (New York: van Nostrand)

Ramo C, Whinnery J R and van Duzer T 1965 *Fields and Waves in Communication Electronics* (New York: Wiley)

Byrd J 2014 *RF Cavities and Linacs* http://uspas.fnal.gov/materials/14UNM/RF_Structures.pdf

Proceedings of linac conferences

Proton Linear Accelerator Conf. (*Chalk River, Canada, 14–17 Sep 1976*) ed S O Schriber Report number AECL-5677

The 1979 Linear Accelerator Conf. (*Montauk, NY, 10–14 Sep 1979*) ed R L Witkover, Report BNL 51134

The 1981 Linear Accelerator Conf. (*Bishop Lodge, Santa Fe, NM, 19–23 Oct 1981*) ed R A Jameson and L S Taylor, Report number LA-9234-C

1984 Linear Accelerator Conf. (*Lufthansa–Schulungszentrum, Seeheim, Fed. Rep. of Germany, 7–11 May 1984*) ed N Angert (GSI Darmstadt publisher)

1986 Linear Accelerator Conf. (*Stanford Linear Accelerator Center, Stanford University, Stanford, CA, 2–6 Jun 1986*) SLAC-Report-303, September 1986

1988 Linear Accelerator Conf. (*Continuous Electron Beam Accelerator Facility, Newport News, VA, 3–7 Oct, 1988*)

1990 Linear Accelerator Conf. (*Albuquerque, NM, 10–14 Sep, 1990*) Report LA-12004-C, March 1991

1992 Linear Accelerator Conf. (*Ottawa, Ontario, Canada, 24–28 Aug, 1992*) AECL-10728

18th Int. Linac Conf. (*Geneva, Switzerland, 26–30 Aug, 1996*) ed C Hill and M Vretenar, CERN 96-07, November 1996

XIX International Linear Accelerator Conf. (*Chicago, IL, 25–28 Aug, 1998*) ANL-98/28, 1999

http://physics.bu.edu/neppsr/2007/TALKS-2007/Accelerators_Barletta.pdf

IOP Publishing

Giuseppe Dattoli, Andrea Doria, Elio Sabia and Marcello Artioli

Chapter 3

Free-electron-based generators of electromagnetic radiation

3.1 An introduction to the process of emission/absorption of photons by free electrons

We reconsider here what has been discussed in chapter 1, with the perspective of introducing generators of coherent radiation that employ beams of free electrons. These devices are designed to produce intense and coherent sources of radiation, operating over a wide range of wavelengths. From the historic point of view, electromagnetic radiation has been generated by means of a field induced by moving electric charges. In 1887 Rudholf Hertz demonstrated that the fields generated by a spark–gap closure can propagate in the free space and induce a current on a coil; this was the first evidence of an electromagnetic wave. Since then, many devices have been designed and constructed for the generation of radiation in the microwave region of the spectrum (300 MHz–300 GHz) based on the interaction of free electron beams with 'structures' of different nature, until lasers appeared in 1960. In the following we will deal with the free-electron-based generators, which will further be detailed in the forthcoming chapters 4 and 5.

The prerequisite for any radiation process is the emission or the absorption of photons (the quanta of the electromagnetic field) by a charge. Any physical mechanism cannot, however, occur if fundamental rules (energy, momentum and charge conservation) are violated. This is indeed what prevents free charged particles from radiating or absorbing radiation in a vacuum.

The above statement can be better appreciated by exploiting the geometrical picture provided by the energy momentum plot in figure 3.1, in which the dispersion relations for electrons (labelled with subscript *e*) and photons (subscript *ph*) are drawn. We have plotted a two-dimensional curve, but we should consider the

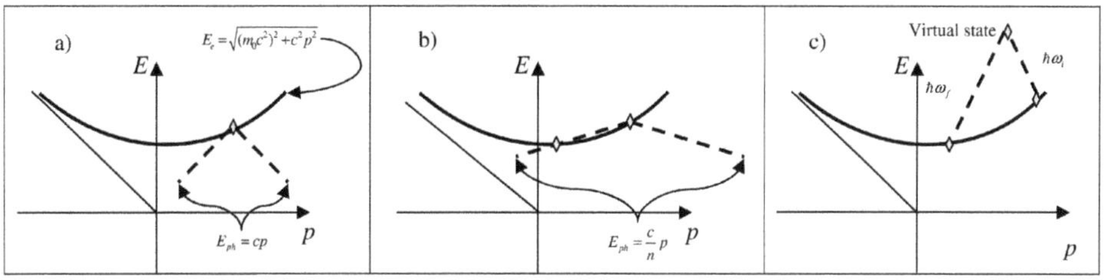

Figure 3.1. The energy–momentum diagram of the electron–photon interaction: (a) in vacuum emission, (b) Cerenkov emission and (c) Compton scattering by means of a virtual state.

corresponding rotation surfaces (a hyperboloid for electrons and a cone for photons).

In figure 3.1(a) it is shown that, from the geometrical point of view, the process of photon emission should be viewed as the subtraction of the photon dispersion relation (dashed line) from that of the electron; energy and momentum are indeed carried off from the electron state. However, since free electron states must lie on the same dispersion curve, the emission process occurs only if photon and electron curves intercept, which is evidently impossible in a vacuum.

On the other side, in the case of a medium with a refractive index n, the photon dispersion relation changes. The slope is c/n and the associated cone is larger. When the photon dispersion is subtracted from that of the electron, we find what is shown in figure 3.1(b), showing the existence of an interception corresponding to the final state of the electron. An electron moving in a medium with index of refraction n can therefore radiate, as occurs in the Cerenkov process.

In figure 3.1(c) we apply the same point of view to Compton scattering. The transition occurs through a 'virtual state', in which the electron 'absorbing' a photon is promoted to a state lying on a dispersion curve with higher energy and then 'relaxes' back to the original curve.

We will see in the following that this last point of view can be adopted to treat the emission by relativistic electrons moving in an intense magnetic field. This field, can indeed be regarded as an ensemble of virtual photons and, from the energy–moment conservation laws, the absorbed and emitted photon are simply connected by the electron kinetic energy.

We will provide below a simple mathematical derivation corroborating what was heuristically discussed with the aid of figure 3.1. We will analytically show that the energy and momentum conservation laws do not allow any emission or absorption process of photons by a particle if the environment is the vacuum.

We proceed by absurdum assuming that the process of emission in a vacuum by an electron with energy E_1 and momentum $\vec{P}_1$ is allowed. When this electron emits a photon of energy $\hbar\omega$ and momentum $\hbar\vec{k}$ it obtains the final state labelled by E_2 and $\vec{P}_2$.

From the energy conservation law we obtain

$$E_1 = E_2 + \hbar\omega,$$

but as $E = m_0c^2\gamma$ we obtain

$$m_0c^2\gamma_1 = m_0c^2\gamma_2 + \hbar\omega \Rightarrow \hbar\omega = m_0c^2(\gamma_1 - \gamma_2).$$

We can, therefore, link the relativistic factor γ to the momentum P as $|P| = m_0v\gamma \Rightarrow \gamma = |P|/(m_0v)$ and thus

$$\hbar\omega = m_0c^2\left(\frac{|P_1|}{m_0v_1} - \frac{|P_2|}{m_0v_2}\right) = c^2\left(\frac{v_2|P_1| - v_1|P_2|}{v_1v_2}\right).$$

Without losing any generality, we can consider the photon energy to be small if compared to that of the particle, which does not change its relativistic velocity significantly during the absorption/emission process, namely $v_2 \approx v_1 = v \approx c$, thus obtaining

$$\hbar\omega \approx v(|P_1| - |P_2|). \tag{3.1}$$

On the other side, the momentum conservation law written in vector form states that

$$\vec{P}_1 = \vec{P}_2 + \hbar\vec{k}.$$

Having assumed that the photon carries a momentum negligible with respect to that of electrons, we find

$$|\vec{P}_1| \cong \cos\theta|\hbar\vec{k}| + |\vec{P}_2|,$$

where θ is the angle formed by the photon momentum with the direction of $\vec{P}_1$. Furthermore, since

$$E_{Ph} = cP_{Ph} \Rightarrow \hbar\omega = c\hbar|k| \Rightarrow |k| = \frac{\omega}{c},$$

the momentum conservation law becomes (the vector symbol has been suppressed as it is unnecessary)

$$|P_1| - |P_2| = \hbar\frac{\omega}{c}\cos\theta. \tag{3.2a}$$

The comparison between (3.1) and (3.2a) yields

$$\cos\theta = \frac{1}{\beta}. \tag{3.3}$$

Equation (3.3) is a clear nonsense because for all particles (except tachyons) $v < c$ and thus $\beta < 1$ gives for the cosine a value greater than unity. The result (3.3) tells us that the original hypothesis of absorption or emission of a photon in vacuum is false.

As already remarked, things change if the process occurs in a medium having a refractive index $n > 1$, in this case the photon will propagate at a reduced speed equal to c/n. Under these conditions the photon energy and momentum read

$$E_{Ph} = \frac{c}{n}P_{Ph}$$

or more explicitly

$$\hbar\omega = \frac{c}{n}\hbar k,$$

which gives $k = n\frac{\omega}{c}$.

Equation (3.2*a*) now can be rewritten as

$$|P_1| - |P_2| = \hbar n\frac{\omega}{c}\cos\theta. \tag{3.2b}$$

A new comparison between (3.2*b*) and (3.1) leads to a new condition for the emission/absorption process by a particle in a medium

$$\cos\theta = \frac{1}{n\beta}. \tag{3.4}$$

Equation (3.4) provides the so-called Cerenkov condition, which can be satisfied for a proper combination of the particle energy and the dielectric medium characteristics.

It is moreover useful to underline how the photon phase velocity, along the particle momentum P_1, is defined. According to the usual definition of phase velocity ($v_f = \frac{\omega}{k}$) we find

$$v_f = \frac{\omega}{|k|\cos\theta} \Rightarrow v_f = \frac{\omega}{n\frac{\omega}{c}\cos\theta} = \frac{c}{n}\frac{1}{\cos\theta}$$

or, from (3.4)

$$v_f = c\beta = v.$$

This last expression tells us that the Cerenkov emission occurs when the particle and the photon have the same velocity and therefore have a constant phase relation.

A simultaneous absorption and emission process can occur if we admit the existence of a virtual state (see figure 3.1(c)). This state allows the particle in state 1 to absorb a photon and, at the same time, to emit a second one before approaching the final state 2.

From the energy conservation law we have

$$E_1 + \hbar\omega_1 = E_2 + \hbar\omega_2. \tag{3.5}$$

From the momentum conservation law we have

$$P_1 - \frac{\hbar\omega_1}{c} = P_2 + \frac{\hbar\omega_2}{c} \Rightarrow P_1 = P_2 + \frac{\hbar\omega_1}{c} + \frac{\hbar\omega_2}{c},$$

where the minus sign indicates absorption.

As $\frac{E}{c}\frac{v}{c} = P$, the previous expression can be written as

$$\frac{E_1}{c}\beta_1 = \frac{E_2}{c}\beta_2 + \frac{\hbar\omega_1}{c} + \frac{\hbar\omega_2}{c} \Rightarrow E_1\beta_1 - \hbar\omega_1 = E_2\beta_2 + \hbar\omega_2. \tag{3.6}$$

Making a proper addition and subtraction of (3.5) and (3.6) we obtain

$$\begin{cases} E_1 + \beta_1 E_1 = E_2 + \beta_2 E_2 + 2\hbar\omega_2 \\ E_1(1 - \beta_1) + 2\hbar\omega_1 = E_2(1 - \beta_2) \end{cases} \Rightarrow \begin{cases} (1 + \beta_1)E_1 = (1 + \beta_2)E_2 + 2\hbar\omega_2 \\ (1 - \beta_1)E_1 = (1 - \beta_2)E_2 - 2\hbar\omega_1 \end{cases},$$

which are more conveniently cast in the form

$$\omega_2 = \frac{c}{2\,\bar{\lambda}_e}[\gamma_1(1 + \beta_1) - \gamma_2(1 + \beta_2)],$$
$$\omega_1 = \frac{c}{2\,\bar{\lambda}_e}[\gamma_2(1 - \beta_2) - \gamma_1(1 - \beta_1)],$$

where $\bar{\lambda}_e$ is the electron Compton wavelength. By keeping the ratio $\frac{\omega_2}{\omega_1}$ we find

$$\frac{\omega_2}{\omega_1} = \frac{\gamma_1(1 + \beta_1) - \gamma_2(1 + \beta_2)}{\gamma_2(1 - \beta_2) - \gamma_1(1 - \beta_1)}.$$

We use now the approximations

$$1 + \beta_{1,2} \cong 2,$$
$$\gamma_{1,2}(1 - \beta_{1,2}) \cong \frac{1}{2\gamma_{1,2}},$$

thus finding

$$\frac{\omega_2}{\omega_1} \cong 4\frac{\gamma_1 - \gamma_2}{\left(\frac{1}{\gamma_2} - \frac{1}{\gamma_1}\right)} = 4\,\gamma_1\gamma_2.$$

We assume a small energy loss by the particle during the interaction (namely $\gamma_2 \approx \gamma_1$), therefore we end up with

$$\omega_2 = 4\,\gamma_1^2\omega_1, \tag{3.7}$$

which is a fairly important result, stating that in a Compton backscattering process the frequency of the backscattered photon is 'Doppler upshifted' by a factor of $4\,\gamma_1^2$.

The following exercise can provide further understanding of the energy transfer mechanism. Assume that initially a photon and an electron are propagating in opposite directions along the z-direction. Denoting by $\hat{e}_z$ the unit vector in the z-direction, we specify the initial electron velocity as $-\hat{e}_z v_i$. The photon is then scattered by an angle θ with respect to the initial velocity and the electron by an angle φ. Use the previous data to state that

$$\omega_2 = \omega_1\frac{\gamma_1(1 + \beta_1)}{\gamma_1(1 + \beta_1\cos(\theta)) + \frac{\bar{\lambda}_e\omega_1}{c}(1 - \cos(\theta))}.$$

Show furthermore that

$$\frac{\omega_2 - \omega_1}{\omega_1} = \frac{\left(\gamma_1\beta_1 - \frac{\lambdabar_e\omega_1}{c}\right)(1 - \cos(\theta))}{\gamma_1(1 + \beta_1\cos(\theta)) + \frac{\lambdabar_e\omega_1}{c}(1 - \cos(\theta))}.$$

Explain the relevant physical meaning, in particular why, if $\gamma_1\beta_1 = \frac{\lambdabar_e\omega_1}{c}$, there is no energy transfer between electrons and photons. Finally recover (3.7) (*hint*: assume $\frac{\lambdabar_e\omega_1}{c} \ll 1$ and $\theta = \pi$).

3.2 Electron-field energy exchange

Metallic walls, such as those found in waveguides and microwave resonant cavities, change the characteristic dispersion relation of the electromagnetic radiation, introducing longitudinal components of the electric field of the wave and thus reducing the group velocity of the radiation. This allows the absorption and emission process as depicted in figure 3.1(b). Generally speaking, in the classical picture, the energy exchange (absorption or emission) occurs between the electric field associated with the electromagnetic wave and the charged particle (electron); this is due to the well-known expression of the Lorentz force

$$\vec{F} = e\,\vec{E} + e(\vec{v} \times \vec{B}). \tag{3.8}$$

The work done (positive or negative) by the fields on a particle that shifts by a quantity $\boldsymbol{s}$ is known to be

$$W = \vec{F} \cdot \vec{s} = e\vec{E} \cdot \vec{s} + e(\vec{v} \times \vec{B}) \cdot \vec{s} = e\vec{E} \cdot \vec{s} = eE_0 s\Re[\exp(i\omega t + \varphi_p)]. \tag{3.9}$$

The work associated with the magnetic field is zero because it is perpendicular to the electron trajectory. The power exchanged (not yet the power emitted to electromagnetic radiation) is then

$$P = \vec{F} \cdot \vec{v} = \vec{J} \cdot \vec{E} = JE_0\Re[\exp(i\omega\, t + \varphi_p)], \tag{3.10}$$

with $\vec{J}$ the electron current (actually we have used the simple choice of current density, therefore the rhs of (3.10) can be viewed as an intensity), ω the frequency of the electromagnetic field and φ_p the particle phase relative to the electromagnetic wave.

A superficial analysis of (3.10) allows the conclusion that is enough to increase the electron current in order to increase the output power. This is only partially true; looking at (3.9) and (3.10) we see how important the global phase term is and in particular the particle phase φ_p with respect to the field. This means that if we add the contributions of many electrons in a beam we must consider the interference mechanism to evaluate the total power exchanged

$$P = eE_0\sum_i v_i\Re[\exp(i\omega t + \varphi_{p_i})]. \tag{3.11}$$

Equation (3.11) tells us that the power exchanged will increase significantly when the electrons, or a great number of them, have the same or similar phase φ_{p_i}. This happens if they are bunched with respect to the wavelength of the electromagnetic field ($\lambda = 2\pi/\omega$).

3.3 Coherent generation of radiation by a charged particle beam

In the following we will reconsider some of the points discussed in chapter 1 from a different perspective, so we will review some concepts and even repeat formulae for the sake of continuity. The reader can skip some parts of this section if they are sufficiently well understood from the content of the previous chapter.

We have already stressed that a charged particle emits radiation if it undergoes an acceleration or is subject to an 'external' disturbance. From classical electromagnetic theory we know that the emission into a solid angle $d\Omega$ in the frequency interval $d\omega$ is

$$\frac{dI}{d\omega d\Omega} = \frac{e^2\omega^2}{4\pi^2 c}\left|\int_{-\infty}^{\infty} \hat{n} \times (\hat{n} \times \vec{\beta}) e^{i\omega\left[t - \frac{\hat{n}\cdot\vec{r}(t)}{c}\right]} dt\right|^2, \tag{3.12}$$

where $\vec{\beta}$ is the normalized particle velocity vector, $\vec{r}(t)$ is the particle position vector and $\hat{n}$ is the unit vector along the direction connecting the particle with the observer.

If we are interested in the emission by a number N of particles, the intensity is found by summing the contribution of each particle, paying attention to the phase of each contribution. Equation (3.12) therefore becomes

$$\frac{dI}{d\omega d\Omega} = \frac{e^2\omega^2}{4\pi^2 c}\left|\int_{-\infty}^{\infty} \sum_{j=1}^{N} \hat{n} \times \left(\hat{n} \times \vec{\beta}_j\right) e^{i\omega\left[t - \frac{\hat{n}\cdot\vec{r}_j(t)}{c}\right]} dt\right|^2. \tag{3.13}$$

Under the very simplified hypothesis of a quasi-monoenergetic beam, we can put $\vec{\beta}_j \approx \vec{\beta}$ and thus rewrite equation (3.13) as

$$\frac{dI}{d\omega d\Omega} = \frac{e^2\omega^2}{4\pi^2 c}\left|\sum_{j=1}^{N} e^{i\frac{\omega}{\beta c} r_j}\right| \left|\int_{-\infty}^{\infty} \hat{n} \times (\hat{n} \times \vec{\beta}) e^{i\omega\left[t - \frac{\hat{n}\cdot\vec{r}(t)}{c}\right]} dt\right|^2. \tag{3.14}$$

We will address the mono-energetic beam hypothesis later in this book, when a more detailed theory of coherence, developed for the free electron laser (FEL), will be presented.

Equation (3.14) has been written considering that $\vec{r}(t)$ represents the center of the mass of the beam, and r_j refers to the particle position along the orbit with respect to the center of the mass. The analysis of this equation reveals that the second term of the rhs contains the full time dependence of the function and is equivalent to (3.12).

The first summation term is the one that includes the coherence effects and makes the difference between (3.12) and (3.14); it can be rewritten as

$$S(\omega) = \sum_{j=1}^{N} e^{i\frac{\omega}{\beta c}r_j} \cdot \sum_{k=1}^{N} e^{-i\frac{\omega}{\beta c}r_k} = \sum_{j=k=1}^{N} e^{i\frac{\omega}{\beta c}(r_j-r_k)} + \sum_{j,k=1(j\neq k)}^{N} e^{i\frac{\omega}{\beta c}(r_j-r_k)}$$
$$\Rightarrow S(\omega) = N + N(N-1)f(\omega), \tag{3.15}$$

where $f(\omega)$ is defined as

$$f(\omega) = \frac{1}{N(N-1)} \sum_{j,k=1(j\neq k)}^{N} e^{i\frac{\omega}{\beta c}(r_j-r_k)}. \tag{3.16}$$

A proper combination of (3.14) and (3.15) gives us the total radiated intensity from a bunch of N charged particles:

$$\frac{dI}{d\omega d\Omega} = [N + N(N-1)f(\omega)]P(\omega), \tag{3.17}$$

where $P(\omega)$ is the power radiated by the single particle as expressed by (3.12).

The function $f(\omega)$ in (3.16) describes the coherence of the emitted radiation; its values range from 0 to 1. The limit $f(\omega) = 0$ indicates an incoherent emission from the bunch that means that the power emitted by N particles is just the sum of the single N contributions (see (3.17)). The upper limit $f(\omega) = 1$ represents the total coherence, when the power emitted by N charged particles is given by N^2 times the emission of the single particle. This result is fully equivalent to the emission of a single particle having a charge $Q = Ne$, because, looking at (3.12), power is proportional to $Q^2 = N^2e^2$.

An appropriate expression of the coherence function $f(\omega)$ strongly depends on the particle distribution in the bunch and the more general problem finds a solution in a statistical physics framework involving an ensemble average over all the particles configurations.

3.4 A single cavity as a basic structure

In order to bunch the electrons we need to induce a velocity modulation in the beam and among the possible techniques, the most elementary involves a geometry realized with metal components, which exploits, again, the interaction between a charged particle and an electromagnetic wave. A solution is provided by, e.g., the gridded gap sketched in figure 3.2.

The velocity of the beam entering the gap is u_0 and is modified by an amount Δu as the beam passes through the gap. The time and position at the entrance to the gap are t_0 and $z = 0$, respectively. The width of the gap is ℓ and the voltage applied to the gap is $V_1 f(z)\sin(\omega t)$, where $f(z)$ is the field shape factor. The equation of motion for electrons in the gap is

$$\frac{d^2z}{dt^2} = -\frac{e}{m}E = \frac{e}{m}\frac{V_1}{\ell}f(z)\sin(\omega t). \tag{3.18}$$

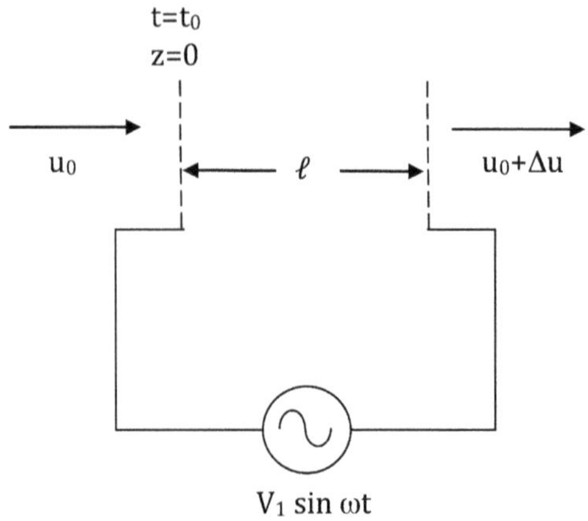

Figure 3.2. Sketch of a single gridded cavity gap, excited by a sinusoidal longitudinal field.

Now we can manipulate (3.18) by multiplying both sides by $2dz$, consequently obtaining an equation for the square of the velocity as follows:

$$2\frac{dz}{dt}\frac{d^2z}{dt^2}dt = \frac{d}{dt}\left(\frac{dz}{dt}\right)^2 dt = 2\frac{V_1}{\ell}\frac{e}{m}f(z)\sin(\omega t)dz. \tag{3.19}$$

Integrating this equation from $z = 0$ to the end of the gap and considering that the approximation of weakly perturbed motion ensures that $t \approx z/u_0$, and denoting the exit velocity by u and defining the beam propagation factor $\beta_e = \omega/u_0$, (3.19) leads to

$$\frac{1}{2}m(u^2 - u_0{}^2) = \frac{eV_1}{\ell}\int_0^\ell f(z)\sin(\beta_e z)dz = eMV_1. \tag{3.20}$$

Equation (3.20) expresses the change in kinetic energy of an electron after going through a narrow gap with a voltage V_1 across it. The modulation coefficient M modifies the potential V_1 yielding an 'effective voltage' $V_{\text{eff}} = MV_1 = \int_0^\ell E_z(z)\sin(\beta_e z)dz$ which acts on the electrons. The 'modulation coefficient', or coupling coefficient, is then found to be

$$M(\beta_e) = \frac{1}{V_1}\int_0^\ell E_z(z)\sin(\beta_e z)dz = \frac{\int_0^\ell E_z(z)\sin(\beta_e z)dz}{\int_0^\ell E_z(z)dz}. \tag{3.21}$$

Since the previous equation is the superposition of the electric field with an oscillating function, the coefficient M is itself an oscillating function. The result is that we have a periodic modulation of the electron energy along the z-direction, as can be seen by analyzing (3.20). A derivation of the coupling coefficient, for a gridded gap, can be obtained by solving (3.18); assuming a slowly varying potential through the gap, we can neglect the form factor $f(z)$, consequently (3.20) can be easily solved, thus obtaining

$$u(\ell) = u_0 - \frac{eV_1}{\omega m\ell}(\cos(\omega t_2) - \cos(\omega t_1)), \tag{3.22}$$

where t_1 and t_2 are the entering and exit gap times, respectively. Defining as t_0 the instant at which the electron is at the center gap ($t_1 = t_0 - \ell/(2u_0)$; $t_2 = t_0 + \ell/(2u_0)$) expression (3.22) can be rewritten as

$$u(\ell) = u_0 + \frac{2eV_1}{\omega m\ell}\sin\left(\frac{\omega\ell}{2u_0}\right)\sin(\omega t_0), \tag{3.23}$$

where u_0 is the electron velocity entering the gap that can be expressed, in terms of the DC voltage of the beam V_0, as $u_0 = \sqrt{2eV_0/m}$. A proper expression of the coupling coefficient, for this specific case, can be deduced from (3.23), and defined as

$$M = \frac{\sin(\omega\ell/2u_0)}{\omega\ell/2u_0}. \tag{3.24}$$

Equation (3.23) can be accordingly written in its final and most known form

$$u(\ell) = u_0 + \frac{eMV_1}{mu_0}\sin(\omega t_0) = u_0\left(1 + \frac{MV_1}{2V_0}\sin(\omega t_0)\right). \tag{3.25}$$

This relation shows, as is expected, that when an electron beam is passed through a modulating electric field, some electrons have their velocities increased, some have their velocities decreased, and some have their velocities unchanged. After the beam leaves the field, those electrons that have had their velocity increased gradually overtake the slower electrons. This process is illustrated in figure 3.3, which is known as an Applegate diagram. The axial positions of electrons are plotted as a function of time, and so the slopes of the lines represent electron velocities. The trajectories of electrons leaving the modulating field when it is decelerating (indicated at the

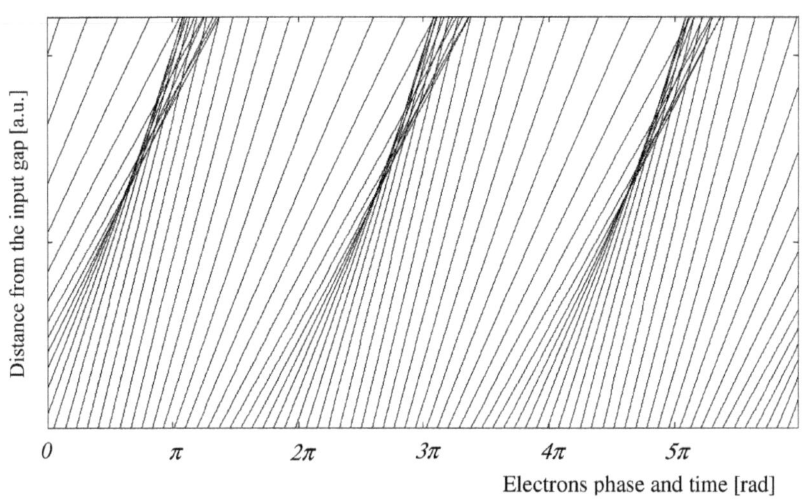

Figure 3.3. Applegate diagram; slope is related to the electron velocity after the modulating field. The difference in velocities (and therefore in slopes) generates regions where trajectories cross each other, resulting in a bunching of the electrons.

bottom of the figure), are represented by lines of reduced slope. Similarly, the trajectories of accelerated electrons are represented by lines of increased slope. In figure 3.3, space charge forces between electrons have been ignored, so the electron trajectories are shown to cross. In practice, at small and medium drive signal levels, space charge forces prevent electron trajectories from crossing; at large signal levels, electron trajectories do cross. As a result of the convergence of electron trajectories illustrated by the Applegate diagram, electron bunching is said to occur. The travelling electron bunches constitute an RF current in the electron beam as requested for a 'correct' energy exchange between electrons and electromagnetic radiation.

The coupling coefficient is an important parameter to be evaluated when studying and designing an electron-based generator of radiation or, more generally, in a project dealing with any device where a charged particle, or a distribution of them, interacts with an electromagnetic field. The coupling coefficient is at the basis of the dynamics of the particles and electromagnetic field in a specific cavity geometry, and it requires a deeper analysis before we proceed to the description of radiation sources.

3.5 General form of the electron-field coupling coefficient in cylindrical cavities

In order to determine the field associated with an electron beam passing through a cavity with a cylindrical geometry, we have to solve the Maxwell equations, including the current term. The pivotal equation is then the Helmholtz equation for the z-component of the vector potential $A_z(r, z)$

$$\nabla^2 A_z + k_0^2 A_z = -\mu_0 J \quad \text{with} \quad k_0 = \frac{\omega}{c}. \tag{3.26}$$

The electric field is related to the vector potential by the following expression, which states in the Lorentz gauge

$$E_z = -i\omega A_z - i\frac{c^2}{\omega}\nabla\nabla \cdot A_z = -i\frac{k_0 A_z + \partial^2 A_z/\partial z^2}{\omega/c^2} = -i\frac{c^2}{\omega}A_z\left(k_0^2 - k^2\right), \tag{3.27}$$

where we have assumed that the field variations along z follow $\exp(-ikz)$. The evaluation of the current J is more complicated and tedious due to the presence of space-charge modulations that must be taken into account for a general discussion.

Let us consider a DC current $J_0 = -\rho_0 v_0$ to which we superimpose an AC current modulation J denoted by a charge density ρ and a velocity v. We can, moreover, consider the AC modulation of the electron density expressed by the value n over a DC amplitude N. We can thus conclude that the electrons' motion is governed by the following equation:

$$(N + n)m\frac{d\vec{v}_T}{dt} = (\rho - \rho_0)[\vec{E} + \vec{v}_T \times \vec{B} + \vec{v}_T \times \vec{B}_0] \quad \text{where} \quad \vec{v}_T = \vec{v}_0 + \vec{v}. \tag{3.28}$$

The guiding magnetic field $\vec{B}_0$ is externally imposed is thus more intense with respect to the radiation associated magnetic field $\vec{B}$, which can be neglected in (3.28).

The total time derivative in (3.28) is expressed as

$$\frac{dv_T}{dt} = \frac{\partial v_T}{\partial t} + (v_T \cdot \nabla)v_T.$$

Due to the reasonable hypothesis on AC and DC relative amplitudes we can consider $v \ll v_0$ and v_0 to be almost constant. The second term on the right hand side of the previous equation becomes

$$[(v + v_0) \cdot \nabla](v + v_0) \approx (v + v_0) \cdot \nabla v \approx (v_0 \cdot \nabla)v.$$

The equations of motion (3.28) are therefore written as

$$(N + n)m\left[\frac{\partial \vec{v}}{\partial t} + (v_0 \cdot \nabla)\vec{v}\right] = (\rho - \rho_0)[\vec{E} + \vec{v} \times \vec{B}_0], \qquad (3.29)$$

in which the force contribution from v_0 does not play any role since, according to figure 3.4, the electron beam propagating along the z-axis, with negligible radial components, is such that $\vec{v}_0 \times \vec{B}_0 = 0$.

If we limit our analysis to the small signal conditions (namely for field intensities not significantly perturbing the electron motion), we can conclude that the terms containing ρ and n in (3.29) are multiplied by other AC quantities and can thus be neglected. The equation of motion finally reads

$$\frac{\partial \vec{v}}{\partial t} + (\vec{v}_0 \cdot \vec{\nabla})\vec{v} = -\eta(\vec{E} + \vec{v} \times \vec{B}_0) \quad \text{where} \quad e = \frac{\rho_0}{N} \quad \text{and} \quad \eta = \frac{e}{m}. \qquad (3.30)$$

Assuming a cylindrical geometry, and a long interaction region, the result is that the magnetic field B_0 extends to infinity and the transverse components of v vanish. Considering a time dependence of $\exp(i\omega t)$, (3.30) becomes

$$i\omega v + v_0\frac{\partial v}{\partial z} = -\eta E_z \Rightarrow (i\omega - ikv_0)v = \eta E_z. \qquad (3.31)$$

Other relations involving the current density J are the 'continuity equation'

$$\vec{\nabla} \cdot \vec{J} = -i\omega\rho \Rightarrow \frac{\partial J}{\partial z} = -i\omega\rho \Rightarrow ikJ = i\omega\rho, \qquad (3.32)$$

and the equation that defines the total current

$$J_0 + J = (-\rho_0 + \rho)(v_0 + v) \approx -\rho_0 v_0 + (\rho v_0 - \rho_0 v),$$

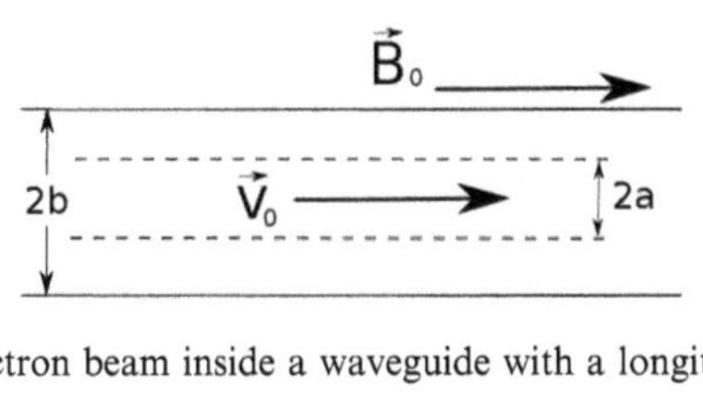

Figure 3.4. Geometry of an electron beam inside a waveguide with a longitudinal confining magnetic field.

having neglected the second order terms in ρv. At the end we obtain for the current density AC and DC components

$$\Rightarrow \begin{cases} J_0 = -\rho_0 v_0 \\ J = \rho v_0 - \rho_0 v \end{cases}. \tag{3.33}$$

Combining (3.32) and (3.33) to eliminate ρ we obtain

$$v = \frac{k v_0/\omega - 1}{\rho_0} J. \tag{3.34}$$

Introducing, for convenience, the term $\beta_0 = \omega/v_0$, which indicates the DC electron-beam momentum, and the plasma frequency defined as

$$\omega_p^2 = \frac{\rho_0 \eta}{\varepsilon_0}, \tag{3.35}$$

we can combine (3.34) and (3.31) to obtain an expression for the current density modulation J in terms of the associated electric field E_z,

$$J = -i\frac{\omega_p^2}{\omega}\frac{\beta_0 \varepsilon_0 E_z}{(\beta_0 - k)^2}. \tag{3.36}$$

Substituting (3.27) in (3.36), and remembering that $c^2 = 1/\mu_0\varepsilon_0$, we obtain

$$\mu_0 J = -\left(\frac{\omega_p}{\omega}\right)^2 \beta_0 \frac{k_0^2 - k^2}{(\beta_0 - k)^2} A_z. \tag{3.37}$$

Inserting the current density expression (3.37) into the Helmholtz equation (3.26), we obtain a set of equations for A_z

$$\begin{cases} \nabla_t^2 A_z + p^2 A_z = 0 & 0 \leqslant r \leqslant a \\ \nabla_t^2 A_z - h^2 A_z = 0 & a \leqslant r \leqslant b \end{cases} \quad \left(\nabla^2 \to \nabla_t^2 - k^2\right), \tag{3.38}$$

where

$$\begin{cases} p^2 = -k^2 + k_0^2 + \left(\frac{\omega_p}{\omega}\right)^2 \beta_0{}^2 \frac{k^2 - k_0^2}{(\beta_0 - k)^2} \\ \quad = -\left(k^2 - k_0^2\right)\left[1 - \left(\frac{\omega_p}{\omega}\right)^2 \left(\frac{\beta_0}{\beta_0 - k}\right)^2\right] \\ h^2 = k^2 - k_0^2 \end{cases}. \tag{3.39}$$

The equations of (3.38), once expressed in cylindrical coordinates, become Bessel equations of order one. Their solutions are Bessel functions with real or imaginary arguments p and h depending on the sign of the expression $(k^2 - k_0^2)$. The evolution of the space charge current, as described by the equation of motion (3.29), is such that it behaves like a slow-wave and, consequently, we can assume $k \approx \beta_0 \gg k_0$.

This condition will be very useful when calculating the power exchanged between the electron beam and radiation, because it ensures a better synchronism.

The slow-wave condition guarantees that the generalized momenta (3.39) will be real and, therefore, the Bessel solutions of (3.38) result as follows:

$$\begin{cases} A_z \approx J_0(pr) & 0 \leqslant r \leqslant a \\ A_z \approx I_0(hr) & a \leqslant r \leqslant b \end{cases}, \tag{3.40}$$

where $I_0(hr)$ is the modified Bessel function used instead of the conventional Bessel function $J_0(ihr)$ with an imaginary argument. A similar behavior can be obtained in the case of an unfocused beam having, again, a space-charge wave dependence $\exp(-ikz)$ along the propagation direction z.

From (3.33) the AC current density module J can be expressed, in terms of radial J_r and longitudinal J_z components, as

$$\begin{cases} J_r = -\rho_0 v_r = \sigma E_r \\ J_z = -\rho_0 v_z + \rho v_0 = \sigma E_z + \rho v_0 \end{cases}, \tag{3.41}$$

where the effective conductivity σ can be derived from (3.31) and can be thus expressed as

$$\sigma = -i\frac{\eta\rho_0}{\omega - kv_0} = -i\frac{\varepsilon_0\omega_p^2}{\omega - kv_0}, \tag{3.42}$$

having made use of (3.35). The continuity equation (3.32) must now be used in vector form, because of the presence of radial and longitudinal components of the electric field. In cylindrical coordinates it can be expressed as

$$\sigma\left(\frac{1}{r}\frac{\partial}{\partial r}rE_r - ikE_z\right) - ikv_0\rho = -i\omega\rho, \tag{3.43}$$

having made use of (3.41). The term inside the parentheses is the divergence of the electric field which, due to the Gauss theorem, can be expressed as $\vec{\nabla} \cdot \vec{E} = \frac{\rho}{\varepsilon_0}$. Therefore, (3.43) becomes

$$\left(\frac{\sigma}{\varepsilon_0} - ikv_0 + i\omega\right)\rho = 0. \tag{3.44}$$

In order to solve general problems, without limiting the values of the momentum k while satisfying different boundary conditions, (3.44) can be fulfilled only for $\rho = 0$. In conclusion our electron beam must have a zero AC space-charge density.

The Helmholtz equation for the electric field can be derived from the curl Maxwell's equations considering a current density expressed by (3.41)

$$\vec{J} = i\omega\varepsilon_0\frac{\sigma}{i\omega\varepsilon_0}\vec{E}. \tag{3.45}$$

The curl equations are

$$\begin{cases} \nabla \times \vec{E} = -i\omega\eta_0\vec{H} \\ \nabla \times \vec{H} = i\omega\varepsilon_0\vec{E} + \vec{J} = i\omega\varepsilon_0\vec{E}\left(1 + \dfrac{\sigma}{i\omega\varepsilon_0}\right) = i\omega\varepsilon\vec{E} \end{cases}, \tag{3.46}$$

where we have introduced the effective permittivity ε as

$$\varepsilon = \varepsilon_0\left(1 + \frac{\sigma}{i\omega\varepsilon_0}\right) = \varepsilon_0\left(1 - \frac{\omega_p^2}{\omega(\omega - kv_0)}\right), \tag{3.47}$$

having made use of (3.42) for the effective conductivity.

Combining the two equations of (3.46) we end up with the equation for the longitudinal electric field component E_z

$$\nabla^2{}_t E_z + \omega^2\eta_0\varepsilon E_z - k^2 E_z = 0,$$

and therefore

$$\begin{cases} \nabla^2{}_t E_z - p^2 E_z = 0 \\ p^2 = k^2 - \eta_0\varepsilon_0\omega^2\left[1 - \dfrac{\omega_p^2}{\omega(\omega - kv_0)}\right] \end{cases}. \tag{3.48}$$

The generalized momentum p can be cast in the form

$$p^2 = k^2 - k_0^2\left[1 - \left(\frac{\omega_p}{\omega}\right)^2 \frac{\beta_0{}^2}{(\beta_0 - k)^2}\right], \tag{3.49}$$

having made use of $\beta_0 = \omega / v_0$.

As previously discussed, also in this case, the radiation behaves as a slow-wave and thus $k \approx \beta_0 \gg k_0$. Under this hypothesis the momentum p is real and the solutions for the electric field E_z, as evaluated from (3.48), are

$$\begin{cases} E_z \approx I_0(pr) \quad 0 \leqslant r \leqslant a \\ E_z \approx K_0(hr) \; r \geqslant a \qquad \left(h^2 = k^2 - k_0^2 \text{ outside the beam radius}\right) \end{cases}, \tag{3.50}$$

where K_0 is a modified Bessel function of the second kind.

We have now all the elements for a general evaluation of the coupling coefficient for an electron beam in a radiofrequency (RF) cavity. As reported in section 3.4 (see (3.21)) the coupling coefficient as a function of the electron-beam momentum $\beta_0 = \frac{\omega}{v_0}$ is

$$M(\beta_0) = \frac{\int_{-\infty}^{\infty} E_z e^{i\beta_0 z} dz}{\int_{-\infty}^{\infty} E_z dz} = \frac{V_{\text{eff}}}{E_0 V_1}, \tag{3.51}$$

where V_{eff} expresses the effective voltage through the gap.

The electric field inside the interaction gap cavity, as expressed by (3.18), contains a longitudinal behavior $f(z)$ that is a continuous function. The axial component E_z is completely specified at the beam limit surface $r = a$ and, due to the continuity of $f(z)$, it must be expressed as a summation over the momentum contributions k as

$$E_z(z, t) = E_0 f(z)e^{i\omega t} = E_0 e^{i\omega t}\int_{-\infty}^{\infty} g(k)e^{-ikz}dk. \tag{3.52}$$

Due to the generic value of the radial position r of the electrons in the cylindrical cavity, we must take into account the solutions found in (3.40) and (3.50); in the case of an unfocussed beam a general formulation of (3.52) is

$$E_z(r, z, t) = E_0 e^{i\omega t}\int_{-\infty}^{\infty} \frac{I_0(pr)}{I_0(pa)} g(k)e^{-ikz}dk, \tag{3.53}$$

where $g(k)$ expresses the Fourier transform of the longitudinal electric field profile $f(z)$

$$g(k) = \frac{1}{2\pi}\int_{-\infty}^{\infty} f(z)e^{ikz}dz. \tag{3.54}$$

Let us now consider the electrons interacting with the field. For the determination of J, as expressed in (3.36) and (3.45), we have considered a DC current with velocity v_0 and a momentum $\beta_0 = \frac{\omega}{v_0}$. In so doing we can write for the phase

$$\omega = \beta_0 v_0 = \beta_0 \frac{z}{t} \Rightarrow \omega t = \beta_0 z.$$

Accordingly, for (3.53) the electric field at the position r of the current is

$$E_z(r, z) = E_0\int_{-\infty}^{\infty} \frac{I_0(pr)}{I_0(pa)} g(k)e^{i(\beta_0 - k)z}dk. \tag{3.55}$$

By integrating (3.55) over z we obtain the effective potential

$$V_{\text{eff}}(r) = \int_{-\infty}^{\infty} E_z(r, z)dz = E_0\int_{-\infty}^{\infty}\int_{-\infty}^{\infty} \frac{I_0(pr)}{I_0(pa)} g(k)e^{i(\beta_0 - k)z}dkdz. \tag{3.56}$$

The use of the integral representation of the Dirac delta distribution function

$$\delta(\beta_0 - k) \equiv \frac{1}{2\pi}\int_{-\infty}^{\infty} e^{i(\beta_0 - k)z}dz$$

yields (3.56) in the form

$$V_{\text{eff}}(r) = 2\pi E_0 g(\beta_0)\frac{I_0(p_0 r)}{I_0(p_0 a)}. \tag{3.57}$$

In (3.57) the momentum p_0 is expressed by (3.49), or equivalently by (3.39), where $k = \beta_0$. As can be seen, the diverging term in p for $k \Rightarrow \beta_0$ is kept under control if the field frequency $\omega \gg \omega_p$, which is a typical condition in microwave tubes.

Table 3.1. Sketch of the field configuration inside the resonant cavity gap, together with the related coupling coefficient term.

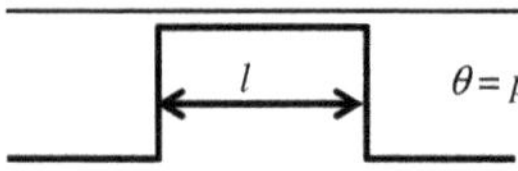 $\theta = pl/2$	$\mu_d = \dfrac{\sin\theta}{\theta}$
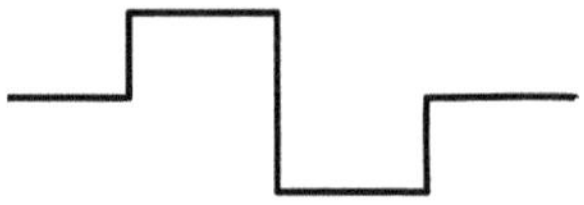	$\mu_d = \dfrac{1 - \cos\theta}{\theta}$
	$\mu_d = \dfrac{2\sin\theta - 4\sin(\theta/2)}{\theta^2}$
	$\mu_d = \dfrac{\pi\sin\theta}{\pi^2 - \theta^2}$
	$\mu_d = \dfrac{(\pi/2)\cos\theta}{(\pi/2)^2 - \theta^2}$
	$\mu_d = \dfrac{\theta\sin\theta}{\pi^2 - \theta^2}$
	$\mu_d = \dfrac{\theta\cos\theta}{(\pi/2)^2 - \theta^2}$

Finally the general expression for the coupling coefficient (3.51) becomes

$$M(\beta_0)\frac{2\pi}{V_1}g(\beta_0)\frac{I_0(pr)}{I_0(pa)} = \eta_r\eta_d\frac{l}{V_1}, \tag{3.58}$$

where l is the RF cavity gap length and

$$\begin{cases} \eta_r = \dfrac{I_0(pr)}{I_0(pa)} \\ \eta_d = \dfrac{2\pi}{l}g(\beta_0) \end{cases}. \tag{3.59}$$

The μ_d parameter is a function of the field distribution $f(z)$ in the gap and, hence, varies from cavity type to cavity type. In table 3.1, various field configurations, together with the μ_d parameters, are listed.

3.6 The klystron

In this section we reconsider the klystron concept introduced in previous chapters. For reasons of convenience and continuity, we emphasize the already discussed issues and invite the reader to reconsider them from the point of view developed so far in this chapter.

When the gridded gap is coupled to a resonant cavity we can excite a specific frequency, in the microwave spectral region, due to the energy exchange mechanism. The central emitted frequency has a bandwidth related to the quality factor of the resonant cavity.

The simplest device exploiting the velocity modulation in a gridded gap is the klystron. It uses two coupled cavities as illustrated in the schematic layout of figure 3.5. The electron beam is generated by an electron gun and passes through the grids of the first cavity: the input cavity. In this case an RF signal is coupled into the cavity by means of a coaxial cable; the voltage developed across the cavity induces the velocity modulation to the beam as described by (3.23)–(3.25) in section 3.4. After the first cavity we have a drift space where the velocity modulation is transformed into a current modulation or bunching. Since there are no accelerating fields in the drift space between the two cavities, each electron moves at a constant velocity $u(\ell)$ given by (3.25) for its particular value of t_0 (the time at which the electron transits at the center of the first gap).

Assuming a separation d between the centers of the two cavity gaps, the arrival time t of a particular electron (t_0 is the time it exits the first cavity) at the output cavity end is given by $t - t_0 = d/u(\ell)$. Multiplying this relation by ω, to discuss in terms of radians, and making use of (3.25) to express the electron velocity, we can obtain an expression for the arrival time difference

$$\omega t - \omega t_0 = \frac{\omega d}{u_0\left(1 + \frac{MV_1}{2V_0}\sin(\omega t_0)\right)} = \theta\frac{\left(1 - \frac{M}{2}\frac{V_1}{V_0}\sin(\omega t_0)\right)}{1 + \left(\frac{M}{2}\frac{V_1}{V_0}\sin(\omega t_0)\right)^2}$$
$$\approx \theta\left(1 - \frac{M}{2}\frac{V_1}{V_0}\sin(\omega t_0)\right) = \theta - X\sin(\omega t_0). \tag{3.60}$$

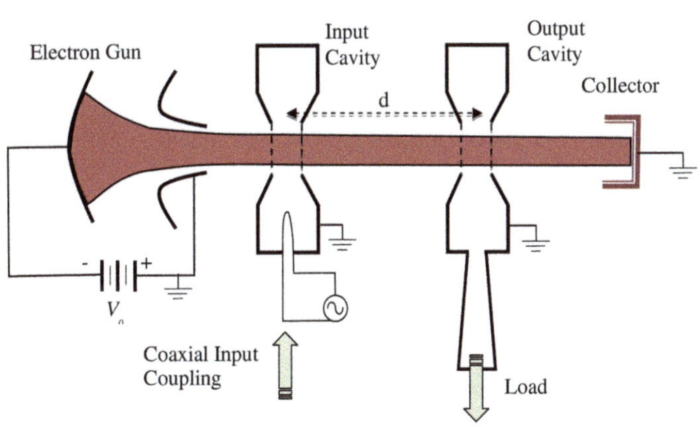

Figure 3.5. Layout of a two-cavity klystron amplifier.

Equation (3.60) has been deduced within the approximation that the input cavity voltage is much less than the electron-beam voltage ($V_1/V_0 \ll 1$) that is true in most cases, except for very high-power devices. Moreover, we have introduced the following important parameters: $\theta = \omega d/u_0$ is the DC transit angle between the cavities and $X = (M/2)(V_1/V_0)\theta$ is the bunching parameter introduced in chapter 1.

In figure 3.6 we show the curves related to the arrival time ωt as a function of the input cavity departure time ωt_0, calculated from (3.60). They refer to different values of X. It is worth noting that, for values greater than one, the departure time is a multivalued function of the arrival time; this occurs for electrons close to the reference electron at the bunch center ($\omega t_0 \approx 0$). Under these conditions electrons may overtake each other. In contrast, the arrival time is always a monotonic function of the departure time.

The role of the bunching parameter X is crucial when solving the equations for the current $i(t) = dq/dt$, where we have indicated the amount of charge dq leaving the buncher and arriving at the second cavity in the time interval dt.

The electric charge amount dq is proportional to the charge imparted in the interval dt_0, as can be deduced from figure 3.6, at least for bunching parameter values $X < 1$. The result is that we can write this proportionality as follows: $dq = -I_0 dt_0$ (the minus refers to the fact that the electrons, which we consider, have a negative charge and I_0 is a positive number). The result is that the electron's current–time behavior is given by

$$i(t) = -I_0 \frac{dt_0}{dt}. \tag{3.61}$$

The derivative expressed in (3.61) can be directly calculated by solving equation (3.60) and is related to the slope of the function $t(t_0)$ plotted in figure 3.6. High

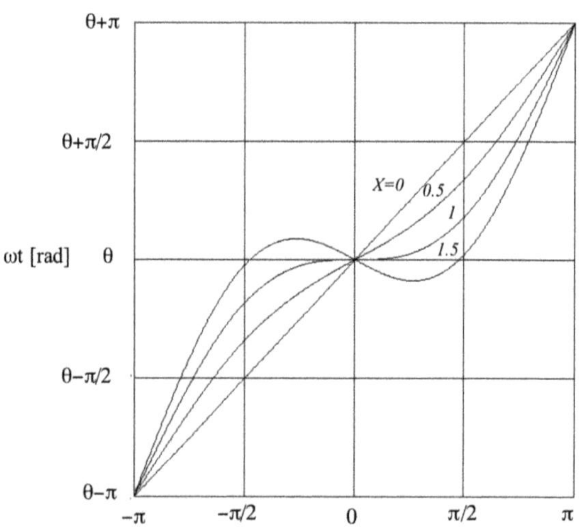

Figure 3.6. Output cavity arrival time ωt as a function of the departure time ωt_0 from the input cavity for various values of the bunching parameter X.

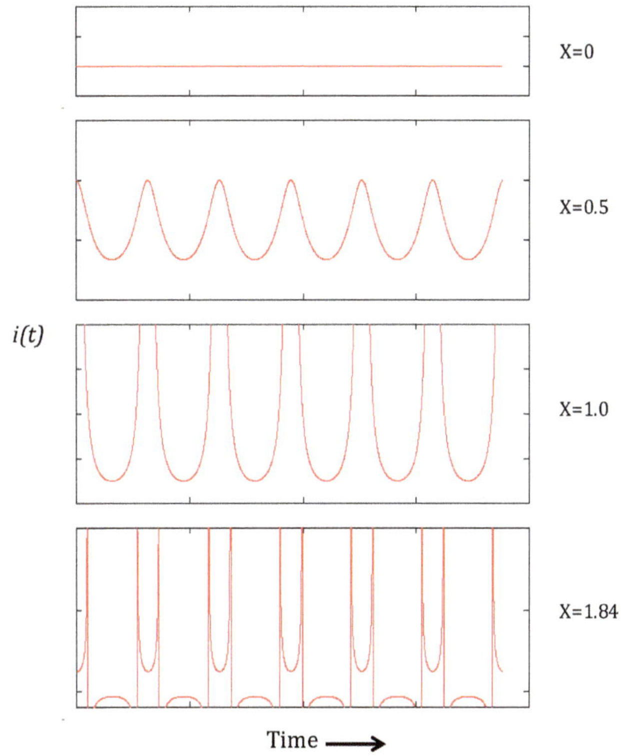

Figure 3.7. Plot of the beam current time shape at the output cavity. For bunching parameter X greater than or equal to unity, infinite peaks occur at the points corresponding to zero slope in figure 3.6.

current peak values occur at any arrival time t for which the curve of (3.60) has a zero-slope. The intensity of the peak is directly related to the value of the parameter X when it ranges between $X = 0$ and $X = 1$ (see figure 3.7). The infinite value, corresponding to $X = 1$, indicates that the charge that left the buncher at a time t_0, arrives at the second cavity in a zero time interval ($dt/dt_0 = 0$).

For values of the bunching parameter exceeding unity we have a multiple contribution for the slope of (3.61) because, as already discussed, (3.60) is a multivalued function and there are three values of slope for a given arrival time. A general form for the expression (3.61) is thus

$$i(t) = -I_0\left[\left|\frac{dt_0}{dt}\right|_1 + \left|\frac{dt_0}{dt}\right|_2 + \left|\frac{dt_0}{dt}\right|_3\right]. \tag{3.62}$$

The absolute values indicate that the contribution to the total current is independent of the sign of the slopes in figure 3.6. The physical meaning is that any charge increment dq has the same sign regardless of the order of the electrons' arrival time.

Due to the periodic nature of the current, as a function of time, we are allowed to expand it in terms of Fourier components (here we repeat what is derived in section (1.3), equations (1.39)–(1.47))

$$i(t) = -I_0 + \sum_{n=1}^{\infty}[a_n \cos(n(\omega t - \theta)) + b_n \sin(n(\omega t - \theta))],$$

where

$$a_n = \frac{1}{\pi}\int_{\theta-\pi}^{\theta+\pi} i(t)\cos n(\omega t - \theta)d(\omega t) \tag{3.63}$$

and

$$b_n = \frac{1}{\pi}\int_{\theta-\pi}^{\theta+\pi} i(t)\sin n(\omega t - \theta)d(\omega t),$$

where, again, θ is the DC transit angle between the two cavities. Making use of (3.61), and referring to figure 3.6, the integration limits of the new variable t_0, are $-\pi$ to π. Moreover, substituting (3.60) into (3.63), we obtain for the Fourier coefficients

$$\begin{aligned} a_n &= -\frac{I_0}{\pi}\int_{-\pi}^{\pi} \cos n[\omega t_0 - X\sin(\omega t_0)]d(\omega t_0) \\ b_n &= -\frac{I_0}{\pi}\int_{-\pi}^{\pi} \sin n[\omega t_0 - X\sin(\omega t_0)]d(\omega t_0) \end{aligned}. \tag{3.64}$$

It is worth noting that the coefficients b_n in (3.64) are vanishing since the integrand is an odd function of ωt_0. The expression of a_n in (3.64) is a known integral form of the Bessel function of the first kind and nth order, which can indeed be written as $a_n = 2I_0J_n(nX)$. The current expressed in (3.63) thus becomes

$$i(t) = -I_0 - 2I_0\sum_{n=1}^{\infty}J_n(nX)\cos n(\omega t - \theta). \tag{3.65}$$

Equation (3.65) shows that the various harmonics in the bunched beam have amplitudes proportional to the Bessel function J_n of order n, where n is the harmonic number. The Bessel functions behaviors are reported in figure 3.8. The abscissa is proportional to the DC transit angle θ, and therefore to the transit time between the bunching and emitting cavities. This time can be adjusted, changing the cavity distance, in order to maximize the current amplitude for any of the harmonics. For an amplifier, where the output cavity amplifies the same frequency of the buncher, we can limit our analysis to the first harmonics ($n = 1$); in this case the maximum for the bunched current is obtained at $X = 1.84$. In other words, with such a value of the bunching parameter we obtain a peak for the fundamental component at the output emitting cavity.

A deeper investigation of figure 3.8 tells us that the klystron could be used as a harmonic generator simply adjusting the cavity separation (proportional to θ and therefore to X) in order to have a transit time corresponding to the peak of one of the higher order Bessel functions. Moreover, due to the fact that all these peaks are quite

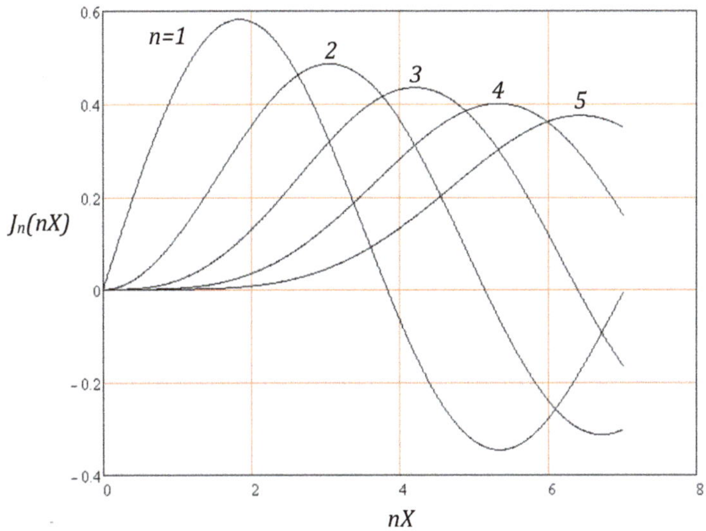

Figure 3.8. The Bessel functions of the first kind of various orders n.

as large as the fundamental one, and well separated, the klystron can be an efficient harmonic generator.

We have so far demonstrated how, in a klystron, the first cavity and a proper drift space generate a bunched current as reported in (3.65). The system is then coupled to a second resonant cavity, which is excited by the electron current (3.65), which induces another current on the metallic walls of this second cavity. This output cavity is connected to a load by means of a waveguide. In the output cavity, which can be called the 'emitter', the amplification and power extraction phenomena occur. To understand the mechanism we can go back over the analysis of the first gridded gap and resonant cavity as reported in sections 3.4 and 3.5. In fact, the induced current is also modulated in the output cavity, by the cavity itself, and by means of a modulation coefficient M_2 having the same form of (3.21) as shown in table 3.1. This coefficient, associated with the bunched current (3.65), gives an induced current that is M_2 times the fundamental component of the beam current at the gap center, both in magnitude and in phase

$$I(t) = -2I_0M_2\sum_{n=1}^{\infty}J_n(nX)\cos\left[n\omega\left(t - \frac{d}{u_0}\right)\right]. \tag{3.66}$$

The important result that can be deduced from this analysis is that the induced current $I(t)$ is independent of the loading of the output cavity. A second important result is that the bunched beam acts, in fact, as a current source. At resonance the phase of the oscillations in the output cavity is such that the maximum decelerating voltage appears across the cavity when the bunch of electrons is crossing the gap. This happens because the output cavity, and the load, act as a resistance connected between the grids, so that when the induced current reaches a maximum, the voltage across the cavity (the gap) is also a maximum. This explains the transfer of energy

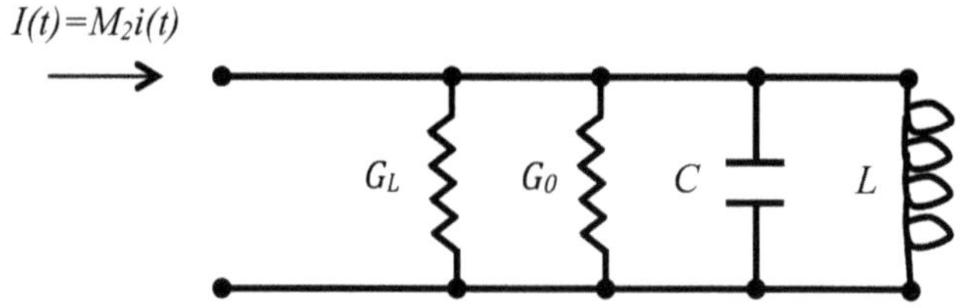

Figure 3.9. Equivalent circuit for the output cavity and its load.

from the electron beam to the cavity. The power extracted can be estimated by means of the product of the current (3.66) and the voltage across the output cavity. To be more explicit we can refer to a schematic equivalent circuit of the output cavity as reported in figure 3.9.

Figure 3.9 tells us that the output cavity can be treated as an *LCG* circuit where the conductance G_0 represents the cavity losses, and the inductance L and the capacitance C are related to the resonant frequency of the cavity by means of the relation $\omega_0 = 1/\sqrt{LC}$. When an external load is connected to the cavity, its conductance G_L is represented in shunt with G_0. The induced current, expressed by (3.66), flows into the circuit through all its components and, therefore, delivers a power to the external load that can be expressed by the known circuit formulas

$$P_0 = \frac{1}{2}\frac{|I(t)|^2 G_L}{(G_0 + G_L)^2}. \tag{3.67}$$

A more rigorous analysis, which goes beyond the purpose of this book, should include the beam-loading in the equivalent circuit in terms of beam conductance, G_b, to be added to G_0 and G_L.

3.7 The travelling wave amplifier

A simple improvement of the two-cavity klystron amplifier is a multi-cavity geometry klystron, which ensures a higher gain and efficiency in the microwave region. The simplest improved model is a three-cavity klystron, such as the one illustrated in figure 3.10(a). The working principle is quite simple: in the first cavity the electron beam undergoes a velocity modulation induced by the electric field associated with the input signal. After a drift space the velocity modulation is transformed into a space modulation in such a way that the bunched beam enters into the intermediate cavity inducing a current on its metallic walls. This induced current produces a voltage, on the second cavity gap, that is much higher with respect to that of the input cavity. This voltage impresses a further velocity modulation on the beam that generates a further current modulation and bunching for the output, and final, cavity. The induced current in this cavity causes the RF power to be delivered to the load.

The higher number of cavities of this improved source reduces the overall spectral bandwidth of the device, but there is a direct linkage among the bandwidth, the quality factor parameter Q and the gain in klystrons. The Q-factor is introduced to evaluate the aptitude of a resonant cavity, or to a number of them, to store energy. It is

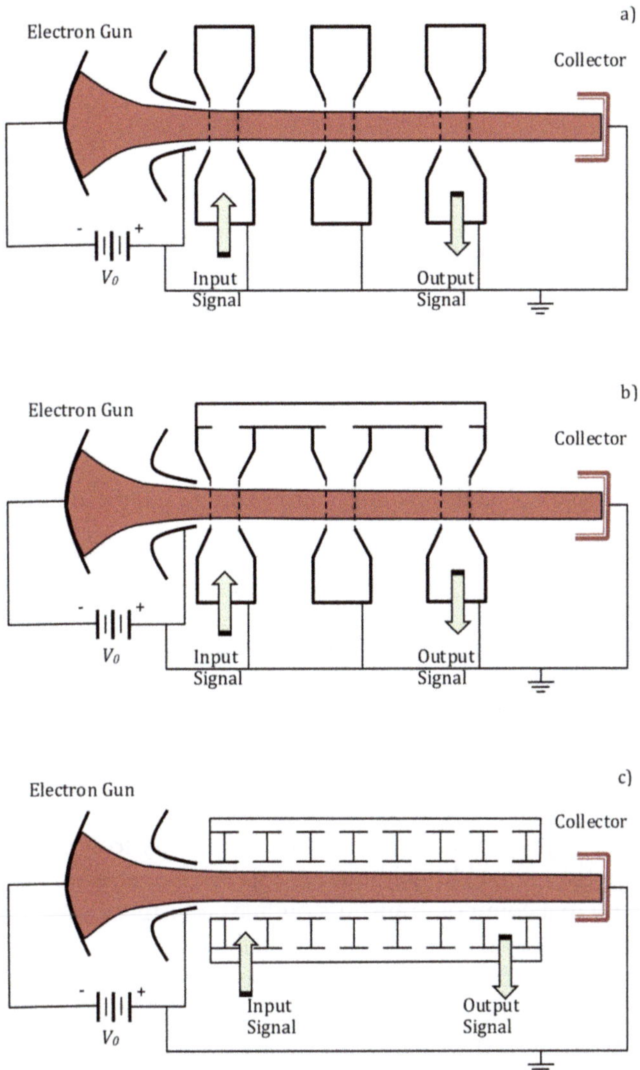

Figure 3.10. Advancement from a multi-cavity klystron to a TWT. (a) Multi-cavity klystron. (b) Multi-cavity klystron with cavities coupled by a transmission line. (c) Travelling wave tube amplifier.

conventionally defined as $Q = 2\pi$(energy stored at resonance)/(energy lost per cycle). Let us consider, as an equivalent circuit, a simplified layout of a resonant cavity, such as the one sketched in figure 3.11. Since the energy that is stored in the capacitance C, that is charged at the peak voltage V_{MAX}, is written as $1/2CV_{MAX}{}^2$, and considering that the frequency at resonance ($\omega_0 = 2\pi f_0$) occurs when $\omega C = 1/(\omega L)$, we can write

$$Q = 2\pi \frac{1/2CV^2{}_{MAX}}{\frac{V_{MAX}{}^2}{2R}\frac{1}{f_0}} = \omega_0 CR = \frac{R}{\omega_0 L}. \qquad (3.68)$$

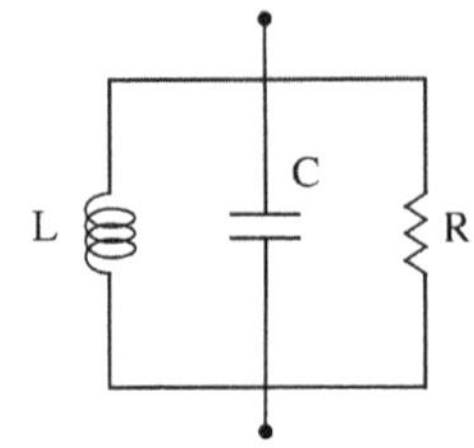

Figure 3.11. Resonant parallel equivalent circuit.

It is known that if an alternate current excites the circuit of figure 3.11, the voltage across it falls at $1/\sqrt{2}$ of its maximum if the frequency satisfies the following relation

$$\left|\omega C - \frac{1}{\omega L}\right| = \frac{1}{R}. \tag{3.69}$$

Multiplying both sides of (3.69) by R/Q derived by (3.68) we obtain

$$\left|\frac{\omega}{\omega_0} - \frac{\omega_0}{\omega}\right| = \frac{1}{Q} = \left|\frac{(\omega - \omega_0)(\omega + \omega_0)}{\omega\omega_0}\right| \approx \frac{2\Delta f}{f_0}.$$

This equation plainly states an inverse proportionality between the quality Factor Q and the bandwidth. However, for high-power sources, it is usually required, by definition, to have high values of Q for the resonant cavities. The result is that the bandwidth of these devices ends up being quite small. Moreover, the gain, defined as the ratio of the input voltage with respect to that developed across the load resistance, is clearly proportional to the load resistance itself, and thus, as results from (3.68), it is also proportional to the Q-factor. A way to overcome the problem and thus increase the bandwidth without an appreciable reduction of the gain, is to couple each cavity to a common transmission line, adjusting either the line length or the beam voltage in a way that the transmission line current arrives at the cavities in phase with the current induced by the electron beam in the resonant cavities.

In figure 3.10(a)–(b) we show the difference between the layouts of a multi-cavity klystron and a klystron whose cavities are coupled by transmission lines. A natural step further is to introduce a larger number of cavities which should be set at a distance smaller than a quarter plasma wavelength (related to the frequency at which the mutual repulsion forces cause the electrons to oscillate, and expressed by (3.35)). This condition corresponds to having an electron beam that becomes only partially bunched between successive cavities. The result is that the gain per cavity is reduced with respect to the klystron while, due to the increased number of cavities, the gain per unit length is increased. Moreover the transmission line loading has the effect of lowering the values of the Q parameter of the cavities, increasing the bandwidth, without a corresponding power loss. In figure 3.10(c) a sketch of a multi-cavity device with a transmission line connecting them is reported; this device is a travelling wave tube (TWT) amplifier.

The TWT is an electromagnetic source based on a periodic structure and it falls into the wider category of the slow-wave structures. These structures present an

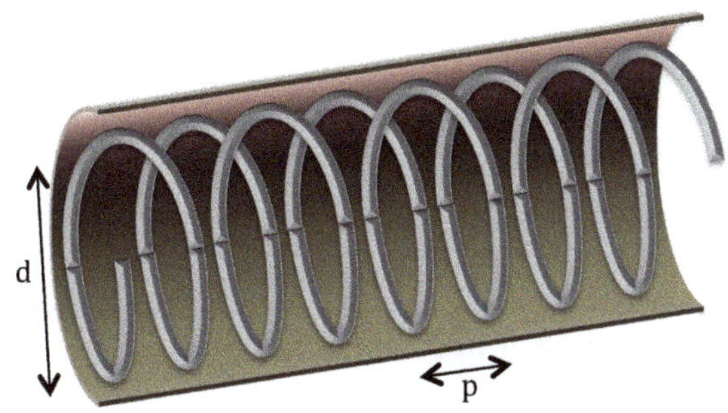

Figure 3.12. Sketch of a helical slow-wave structure.

important feature: they sustain waves with phase velocities smaller than the velocity of light. A heuristic way to understand this phenomenon is the analysis of the helix slow-wave structure reported in figure 3.12. The helical wire and the metallic waveguide form a double conductor transmission line with diameter d and period (pitch) indicated with p. The electromagnetic field moves along the wire with a speed close to the light velocity in vacuum c; but the resultant axial phase velocity is much smaller than this value. An approximate value can be obtained by geometrical considerations from figure 3.12, giving

$$v_{ph} = c\frac{p}{\sqrt{p^2 + (\pi d)^2}}.$$

A first evident advantage of the slow-wave structures is that an electron beam can travel along the cylindrical axis with a velocity that can be locked with that of the electromagnetic wave.

Due to the above analysis we can conclude that all the periodic guiding structures are slow-wave structures; but there are other categories of slow-wave devices based on the combination of dielectric and metallic materials. We will come to this point in the following sections of this chapter, when describing some FEL devices.

The solutions of electromagnetic waves in periodic structures involve a more general theorem of the differential equations branch of mathematics, Floquet's theorem. We cannot give a complete and rigorous proof of the theorem, but again a heuristic derivation will be sufficient for the purposes of this book. The basic assumption, which can be easily accepted, is that the electric field solution (or equivalently the magnetic field) in a periodic structure, with period L, assumes the following form

$$\vec{E}(x, y, z) = \vec{E}_p(x, y, z)e^{-i\beta_0 z}, \tag{3.70}$$

where $\vec{E}_p(x, y, z)$ is a periodic function of z with period L. If we replace z with $(z - L)$ in (3.70) we obtain

$$\vec{E}(x, y, z - L) = \vec{E}_p(x, y, z - L)e^{-i\beta_0(z-L)}. \tag{3.71}$$

Due to the periodic nature of $\vec{E}_p(x, y, z)$, we obtain for (3.71)

$$\vec{E}(x, y, z - L) = \vec{E}_p(x, y, z)e^{-i\beta_0 z}e^{i\beta_0 L}. \tag{3.72}$$

Making use of (3.70) we can obtain the Floquet's theorem expression

$$\boxed{\vec{E}(x, y, z - L) = \vec{E}(x, y, z)\exp(i\beta_0 L)}. \tag{3.73}$$

Equation (3.73) states that the steady state solution of an electromagnetic field in a periodic structure has the property that the fields in adjacent cells differ only by a multiplicative phase constant. Because of the periodic nature of the electromagnetic field, we are allowed to express it in terms of Fourier components as

$$\vec{E}(x, y, z) = \sum_{-\infty}^{\infty} \vec{E}_n(x, y)\exp(-i\beta_n z), \tag{3.74}$$

where $\beta_n = \beta_0 + 2\pi n/L$ is the wave momentum at the harmonic n. Equation (3.74) states that the field is composed of a series of space harmonics and that it is physically impossible to have a single harmonic propagating in a periodic structure. Each space harmonic propagates along z with a different phase velocity given by

$$v_{ph} = \frac{\omega}{\beta_n}. \tag{3.75}$$

Equation (3.75) implies another important result, i.e. it is not possible for an 'observer', moving with the electron beam, to always be in the same phase of the total field (3.74). An 'observer' can, in principle, be synchronized with only one of the space harmonics that constitutes the field. The other harmonics will have continuously changing phases as seen by the 'observer'. This means that the observer, after a proper time average of the fields that he sees, will obtain a net contribution only from the synchronous harmonic, the other harmonics being contributions compensated and therefore negligible.

A useful tool to better understand the periodic structures and devices is the Brillouin diagram relevant to the dispersion relation ($\omega - \beta$) curve, obtained by solving the homogeneous Helmholtz equation (3.26) under Floquet's hypothesis (3.73) and the Fourier expansion (3.74). The curve behavior strongly depends on the geometry of the device, but some general elements are common to all of them and it is useful to analyze them in detail.

Without entering too much into the calculation details, we can discuss a TWT device with the geometry reported in figure 3.13. It consists of a parallel plane waveguide with a periodic corrugation given by thick teeth.

The dispersion relation for this geometry is

$$\frac{\cot(k_0 h)}{k_0 h} = \frac{s}{L}\sum_{n=-\infty}^{\infty}\left[\frac{\sin\left(\frac{\beta_n s}{2}\right)}{\left(\frac{\beta_n s}{2}\right)}\right]^2 \frac{\cot h(\gamma_n d)}{\gamma_n h}, \tag{3.76}$$

where k_0 is the radiation momentum in the vacuum as expressed by (3.26) (i.e. the radiation frequency normalized to the light velocity), β_n is the nth space harmonic

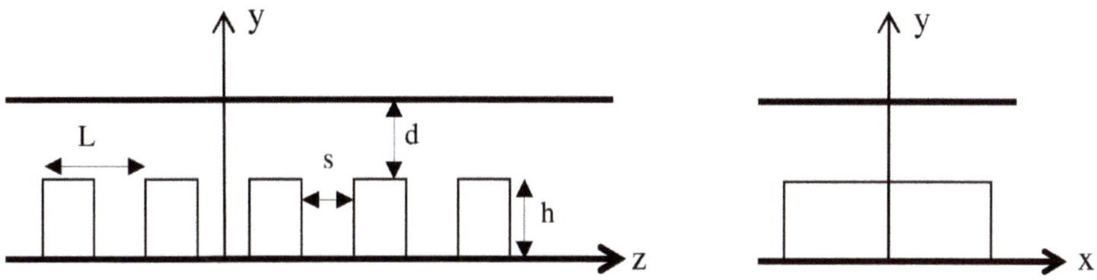

Figure 3.13. Corrugated periodic slow-wave structure inside a parallel plane waveguide.

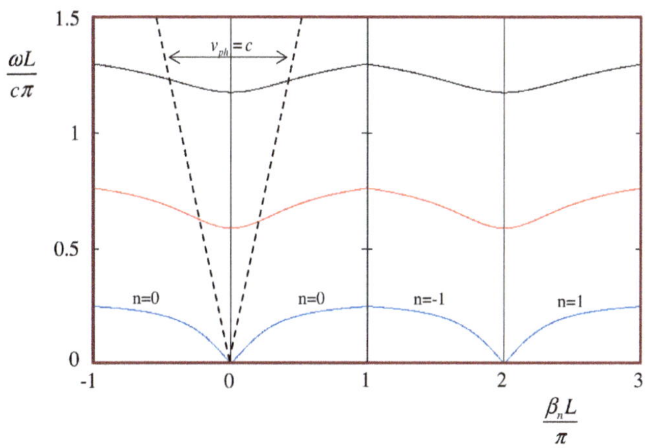

Figure 3.14. The Brillouin diagram of the periodic slow-wave structure of figure 3.13.

momentum (see (3.74)) and $\gamma_n = \sqrt{\beta_n^2 - k_0^2}$ is the cut-off momentum of the periodic slow-wave structure.

Due to the transcendental nature of (3.76), we should expect a multi-branch solution for any of the harmonics n. Equation (3.76) can be solved numerically, and the result is reported in figure 3.14, which is the Brillouin diagram we were looking for.

Many interesting results can be deduced from the analysis of figure 3.14. First of all, the study of the group velocity (defined as dispersion relation curve slope $v_g = \partial\omega/\partial\beta$) reveals that when the momentum is in the range $n\pi \leqslant \beta_n L \leqslant (n+1)\pi$, for $n = 0, \pm 2 \ldots$, we always have positive values for the group velocity for any of the propagating modes of the specific branch. This means that the radiation propagates in the forward direction, and thus concords with the electron beam. In contrast, when the momentum is in the range $n\pi \leqslant \beta_n L \leqslant (n+1)\pi$, but for $n = 1, \pm 3 \ldots$, we have the opposite situation in which the group velocity is negative and the radiation propagates in the reverse direction with respect to the electron beam.

Other interesting features arise from the analysis of the phase velocity (3.75). In figure 3.14 we have reported two lines indicating phase velocity values corresponding to $v_{ph} = \pm c$; it is evident that all the modes, except the fundamental ones, may have phase velocities higher or lower with respect to the light velocity in vacuum. As the fundamental modes of any space harmonics, we always have $v_{ph} \leqslant c$. Another important, and unique, feature of the periodic slow-wave structure is evident if we consider the negative space harmonics ($n = -1, -2\ldots$) with positive branches, as, for instance when $\pi \geqslant \beta_n L \geqslant 2\pi$ (see figure 3.14). In these cases we have a positive phase velocity, but a negative group velocity. This means that the electron bunch, considered as an 'observer', can remain in synchronism with a wave that transports the energy in the opposite direction. This is the basis of the backward wave oscillator, which will be described in the next section.

A final property to be mentioned is that such devices, due to the fundamental mode aspect, do not present any cut-off frequency, and, therefore, can propagate electromagnetic radiation power with frequencies ranging down to zero.

In order to realize an effective TWT we have to assemble the coupled cavities that constitute the slow-wave structure, such that the backward energy contributions have a phase which means they cancel each other. We can define, for the electromagnetic radiation, a phase shift per period as $\theta_1 = \beta_0 L$, where β_0 is the phase shift per unit length. As far as the electrons are concerned, they travel from one cavity to another in a time $T = L/u_0$ (we are considering a mono-energetic beam). An optimal interaction between the electromagnetic field and the electron beam requires that an electron 'sees' the same phase of the radiation as it passes from one cavity to another. In other words the phase change in a cavity after a time interval T must be $\theta_1 + 2n\pi$. Hence

$$\omega = \frac{\theta_1 + 2n\pi}{T} = \frac{(\beta_0 L + 2n\pi)u_0}{L}. \tag{3.77}$$

Defining the electron associated momentum as $\beta_e = \omega/u_0$ we obtain

$$\beta_e = \beta_0 + \frac{2n\pi}{L} = \beta_n. \tag{3.78}$$

Equations (3.75) and (3.78) state an important result $u_0 = v_{ph}$. In other words travelling wave amplification is obtained when the DC electron-beam velocity is approximately equal to the phase velocity of the fundamental or any of the space harmonics with positive group velocity v_g. A large bandwidth operation, in TWTs, is possible when the slow-wave structure is to be designed in such a way that (3.78) holds for a wide range of frequencies and an extensive space harmonic number.

In conclusion the interaction process can be summarized in four conceptual steps:

1. The electric field associated with the input electromagnetic signal exerts a force on the DC current electron beam causing the electrons to modulate their velocities, with the beginning of bunch formation around the zero-crossing of the electromagnetic wave.
2. If the electron-beam velocity has been set slightly higher than the wave one, the bunch centroids will drift toward the retarding electric field; thus more

electrons are in the retarding field than in the accelerating one and so a net transfer of energy occurs from the beam to the field.

3. The process of step 2 continues along the structure, more electrons enter the retarding phase and the increasing amplitude of the axial electric field makes the bunches more and more compact.
4. An appreciable fraction of the kinetic energy of the electrons has been converted into electromagnetic energy; the amplified wave is then propagated into the output waveguide and to the load.

In conclusion we can say that the velocity modulation produced in the beam by the input electric field and the subsequent conversion to density modulation is used to transfer energy from the beam to the radiation, just as we have described for the klystron. However, unlike the klystron, in the TWT this process occurs simultaneously and continuously along the whole length of the slow-wave structure.

3.8 The backward wave oscillator

Another important electron based radiation source is the backward wave oscillator (BWO) which, as discussed for the TWT, is based on the slow-wave structure geometry. We have already mentioned, among the properties of the Brillouin diagram of figure 3.14, that in a slow-wave we can set the parameters in a way in which we can have a positive phase velocity and, at the same time a negative group velocity, or vice versa. Let us stress again that this means that an electron can remain in synchronism with a wave that is actually transporting energy in the opposite direction. This very interesting property of periodic slow-wave structures has no parallel in smooth waveguides and is at the basis of the BWO.

To better explain some of their features let us briefly introduce one of the most common slow-wave structures used for BWOs (as well as for TWTs), the tape helix. A schematic layout of a BWO in the amplifier configuration is shown in figure 3.15, where the most evident difference to TWTs is the fact that the microwave input signal is in the electron-beam collector region, while the radiation output is placed near the electron gun.

In section 3.5 we discussed the behavior of electromagnetic fields in the presence of electron currents and in cylindrical structures. Equation (3.50), together with the definitions of h and p reported in (3.39) and (3.49), indicates that in slow-wave

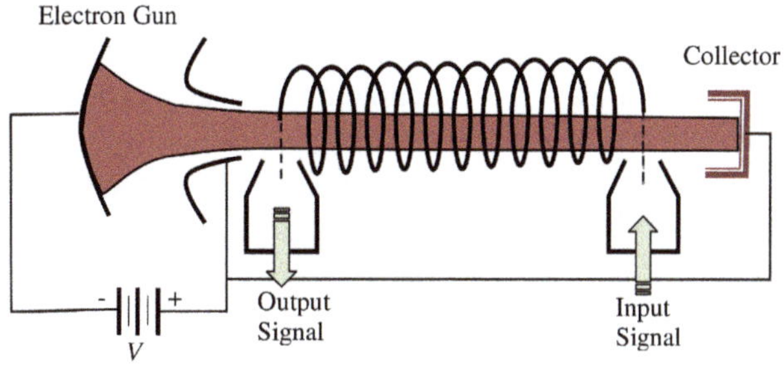

Figure 3.15. Layout of a BWO amplifier.

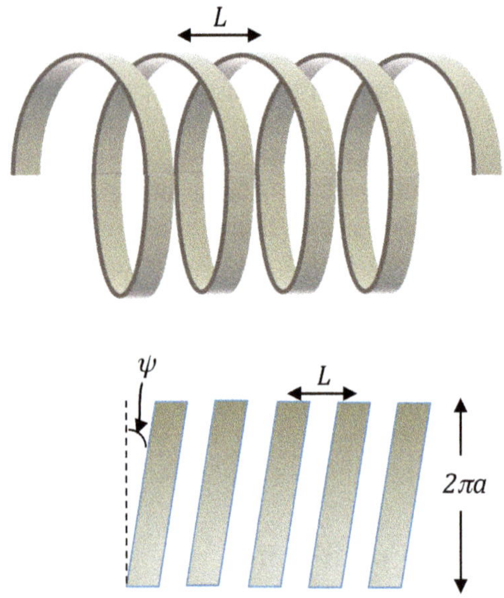

Figure 3.16. Tape helix geometry for a BWO.

structures the electric and magnetic fields are proportional to the modified Bessel functions I_n and K_n (n being again the harmonic number not considered in the analysis of section 3.5).

Considering the geometry reported in figure 3.16 it can be shown that the dispersion relation, obtained by imposing the boundary conditions to the electric and magnetic field, can be expressed for the fundamental harmonic as

$$\frac{K_1(ha)I_1(ha)}{K_0(ha)I_0(ha)} = \frac{(ha)^2 \tan^2\psi}{(k_0a)^2}. \tag{3.79}$$

As stated in section 3.5, slow-wave modes are such that $\beta_n \approx \omega/v_0 \gg k_0$, where ω/v_0 is the DC electron-beam momentum, in order to ensure a better synchronism between beam and fields; this implies $h_n \approx \beta_n$. Moreover, for high values of the argument, it can be easily directly demonstrated that Bessel's ratio, at the lhs of (3.79), approaches unity. After these considerations we can obtain the fundamental relation of this device that concerns the phase velocity and states

$$k_0 = \frac{\omega}{c} = \beta_n \tan\psi \Rightarrow v_{ph} = \frac{\omega}{\beta_n} = c\tan\psi. \tag{3.80}$$

In other words, any mode always has the same phase velocity that only depends on the geometry of the helix in terms of the pitch angle ψ. Equation (3.80) suggests another interesting result: the group velocity is equal to the phase velocity and, due to its independence from the mode index, there is no frequency dispersion in such a structure.

The Brillouin diagram for the structure in figure 3.16 is shown in figure 3.17. The dispersion relation is now a simple straight line, as deduced by (3.80), which has the same slope for any of the modes. In figure 3.17 the modes are shown and numbered with the order of the space harmonics, moreover the lines indicating the phase

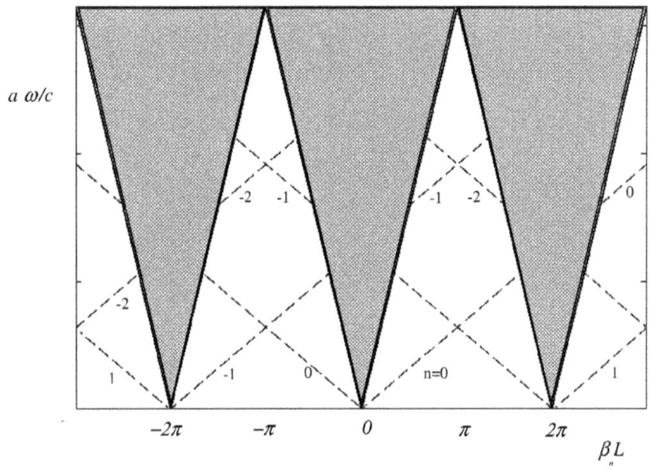

Figure 3.17. Brillouin diagram of the helix tape periodic structure.

velocity equal to the speed of light c are also shown. In order to complete the analysis of this diagram we must mention the presence of the so-called 'forbidden regions' where electromagnetic wave propagation is not possible without losses. These regions are delimited by the lines

$$k_0 = \omega/c = \pm\beta_n = \pm\left(\beta_0 + \frac{2\pi n}{L}\right), \tag{3.81}$$

which are the above-mentioned lines for which $v_{ph} = c$. Recalling the propagation constant expression (see the second equation of (3.39))

$$h_n = \sqrt{\left(\beta_0 + \frac{2\pi n}{L}\right)^2 - \left(\frac{\omega}{c}\right)^2}. \tag{3.82}$$

We understand that in the forbidden regions, where $\omega/c \geqslant \beta_n$, the propagation constant assumes imaginary values. As the field is proportional to the modified Bessel function K_n, and as the asymptotic behavior is $K_n(hr) \approx \sqrt{\pi/(2hr)}\exp(-hr)$, we can conclude that in the forbidden region the propagation has an evanescent behavior due to the losses.

It is possible to realize a backward wave interaction in a slow-wave scheme by having a helix voltage V_h (see figure 3.15) adjusted to a value lower than that normally used in TWTs. The electron velocity v_0 associated with the voltage is such that $\beta_e L \approx \omega L/v_0$, giving a beam-line, shown as OA in the Brillouin diagram of figure 3.18. This line intercepts the $n = -1$ space harmonic line at a point where the synchronism is established. Note that the group velocity of this space harmonic (i.e. the slope of the dispersion relation) is opposite to the direction of the electrons as expected in a BWO.

A detailed study of Klystrons TWTs and BWOs goes beyond the purposes of this book and thus, after this quick overview of the essential features underlying the operating mechanisms of these electromagnetic sources, we present an analysis of the relevant device performance.

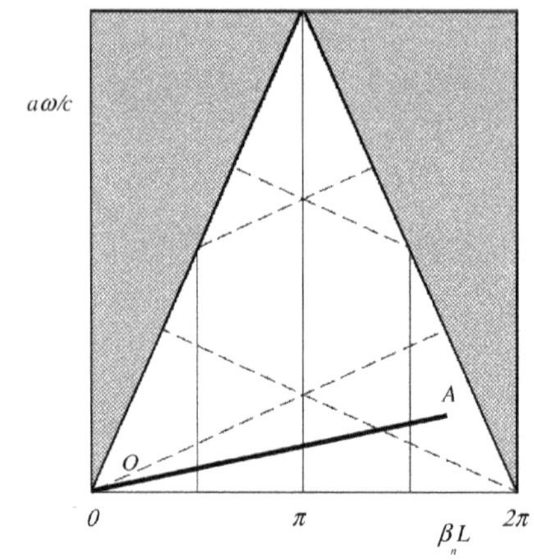

Figure 3.18. Brillouin diagram reporting the synchronism in a BWO.

3.9 Electron-based source characteristics and limitations

We have up to now illustrated the theory at the basis of the electron-based sources of electromagnetic radiation. The devices we have described (klystron, TWT and BWO) are the popular sources used in a wide spectral range. They can operate with continuity, from a few hundred MHz up to about 100 GHz. They can offer a wide variety of applications, from telecommunications, particle acceleration, plasma generation, plasma heating and so on; this type of device is associated with a series of features that can be exploited.

First, it is worth mentioning the coherence related to the bunching emission mechanism, expressed by (3.17), and by the oscillatory nature of the current drive expressed by (3.25) and (3.66). This property is suitable for any application that requires a stable phase relation such as, for instance, telecommunications.

Another important feature is the high power that these sources may generate. This characteristic has manifold explanations. The first is again related to the coherent nature of the emission process; the N^2 term in (3.17) tells us that we have a quadratic proportionality of the electromagnetic emitted power, with respect to the electron-beam current. A second aspect allowing these devices to be extremely powerful is the fact that the radiation travels in a low-loss evacuated waveguide and does not pass through any material, as happens for conventional laser sources. This fact, which is common to all the free-electron-based sources of electromagnetic radiation, puts very high limits on the energy and power generation, which are mainly related to the ohmic losses of the guiding structure. The high power, which can range up to the megawatt level in quasi-continuous wave operation, has found extensive applications in industrial use, for plasma generation, or in the big scientific laboratories, for high-energy electron-beam acceleration or for plasma heating in fusion research. A final aspect regarding the power is the relatively high efficiency that these systems present. Efficiency is figure of merit concerning the capability of a device to transform the input power into output power; the described sources have intrinsic efficiencies of the order of 30%. Moreover, the efficiency is strictly related to the dynamic process of the energy exchange between the electrons and electromagnetic field, and also to the saturation mechanism that stops the process at a certain point due to the degeneration of the quality parameters. Both the dynamics and the saturation mechanisms will be described in the forthcoming sections.

In spite of the extraordinary variety of the design, manufacturing, applicability and spectral range of klystrons, TWTs, BWOs and their variants, such as the Magnetron for instance, these sources all suffer from limitations in extending the emitted frequencies toward the far infrared (FIR) and infrared (IR). This is due to the fact that the resonant cavities, at the basis of the velocity modulation process and bunching emission process, have dimensions comparable to the wavelength of the radiation generated. This statement is easily understood looking at the coupling coefficient M, given by (3.24), which is a sinc function of a variable proportional to $\omega\ell$, where ℓ is the cavity dimension. If ℓ were too large with respect to the wavelength, the sinc function would approach zero values. Similar conclusions can be achieved by examining table 3.1, where all the μ_d parameters are expressed in terms of the variable $\theta = p\ell/2$ which is significant only when it assumes small values.

To overcome this drawback—which made the realization of a klystron or a TWT emitting radiation at, say, $\lambda = 10\ \mu\text{m}$ practically impossible—we had to wait for the invention of the laser, which occurred in 1960. A detailed analysis of the physical mechanism of laser devices goes beyond the purposes of the present book, but a few words are worth spending in order to introduce the free electron laser sources in the forthcoming sections of this chapter. A laser is based upon three fundamental elements:

1. A 'pumping system' that acts as an energy reservoir and is capable of delivering that energy by means of different mechanisms, mainly electrical and optical.
2. The 'active medium' which receives the energy from the pumping system. It is based on a quantum level system (gas, liquid or solid) that reaches an excited state during the pumping process, and then relaxes down to a ground level by means of emission of photons (electromagnetic quanta). The overall aim of the active medium is to transform the energy of the pumping system into electromagnetic radiation energy.
3. The 'optical cavity', the tool providing order among the all phases of the emitted photons by means of the stimulated emission mechanism that will be described further on.

The main advantage of the laser systems is that they can operate at any frequency provided that a suitable active medium, with proper transition levels, could be found. After 1960 many teams in the world began an unconventional 'race' in materials science to achieve lasing action in new active media. The result is that now we have laser sources ranging from the FIR spectral region up to the x-ray region, but with a huge variety of performances in terms of power, pulse duration, repetition rate, efficiency, spectral bandwidth, time and space coherence.

Together with the benefits there are some weaknesses associated with 'conventional' laser sources. The main drawback is that they cannot be designed and realized for any required set of parameters, in particular concerning the emitted frequency, the spectral bandwidth and the power. This happens because any active medium has a specific set of quantum levels that regulate, in particular, the emitted photon energy (and therefore the frequency of the radiation) and the spectral bandwidth (related to the quantum lifetime of the occupied levels). Moreover, any material has a particular and definite damage threshold that limits the power that can flow through it, and therefore the radiation power that it can generate if used as the laser active medium. It is interesting to stress that the difficulties of the laser devices are, at the same time, the advantages of the electron-based radiation sources, as discussed above.

To overcome the weaknesses of the electron-based sources and of the laser systems and, at the same time, take advantage of all the positive characteristics of both these categories of radiation emitting tools, a new device appeared on the scene in 1977. As mentioned above, this is called the free electron laser (FEL) and, as implicit in its name, is a laser based on free charged particles (electrons not bound to quantum orbitals) that interact with electromagnetic radiation. As explained in section 3.1 a particle cannot emit or absorb a photon in a vacuum, but if it can exchange momentum with an 'external entity' the process is allowed. Together with the described emission associated with a dielectric medium (for the Cerenkov effect) and with a virtual state (for the Compton scattering), we can consider a strong magnetic field as the required 'external entity'. This is the so-called synchrotron radiation emission process.

All the electromagnetic calculations presented in the following sections are expressed in the convention of Gaussian units. The reason for this choice is related to the fact that most of the FEL literature has been produced following this system and a proper comparison can therefore be facilitated.

3.10 Radiation by moving charges and synchrotron radiation emission

In order to evaluate the behavior of the radiation emitted by a moving charge, we have to consider the non-homogeneous Maxwell's equation in a vacuum, expressed as follows

$$\begin{cases} \vec{\nabla} \cdot \vec{E} = 4\pi\rho \\ \vec{\nabla} \times \vec{B} - \dfrac{1}{c}\dfrac{\partial \vec{E}}{\partial t} = \dfrac{4\pi}{c}\vec{J}, \end{cases} \qquad \begin{cases} \vec{\nabla} \cdot \vec{B} = 0 \\ \vec{\nabla} \times \vec{E} - \dfrac{1}{c}\dfrac{\partial \vec{B}}{\partial t} = 0. \end{cases} \tag{3.83}$$

A more convenient and synthetic form for (3.83) is their covariant formulation, which has the important advantage of being Lorentz transformation invariant. The two non-homogeneous equations of (3.83) can be recast in the following form

$$\partial_\alpha F^{\alpha\beta} = \frac{4\pi}{c} J^\beta, \tag{3.84}$$

where the antisymmetric tensor $F^{\alpha\beta}$ is expressed by

$$F^{\alpha\beta} = \begin{pmatrix} 0 & -E_x & -E_y & -E_z \\ E_x & 0 & -B_z & B_y \\ E_y & B_z & 0 & -B_x \\ E_z & -B_y & B_x & 0 \end{pmatrix}. \tag{3.85}$$

The four-vector J^β and the four-vector derivative operator are, respectively,

$$J^\beta \equiv \left(c\rho, \vec{J}\right); \qquad \partial_\alpha \equiv \left(\frac{\partial}{\partial x^0}, -\vec{\nabla}\right). \tag{3.86}$$

A further useful formulation of Maxwell's equations is by means of the vector $\vec{A}$ and scalar Φ potentials of the electric and magnetic fields. We have already introduced the vector potential in (3.26), in order to express the Helmholtz equation for the evaluation of the coupling coefficient between electrons and fields in cylindrical cavities. However, the most general way to express the fields $\vec{E}$ and $\vec{B}$ in terms of their potentials is

$$\begin{cases} \vec{E} = -\dfrac{1}{c}\dfrac{\partial\vec{A}}{\partial t} - \vec{\nabla}\Phi \\ \vec{B} = \vec{\nabla} \times \vec{A} \end{cases}. \tag{3.87}$$

Equation (3.87) can be considered as a definition 'inspired' by the two homogeneous electromagnetic equation (3.83) ($\vec{\nabla} \cdot \vec{B} = \vec{\nabla} \cdot (\vec{\nabla} \times \vec{A}) \equiv 0$ always, for instance). Introducing the quadrivector potentials as

$$A^\alpha \equiv (\Phi, \vec{A}) \tag{3.88}$$

we can write (3.87) in covariant form by means of the field strength tensor (3.85) as follows:

$$F^{\alpha\beta} = \partial^\alpha A^\beta - \partial^\beta A^\alpha. \tag{3.89}$$

Substituting (3.89) into (3.84) we obtain

$$\partial_\alpha \partial^\alpha A^\beta - \partial_\alpha \partial^\beta A^\alpha = \frac{4\pi}{c} J^\beta. \tag{3.90}$$

It is worth mentioning that the four-vector potential it is not uniquely determined, but it may differ by a transformation, $A'_\mu = A_\mu + \partial\Lambda(x)/\partial x_\mu$, where $\Lambda(x)$ is an arbitrary scalar function. The above transformation, called 'gauge transformation', leaves (3.89) unaltered, as can be proven by a direct substitution. The freedom in the choice of A_μ allows us to use the following condition known as the 'Lorentz gauge'

$$\partial_\alpha A^\alpha \equiv 0. \tag{3.91}$$

Under this gauge condition (3.90) becomes

$$\partial_\alpha \partial^\alpha A^\beta = \frac{4\pi}{c} J^\beta. \tag{3.92}$$

Equation (3.92), with the condition (3.91), yields the electromagnetic potential A^β in terms of moving charges expressed by the quadrivector J^β. The equation can be recast in terms of the vector potential $\vec{A}$ and scalar potential Φ and in terms of the charge density ρ and current density $\vec{J}$. By means of (3.86) and (3.88) we obtain

$$\begin{cases} \nabla^2 \vec{A} - \dfrac{1}{c^2}\dfrac{\partial^2 \vec{A}}{\partial t^2} = -\dfrac{4\pi}{c}\vec{J} \\ \nabla^2 \Phi - \dfrac{1}{c^2}\dfrac{\partial^2 \Phi}{\partial t^2} = -4\pi\rho \end{cases} \tag{3.93}$$

for the electromagnetic potentials equations and

$$\frac{1}{c}\frac{\partial \Phi}{\partial t} + \vec{\nabla} \cdot \vec{A} = 0 \tag{3.94}$$

for the Lorentz gauge condition.

A detailed and rigorous derivation of the solutions of (3.93) goes beyond the purposes of this book. We will present an effective heuristic method based on the linearity of the equations that allows us to divide the space-volume into infinitesimal elements and calculate the real field as the sum over all these contributions. Let us consider the scalar field first. The charge density ρ in an infinitesimal volume, at a distance $\vec{R}$ from the origin of the coordinates, at the time instant t, can be written as $\rho = dq(t)\delta(\vec{R})$. Thus the scalar potential (3.93) becomes

$$\nabla^2 \Phi - \frac{1}{c^2}\frac{\partial^2 \Phi}{\partial t^2} = -4\pi dq(t)\delta(\vec{R}). \tag{3.95}$$

At the origin of the coordinates ($\vec{R} \equiv 0$), (3.95) becomes a homogeneous wave equation having clear central symmetry behavior. The general solution depends only on t and $R = |\vec{R}|$. The solution of the homogeneous equation (3.95) can therefore be written as

$$\Phi = \frac{\chi\left(t - \dfrac{R}{c}\right)}{R}. \tag{3.96}$$

The function χ is, at the moment, an arbitrary function that must be chosen in order to satisfy the non-homogeneous equation (3.95) at the coordinate origin as well. By inspecting (3.96) we can deduce that at the origin ($R \to 0$) the potential Φ diverges and we can therefore neglect the time derivative $\partial^2/\partial t^2$ in (3.95) with respect to ∇^2.

In so doing (3.95) reduces to the Poisson equation whose solution is provided by the Coulomb law. This allows us to write $\chi(t) = dq(t)$. We eventually end up with

$$\Phi = \frac{dq\left(t - \frac{R}{c}\right)}{R}. \tag{3.97}$$

The general solution of (3.95) can be obtained by considering an arbitrary distribution of the charges $\rho(x, y, z, t)$, related to the infinitesimal volume element considered by $dq = \rho dV$, and summing up, by means of an integration, over all the contributions in the space, thus ending up with

$$\Phi(\vec{r}, t) = \int \frac{1}{R}\rho\left(\vec{r}', t - \frac{R}{c}\right)dV' + \Phi_0. \tag{3.98}$$

In the above equation we have added Φ_0 in order to obtain the most general expression for the scalar potential. In the previous equation dV' denotes the elementary integration volume, $\vec{r}'(x', y', z')$ and $\vec{r}(x, y, z)$ the origin and observation vectors, and $R = |\vec{r} - \vec{r}'|$ is the origin distance with respect to the observation point.

Equation (3.98) can also be written as

$$\Phi(\vec{r}, t) = \iint \frac{\rho(\vec{r}', \tau)}{|\vec{r} - \vec{r}'|}\delta\left(\tau - t + \frac{|\vec{r} - \vec{r}'|}{c}\right)d\tau dV' + \Phi_0. \tag{3.99a}$$

An analogous solution can be found for the vector potential

$$\vec{A}(\vec{r}, t) = \iint \frac{1}{c}\frac{\vec{J}(\vec{r}', \tau)}{|\vec{r} - \vec{r}'|}\delta\left(\tau - t + \frac{|\vec{r} - \vec{r}'|}{c}\right)d\tau dV' + \vec{A}_0. \tag{3.99b}$$

Let us now assume that a charge q is moving along a trajectory $\vec{r} = \vec{r}_0(t)$, with a velocity $\vec{v}(t) = \dot{\vec{r}}_0(t)$. The charge and the current densities can be expressed by

$$\begin{cases} \rho(\vec{r}, t) = q\delta(\vec{r} - \vec{r}_0(t)) \\ \vec{J}(\vec{r}, t) = q\vec{v}(t)\delta(\vec{r} - \vec{r}_0(t)) \end{cases}. \tag{3.100}$$

Introducing the notation $\vec{R}(\tau) = \vec{r} - \vec{r}_0(\tau)$ and $R(\tau) = |\vec{R}(\tau)|$, substituting (3.100) in (3.99), integrating over the coordinates $\vec{r}'$ and neglecting the homogeneous solutions, we obtain

$$\begin{cases} \Phi(\vec{r}, t) = q\int \frac{1}{R(\tau)}\delta\left(\tau - t + \frac{R(\tau)}{c}\right)d\tau \\ \vec{A}(\vec{r}, t) = \frac{q}{c}\int \frac{\vec{v}(\tau)}{R(\tau)}\delta\left(\tau - t + \frac{R(\tau)}{c}\right)d\tau \end{cases}. \tag{3.101}$$

A wise interpretation of (3.101) tells us that the potentials Φ and $\vec{A}$, evaluated at the observation point $P(\vec{r})$ and at the time instant t, are related to the motion state of the charge q, at any previous time τ, for which the propagation time of the electromagnetic signal obeys the following equation:

$$\tau + \frac{R(\tau)}{c} = t. \tag{3.102}$$

The time instant τ is commonly known as the retarded time: $t_{\text{ret}} = \tau$. Introducing a new variable as $t' = \tau - t + [R(\tau)/c]$, we can easily verify that

$$dt' = \left[1 - \frac{1}{c}\vec{n}(\tau) \cdot \vec{v}(\tau)\right]d\tau, \tag{3.103}$$

from the definition of $R(\tau)$ and $\vec{v}(\tau)$, having introduced the unit vector $\vec{n}(\tau) = \vec{R}(\tau)/R(\tau)$ indicating the direction between the observation point $P(\vec{r})$ and the moving charge q.

According to the previous definitions (3.101) may be written as

$$\begin{cases} \Phi(\vec{r}, t) = q\displaystyle\int \frac{\delta(t')}{R(\tau)\left[1 - \frac{1}{c}\vec{n}(\tau) \cdot \vec{v}(\tau)\right]}dt' \\ \vec{A}(\vec{r}, t) = \dfrac{q}{c}\displaystyle\int \frac{\vec{v}(\tau)\delta(t')}{R(\tau)\left[1 - \frac{1}{c}\vec{n}(\tau) \cdot \vec{v}(\tau)\right]}dt' \end{cases}. \tag{3.104}$$

The integration over t' clearly indicates that a non-zero result only occurs at $t' = 0 \;\Leftrightarrow\; \tau = t_{\text{ret}}$. Equation (3.104) can be hence written as

$$\begin{cases} \Phi(\vec{r}, t) = \dfrac{q}{\left[R(t_{\text{ret}}) - \frac{\vec{R}(t_{\text{ret}}) \cdot \vec{v}(t_{\text{ret}})}{c}\right]} \\ \vec{A}(\vec{r}, t) = \dfrac{q\vec{v}(t_{\text{ret}})}{c\left[R(t_{\text{ret}}) - \frac{\vec{R}(t_{\text{ret}}) \cdot \vec{v}(t_{\text{ret}})}{c}\right]} \end{cases}. \tag{3.105}$$

Equations (3.105) are known as the Lienard–Wiechert potentials. Another common way to write the above potentials follows after the introduction of the following notation

$$\vec{\beta} \equiv \frac{\vec{v}}{c}; \quad k \equiv 1 - \vec{n} \cdot \vec{\beta}. \tag{3.106}$$

Equation (3.105) can be thus cast in the form

$$\begin{cases} \Phi(\vec{r}, t) = \left[\dfrac{q}{kR}\right]_{\text{ret}} \\ \vec{A}(\vec{r}, t) = \left[\dfrac{q\vec{\beta}}{kR}\right]_{\text{ret}} \end{cases}, \tag{3.107}$$

where the square brackets indicate that their content must be calculated at the 'retarded time' such that $t = t_{\text{ret}} + R(t_{\text{ret}})/c$. The quantitative weight of the parameter k in (3.106) increases at relativistic velocities; for small velocities $k \approx 1$ and (3.107) tends to the result of static or quasi-static electromagnetic theory.

The electric and magnetic fields associated with a moving charged particle can be calculated from the retarded potentials (3.107) by means of (3.87), or by means of the covariant equation (3.89). The calculation is straightforward, but long and tedious; the results are summarized below:

$$\begin{cases} \vec{E}(\vec{r}, t) = q\left[\dfrac{(1-\beta^2)(\vec{n}-\vec{\beta})}{k^3 R^2}\right]_{\text{ret}} + \dfrac{q}{c}\left[\dfrac{\vec{n}}{k^3 R}\times\left\{(\vec{n}-\vec{\beta})\times\dot{\vec{\beta}}\right\}\right]_{\text{ret}}. \\ \vec{B}(\vec{r}, t) = [\vec{n}\times\vec{E}(\vec{r}, t)]_{\text{ret}} \end{cases} \tag{3.108}$$

It is important to stress how the electric field proposed in (3.108) is composed of two terms. The first term is associated with the charged particle velocity only, it falls off as $1/R^2$ and it represents a generalization of the Coulomb law. The direction of this electric field component always points to the particle current from the observation point; moreover this is the only term contributing to the fields for charged particles moving at constant velocity. The second term arises for accelerated charges only and falls off as $1/R$ and always has a direction perpendicular to the versor $\vec{n}$ connecting the observer with the charge q.

Before applying the above theoretical discussion to a specific radiation emission mechanism, of great interest in FELs, let us evaluate the energy emitted by the electromagnetic field expressed by (3.108). The instantaneous energy flux can be calculated by means of the Poynting vector, defined as $\vec{S} = c/(4\pi)(\vec{E}\times\vec{B})$, which in our case becomes $\vec{S} = c/(4\pi)\vec{n}\,|\vec{E}(\vec{r}, t)|^2_{\text{ret}}$. By the Poynting theorem it follows that

$$\frac{d\mathcal{E}}{dt d\Omega} = \frac{c}{4\pi}\left|[R\vec{E}(\vec{r}, t)]_{\text{ret}}\right|^2 = \frac{q^2}{4\pi c}\left|\left[\frac{\vec{n}\times\left\{(\vec{n}-\vec{\beta})\times\dot{\vec{\beta}}\right\}}{k^3}\right]_{\text{ret}}\right|^2. \tag{3.109}$$

In writing (3.109) we have restricted our analysis to the far field region, therefore neglecting the $1/R^2$ term of the radiation field (3.108). An examination of (3.109) suggests that in the non-relativistic case $k \to 1$ (see (3.106)) we can write

$$\frac{d\mathcal{E}}{dt d\Omega} = \frac{q^2}{4\pi c^3}|\dot{\vec{v}}|^2 \sin(\Theta)^2 \Rightarrow \frac{d\mathcal{E}}{dt} = \frac{2}{3}\frac{q^2}{c^3}|\dot{\vec{v}}|^2, \tag{3.110}$$

where Θ is the angle between $\dot{\vec{v}}$ and $\vec{n}$. This is the well-known Larmor formula. In the relativistic regime ($\beta \to 1$), the radiated energy increases noticeably (see (3.106)). Furthermore, denoting by θ the angle between $\vec{n}$ and $\vec{v}$ we can write

$$k = 1 - \frac{v}{c}\cos\theta \approx 1 - \frac{v}{c} + \frac{\theta^2}{2},$$

where we have used an expansion for small angles, since $\theta \approx \sqrt{1-\beta^2} = 1/\gamma$, since, for relativistic velocities, the radiation is emitted mostly in the same direction as the charge motion.

In order to evaluate the spectral distribution and the properties of the radiation emitted by moving charges, it is necessary to manipulate (3.109). The use of the Fourier transform and Parseval's theorem allows us to write

$$\begin{aligned}\frac{d\mathcal{E}}{d\omega d\Omega} &= \frac{c}{4\pi^2}\left|\int [R\vec{E}(\vec{r},t)]_{\text{ret}} e^{i\omega t} dt\right|^2 \\ &= \frac{q^2}{4\pi^2 c}\left|\int \left[\frac{\vec{n}\times\left\{(\vec{n}-\vec{\beta})\times\dot{\vec{\beta}}\right\}}{k^3}\right]_{\text{ret}} e^{i\omega t} dt\right|^2 .\end{aligned} \tag{3.111}$$

Equation (3.111) must, again, be evaluated at the 'retarded time' and we are thus allowed to operate the same variable transformation expressed in (3.103), $dt = kd\tau$. Moreover, due to the far field condition already evocated, we can perform the expansion $R(\tau) \approx |\vec{r}| - \vec{n}\cdot\vec{r}_0(\tau)$, valid for $|\vec{r}_0| \ll |\vec{r}|$. Omitting the constant phase term $\exp(i\omega t)$ we can write (3.111) as

$$\frac{d\mathcal{E}}{d\omega d\Omega} = \frac{q^2}{4\pi^2 c}\left|\int \left[\frac{\vec{n}\times\left\{(\vec{n}-\vec{\beta})\times\dot{\vec{\beta}}\right\}}{k^2}\right] e^{i\omega\left(\tau - \vec{n}\cdot\frac{\vec{r}_0(\tau)}{c}\right)} d\tau\right|^2 . \tag{3.112}$$

After an integration by parts we find

$$\frac{d\mathcal{E}}{d\omega d\Omega} = \frac{q^2\omega^2}{4\pi^2 c}\left|\int [\vec{n}\times(\vec{n}\times\vec{\beta})] e^{i\omega\left(\tau - \vec{n}\cdot\frac{\vec{r}_0(\tau)}{c}\right)} d\tau\right|^2 . \tag{3.113}$$

We have now an appropriate physical and mathematical tool-set to deal with the radiation mechanism underlying 'magnetic based' FELs, namely 'synchrotron radiation emission'. Let us consider a charged particle, with mass m and charge q, moving in an external magnetic field $\vec{B}$ (not to be confused with the radiation magnetic field in (3.108)) which is uniform and constant. The Lorentz four-vector force equation is written, with $\vec{E} \equiv 0$,

$$\frac{d}{dt}(\gamma m\vec{v}) = \frac{q}{c}\vec{v}\times\vec{B}; \quad \frac{d}{dt}(\gamma mc^2) = q\vec{v}\cdot\vec{E} \equiv 0. \tag{3.114}$$

The second equation of (3.114) tells us that the absence of the electric field implies that $\gamma = \text{const}$ and therefore

$$m\gamma\frac{d\vec{v}}{dt} = \frac{q}{c}\vec{v}\times\vec{B} \Rightarrow \begin{cases}\dfrac{d\vec{v}_{//}}{dt} = 0 \\ m\gamma\dfrac{d\vec{v}_{\perp}}{dt} = \dfrac{q}{c}\vec{v}_{\perp}\times\vec{B}\end{cases} . \tag{3.115}$$

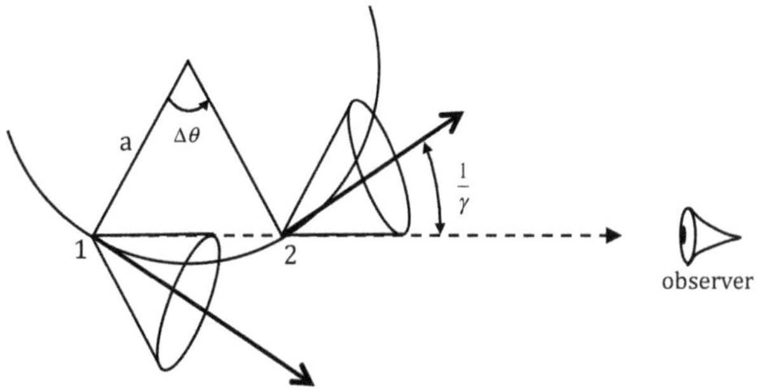

Figure 3.19. Emission cones of a moving charge along a circular trajectory.

The charge particle velocity component parallel to the magnetic field remains constant and, as a consequence, also the orthogonal component is constant ($|v_{\perp}| = \text{const}$). To find the total power emitted we can make use of the Larmor equation (3.110) writing it in the Lorentz invariant form, considering only the vector part of the four-vector:

$$\frac{d\mathcal{E}}{dt} = \frac{2}{3}\frac{q^2}{m^2c^3}\left|\frac{d\vec{P}}{dt}\right|^2. \tag{3.116}$$

Making use of (3.115) we obtain

$$\frac{d\mathcal{E}}{dt} = \frac{2}{3}\frac{q^4}{m^2c^5}\gamma^2\,|\vec{B}|^2|\vec{v}_{\perp}|^2. \tag{3.117}$$

A cursory look to (3.117) ensures that the emission process is more significant for light particles with high energies due to the quadratic behavior with respect to the relativistic factor γ. We already discussed, while analyzing (3.109), how, for a highly relativistic charged particle, the emitted radiation is concentrated in a narrow set of forwarding directions, with an angular aperture of about $1/\gamma$. This fact has a valuable consequence for the spectral properties of the emitted radiation, because an observer will see a pulse of radiation for a time interval much shorter than the gyration period. The spectrum will thus result in much spread over a broad region. Referring to figure 3.19, the observer will see the pulse during the particle passage from point 1 to point 2. The trajectory is $\Delta s = a\Delta\theta$ where a is the radius of curvature. From figure 3.19 we also have $\Delta\theta = 2/\gamma$ so that $\Delta s = 2a/\gamma$. From the equation of motion (3.115) we therefore have

$$\gamma m\frac{\Delta v}{\Delta t} = \gamma m\frac{v\Delta\theta}{\Delta s/v} = \frac{q}{c}\vec{v}\times\vec{B} \Rightarrow \frac{\Delta\theta}{\Delta s} = \frac{qB}{\gamma mcv} = \frac{\omega_B}{v} = \frac{1}{a}.$$

In the above expression we have introduced the frequency of rotation ω_B also defined as cyclotron frequency,

$$\omega_B = \frac{qB}{\gamma mc}. \tag{3.118}$$

From the above considerations we can conclude that

$$\Delta s \approx \frac{2v}{\gamma\omega_B} = v(t_2 - t_1) \Rightarrow (t_2 - t_1) \approx \frac{2}{\gamma\omega_B}. \qquad (3.119)$$

Equation (3.119) expresses the time at which the particle passes from point 1 to point 2. On the other end the observer will 'see' different arrival times $t^0{}_1$ and $t^0{}_2$, with a difference reduced by the time the radiation required to move a distance Δs: i.e. $\Delta s/c$. From (3.119)

$$\Delta t^0 = (t_2 - t_1) - \frac{\Delta s}{c} = (t_2 - t_1)\left(1 - \frac{v}{c}\right) = \frac{2}{\gamma\omega_B}\left(1 - \frac{v}{c}\right) \approx \frac{1}{\gamma^3\omega_B}, \qquad (3.120)$$

where we have considered a relativistic particle for which $(1 - v/c) \approx 1/(2\gamma^2)$.

Equation (3.120) is a very important result: it states that the pulse duration observed is smaller, by a factor γ^3, with respect to the gyration period. Due to the Fourier theorem we shall expect a fairly broad spectrum, cutting at frequencies of the order of $1/\Delta t^0$. If we introduce the 'critical frequency' as

$$\omega_C \equiv \frac{3}{2}\gamma^3\omega_B = \frac{3}{2}\gamma^3\frac{v}{a}, \qquad (3.121)$$

we can expect the spectral power to extend up to about ω_C before decreasing.

A complete and detailed spectral analysis of the synchrotron radiation really goes beyond the commitments of this book, but now we have all the tools to understand some of the basic steps. The relevant equation to this purpose is (3.113). Referring to figure 3.20 we can assume that the orbit lying in the (x, y)-plane, an observing direction $\vec{n}$ forming an angle θ with the x-axis. The two polarizations considered are $\vec{\varepsilon}_\perp$, along the y-axis and $\vec{\varepsilon}_{//} = \vec{n} \times \vec{\varepsilon}_\perp$.

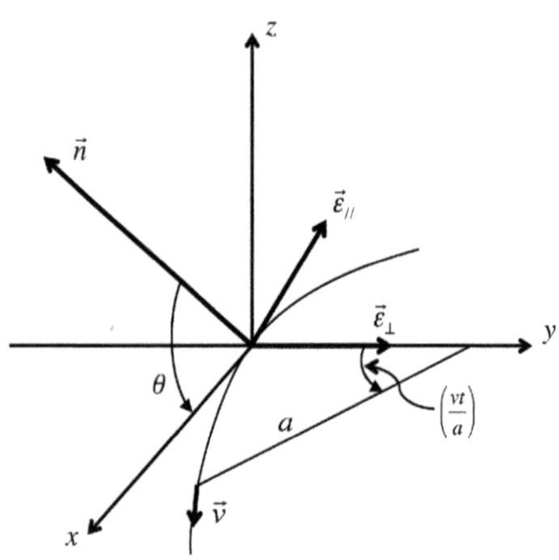

Figure 3.20. Geometry of the synchrotron radiation emission. At the time $t = 0$ the particle velocity is along the x-axis and a is the trajectory radius of curvature.

We can accordingly write the vector and products in (3.113) as

$$\begin{cases} \vec{n} \times (\vec{n} \times \vec{\beta}) = \beta\left[-\vec{\varepsilon}_{\perp} \sin\left(\frac{v\tau}{a}\right) + \vec{\varepsilon}_{//}\cos\left(\frac{v\tau}{a}\right)\sin\theta\right] \\ \omega\left(\tau - \frac{\vec{n}\cdot\vec{r}_0(\tau)}{c}\right) = \omega\left[\tau - \frac{a}{c}\cos\theta \sin\left(\frac{v\tau}{a}\right)\right] \\ \qquad \approx \frac{\omega}{2\gamma^2}\left[(1+\gamma^2\theta^2)\tau + \frac{c^2\gamma^2\tau^3}{3a^2}\right] \end{cases} . \quad (3.122)$$

In the second equation of (3.122) we have expanded the trigonometric functions for small θ angles and short times τ. In the relativistic regime $\beta \to 1$ ($v \to c$), (3.113) takes the concise form

$$\frac{d\mathcal{E}}{d\omega d\Omega} = \frac{d\mathcal{E}_{//}}{d\omega d\Omega} + \frac{d\mathcal{E}_{\perp}}{d\omega d\Omega}$$
$$= \frac{q^2\omega^2}{4\pi^2 c}\left|\int\left(\theta + \frac{c\tau}{a}\right)\exp\left\{\frac{i\omega}{2\gamma^2}\left[(1+\gamma^2\theta^2)\tau + \frac{c^2\gamma^2\tau^3}{3a^2}\right]\right\}d\tau\right|^2 . \quad (3.123)$$

The integrals in (3.123), over the 'retarded time', can be identified as the Airy integrals yielding the modified Bessel functions as

$$\frac{d\mathcal{E}}{d\omega d\Omega} = \frac{q^2\omega^2}{3\pi^2 c}\left\{\theta^2\left[\frac{a^2(1+\gamma^2\theta^2)}{\gamma^2c^2}\right]K^2_{\frac{1}{3}}(\eta) + \left[\frac{a(1+\gamma^2\theta^2)}{\gamma^2 c}\right]^2 K^2_{\frac{2}{3}}(\eta)\right\}, \quad (3.124)$$

where we have introduced the variable

$$\eta \equiv \frac{\omega a}{3c\gamma^3}(1+\gamma^2\theta^2)^{3/2} = \frac{\omega}{2\omega_C}(1+\gamma^2\theta^2)^{3/2}, \quad (3.125)$$

and we have made use of the definition (3.121) in the relativistic regime with $v \to c$.

Many things can be discovered by the analysis of (3.124). From the known properties of the modified Bessel functions, which we can summarize as follows

$$K_\nu(\eta) \to \begin{cases} -\left[\ln\left(\frac{\eta}{2}\right) + 0.5772 + \dots\right], & \nu = 0 \\ \frac{\Gamma(\nu)}{2}\left(\frac{2}{\eta}\right)^2, & \nu \neq 0 \end{cases} \qquad \eta \ll 1,$$

$$K_\nu(\eta) \to \sqrt{\frac{\pi}{2\eta}}e^{-\eta}\left[1 + 0\left(\frac{1}{\eta}\right)\right] \qquad \eta \gg 1,$$

we can deduce that the energy radiated by the charge in circular motion is negligible for $\eta \gg 1$, due to the exponential factor. From (3.125) this condition is satisfied at large angles θ for small frequencies values ω; at higher frequencies the critical angle,

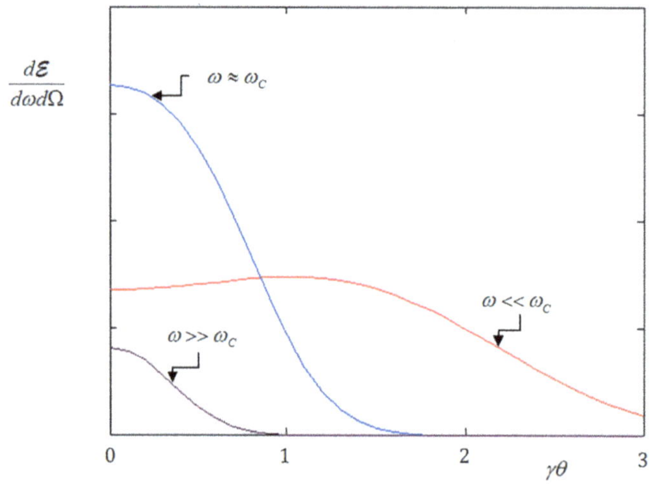

Figure 3.21. Frequency spectra as a function of the angular distribution $\gamma\theta$. The curves correspond to the three different frequency ranges with respect the critical frequency ω_C.

beyond which the radiation emission falls away, becomes smaller. This confirms how the radiation is mainly confined to the plane of motion. At larger frequencies, however, the variable η will be large at all angles; the meaning of the critical frequency of (3.121) is stronger as the limit, above which the emission reduces consistently ($\omega = \omega_C \Rightarrow \eta = 1/2$ for $\theta = 0$). The angular distribution behavior of the radiation is shown, qualitatively, in figure 3.21. The three curves correspond to the distribution for frequencies small, comparable or higher, with respect to the critical frequency ω_C. The curves have been calculated by means of (3.124) and using the natural unit $\gamma\theta$ for the angles.

The frequency distribution of the synchrotron radiation emission can be obtained by the integration of (3.124) over the angles. We have already demonstrated how the radiation is confined in the orbit plane when the charge moves in the relativistic regime. This fact allows us to extend the integral limits to infinity, and gives us the opportunity to make use of the Bessel function properties. A proper integration, which we omit for brevity, gives

$$\begin{aligned}\frac{d\mathcal{E}}{d\omega} &= 2\pi\int_{-\pi/2}^{\pi/2}\frac{d\mathcal{E}}{d\omega d\Omega}\cos\theta d\theta \approx 2\pi\int_{-\infty}^{\infty}\frac{d\mathcal{E}}{d\omega d\Omega}d\theta \\ &= \sqrt{3}\frac{q^2}{c}\gamma\frac{\omega}{\omega_C}\int_{\omega/\omega_C}^{\infty}K_{\frac{5}{3}}(x)dx.\end{aligned} \tag{3.126}$$

The plot of the spectrum of the total energy emitted is shown in figure 3.22. Due to the properties of the integral in (3.126) the behavior of the spectrum in the limit $\omega \ll \omega_C$ reduces to $\omega^{1/3}$. This gives a very broad flat spectrum at frequencies below ω_C. For the high-frequency limit $\omega \gg \omega_C$ we have a trend given by $(\omega/\omega_C)^{1/2}\exp(-\omega/\omega_C)$ corresponding to an exponential damping.

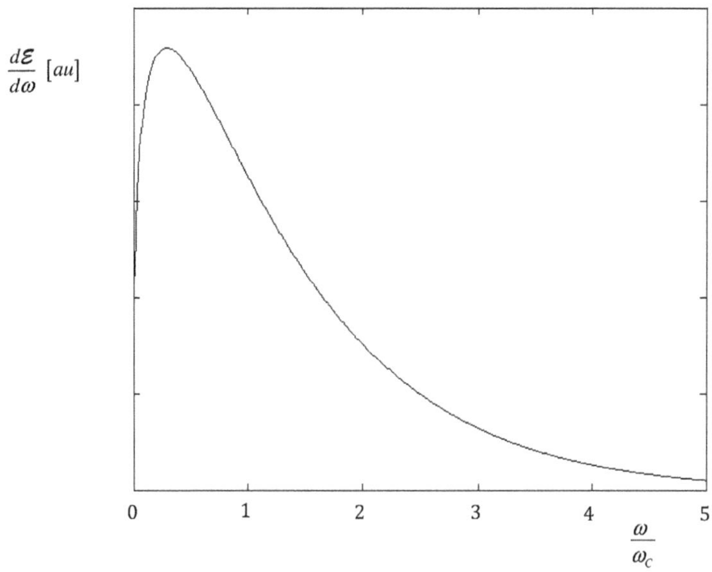

Figure 3.22. Synchrotron radiation spectrum.

Synchrotron radiation emission is a very important phenomenon in circular electron accelerators and it is currently widely exploited in diagnostic sciences, due to its very broad frequency distribution covering frequencies from IR to x-ray regions. Synchrotron radiation, also associated with cyclotron emission, is also an important mechanism observed in hot plasmas both in astronomical objects and in nuclear fusion devices.

3.11 The magnetic undulator

At the end of the previous section we have discussed how the synchrotron radiation spectrum, associated with a charged particle in circular motion, is very broad. Many applications require brighter sources and narrower spectra. To satisfy these requests a specific device, called a magnetic undulator, has been realized. The magnetic field associated with an undulator is such that a charged particle undergoes an oscillating trajectory in the plane orthogonal to the magnetic field. It is easy to understand that if the particle angular excursion, along the undulating trajectory, is smaller than the emission cone width (as described in figure 3.19 of the previous section), the observer will see an emission duration longer with respect to that of (3.120). The result is a much narrower spectrum as expected.

A complete theory of magnetic undulator radiation emission is complicated due to the many possible magnetic field geometries that can be considered. The equation to be used is again the one derived from the Lienard–Wiechert potentials, i.e. (3.113), for which it is necessary to evaluate the electron trajectory $\vec{r}_0(\tau)$. For our purposes let us consider a simplified geometry, such as the one reported in figure 3.23, for which we can consider a sinusoidal magnetic field. From (3.83), in the

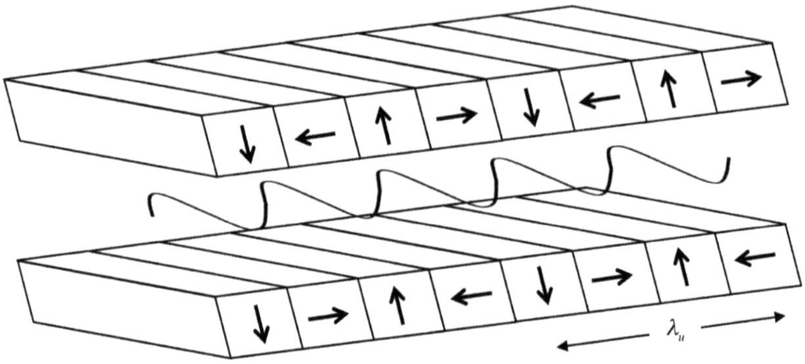

Figure 3.23. Schematic view of a planar magnetic undulator and the associated sinusoidal charged particle trajectory.

absence of currents and for static fields, we have $\vec{\nabla} \times \vec{B} = 0$, which implies $\vec{B} = -\vec{\nabla}\Phi_{\mathrm{MAG}}$, and consequently $\nabla^2\Phi_{\mathrm{MAG}} = 0$. This Laplace equation has a general solution: $\Phi_{\mathrm{MAG}}(y, z) = f(y)\sin(2\pi z/\lambda_u)$ having neglected any variation in the x-direction. A direct substitution in the Laplace equation gives $d^2f/dy^2 - k_u{}^2 f = 0$ (where $k_u = 2\pi/\lambda_u$) which has a general solution: $f(y) = a_1 \sinh(k_u y) + a_2 \cosh(k_u y)$. From the known relation between the a potential and the associated field, and imposing the symmetry with respect to plane $y = 0$, we obtain for the undulator magnetic field

$$B_y = -\frac{\partial\Phi_{\mathrm{MAG}}}{\partial y} = -k_u a_1 \cosh(k_u y)\sin(k_u z) = -B_0 \cosh(k_u y)\sin(k_u z). \quad (3.127)$$

Restricting, for simplicity, our analysis to the plane $y = 0$, we obtain

$$\vec{B} = -B_0 \sin(k_u z)\widehat{\varepsilon_y}. \quad (3.128)$$

The charged particle trajectory can be calculated, as previously done for the synchrotron radiation, by solving (3.115) and using expression (3.128) for the magnetic field:

$$m\gamma\frac{d\vec{v}}{dt} = \frac{q}{c}\vec{v} \times \vec{B} \Rightarrow \begin{cases} \ddot{x}(t) = \dfrac{qB_0}{mc\gamma}\dot{z}(t)\sin(k_u z) \\ \ddot{z}(t) = -\dfrac{qB_0}{mc\gamma}\dot{x}(t)\sin(k_u z) \end{cases}. \quad (3.129)$$

The two equations of (3.129) are coupled, but we can obtain a first order solution just considering that $\dot{z} = \beta c \approx$ const and $\dot{x} \ll \dot{z}$, which implies

$$\begin{cases} z(t) = \beta c t \\ x(t) = -\dfrac{qB_0}{mc\gamma}\dfrac{1}{k_u{}^2\beta c}\sin(k_u\beta c t) \end{cases}. \quad (3.130)$$

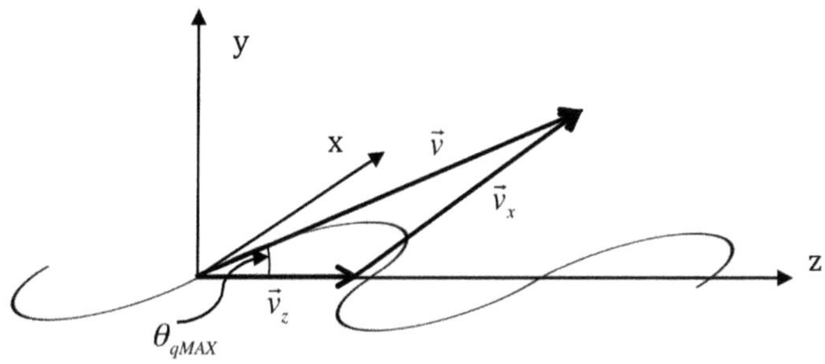

Figure 3.24. Particle angular excursions are harmonic, with maximum excursion K/γ. For $K < 1$ the angular excursions are within the natural radiation cone ($1/2\gamma$), leading to interference effects that are manifested in spectral narrowing. The case of small angular excursions ($K < 1$) is known as the undulator regime. This opposite limit ($K \gg 1$) is known as the wiggler regime.

Introducing the important 'undulator parameter' K defined as

$$K = \frac{qB_0}{mc^2 k_u} \tag{3.131}$$

we obtain a solution $x(t) = -K/(\beta\gamma k_u)\sin(k_u\beta ct)$, which, clearly, is a sinusoidal trajectory, as expected. From (3.130) and referring to figure 3.24, we can deduce the physical meaning of the parameter K.

From the velocities $\dot{z}(t) = \beta c$ and $\dot{x}(t) = K/(\beta\gamma)\cos(k_u\beta ct)\beta c$ we obtain that the tangent with respect to the particle trajectory is

$$\tan(\theta_q) = \frac{dx}{dz} = \frac{\dot{x}(t)}{\dot{z}(t)} = \frac{K}{\beta\gamma}\cos(k_u z), \tag{3.132}$$

therefore the maximum excursion angle, in the relativistic regime, is

$$|\theta_{q\,\mathrm{MAX}}| \approx \frac{K}{\gamma}. \tag{3.133}$$

As discussed when analyzing (3.106), in the relativistic regime, the synchrotron radiation is concentrated in a narrow cone with an angular aperture of $1/\gamma$. This means that, if the undulator parameter $K \leqslant 1$, the particle angular excursion lies within the radiation cone; this is called the 'undulator regime'. In contrast, in the case of strong magnetic fields, when we have $K \gg 1$ the radiation is emitted in a series of bursts separated in space; this case corresponds to the so-called 'wiggler regime'.

The equations of (3.130) are clearly approximated, because the sinusoidal transverse component also implies a longitudinal oscillation. We can, in fact, write

$$\dot{z}(t) = c\sqrt{\beta^2 - \left(\frac{\dot{x}(t)}{c}\right)^2} = c\sqrt{1 - \left(\frac{1}{\gamma^2} + \left(\frac{\dot{x}(t)}{c}\right)^2\right)} \approx c\left(1 - \frac{1}{2\gamma^2} - \frac{\dot{x}(t)^2}{2c^2}\right).$$

Inserting the above calculated expression for the transverse velocity $\dot{x}(t)$ we obtain

$$\begin{aligned}\dot{z}(t) &= c\left[1 - \frac{1}{2\gamma^2} - \frac{K^2}{2\gamma^2}\cos^2(k_u z)\right] \\ &= c\left[1 - \frac{1}{2\gamma^2}\left(1 + \frac{K^2}{2}\right)\right] - \frac{cK^2}{4\gamma^2}\cos(2k_u z).\end{aligned} \tag{3.134}$$

From equation (3.134) we can obtain the average longitudinal velocity by an integration over the undulator period

$$\bar{v}_z = c\left[1 - \frac{1}{2\gamma^2}\left(1 + \frac{K^2}{2}\right)\right] \equiv c\bar{\beta}, \tag{3.135}$$

and therefore the particle trajectory, calculated at the second order, is expressed by the following equations

$$\begin{cases} x(t) = \dfrac{K}{\beta\gamma k_u}\sin(\omega_u t) \\ z(t) = \bar{v}_z t - \dfrac{K^2}{8\gamma^2 k_u}\sin(2\omega_u t) \end{cases} \quad \text{where} \quad \omega_u = c\bar{\beta}k_u. \tag{3.136}$$

Together with the reduced velocity $\bar{\beta}$, expressed by (3.135), we can also obtain the reduced relativistic factor $\bar{\gamma}$. For relativistic particles, in fact, $\bar{\beta} \approx 1 - 1/(2\bar{\gamma}^2)$; from (3.135) we thus have

$$\bar{\gamma} = \frac{\gamma}{\sqrt{1 + \dfrac{K^2}{2}}}. \tag{3.137}$$

We can now use a simple treatment to understand the properties of the radiation emitted by a charged particle passing through a magnetic undulator. Let us analyze the undulator structure as seen in the particle reference frame. In this frame the undulator period will be contracted as $\lambda' = \lambda_u/\gamma$. At the same time the particle, in its co-moving frame, will transversely oscillate like a dipole, thus irradiating at a frequency $\nu' = c/\lambda' = c\gamma/\lambda_u$.

An observer at rest, in the laboratory frame, will see a radiation emitted at a different frequency value due to the Doppler shift. Considering an observation angle θ (between the observer and the velocity vector of the moving frame), the transformed frequency becomes

$$\nu = \frac{\nu'}{\gamma(1 - \bar{\beta}\cos\theta)} = \frac{c}{\lambda_u(1 - \bar{\beta}\cos\theta)}. \tag{3.138}$$

For small θ angles we can expand the cosine and obtain

$$\begin{aligned}\nu &= \frac{c}{\lambda_u\left[1-\bar{\beta}\left(1-\frac{\theta^2}{2}\right)\right]} = \frac{c}{\lambda_u\left[1-\bar{\beta}+\frac{\bar{\beta}\theta^2}{2}\right]}\\ &\approx \frac{c}{\lambda_u\left[\frac{1}{2\bar{\gamma}^2}+\left(1-\frac{1}{2\bar{\gamma}^2}\right)\frac{\theta^2}{2}\right]}\\ &\approx \frac{2c\bar{\gamma}^2}{\lambda_u(1+\bar{\gamma}^2\theta^2)}\end{aligned} \tag{3.139a}$$

having made use of the relativistic approximation $1-\bar{\beta}\approx 1/2\bar{\gamma}^2$ and having considered $\theta^2 \ll \bar{\gamma}^2\theta^2$.

We can finally write the wavelength of the electromagnetic radiation emitted, also making use of (3.137),

$$\lambda = \frac{c}{\nu} = \frac{\lambda_u}{2\bar{\gamma}^2}(1+\bar{\gamma}^2\theta^2) = \frac{\lambda_u}{2\gamma^2}\left(1+\frac{K^2}{2}+\gamma^2\theta^2\right). \tag{3.139b}$$

Many things can be deduced by reading (1.139). The wavelength of the emitted radiation is proportional to the undulator period and reduced by the square of the particle energy. This means that if we think to realize a source of electromagnetic radiation, based on a charged particle beam, passing through a magnetic undulator, we can tailor the emitted wavelength by choosing the undulator period or by changing the particle energy. Another tool is the strength of the magnetic field, which enters in (3.139) by means of the parameter K (see (3.131)). This characteristic overcomes the limitations of the conventional electron-based sources as expressed in section 3.9.

To evaluate the power of the emitted radiation we could follow the rigorous procedure that requires the use of (3.113), where we may insert the particle trajectory (3.136). An easy technique, that exploits the Lorentz transformations, allows us to evaluate the total power emitted. Analyzing (3.136) we realize that the charged particle, in its rest frame, oscillates like a dipole. This fact gives us the opportunity to use the integrated equation (3.110) for the emitted power; we only need to calculate the acceleration in the particle rest frame. Equation (3.136) and the Lorentz transformations of the transverse motion give us

$$x'(t') = \frac{K}{\gamma k_u}\sin(\omega' t') \quad \text{where} \quad \omega' = \bar{\gamma}\omega_u;\ t' = \frac{t}{\bar{\gamma}}.$$

After a double differentiation over the time t' and making use of (3.137), we obtain

$$\ddot{x}'(t') = -\frac{K}{\gamma k_u}\omega'^2\sin(\omega'^2 t') = -\frac{K\gamma k_u c^2}{\left(1+\frac{K^2}{2}\right)}\sin(\omega' t').$$

The average square acceleration, to be used in the Larmor formula (3.110), is therefore

$$|\dot{v}_x'|^2 = \frac{1}{2}\frac{\gamma^2 k_u{}^2 c^4 K^2}{\left(1+\frac{K^2}{2}\right)^2}, \tag{3.140}$$

which leads to an amount of total power, integrated over the observation angles, equal to

$$\frac{d\mathcal{E}}{dt} = \frac{2}{3}\frac{q^2}{c^3}|\dot{v}_x'|^2 = \frac{q^2 c\gamma^2 k_u{}^2 K^2}{3\left(1+\frac{K^2}{2}\right)^2}. \tag{3.141}$$

The complete power distribution, in terms of frequency and angular directions, is obtained as anticipated by means of (3.113) and (3.136). We omit the long algebraic calculations, finally obtaining

$$\begin{aligned}
\frac{d\mathcal{E}}{d\omega d\Omega} &= \frac{8(q\gamma N)^2}{c}\sum_{m=1}^{\infty}\left(\frac{m\xi}{K}\frac{\sin\nu_m/2}{\nu_m/2}\right)^2 \\
&\times\left[\frac{1}{2}\left(\frac{\gamma\theta}{K}\right)^2 A^2{}_{0,m} + \sqrt{2}\left(\frac{\gamma\theta}{K}\right)\cos\phi A_{0,m}A_{1,m} + A^2{}_{1,m}\right], \\
A_{\alpha,m} &= (-1)^{m+\alpha}\sum_{m'=-\infty}^{\infty}(-1)^{m'}J_{m'}(m\xi) \\
&\times\left[J_{m-2m'-\alpha}(m\sqrt{2}\chi\cos\phi) + J_{m-2m'+\alpha}(m\sqrt{2}\chi\cos\phi)\right],
\end{aligned} \tag{3.142}$$

where $\alpha = 0,\ 1; \xi = \frac{1}{2}\frac{K^2}{1+K^2+\gamma^2\theta^2};\quad \chi = \frac{2K\gamma\theta}{1+K^2+\gamma^2\theta^2};$

$$\nu_m = 2\pi N\left(\frac{\nu - m\nu_0}{\nu_0}\right),$$

where ν_0 is the central frequency emitted as expressed by (3.139) and m refers to the harmonic number. The geometry explaining the angles involved in (3.142) is shown in figure 3.25.

Equation (3.142) is not easy to read except for a couple of facts. The first is the presence of sinc function concerning the frequency behavior; this is strictly related to the lengthening of the pulse duration and the consequent restriction of the spectrum. The specific sinc-like behavior is related to the Fourier transformation, over a finite time $N\lambda_u/c$, of the oscillating field. The second fact is related to the presence of the harmonics m directly connected to the fact that the particles oscillate at a double frequency, in the longitudinal direction, with respect to that of the transverse coordinate. All the other characteristics are very difficult to extract at a first glance.

The case of an on-axis observation produces the spectrum that is visible in figure 3.26. The graph, which is plotted as a function the detuning parameter ν_m, has been

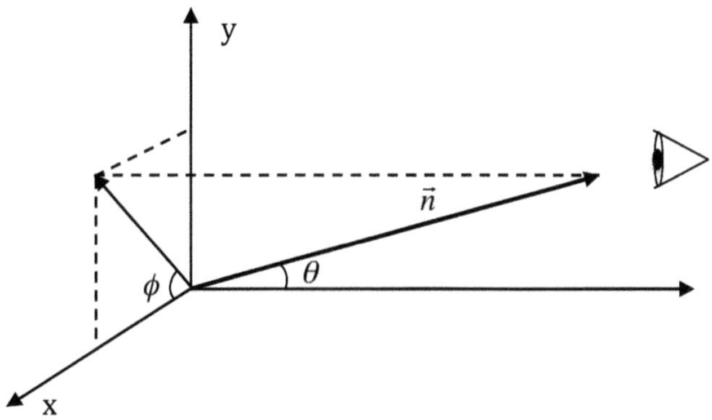

Figure 3.25. Radiation directions and angles in the laboratory reference frame (note that the x-axis should be reversed to have a left-handed vector system).

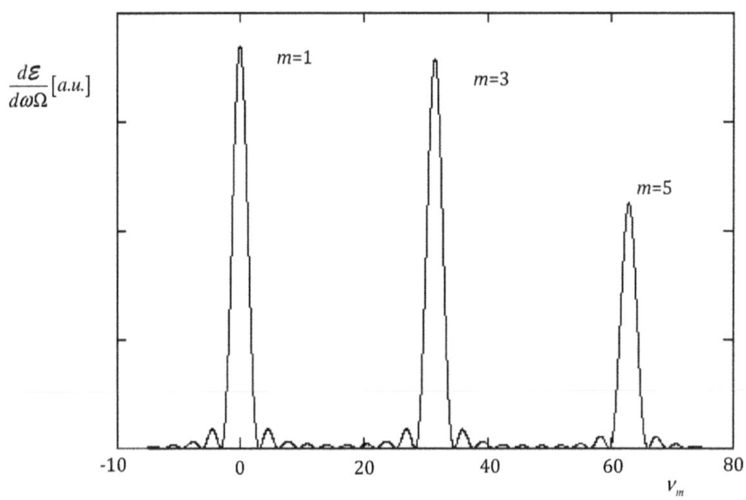

Figure 3.26. On-axis ($\theta = 0$) electromagnetic spectrum of an emitted electron passing through a magnetic undulator with $N = 5$, $\varphi = 1$ and $K = 1$.

calculated considering an undulator parameter $K = 1$, which ensures a high generation of harmonics. In figure 3.26 only the odd harmonics ($m = 1, 3, 5$ in the present case) are visible; this is a general feature for $\gamma\theta = 0$.

The complete spectrum, including the reduced polar angle $\gamma\theta$, is reported in figure 3.27. It is interesting to note how the even harmonics start to appear at increasing values of the parameter $\gamma\theta$. The small amplitudes of the even harmonics with respect to the odd ones is due to the choice of the azimuthal angle that for figure 3.27 is $\varphi = \pi/2$.

As has been mentioned several times, the polar angular distribution for the fundamental harmonic is related to the inverse of the relativistic factor γ, but it is not sensitive to the azimuthal angle φ. The higher harmonics have different angular

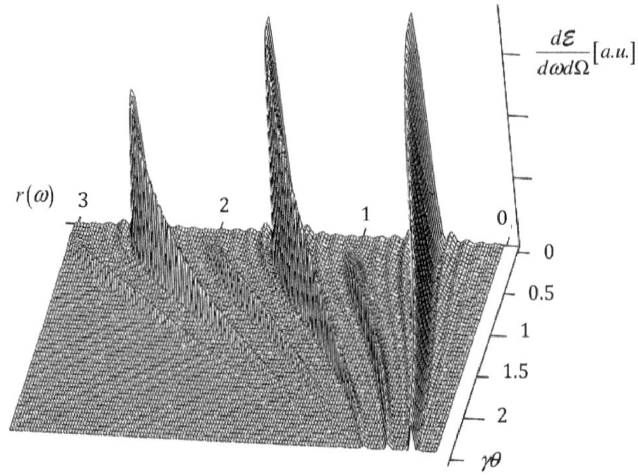

Figure 3.27. Spectral brightness in a three-dimensional plot of the radiation emitted in an undulator as a function of $r = \omega/(2\gamma^2\omega_u)$ and $\gamma\theta(\varphi = \pi/2)$.

behavior depending also on whether they are odd or even. In figure 3.28 we have reported the power distribution in terms of contour plot and corresponding three-dimensional plot, for the various harmonics.

3.12 The gain mechanism

We have anticipated in section 3.9 how FELs share, with conventional lasers, the stimulated emission process, which is the basis for the coherence, both spatial and temporal, of the radiation emitted by all these devices. Stimulated emission, in FELs, is governed by the so-called gain mechanism that establishes the relative energy increase of the electromagnetic radiation field, during one passage inside the magnetic undulator, with respect to the pre-existing field. To be more explicit, if we have an electromagnetic wave present inside the volume of the magnetic field generated by a magnetic undulator, this wave will increase in amplitude if a charged particle interacts (with the appropriate phase) with the electric field associated with the wave and with the magnetic field of the undulator. We already discussed, in section 3.2, how field and charged particles can exchange energy, but in this case the presence of the electromagnetic wave can stimulate the emission (or absorption). This is the stimulated emission process where the field–particle relative phase plays an important role.

We have already introduced the Lorentz force and argued its connection in the particle field energy exchange in equations (3.8) and (3.9), respectively. We can now write them in a more general relativistic form:

$$\begin{cases} \dfrac{d}{dt}(\gamma\vec{\beta}) = \dfrac{q}{m_0 c}[\vec{E}_r + \vec{\beta} \times (\vec{B}_u + \vec{B}_r)] \\ \dfrac{d}{dt}\gamma = \dfrac{q}{m_0 c}\vec{\beta} \cdot \vec{E}_r \end{cases} . \tag{3.143}$$

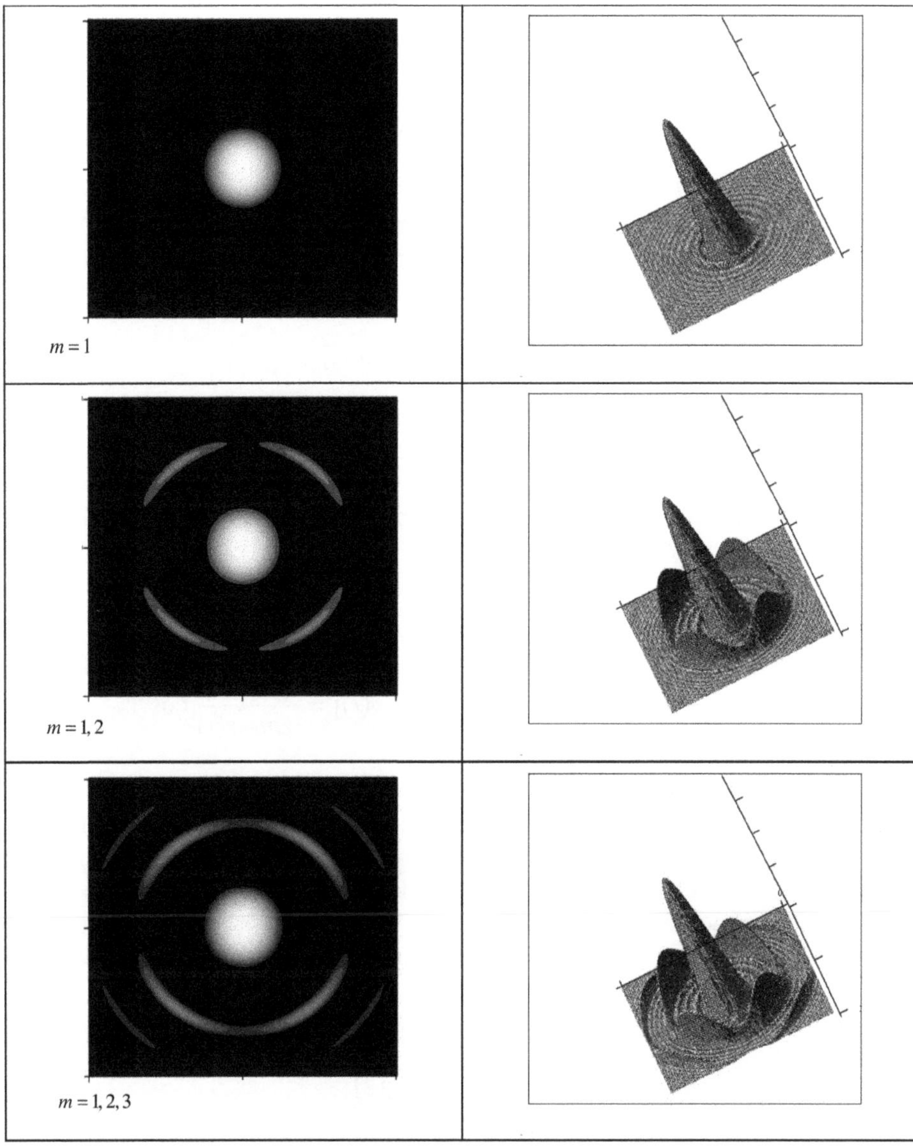

Figure 3.28. Angular distribution of the spectral brightness as level curves and three-dimensional plots. The first set refers to the presence of only the fundamental harmonic ($m = 1$). In the second set the second harmonic is also present, while in the third set a faint third harmonic is visible.

In (3.143) $\vec{E}_r$ and $\vec{B}_r$ are the electric and magnetic fields associated with the electromagnetic wave, $\vec{B}_u$ is the undulator magnetic field and m_0 is the charged particle mass in its rest frame. The vector velocity $\vec{\beta}$ has been previously calculated from (3.129), and is

$$\begin{cases} \beta_x(t) = \dfrac{K}{\gamma}\cos(k_u z(t)) \\ \beta_z(t) = \left[1 - \dfrac{1}{2\gamma^2}\left(1 + \dfrac{K^2}{2}\right)\right] - \dfrac{K^2}{4\gamma^2}\cos(2k_u z(t)) \end{cases}. \tag{3.144}$$

The electric field associated with the co-propagating electromagnetic wave can be easily derived from the vector potential $\vec{A}(\vec{r}, t)$, the solution of the homogeneous Helmholtz equations (3.26) or (3.93). The result is

$$\vec{E}_r = \vec{E}_0(t)\cos[\omega t - kz + \varphi(t)], \tag{3.145}$$

where $k = \omega/c = 2\pi/\lambda$ is the radiation wavenumber and $\varphi(t)$ is the radiation–particle relative phase. Considering the geometry shown in figure 3.24 and an electromagnetic wave propagating in the z-direction, we have a particle–wave coupling in terms of energy exchange only in the x-direction. Under these hypotheses the second equation of (3.143) is written

$$\begin{aligned} \frac{d\gamma(t)}{dt} &= \frac{qKE_0(t)}{m_0c\gamma(t)}\cos(k_u z(t))\cos[\omega t - kz(t) + \varphi(t)] \\ &\approx \frac{qKE_o(t)}{2m_0c\gamma(t)}\cos[(k + k_u)z(t) - \omega t - \varphi(t)] = \frac{qKE_o(t)}{2m_0c\gamma(t)}\cos\Psi. \end{aligned} \tag{3.146}$$

In the manipulation of (3.146) we have neglected a term proportional to $\cos[(k - k_u)z(t) - \omega t - \varphi(t)]$, because its phase term cannot be kept constant except for backward propagating waves and, therefore, does not allow an efficient energy exchange process. The phase term Ψ is also called the 'ponderomotive phase', and it contains the longitudinal coordinate $z(t)$.

From (3.134) we can calculate the longitudinal acceleration, after neglecting the oscillating term that is average to zero:

$$\frac{d^2z(t)}{dt^2} = c\frac{\dot{\gamma}(t)}{\gamma^3(t)}\left(1 + \frac{K^2}{2}\right). \tag{3.147}$$

Inserting (3.146) in (3.147) we obtain

$$\ddot{z}(t) = \frac{qE_0(t)K}{2m_0\gamma^4(t)}\left(1 + \frac{K^2}{2}\right)\cos\Psi = \frac{qB_0E_0(t)\lambda_u}{4\pi\gamma^4(t)(m_0c)^2}\left(1 + \frac{K^2}{2}\right)\cos\Psi. \tag{3.148}$$

From the definition of the ponderomotive phase we also obtain

$$\ddot{\Psi}(t) = (k + k_u)\ddot{z} \approx k\ddot{z} = \frac{4\pi\gamma^2}{\lambda_u}\frac{1}{\left(1 + \frac{K^2}{2}\right)}\ddot{z} \tag{3.149}$$

after (3.139b) and for on-axis observation ($\theta = 0$). The combination of (3.148) and (3.149) leads to the most relevant equation for the FEL dynamics:

$$\ddot{\Psi} = (k + k_u)\frac{qE_0(t)K}{2m_0\gamma^2(t)}\left(1 + \frac{K^2}{2}\right)\cos\Psi = \Omega^2\cos\Psi. \tag{3.150}$$

This is a pendulum-like equation where Ω plays the role of the Rabi frequency, which regulates the Bloch oscillations in two-quantum-level systems. This frequency is proportional to the square root of the radiation associated electric field amplitude, which is, at the same time, a function of the time. Another convenient way to express the pendulum equation (3.150) is obtained easily by means of the dimensionless parameters, such as time $\tau = ct/L$ (where $L = N\lambda_u$is the undulator length) and the optical electric field $|a|$

$$\frac{d^2\Psi}{d\tau^2} = |a|\cos\Psi, \qquad \text{where} \quad |a| = \frac{4\pi qNLKE_0(t)}{m_0(c\gamma(t))^2} \tag{3.151}$$

Equations (3.150) and (3.151) describe the dynamical evolution of the charged particle under the influence of the undulator magnetic field $\vec{B}_0$ and of the electric field associated with the electromagnetic radiation $\vec{E}_r$.

It is clearly understandable that (3.150), or (3.151), cannot completely describe the dynamics of an FEL which must also include the evolution of the electric field and, consequently, the evolution of the electromagnetic intensity. The equation to be considered is (3.90), which describes the evolution of the electromagnetic potentials.

It is useful to use a different gauge to solve the present problem; the most appropriate is, in fact, the Coulomb gauge, also called the transverse gauge. This gauge implies that $\vec{\nabla} \times \vec{A} = 0$, which transforms (3.93) as follows (we do not enter into the calculations details, but the result can be directly proved starting from (3.90)):

$$\begin{cases} \vec{\nabla}\left(\dfrac{\partial\Phi}{\partial t}\right) = 4\pi\vec{J}_z \\ \nabla^2\vec{A} - \dfrac{1}{c^2}\dfrac{\partial^2\vec{A}}{\partial t^2} = -\dfrac{4\pi}{c}\vec{J}_\perp \\ \nabla^2\Phi = -4\pi\rho \end{cases} \tag{3.152}$$

This solution is ideal for an FEL because no static charges ρ are involved ($\Rightarrow\Phi = 0$) and the coupling between the charged current and the electromagnetic field is purely transverse and this fact justifies the inhomogeneous term of the equation for $\vec{A}$ in (3.152).

The current that drives the dynamics and is generally expressed by (3.100) can be written for an ensemble of particles as

$$\vec{J}_\perp = qc\sum_m \vec{\beta}_{m\perp}\delta(\vec{r} - \vec{r}_m(t)), \tag{3.153}$$

where $\vec{r}_m(t)$ represents the trajectory of the mth particle (expressed by (3.136)) and $\vec{\beta}_{m\perp}$ is the transverse velocity calculated from (3.130).

The electromagnetic potential, expressed in the Coulomb-transverse gauge, can be derived from (3.87), which, for $\Phi = 0$ reduces to $\vec{E} = -1/c\partial\vec{A}/\partial t$; the result is

$$\vec{A}(t) = \frac{\vec{E}_0(t)}{k}\sin(kz - \omega t + \varphi(t)) \approx \frac{\vec{E}_0(t)}{k}e^{i(kz-\omega t)}e^{i\varphi(t)}, \tag{3.154}$$

where we have made use of (3.145) and the dispersion relation in a vacuum $k = \omega/c$. Inserting (3.153) and (3.154) into (3.152) we can come to the equation that rules the time behavior of the electromagnetic radiation amplitude $E_0(t)$ in terms of the relative phase $\Psi(t)$. Considering only the slow variations of these two functions, with respect to the radiation period ω^{-1}, we are allowed to neglect the second derivatives of the relation (3.152), finally leading, after some modest algebra, to the following equation:

$$\begin{aligned}&\frac{1}{c}\frac{\partial E_0(t)}{\partial t}e^{i(kz-\omega t)}e^{i\varphi(t)} = -\pi qK\sum_m \frac{1}{\gamma_m}e^{-i(k_u z)}\delta(\vec{r} - \vec{r}_m(t))\\ &\Rightarrow \frac{1}{c}\frac{\partial E_0(t)}{\partial t} = -\pi qK\sum_m \frac{e^{-i\Psi(t)}}{\gamma_m}\delta(\vec{r} - \vec{r}_m(t)).\end{aligned} \tag{3.155}$$

Introducing, as previously done, the dimensionless parameters τ and a, we obtain the compact equation

$$\frac{\partial a(\tau)}{\partial \tau} = -j\langle e^{-i\Psi(\tau)}\rangle \quad \text{where } j = \frac{4N(\pi qLK)^2}{m_0c^2\gamma^3}\Delta_q. \tag{3.156}$$

The dimensionless current density j is proportional to the charged particle density distribution Δ_q and the brackets represent the normalized average over the same particles. The coupled equations (3.150) and (3.155), or their dimensionless compact form (3.151) and (3.156), represent the complete dynamics of an FEL device; their solutions $(E_0(t), \Psi(t))$ or $(a(t), \Psi(t))$ give the needed answers regarding the growth of the electromagnetic signal and its linkage with the phase evolution $\Psi(t)$ and its first derivative $\dot{\Psi}(t)$.

A convenient way to visualize this process is to consider the so-called 'phase-space' $[\Psi(t), \dot{\Psi}(t)]$. In the phase-space the particles are distributed, at $t = 0$ (or $\tau = 0$), with a set of initial conditions $[\Psi_n(0), \dot{\Psi}_n(0)]$; we can consider, as an ideal example, a 'fluid' of charged particles that equally populates all the phases $\Psi(t)$ (in the range $-\pi \rightarrow \pi$), while it might have a Gaussian distribution in the phase velocity $\dot{\Psi}(t)$ that, we will show later, is proportional to the energy distribution of the particles.

In figure 3.29(a) we can see the phase-space at $t = 0$, with the particle initially distributed as described and a separatrix line defined by $\dot{\Psi} = \pm\sqrt{2a[1 - \cos(\Psi)]}$. When the time evolves, the particle fluid starts to feel the ponderomotive potential and undergoes a rotation in the phase-space; at the same time the separatrix grows, due to its relation with the field amplitude, and gradually includes more particles

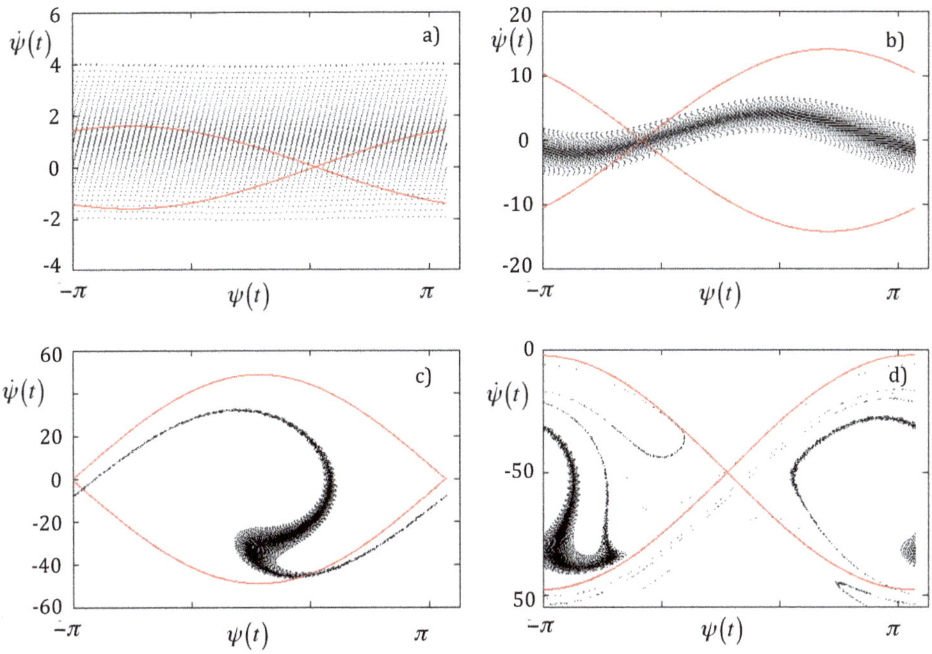

Figure 3.29. Phase-space trajectory evolution of the particle distribution. (a) At the initial time the particles are distributed according to their initial conditions, which in the present case are a uniform distribution in Ψ and a Gaussian distribution in $\dot{\Psi}$. (b) and (c) The system evolves and the particles are trapped in the ponderomotive potential delivering its energy to the electromagnetic field as proved by the increasing values of the separatrix. (d) The process reaches its saturation and the field amplitude stabilizes reaching its maximum value.

(see figure 3.29(b) and (c)). In figure 3.29(d) the process ends at a saturation level and the energy is no longer transferred from the particles to the electromagnetic field; almost all the particles are now trapped in the potential and continue their synchrotron motion.

From the above analysis we can introduce the important parameter called the 'gain'. This parameter is widely used when describing all the generators of coherent radiation, therefore including conventional laser and FELs. The gain can be defined as the amount of energy increase, by the electromagnetic field, during the passage of the charged particles inside the magnetic undulator, with respect a pre-existing, or injected, field energy

$$G(t) = \frac{\Delta\mathcal{E}_r}{\mathcal{E}_r^0} = -m_0c^2\frac{\Delta\gamma(t)}{\mathcal{E}_r^0}. \tag{3.157}$$

The $\Delta\gamma$ term is related to the charged particle energy loss and, therefore, this fact explains the minus sign. The way to evaluate this term passes through a relation derived combining (3.147) and (3.149):

$$\ddot{\Psi}(t) = (k + k_u)c\frac{\dot{\gamma}(t)}{\gamma^3(t)}\left(1 + \frac{K^2}{2}\right). \tag{3.158}$$

By simply integrating (3.158) we obtain

$$\Delta\gamma(t) = \frac{\gamma^3(t)}{c(k + k_u)\left(1 + \frac{K^2}{2}\right)}\Delta\dot{\Psi}(t). \tag{3.159}$$

Equation (3.159) reveals the link between the energy exchange $\Delta\gamma(t)$ between the particles and field, and the phase derivative variation. This result better explains the use of the phase space for the analysis of the FEL dynamics (see figure 3.29). A direct substitution of equation (3.159) into (3.157) gives

$$G = -\frac{m_0 c^2 \gamma^3(t)}{c(k + k_u)\left(1 + \frac{K^2}{2}\right)\mathcal{E}_r^0} N_e \langle \Delta\dot{\Psi}(t) \rangle. \tag{3.160}$$

As previously done for (3.156) we have introduced a normalized average over the N_e particles in order to express the gain of a real charged beam. Equation (3.160) tells us that to evaluate the gain process in an FEL we need to express the average $\langle \Delta\dot{\Psi}(t) \rangle$. If we restrict our analysis to the 'small signal' regime we need to simply solve the pendulum equations (3.150), or (3.151), neglecting (3.155), or (3.156). Moreover, we see that this equation cannot be solved analytically and requires a perturbative approach; this is allowed in the so-called 'small gain' regime. The perturbative analysis involves the Rabi frequency $\Omega^2 \equiv \varepsilon$ as an expansion parameter allowing us to write $\Psi = \Psi_0 + \varepsilon\Psi_1 + \varepsilon^2\Psi_2 + \ldots$. Equation (3.150) can therefore be written as

$$\ddot{\Psi}_0(t) + \varepsilon\ddot{\Psi}_1(t) + \varepsilon^2\ddot{\Psi}_2(t) = \varepsilon\cos(\Psi_0(t) + \varepsilon\Psi_1(t)). \tag{3.161}$$

Equation (3.161) represents the pendulum equation derived up to the second order expansion in the parameter ε.

By equating the rhs terms with the left-hand terms having the same perturbation order we obtain, after simple algebra, the normalized average of the phase derivative:

$$\langle \Delta\dot{\Psi}(t) \rangle = \left(\frac{\Omega L}{c}\right)^4 \frac{c}{4L}\frac{d}{d\nu}\left(\frac{\sin \nu/2}{\nu/2}\right)^2. \tag{3.162}$$

The dimensionless detuning parameter ν is defined in (3.142) and indicates the frequency shift with respect the central emitted frequency defined by (3.139). By inserting (3.162) into (3.160) we have

$$G = -\frac{m_0 c^2 \gamma^3(t)}{c(k + k_u)\left(1 + \frac{K^2}{2}\right)} \frac{N_e}{\mathcal{E}_r^0}\left(\frac{\Omega L}{c}\right)^4 \left(\frac{c}{4L}\right)\frac{d}{d\nu}\left(\frac{\sin \nu/2}{\nu/2}\right)^2. \tag{3.163}$$

A proper manipulation of the above relation makes use of (3.150) for the Rabi frequency Ω, the definition of the Alfven current as $I_0 \equiv m_0 c^3/e$, and it needs the introduction of the total electromagnetic energy in a vacuum as $\mathcal{E}_r^0 = \mathrm{E}_0^2 V/(8\pi)$ and the evaluation of the total number of particles N_e (actually electrons) over the electromagnetic volume V as $N_e/V = I/(ec\Sigma_r)$. We have denoted by I the charged particles current, while Σ_r is the average radiation transverse section. Equation (3.163) eventually becomes

$$G = -\pi \frac{\left(1 + \frac{K^2}{2}\right)K^2}{\gamma^5} \frac{L^3}{2\Sigma_r} \frac{I}{I_0}(k + k_u) \frac{d}{d\nu}\left(\frac{\sin \nu/2}{\nu/2}\right)^2. \tag{3.164}$$

This a result of crucial importance for the design of the FEL device and will be more carefully discussed in forthcoming chapters.

Let us note that it yields a general criterion to assemble the physical parameters of the entire device to obtain a safe lasing. First of all the gain is proportional to the current I and to the third power of the undulator length L. High values of these variables are not always easy to accomplish apart from in big installations. Another important parameter is the normalized beam energy γ that scales with the inverse fifth power. We have already seen, in (3.139), how small wavelength radiation (high-energy photons) can be generated in an FEL by increasing the beam energy, but now we understand that this is not an easy task. An FEL source, operating from the visible to ultraviolet spectral region, or beyond, requires very long undulator and high currents, to compensate the gain reduction related to the γ^5 factor. The extension of the FEL theory to include devices operating at short wavelengths (the x-ray region of the spectrum) will be discussed in forthcoming chapters.

Let us conclude this FEL gain section by discussing the spectral behavior of (3.164). The relevant term is the derivative of the 'sinc' function that, as established in (3.142), is the line shape of the undulator spontaneous emission. This result has been obtained considering the particular geometry of a linear undulator, such as the one described in section 3.11, but it holds in general, as has been demonstrated by Madey in 1979. A sketch of the gain line shape is reported in figure 3.30(b), together with the spontaneous emission line shape (figure 3.30(a)). As we can see, the frequency where we have the maximum of the spontaneously emitted radiation corresponds to a net gain of zero. The FEL, during the laser operation, will therefore emit at a shifted frequency corresponding to a detuning parameter $\nu = 2.6$. Another

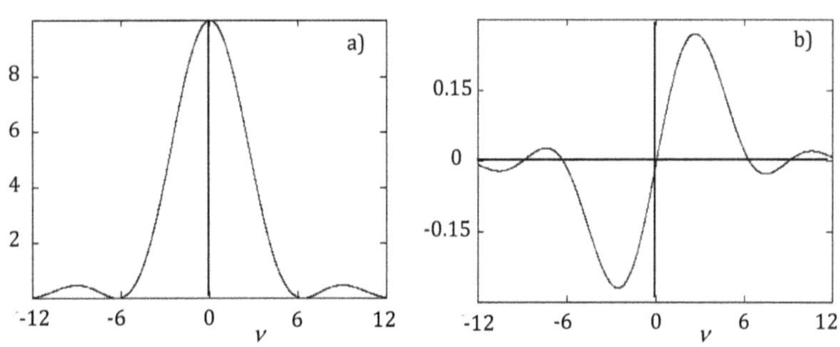

Figure 3.30. Spectral course of the spontaneous emission and gain. (a) Spontaneous emission versus ν behaves like a 'sinc' function (see (3.142)). (b) Gain coefficient versus ν performs like the spontaneous emission derivative (see (3.164)).

important feature, which can be deduced from figure 3.30, is that, due to the asymmetric nature of the gain shape, we have frequency values for which the gain is negative. This means that, under these conditions, the electromagnetic field loses its energy in favor of the charged particle beam that, as a final result, accelerates. This is a unique feature of FELs in comparison with conventional lasers, in which radiation will never return energy to the pumping system.

All the previous discussions are valid in the small-signal and small-gain regimes, which is the most appropriate case for an FEL operating as an oscillator. Different regimes, such as high-gain and the self-amplified spontaneous emission (SASE), present other relevant features and will be described in the forthcoming chapters.

Further reading for chapter 3

General electromagnetic theory

Hertz H 1963 *Electric Waves* (New York: Dover)

Svelto O 1989 *Principles of Lasers* (New York: Plenum)

Pantell R H 1980 Energy and momentum conservation requirements for electron interactions with electromagnetic radiation *Free-Electron Generators of Coherent Radiation, Physics, and Quantum Electronics* vol 7 ed S F Jacobs *et al* (Reading, MA: Addison Wesley)

Jelley J V 1958 *Cerenkov Radiation and its Applications* (Oxford: Pergamon)

Heitler W 1954 *The Quantum Theory of Radiation* (Oxford: Clarendon)

Jackson J D 1975 *Classical Electrodynamics* (New York: Wiley)

Hirschmugl C J, Sagurton M and Williams G P 1991 Multiparticle coherence calculations for synchrotron radiation emission *Phys. Rev.* A **44**(2) 1316–20

Landau L D and Lifshits E M 1994 *The Classical Theory of Fields* (Oxford: Butterworth Heinemann)

Microwave and electron tubes

Barker R J, Booske J H, Luhmann N C and Nusinovich G S 2005 *Modern Microwave and Millimeter Wave Power Electronics* (New York: Wiley)

Gewartowky J W and Watson H A 1965 *Principles of Electron Tubes* (Princeton, NJ: van Nostrand)

Branch G M 1961 Electron beam coupling in interaction gaps of cylindrical symmetry *IRE Trans. Electr. Devices* **8**(3) 193–207

Collin R E 1966 *Foundations of Microwave Engineering* (Singapore: McGraw-Hill)

Synchrotron radiation and related topics

Schwinger J 1949 On the classical acceleration of accelerated electrons *Phys. Rev.* **75**(12) 1912–25

Rybicki G B and Lightman A P 2004 *Radiative Processes in Astrophysics* (Weinheim: Wiley)

Attwood D 1999 *Soft x-ray and Extreme Ultraviolet Radiation* (New York: Cambridge University Press)

Luchini P and Motz H 1990 *Undulators and Free Electron Lasers* (Oxford: Clarendon)

Barbini R, Ciocci F, Dattoli G and Giannessi L 1990 Spectral properties of the undulator magnets radiation: analytical and numerical treatment *Riv. Nuovo Cimento* **13(6)** 1990

Dattoli G, Renieri A and Torre A 1993 *Lectures on the Free Electron Laser Theory and Related Topics* (Singapore: World Scientific)

Dattoli G and Renieri A 1985 Experimental and theoretical aspects of free electron laser *Laser Handbook* vol 4, ed M L Stitch and M Bass (Amsterdam: North Holland)

Ciocci F, Dattoli G, Torre A and Renieri A 2000 *Insertion Devices for Synchrotron and Free Electron Laser* (Singapore: World Scientific)

Doria A 2014 Relativistic electron-based THz Sources: principles of operation and the ENEA experience *Physics and Applications of Terahertz Radiation* ed M Perenzoni and D J Paul (Heidelberg: Springer)

IOP Publishing

Charged Beam Dynamics, Particle Accelerators and Free Electron Lasers

Giuseppe Dattoli, Andrea Doria, Elio Sabia and Marcello Artioli

Chapter 4

Linac-based free electron laser devices: oscillator and single passage operating modes

4.1 Introduction

In the previous chapter we have discussed the physical principles underlying the operation of free electron devices and we have framed them under the common definition of generators of coherent radiation by electrons, not bound in an atomic or molecular system. The notion of the free electron laser (FEL) traces back to the early 1970s, when John M J Madey (1943–2016), at the time a PhD student at Stanford University, conceived the idea that relativistic electrons moving in a magnetic undulator can emit and amplify coherent radiation, suitably stored in an optical cavity, in a laser-like fashion. Madey was the leader of the first two experiments at Stanford University proving FEL amplification and oscillation.

In the past there was also a sterile debate regarding the question whether an FEL is a laser or not, and this meaningless question arises from time to time. Even though the matter is bereft of any substantive scientific content, we dwell on this argument to provide the reader with a more general view regarding the interplay of 'conventional' lasers and FEL systems.

As already stressed, FELs exploit the emission by electrons, whose motion is constrained in an alternating magnetic field and is not the result of a de-excitation of an atomic or molecular ensemble, which have undergone a population inversion. Notwithstanding, as we have already discussed and as will be further stressed in the following sections, both devices can be described using the same sets of macroscopic parameters (gain, saturation intensity, efficiency, etc) they can accordingly be treated on the basis of a unified description, which will be sketched in this introductory section.

doi:10.1088/978-0-7503-1239-4ch4

To make the discussion general and self-contained we will start *ab ovo* going back to the Einstein treatment of the black-body problem and to the relevant distribution law, yielding the equilibrium between radiation at a given frequency and matter at a given temperature T.

The point of view elaborated by Einstein to derive the Planck distribution was prompted both by Planck's original treatment and by the idea put forward by Bohr according to which the 'size' of energy variation at the atomic scale occurs on integer multiple of the Planck constant. The theory of spontaneous and stimulated emission in quantum systems, be they atomic or molecular, is sketched in figure 4.1. There we have considered a system (say atomic) characterized by two energy levels $E1$ and $E2$ (1 being the lower energy state).

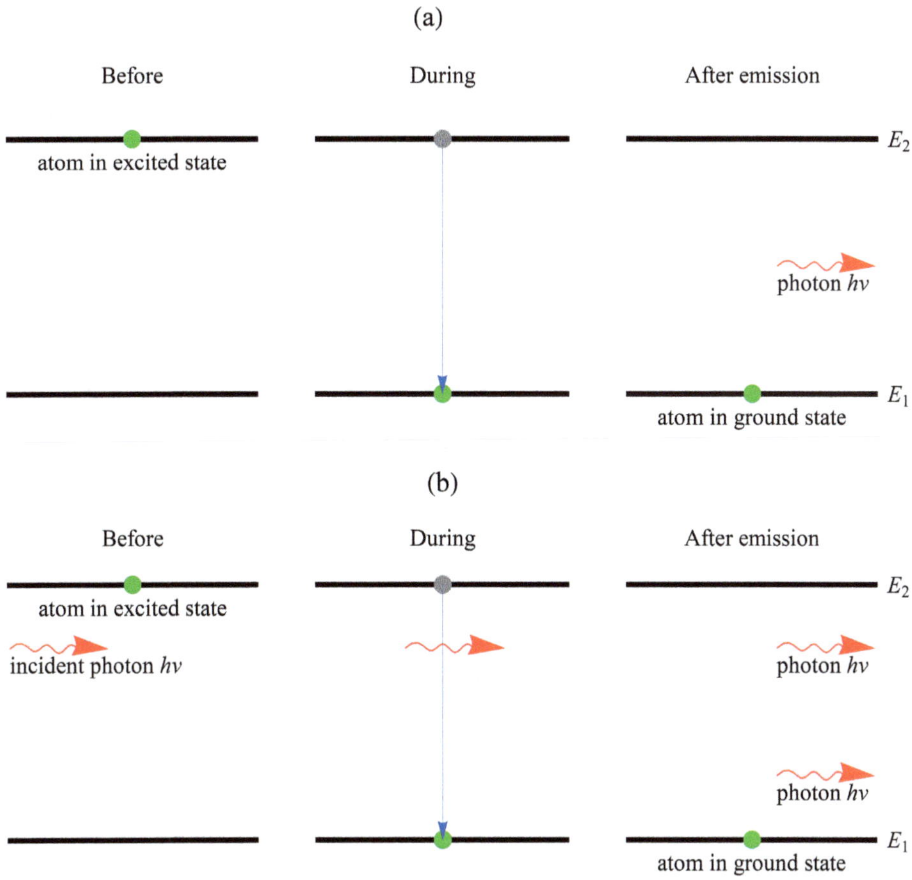

Figure 4.1. Two-level atom interaction: spontaneous (a) and stimulated (b) emission.

The only way the system can interact with the surrounding environment is that of absorbing or emitting a photon with the frequency matching the energy gap, namely

$$\varepsilon = \hbar\omega, \ \omega = \frac{E_2 - E_1}{\hbar}. \tag{4.1}$$

In figure 4.1(a) it is assumed that the system is sitting in the upper energy level (the excited state). The principles of minimum energy and of conservation of energy require that it 'decays' to the lower state spontaneously by emitting a photon of energy given by (4.1).

The other process, illustrated in figure 4.1(b), considers the transition from the upper to the lower level assisted by a photon with energy (4.1) (for a more complete description of two level systems interacting with electromagnetic radiation see section 4.10). The existence of this further de-excitation mechanism was Einstein's key idea paving the way to the concept of stimulated emission along with the conception of the laser (an acronym for 'light amplification (by) stimulated emission (of) radiation') mechanism, as we will see below.

Following the Einstein treatment, we assume that the two processes, spontaneous and stimulated emission, occur with probabilities $A_{2,1}$ and $B_{2,1}$, respectively. It is furthermore supposed that the system under study, consisting of N atoms ($N1$ are in the lower state and $N2$ in the higher state ($N1 + N2 = N$)), is interacting with an external field of intensity $I(\omega)$. We can therefore write the following rate equation:

$$N_1 B_{1,2} I(\omega) = N_2 B_{2,1} I(\omega) + N_2 A_{2,1}. \tag{4.2}$$

which rules the equilibrium between the atoms and the radiation. We have denoted by A and B two coefficients controlling the photon emission from the upper to the lower level and vice versa. In particular, $B_{1,2(2,1)}$ is the probability that the radiation induces a transition from the lower to the upper level (and upper lower), furthermore $A_{2,1}$ controls the so-called spontaneous emission, namely the transition from the higher to the lower state. The roles played by the two coefficients are significantly different and it is worth stressing that the physical process controlled by A does not depend on the radiation intensity.

Solving for $I(\omega)$ we find

$$I(\omega) = \frac{A_{2,1}}{\frac{N_1}{N_2} B_{1,2} - B_{2,1}}. \tag{4.3}$$

We can complete our analysis with the following further remarks:

a) Assume that $B_{1,2} = B_{2,1}$, namely that the probability of inducing a transition from the lower to the upper levels and vice versa are identical.

b) Note that according to the Boltzmann law $\frac{N_1}{N_2} = e^{-\frac{\hbar\omega}{KT}}$.

c) Exploit the Wien law to set $A_{2,1} = \frac{\hbar\omega^3}{\pi^2 c^2} B_{2,1}$.

We accordingly find from (4.3)

$$I(\omega) = \frac{1}{\pi^2 c^2}\frac{\hbar\omega^3}{e^{-\frac{\hbar\omega}{KT}} - 1}, \tag{4.4}$$

which is the Planck distribution law.

It is now worth stressing how the previous point of view, allowing the derivation of Planck's law, is important to understand the laser process itself. We consider therefore figure 4.2, in which we have shown a large number of two-level systems, all sitting in the state with larger energy.

This particular condition realizes that which, in laser physics, is called an active medium with an inverted population; it is therefore evident that the stimulated emission will cause the de-excitation of the medium with a consequent amplification of the incoming field.

Let us therefore consider a medium consisting of $N = N_1 + N_2$ two-level systems, supposed to interact with an electromagnetic field with constituent photon energies matching the energy difference between the 1, 2 states. Assuming that the wave moves along the z-direction (see figure 4.3), we can evaluate the variation of the number of photons using the following differential equation, which is a naïve consequence of (4.2):

$$\frac{dn}{dz} = (N_2 - N_1)b_{1,2}n + a_{2,1}N_2, \tag{4.5}$$

where the Einstein coefficients have just been rescaled by keeping account of the change in physical dimensions from field intensity to number of photons.

The differential equation in (4.5) can be solved by keeping the level population constant during the interaction, thus finding

$$\begin{aligned} n(z) &= n_o \exp(\alpha z) + \beta\frac{\exp(\alpha z) - 1}{\alpha}, \\ \alpha &= (N_2 - N_1)\, b_{2,1},\ \beta = a_{2,1}\, N_2, \end{aligned} \tag{4.6}$$

with $n_o$being the number of the photons of the seed field.

Albeit simple, the previous solution is extremely useful, it is indeed provided by two contributions, one of which (the spontaneous part) is independent of the number

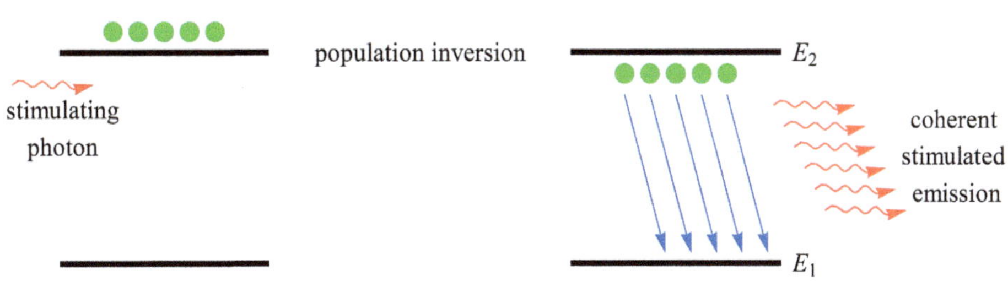

Figure 4.2. Stimulated emission and consequent amplification of the incoming photon

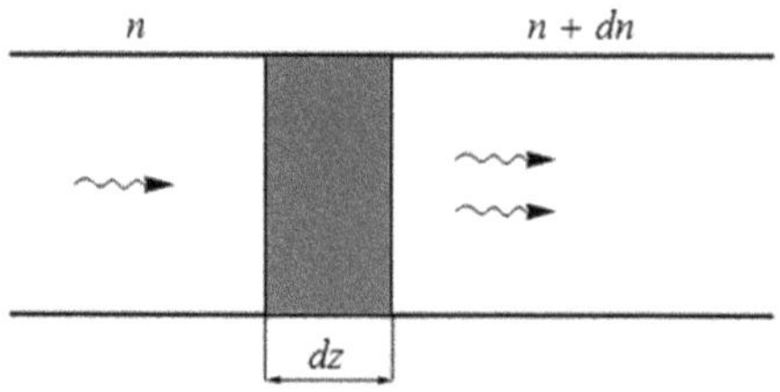

Figure 4.3. Radiation increment.

of initial photons in the input field. If we assume that the stimulated part is dominant with respect to the other and if $N_2 < N_1$, namely $\alpha < 0$, we find

$$n(z) = n_o \exp(-|\alpha|z), \tag{4.7}$$

namely the process described by (4.5) is nothing other than an absorption.

For $N_2 = N_1$ the number of photons remains constant during the wave propagation along z. It is finally evident that an amplification of the wave is possible only for $N_2 > N_1$, namely when a population inversion has been realized.

We can now answer the following question: what are the elements allowing the realization of a laser? According to the previous discussion we can quote at least two items:

1) The existence of an 'active' medium with an 'inverted population'.
2) An environment allowing the growth, storage and the feedback of the radiation.

In figure 4.4 we have sketched such an assemblage of devices representing what we will refer to as a laser oscillator.

The cavity plays a manifold role, one of which is the selection of the transverse mode. The spontaneous emission is indeed an isotropic process and therefore the radiation, going back and forth, selects modes propagating along the cavity axis and confined inside the cavity itself, as shown in figure 4.5, where we have sketched the round trip pattern leading to the spatial coherence.

The radiation moves back and forth inside the cavity and re-interacts with the active medium, part of the radiation is coupled out through the reflective mirror, while the other part is reflected.

Having assumed that some gain mechanism exists, we expect that at any reflection back and forth (a round trip) of the light in the optical cavity, the intra-cavity radiation intensity increases and that only a fraction of it is transmitted outside. The evolution process should be characterized by the supposed gain mechanism (including the losses) and by another effect, determining an *intensity dependent feedback to the gain*, defining the saturation mechanism. The expression quoted in cursive characters is a rather obscure statement pointing out that the laser gain is counteracted by the increase of the intensity, which determines the slow-down of the growth.

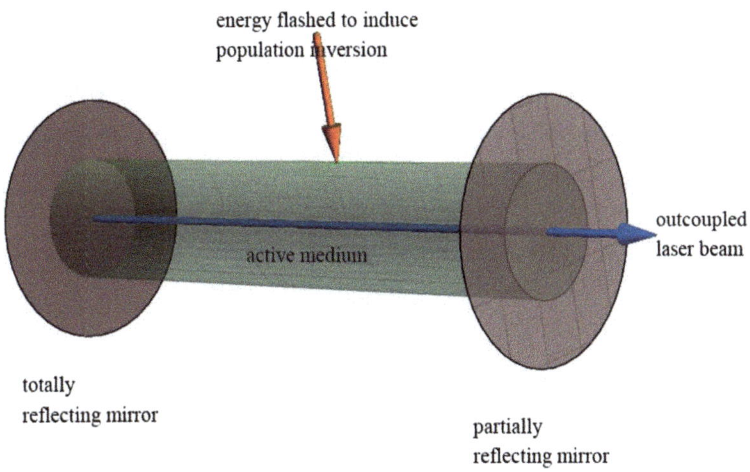

Figure 4.4. Laser oscillator and its components.

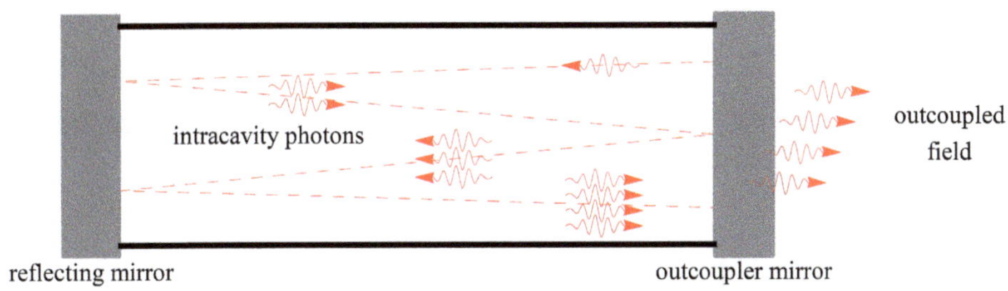

Figure 4.5. Selection of a transverse mode in an optical cavity.

Gain and losses are the two processes concurring with the fulfillment of the so-called equilibrium condition, which occurs when the gain, reduced by the previously quoted mechanism, equals the cavity losses.

The power density growth (per round trip) inside the cavity (intra-cavity) can be accounted for by the following rate equation:

$$I_{n+1} = (1 - \eta)(G(I_n) + 1)I_n, \tag{4.8}$$

where n represents the number of round trips.

The solution of (4.8) requires an assumption on the explicit dependence of the gain $G(I_n)$ on the power intensity at the nth round trip. Before proceeding we make a distinction between small signal and saturated gain. By small signal we mean the gain unaffected by intensity corrections, while in the other regime the intensity dependence may be more or less significant according to the value of some threshold to be specified.

Denoting by g the small signal gain, we solve (4.8) assuming the following saturated gain intensity dependence

$$G(I) = \frac{g}{1 + \frac{I}{I_s}}, \tag{4.9}$$

where we have denoted by I_s the previously quoted intensity threshold. It is a further key quantity of the laser dynamics, called 'the saturation intensity' in the jargon of laser physics and defined as the value of the intra-cavity power density halving the small signal laser gain. When $\frac{I}{I_s} \ll 1$, the non-linear effects associated with the saturation mechanism itself are not significant and become crucial for values larger than unity.

Equation (4.8) is a difference equation (see section 4.10) which can be transformed into a first order differential equation after setting

$$I_{n+1} - I_n \cong T_R \frac{dI}{d\tau}. \tag{4.10}$$

Thus obtaining

$$\frac{dI}{d\tau} = [(1 - \eta)G(I) - \eta]\, I, \tag{4.11}$$

where τ is a non-dimensional time, linked to physical time and to the cavity length L_c by $\tau = \frac{t}{T_R}$, $T_R = \frac{2\,L_c}{c}$, with T_R being the round trip period, namely the time in which the light makes a full back and forth journey inside the optical cavity.

It is convenient to cast (4.11) in the form

$$\begin{gathered} \frac{dX}{\left[(1 - \eta)\frac{1}{1 + X} - r\right]X} = g d\tau, \\ X = \frac{I}{I_s},\ r = \frac{\eta}{g}, \end{gathered} \tag{4.12}$$

and it is evident that for $X \ll 1$, namely far from saturation, the relevant solution is almost trivial and is written

$$X = X_0 \exp[[(1 - \eta)\, g - \eta]\tau], \tag{4.13}$$

On the other side the equilibrium condition, namely $\frac{dX}{d\tau} = 0$, is satisfied by

$$(1 - \eta)\frac{1}{1 + X_E} - r = 0, \tag{4.14}$$

yielding the intra-cavity equilibrium intensity

$$I_E = \left[\frac{1 - \eta}{\eta} g - 1\right] I_s. \tag{4.15}$$

The losses η are given by the sum of active η_a and passive η_p cavity losses. The former are understood as a means to extract the optical power stored inside the cavity, the latter as an unwanted effect determining loss of power, due to heating, absorption, etc.

It is also evident that if we assume the passive losses to be negligible and $\eta \cong \eta_a \ll 1$, we find that the power extracted from the cavity at equilibrium is

$$I_o = \eta I_E \cong g I_s. \tag{4.16}$$

This last identity yields a further hint on the importance of the saturation intensity as a macroscopic quantity of practical interest. Along with the small signal gain coefficient it allows, indeed, the derivation of the output power.

The solution of (4.8), without any approximation, can be written as

$$I_n = I_0 \frac{[(1-\eta)(g+1)]^n}{1 + \frac{I_0}{I_E}([(1-\eta)(g+1)]^n - 1)}, \tag{4.17}$$

where I_0 is the initial value of the intra-cavity intensity. We did not include spontaneous emission contributions in (4.17).

The associated behavior versus round trip number is shown in figure 4.6, which displays an S-shaped form, providing a sort of signature for the processes we are interested in. We underline that at the beginning the growth displays an exponential behavior, which is then counteracted by the gain saturation mechanism and the stationary condition is asymptotically reached, when the gain gets closer and closer to the cavity losses.

In the following we will make large use of S-shaped-like curves. A thorough analysis of the underlying dynamics, known as logistic behavior, is given in section 4.10, where we will provide further elements of discussion.

We will show in the forthcoming sections that FELs are also characterized by such a paradigmatic behavior.

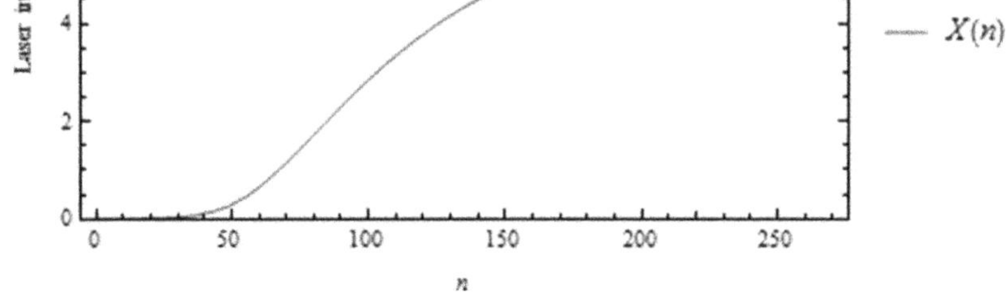

Figure 4.6. Laser intensity (a.u.) versus round trip number for g= 15 %, η = 2%.

4.2 Free electron laser

In the previous chapter we have framed the FEL devices within the family of coherent radiation devices, originated by the klystron. The unifying mechanism is that of an electron-beam density bunching, induced by an appropriate energy modulation. In the case of an FEL the energy modulation is realized through an undulator magnet, providing a transverse component of the electron motion and the consequent coupling to a co-propagating electromagnetic wave. We will look in this section at the FEL process from a slightly different point of view.

The electrons in a magnet emit brehmsstrahlung radiation and the same effect occurs in an undulator magnet specified by the following oscillating magnetic field:

$$\vec{B} = B_0\left(0,\ \sin\left(\frac{2\pi\, z}{\lambda_u}\right),\ 0\right), \tag{4.18}$$

with vertical polarization. We have denoted by B_0 the on-axis peak field, while λ_u denotes the magnet periodicity.

Although the magnetic intensity distribution in (4.18) does not fulfill the Maxwell equations, we can safely assume that it approximates, close to the undulator axis, the exact form of the field generated by the arrangement of N–S magnets placed as shown in figure 4.7, in which the electrons, once injected, execute in the x–z plane an oscillating trajectory. The undulator impresses therefore a transverse component to the, previously full longitudinal, electron motion.

The relevant dynamics, in the absence of other fields but that static of the undulator, is described through the Lorentz force

$$\frac{d}{dt}\vec{p} = -e\,\vec{v}\times\vec{B}. \tag{4.19}$$

The use of the energy and velocity conservation allows the derivation of the transverse velocity component which will be simply provided at the expense of the longitudinal part (see section 4.10) according to the identity

$$\begin{aligned} v_z &\cong c\left[1 - \frac{1}{2\gamma^{*2}}\right], \\ \gamma^* &= \frac{\gamma}{\sqrt{1+\frac{K^2}{2}}},\quad K = \frac{eB_0\lambda_u}{2\pi m_e c}. \end{aligned} \tag{4.20}$$

The parameter K, usually referred to as undulator strength, has as we will see later a pivotal role in the physics of FELs; here it has been introduced to show how the Lorentz force experienced by the electrons inside the magnet modifies the longitudinal velocity. The K parameter is given in SI units, very often it is written in cgs as $K = \frac{eB_0\lambda_u}{2\pi m_e c^2}$ and in practical units reads $K = \frac{\lambda_u[\text{cm}]B_0[\text{KG}]}{10.71}$ (the reader is invited to make the appropriate transformations from one system of units to the other).

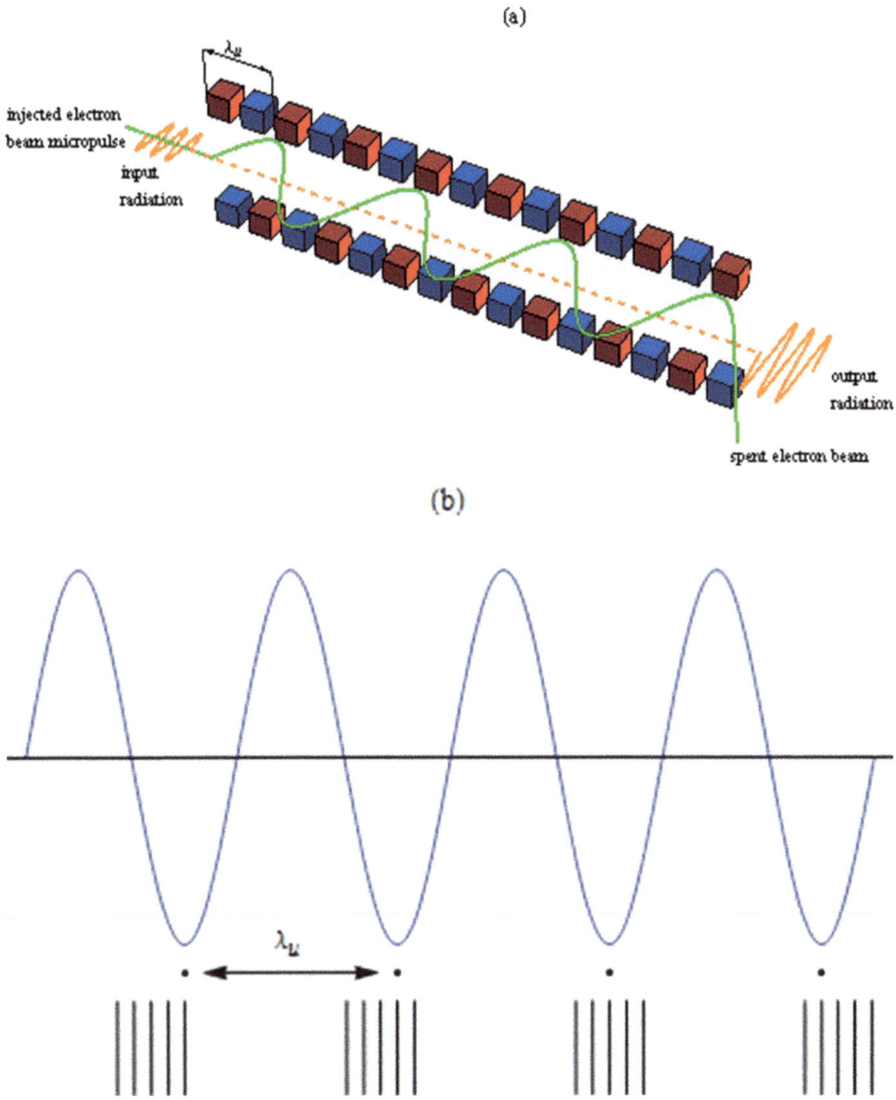

Figure 4.7. (a) Electrons moving in an undulator magnet, together with an input radiation. (b) Inside the magnet electrons and radiation move at different velocities, at each period electrons are left behind the wave by a slight amount δ, constituting the radiation phase advance.

We have already seen that the prerequisite for a laser operation is a frequency selection mechanism, which in the case of FEL is less evident than in a quantized atomic or molecular system, where it is naturally ensured by the natural energy gap.

To recover such a mechanism, within the present purely classical framework, we note that (see figure 4.7), when the electron(s) emit(s) radiation owing to the different velocities, the latter, moving at greater velocity, will overcome the electrons, after each undulator period λ_u by a quantity

$$\delta \cong (c - v_z)\frac{\lambda_u}{c}. \tag{4.21}$$

Such a quantity represents a phase advance and in order to have a positive interference with the radiation emitted at a later time the following Bragg-like condition should be satisfied:

$$\delta = n\lambda. \tag{4.22}$$

Where n is an integer, from (4.21) and (4.22) we finally obtain the expression of the wavelength, in terms of the electron energy, period length and strength parameter,

$$\lambda = \frac{\lambda_u}{2n\gamma^2}\left(1 + \frac{K^2}{2}\right). \tag{4.23}$$

We will say that $n = 1$ represents the fundamental harmonic, larger values provide higher order harmonics.

We limit ourselves, for the moment, to the case $n = 1$ and we will try to further clarify the relevant characteristics, by evaluating, e.g., the relative bandwidth.

The emitted pulse at the end of the undulator can be safely assumed as a kind of step function with a length $N\delta$, the relevant spectral distribution is provided by the associated Fourier transform (see figure 4.8) and with the pulse duration given by

$$\Delta\tau \cong \frac{N\delta}{c}. \tag{4.24}$$

We find from the Parseval inequality

$$\Delta\omega\ \Delta\tau \cong \pi. \tag{4.25}$$

The line relative bandwidth is therefore obtained from the last two equations as

$$\frac{\Delta\omega}{\omega} = \frac{1}{2N}. \tag{4.26}$$

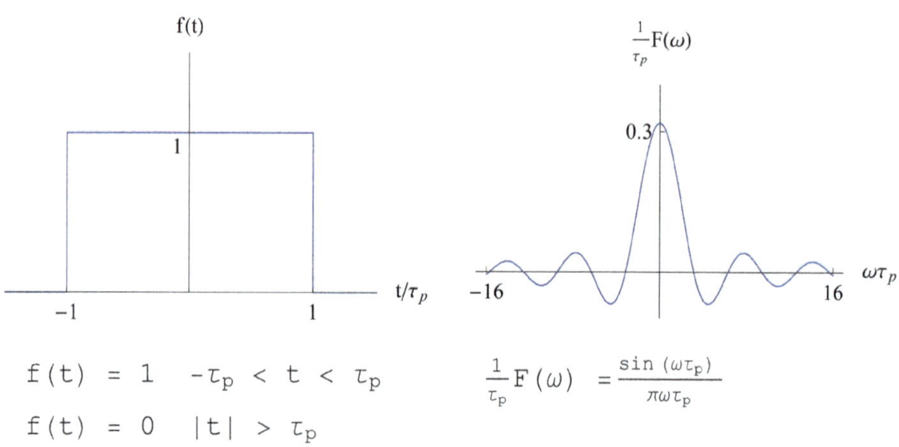

Figure 4.8. Step function and relevant Fourier transform.

The spectral profile can accordingly be written as

$$s(\nu) \propto \left[\frac{\sin\left(\frac{\nu}{2}\right)}{\frac{\nu}{2}} \right]^2, \qquad \nu = 2\pi N \frac{\omega - \omega_0}{\omega_0}, \tag{4.27}$$

where

$$\omega_0 = \frac{2\pi c}{\lambda} = 2\frac{\gamma^2}{1 + \frac{K^2}{2}}\omega_u, \qquad \omega_u = \frac{2\pi c}{\lambda_u}, \tag{4.28}$$

is the resonance frequency. The parameter ν, called detuning, is exploited to define the spectral line shape with respect to the central resonant frequency (see figure 4.9). As ν is a dimensionless quantity the FWHM of the spectral distribution is $\Delta\nu \cong \pi$.

We have so far specified, using simple kinematics and wise physical intuition, how the operating frequency and the relevant bandwidth of an FEL device is determined in terms of the electron energy and undulator parameters. According to our description the undulator is an array of magnets which forces electrons to emit radiation at a given frequency with a sharp spectral width. It is the last born of a family of *insertion devices* (ID), which are magnetic elements inducing the emission of synchrotron radiation (SR). The archetype of ID is the bending magnet, which was the first device that allowed the generation of SR, initially viewed as an annoyance since it appeared to be a limiting factor for reaching higher energies in circular accelerating devices. Later SR became a research tool, thus the evolution of the associated ID proceeded, as shown in figure 4.10.

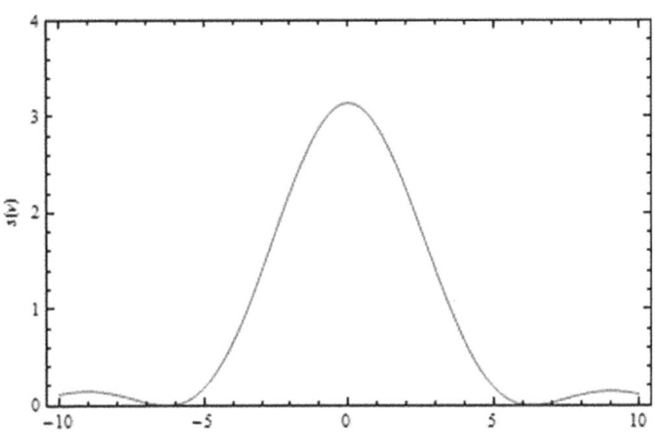

Figure 4.9. Spontaneous emission line shape spectrum.

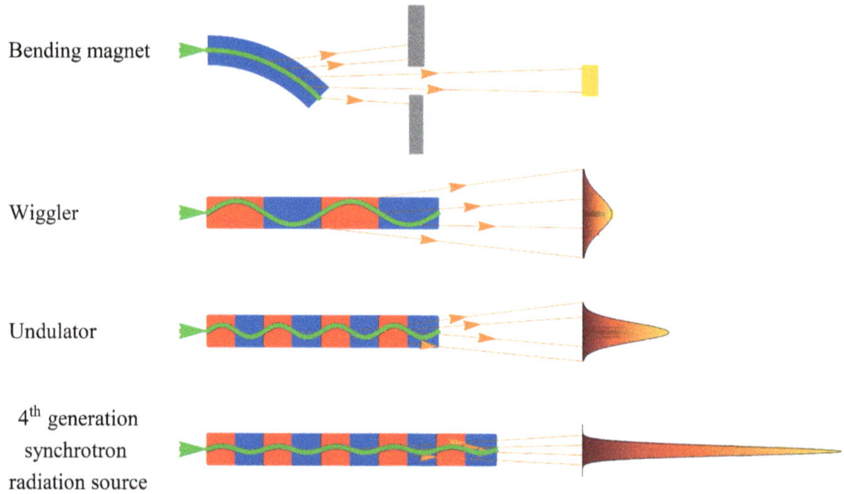

Figure 4.10. Evolution of IDs.

The wiggler, viewed as a chain of bending magnets, was conceived to reduce the emission bandwidth which was extremely large in the case of bending magnets. The next ID was the just described undulator and finally came the FEL operating in self-amplified spontaneous emission (SASE).

The evolution of these devices has been characterized by two significant elements: the narrowing of the emission band and the increase of the intensity of the emitted radiation. We will see in the forthcoming parts of this chapter the role played in this evolution process by the improvement of the qualities of both the electron beam and magnets.

In between the undulator and SASE there was the development of a technology which can be framed within the context of ID and is the FEL oscillator (FEL-O), which is sketched in figure 4.11.

An FEL-O is essentially a laser oscillator, since it fits the previously described paradigm. We can indeed recognize a process of spontaneous emission by free electrons, radiation stored in an optical cavity interacting with a fresh electron beam and inducing bunching, and a consequent coherent emission, which lasts until saturation of the gain occurs.

The FEL gain mechanism (see also chapter 3) can be understood by following a straightforward picture, based on the contents of figure 4.12 where we have drawn the elementary process of absorption and emission of a photon by an electron moving in a magnetic field.

If we watch at the process in a reference frame where the electron motion is non-relativistic, the undulator field can be viewed as an electromagnetic wave. The electrons therefore scatter the 'undulator photons' into real photons, moving in the opposite direction with an almost equivalent frequency. If these photons are fed back to re-interact with the electrons (e.g. by means of a storing cavity) the interaction process is that of an electron moving two counter-propagating waves.

It should now be noted that, in our picture, we have considered the electromagnetic field as being limited to just two modes (forward and back-propagating),

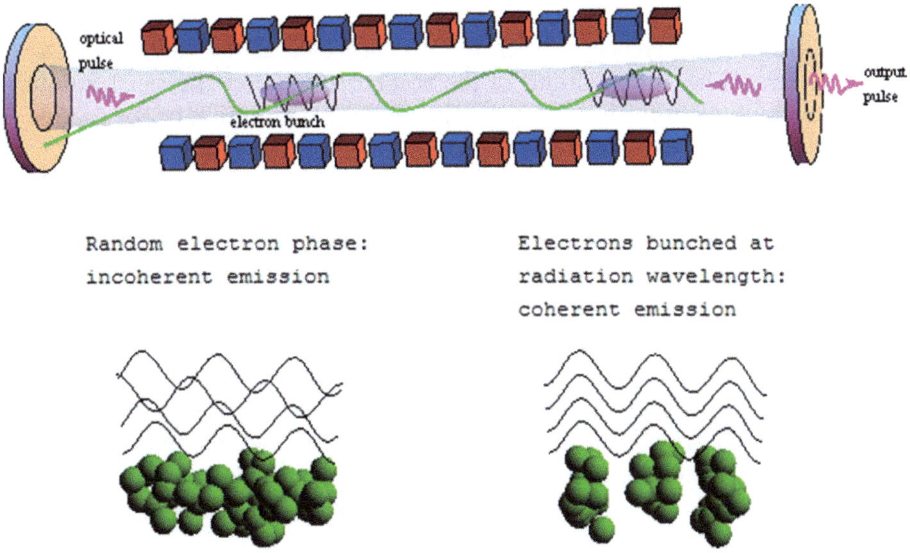

Figure 4.11. FEL oscillator.

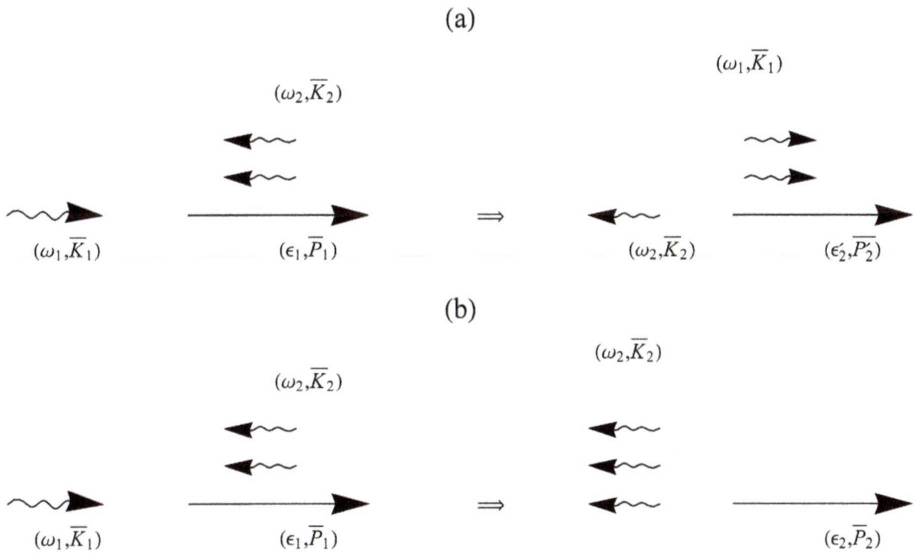

Figure 4.12. The elementary process of emission and absorption of a photon by an electron moving in a magnetic field.

this is again a kind of two-level model. Within such a picture the electrons have only two chances:

a. Scatter the counter-propagating undulator photon into the forward directions.
b. Scatter the co-propagating photon into the undulator mode.

We can consider the mechanisms ‘a’ and ‘b’ as emission and absorption processes.

What causes the difference between the two processes is the fact that in the first case the electron loses momentum, in the second it acquires the same amount of momentum. In the non-relativistic frame, hosting the description of the process, the amount of gained or lost linear momentum is

$$\Delta p = 2\hbar k. \tag{4.29}$$

The line shape of the emission and absorption processes are the same, with the significant difference of a Doppler shifted frequency due to the momentum gain or loss.

We associate with process ‘a’ a photon amplification (photons are created at the expense of the electron momentum), vice versa process ‘b’ corresponds to a destruction of photons with a consequent increase of the electron momentum. This qualitative discussion will be put on rigorous mathematical ground in section 4.10, where we will provide a close analogy to a two-level atomic maser. In terms of the detuning parameter the amplification process will be characterized by a down-shifted line frequency (the electrons have lost energy), while for the opposite effect we find an upshift.

In conclusion, the gain can be viewed as a balance between the two effects, namely

$$G(\nu) \propto s(\nu_-) - s(\nu_+), \tag{4.30}$$

where $\nu_\pm$ refers to the up- and down-shifted detuning differing by the Doppler shift induced by the momentum variation fixed by (4.29).

We have accordingly

$$\nu_\pm \cong \nu + o\left(\frac{\Delta p}{m_e c}\right), \tag{4.31}$$

thus finding in conclusion

$$G(\nu) \propto -\frac{\partial}{\partial \nu} s(\nu). \tag{4.32}$$

This is just a ‘single electron’ process (the reader may raise serious doubts on how to reconcile such a point of view with that discussed in the previous chapters, which have been based on the bunching mechanism, which is rather difficult to imagine using one electron only), however, by taking into account the full electron beam and the undulator geometry we end up with

$$\begin{aligned} G(\nu) &= -\pi g_0 \frac{\partial}{\partial \nu} s(\nu), \\ g_0 &= 4\pi \frac{|J|}{I_0}\left(\frac{N}{\gamma}\right)^3 (\lambda_u K^* f_b)^2, \\ f_b &= J_0\left(\frac{1}{2}\frac{K^{*2}}{1+K^{*2}}\right) - J_1\left(\frac{1}{2}\frac{K^{*2}}{1+K^{*2}}\right), \\ K^* &= \frac{K}{\sqrt{2}}. \end{aligned} \tag{4.33}$$

The gain curve is reported in figure 4.13 and it should be noted that the maximum gain does not occur at the maximum of the spontaneous emission (namely at $\nu_0 \cong 0$) but at $\nu_0 \cong 2.6$. The quantity g_0 is called the small signal gain coefficient, J denotes the current density and f_b is the Bessel factor, which includes the effect of a non-perfectly sinusoidal trajectory of electrons inside the undulators with linear polarization.

It is worth noting the antisymmetric nature of the gain curve itself, according to which the gain can be either negative or positive. This point will be better clarified later.

We can now use what we have learned in the introductory section of this paper to introduce along with gain the other laser pivotal quantity, namely the FEL saturation intensity. The FEL process is the mechanism according to which electron-beam power is transformed into coherent electromagnetic power. We insist again on an antisymmetric gain shape and note that the electron energy loss brings the gain process toward the negative gain region; when the energy loss is such, the gain process stops and the system reaches a kind of steady state.

Such an energy loss can be obtained from a cursory look at the gain curve and note that from the maximum gain to the zero detuning we have

$$\Delta \nu \cong 4\pi N \frac{\Delta \gamma}{\gamma} \cong \pi, \tag{4.34}$$

which yields, in terms of fractional energy variation,

$$\frac{\Delta \gamma}{\gamma} \cong \frac{1}{4N}, \tag{4.35}$$

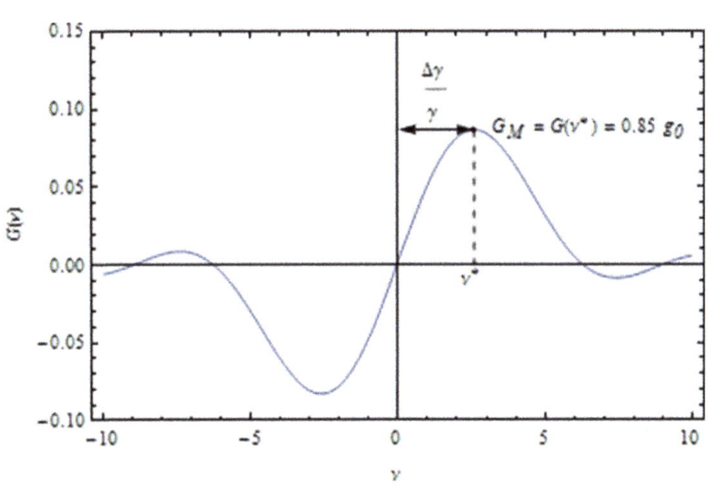

Figure 4.13. FEL gain curve and position of the peak gain.

which provides the power delivered for the electron beam to the laser field as

$$P_{\mathrm{FEL}} \cong \frac{P_{e-b}}{4N}. \tag{4.36}$$

We can accordingly define the FEL efficiency as

$$\chi = \frac{P_{\mathrm{FEL}}}{P_{e-b}} \cong \frac{1}{4N}. \tag{4.37}$$

Assuming for the FEL oscillator the same dynamics leading to (4.16) we find

$$\begin{aligned} &\chi I_{e-b} \cong g_0 I_s \\ &\rightarrow I_s = \frac{\chi I_{e-b}}{g_0} = \frac{I_{e-b}}{4Ng_0}, \\ &I_{e-b} = JE, \end{aligned} \tag{4.38}$$

where I_{e-b} is the electron-beam power density. In practical units we obtain for the saturation intensity

$$I_S[\mathrm{MW/cm^2}] = 6.9312 \times 10^2 \frac{1}{2}\left(\frac{\gamma}{N}\right)^4 \left(\lambda_u[\mathrm{cm}]\, K^* f_b\right)^{-2}. \tag{4.39}$$

In this section we have provided arguments allowing the understanding of an FEL-O using a point of view very much similar to that of a conventional laser, this method will be further corroborated in forthcoming sections.

4.3 The FEL small signal equation low and high-gain regimes

We derive here the FEL gain by the use of a slightly more elaborate procedure. In the previous section we have used the spontaneous emission line shape to derive the FEL gain. We have dealt with a real function, but to properly describe the emission process we should use complex amplitudes. To this aim we introduce the following complex function

$$\begin{aligned} s(\nu, \tau) &= \int_0^\tau e^{-\imath\nu\xi}\, d\xi, \\ \tau &= \frac{z}{L_u}, \end{aligned} \tag{4.40}$$

whose square modulus yields the spontaneous emission line shape.

The dimensionless variable τ is linked to the longitudinal coordinate z along the undulator and is in turn linked to the time by $z \cong \beta_z ct$. The equation yielding the evolution of the FEL complex amplitude $a(\tau)$ (left unspecified for the moment) can be simply written as

$$\frac{d}{d\tau}a(\tau) = -a(\tau)\pi g_0 \frac{\partial}{\partial\nu} s(\nu, \tau), \tag{4.41}$$

and thus, more conveniently, cast in the form

$$\frac{d}{d\tau}a(\tau) = i\pi g_0 \int_0^{\tau} \xi e^{-i\nu\xi} a(\tau) d\xi. \tag{4.42}$$

Let us now note that (4.41) and (4.42) are the transcription of the photon balance mechanism, introduced in the previous section. According to (4.42) the process is assumed to occur once only in a well-defined position inside the undulator. By taking into account that the process may occur everywhere (at different positions and hence at different times) and that the field also brings a memory of the previous interactions, it is more correct to write (4.42) in the following Volterra-like form

$$\frac{d}{d\tau}a(\tau) = i\pi g_0 \int_0^{\tau} \xi e^{-i\nu\xi} a(\tau - \xi) d\xi, \tag{4.43}$$

where the convolution integral on its rhs accounts for delocalization and memory effects.

There are different ways of treating the previous equation, which can be solved using a plethora of methods. Regarding the gain it will be defined as the fractional variation of the square modulus of the dimensionless amplitude at the end ($\tau = 1$) and at the beginning ($\tau = 0$) of the interaction, namely

$$G(\nu) = \frac{|a(1)|^2 - |a(0)|^2}{|a(0)|^2}. \tag{4.44}$$

Using, e.g., the small signal gain coefficient as a perturbation coefficient to solve the above equation we end up with the following gain dependence on g_0 and on the detuning parameter

$$\begin{aligned} G(g_0, \nu) &\cong \sum_{s=1}^{3} g_0^s g_s(\nu) \\ g_1(\nu) &= \frac{2\pi}{\nu^3}[2(1 - \cos(\nu)) - \nu \sin(\nu)] \\ g_2(\nu) &= \frac{\pi^2}{3\nu^3}[84(1 - \cos(\nu)) - 60\nu \sin(\nu) + 3\nu^2 + 15\nu^2 \cos(\nu) + \nu^3 \sin(\nu)] \\ g_3(\nu) &= \frac{\pi^3}{60\nu^9}[11520(1 - \cos(\nu)) - 9000\nu \sin(\nu) + 360\nu^2 + 2880\nu^2 \cos(\nu) \\ &\quad + 480\nu^3 \sin(\nu) - 20\nu^4(1 + 2\cos(\nu)) - \nu^5 \sin(\nu)]. \end{aligned} \tag{4.45}$$

The plot of the gain $G(g_0, \nu)$ versus the detuning parameter for different values of g_0 are shown in figure 4.14.

The physical meaning of the curve is quite transparent. The low gain case (see figure 4.14(a)) exhibits the anti-symmetric shape discussed in the previous section. When g_0 is small (namely not larger than 10%) the approximation (4.41) of the FEL

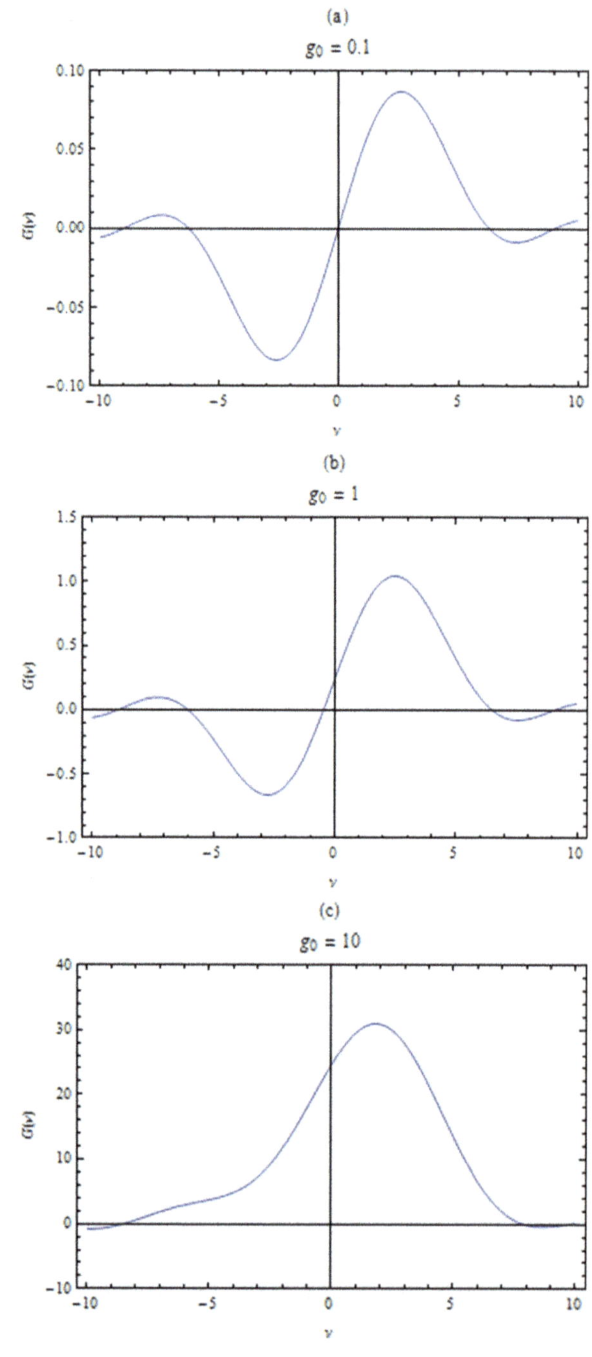

Figure 4.14. FEL gain for three different values of the small signal gain coefficient: (a) $g_0 = 0.1$, (b) $g_0 = 1$ and (c) $g_0 = 10$.

gain equation is rather accurate; when it increases the cumulative contributions due to the integral form (4.43) provides a progressive distortion of the low gain antisymmetric shape, as shown in figure 4.14(b) and (c).

The maximum gain is a function of the small signal gain coefficient and is provided by the simple identity

$$G_M = G(g_0) \cong 0.85 g_0 + 0.192 g_0^2 + 4.23 \times 10^{-3} g_0^3. \tag{4.46}$$

In the forthcoming sections we will provide further solutions of the FEL integral equation, by discussing different gain regimes. The previous results are, however, sufficient for the next considerations regarding the FEL oscillators.

We have introduced the dimensionless amplitude $a(\tau)$, but we did not specify how it should be effectively defined. We first note that its square modulus should be associated with a kind of dimensionless intensity; it is therefore almost natural to express $(a(\tau)|^2$ in terms of the saturation intensity as

$$|a(\tau)|^2 = \kappa \frac{I}{I_s}, \tag{4.47}$$

where the proportionality coefficient is $\kappa = 8\pi^2$.

The use of the amplitude $a(\tau)$, usually referred as the Colson amplitude[1], is extremely useful to understand the regime of operation, and in particular the landmark between the small signal and saturated regime, fixed by

$$|a(\tau)|^2 < \kappa \to I < I_s. \tag{4.48}$$

In the following section the previous, apparently naïve remarks, will be crucial in the understanding of the FEL oscillator dynamics.

4.4 FEL oscillators: a preliminary analysis

In the previous sections we have introduced the crucial concepts of gain and saturation intensity for FELs, we have seen how these quantities allow the description of the dynamics of the conventional oscillator. We can therefore apply the previous methodology to the study of the FEL-O intra-cavity intensity evolution.

The point of view underlying our description is that there is no conceptual difference between an ordinary laser and an FEL oscillator, which is in any case provided by an optical cavity with its own losses, by an active medium and by a radiation beam moving back and forth inside the cavity. In order to derive the evolution of the FEL intensity we proceed as follows:

[1] W B Colson is, in the opinion of the authors of this book, the most eminent contributor to the theory of FELs, his clear and profound work has been a guide for a generation of scholars studying FEL physics.

a. Limiting ourselves for the moment to the low gain case, we specify the FEL gain saturation as

$$G(X) = \frac{G_M}{F(X)}, \qquad X = \frac{I}{I_S}, \qquad G_M = 0.85\, g_0, \tag{4.49}$$

where

$$F(X) = 1 + \alpha X + \beta X^2, \qquad \alpha + \beta = 1, \qquad \alpha = 2\left(\sqrt{2} - 1\right), \qquad \beta = 3 - 2\sqrt{2}. \tag{4.50}$$

It should be noted that the saturation function is slightly different from the one exploited in section 4.1 (see (4.9)), this is, however, a technical detail only, without any consequence on the substantive aspects of the relevant physics.

b. We write the round trip evolution using the same rate equation exploited for conventional oscillators, namely

$$I_{r+1} = [(1 - \eta)\, G(X_r) - \eta]\, I_r \tag{4.51}$$

namely a difference equation that can be solved using standard means. It is evident that the intra-cavity intensity grows until it reaches the equilibrium level such that the following condition holds $I_{r+1} = I_r$, and therefore

$$G(X_e) = \frac{\eta}{1 - \eta}, \tag{4.52}$$

which yields the following equilibrium intensity

$$I_e = \left(\sqrt{2} + 1\right)\left(\sqrt{\frac{1 - \eta}{\eta} G_M} - 1\right) I_S. \tag{4.53}$$

In conclusion, the round trip evolution of the FEL intensity is provided by the following equation

$$I_{r+1} = I_0 \frac{[(1 - \eta)\ (G_M + 1)]^r}{1 + \frac{I_0}{I_e}\{[(1 - \eta)(G_M + 1)]^r - 1\}}, \tag{4.54}$$

where I_0 is the seed intensity provided, e.g., by the spontaneous emission. An example of comparison between the prediction of the formula in (4.54) and a numerical simulation with the code PROMETEO (the code was developed by P L Ottaviani (1941–2014) in collaboration with Giuseppe Dattoli and Simonetta Pagnutti) is given in figure 4.15, where we have shown the growth of the intensity (in MW cm^{-2}) and the associated gain reduction. The intensity growth is that of conventional lasers, with the characteristic sigmoidal shape, and such a logistic shape will be discussed further in the forthcoming parts of this chapter.

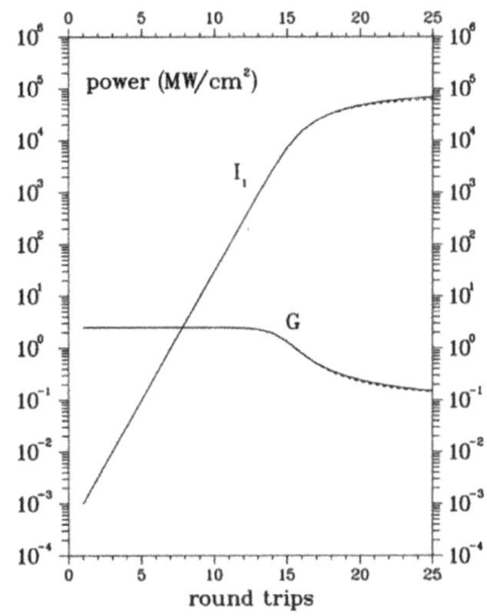

Figure 4.15. (a) Intra-cavity intensity versus the round trip number (continuous line numerical integration, dot line analytical formula); simulation parameters (the gain coefficient is larger than the low gain approximation, but the validity of the formula extends to these cases too): $E = 108.24$ MeV, $\lambda_u = 2.8$ cm, $K = 2.10$, $N = 50$, $g_0 = 2.0$, $P_{e-b} = 2.012 \times 10^6$ MW cm^{-2}, $I_s = 8.08 \times 10^3$ MW cm^{-2}, $\lambda_0 = 1.008$ μm and $\eta = 0.10$. (b) Net gain $\left(G_{\text{net}}(X_{r+1}) = \frac{I_{r+1} - I_r}{I_r} = (1 - \eta)\, G(X_r) - \eta\right)$ versus round trip.

4.5 FEL oscillators: mode locking

In the previous section we discussed a fairly primitive model of an FEL-O, without taking into account that these devices are usually driven by an electron beam delivered by radio-frequency (RF) linacs, and that the structure of the laser beam follows the macro- and micro-patterns induced by the electron beam itself (see figure 4.16 and chapter 2).

The result of this operating mode is the creation of short laser micro-pulses, with a length comparable to that of the electron bunch and providing (at equilibrium) longitudinal mode configurations involving peculiar structures. These structures must be compatible with the boundary conditions imposed by the cavity itself.

We should first recall that the field of the optical cavity is a superposition of longitudinal modes, namely

$$a(t) = \sum_m \alpha_m e^{i\,(\omega_m t + \varphi_m)}, \qquad \omega_m = m\frac{\pi c}{L_c}, \tag{4.55}$$

with ϕ_m being the phase of each mode. If all modes are oscillating in the cavity with random phases, the total intensity averages to zero whenever a fixed phase relation is determined between a number of longitudinal modes; a phase (or mode) locking is realized and the intensity is provided by a sequence of pulses, as shown in figure 4.17.

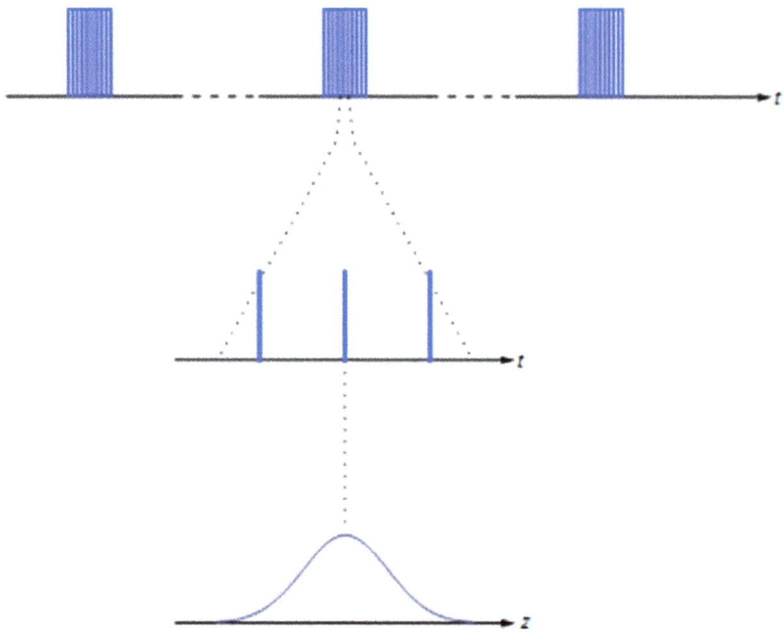

Figure 4.16. Structure of an RF linac electron beam and micro-bunch Gaussian shape.

Before getting into further details let us try to understand the number of longitudinal modes that can be involved in the FEL process. From the gain curve we infer that the frequency width is approximately

$$(\Delta f)_{\text{FEL}} = \frac{\pi c}{N\lambda}, \tag{4.56}$$

with the cavity frequency bandwidth given by

$$(\Delta f)_c = \frac{c}{L_c}.$$

We infer that the number of potentially coupled modes is fixed by

$$N_m = \frac{(\Delta f)_{\text{FEL}}}{(\Delta f)_c} = \frac{\pi L_c}{2\Delta}, \quad \Delta = N\lambda, \tag{4.57}$$

where we have introduced the 'slippage' parameter $\Delta = N\lambda$, whose physical meaning will be discussed later in this section.

Without any coupling mechanism the longitudinal modes of an FEL-O will evolve independently, which means with uncorrelated phases. The device allowing a phase locking in the FEL case is the electron bunch itself. The strength of the coupling can be understood using a fairly simple reasoning which takes into

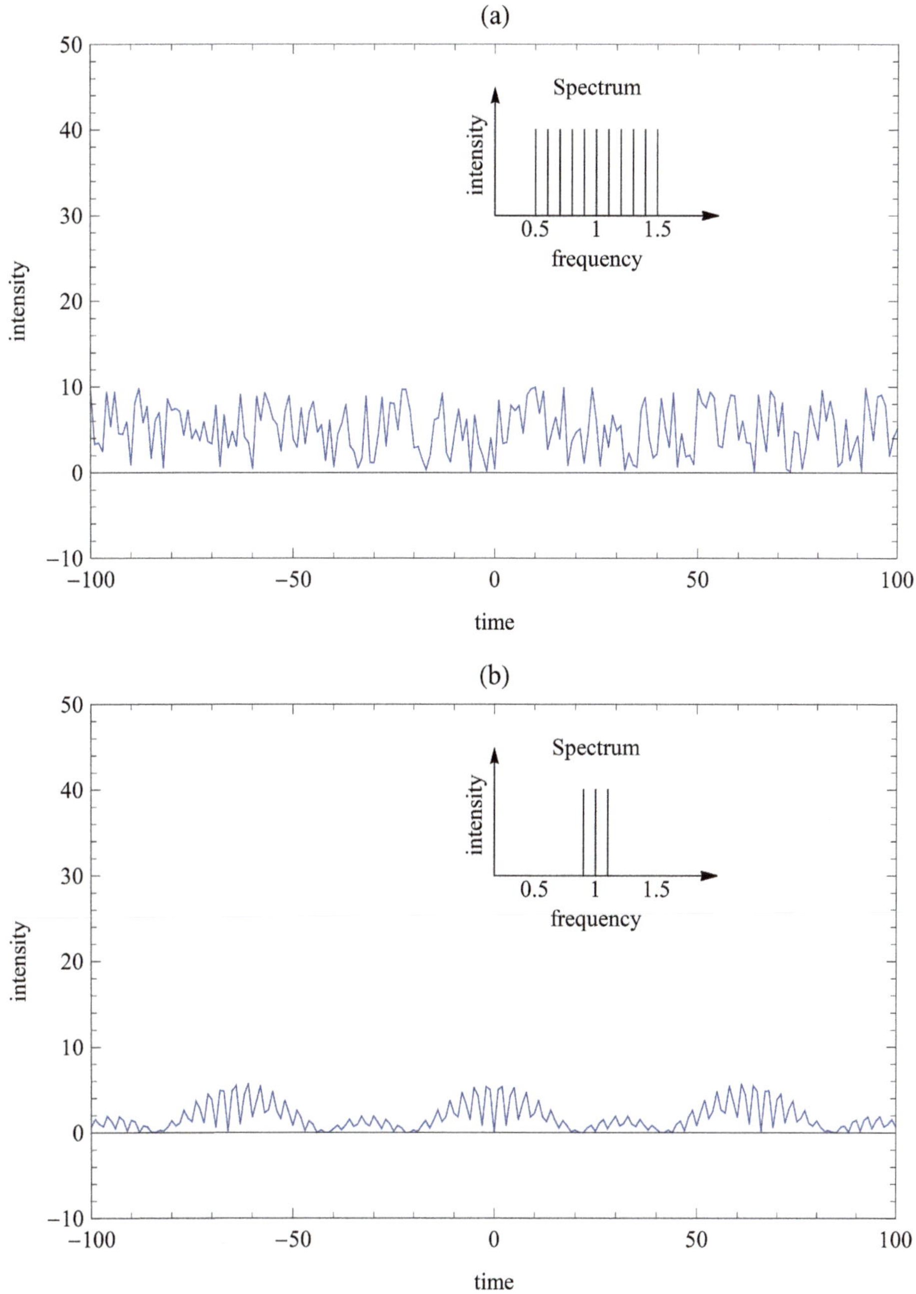

Figure 4.17. (a) Intensity determined by 11 modes with random phase. (b) The same as (a) for three modes all with the same phase and with a relative separation $\delta = \frac{\delta\omega}{\omega_0} = 0.1$. The distance between the peaks is then fixed by $T = \frac{2\pi}{\delta}$.

account the longitudinal distribution of the electron bunch (assumed to be Gaussian), namely

$$f(\zeta) = \frac{1}{\sqrt{2\pi}\,\sigma_\zeta} e^{-\frac{\zeta^2}{2\sigma_z^2}}, \tag{4.58}$$

and by the relevant Fourier transform in the FEL gain space. Denoting by k the wave vector, with k_0 the central frequency associated wave vector, we find after using the detuning ν as the transform variable

$$\begin{aligned} \tilde{f}(\nu) &= \frac{1}{\sqrt{2\,\pi}} \int_{-\infty}^{\infty} f(\zeta)\, e^{-i(k-k_0)\zeta} d\zeta \\ &= \frac{1}{\sqrt{2\,\pi}} \int_{-\infty}^{\infty} f(\zeta)\, e^{-i\frac{\nu}{\Delta}\zeta} d\zeta \propto \frac{1}{\sqrt{2\pi}\,\mu_c} e^{-\frac{\nu^2}{2\,\mu_c^2}}, \\ \mu_c &= \frac{\Delta}{\sigma_\zeta}, \end{aligned} \tag{4.59}$$

where we have introduced the parameter μ_c measuring the strength of the coupling of the longitudinal modes within the laser bandwidth.

At this point we quote the following definition of mode-locking (taken from Wikipedia):

> Mode-locking is a technique in optics by which a laser can be made to produce ultrashort pulses with a width of the order of sub-picoseconds ($< 10^{-12}$s). The basis of the technique is to induce a fixed phase relationship between the modes of the laser's resonant cavity. The laser is then said to be phase-locked or mode-locked. Interference between these modes causes the laser light to be produced as a train of pulses. Depending on the properties of the laser, these pulses may be of extremely brief duration, as short as a few femtoseconds

and conclude that in an FEL-O mode-locking is naturally achieved by the presence of a finite extension of the electron bunch itself.

Let us clarify the physical meaning of the slippage parameter Δ which is simply due to the fact that electrons and photons move at different speeds during their flight inside the undulator (see section 4.2), therefore the radiation tends to be ahead of the electrons by the quantity

$$\begin{aligned} \Delta &= (1 - \beta_z)\, L_u, \\ 1 - \beta_z &= \frac{1}{2\,\gamma^2}\left(1 + \frac{K^2}{2}\right), \\ L_u &= N\lambda_u. \end{aligned} \tag{4.60}$$

The slippage length measures therefore the distance covered by the radiation over the electrons in the time the photons and electrons travel in an undulator of length L_u.

It is not difficult to figure out how the FEL integral equation should be modified to account for the short pulse effects. Denoting indeed by $a(\zeta, \tau)$ the Colson dimensionless amplitude and with $g_0(\zeta)$ $(=\sqrt{2\pi}g_0 f(\zeta))$ the small signal gain coefficient, including the current longitudinal profile, we end up with

$$\frac{\partial}{\partial \tau}a(\zeta, \tau) = i\pi g_0(\zeta + \Delta\tau)\int_0^\tau \tau' a(\zeta + \Delta\tau', \tau - \tau')\, e^{-i\nu\tau'} d\tau', \tag{4.61}$$

which displays in mathematical terms what has been discussed regarding the slippage mechanism. It is worth noting that we have here two distinct variables (ζ, τ) the first associated with the electron (or optical) bunch longitudinal distribution coordinate, the second with the undulator coordinate z. Accordingly (4.61) describes the slow motion of the optical packet with respect to the electron bunch.

Within the present context the role of the coupling parameter $\mu_c = \frac{\Delta}{\sigma_\zeta}$ becomes clear. It measures indeed the relevance of the slippage with respect to the electron bunch length; when $\mu_c \cong 0$ (namely a continuous electron beam) equation (4.61) reduces to the case discussed in the previous section. For $\mu_c \neq 0$ many longitudinal modes are coupled and the gain formulae discussed in the previous section should be modified (as discussed later in this chapter).

Before closing this section it is necessary to quote a further important aspect of FEL physics (not only peculiar to oscillators and discovered for the first time on an FEL-O) known as *lethargy*.

We have remarked that radiation tends to slip over the electrons because of the relative different velocities. This is strictly true if the two devices are not interacting.

According to the previous discussion, the process, giving rise to the FEL gain and to the associated dynamics, displays a dispersive nature since it involves both a phase and intensity variation of the field interacting with the electrons. We can expect therefore the existence of a group refractive index linked to the phase refractive index n by

$$n_g = n\left(1 - \frac{\lambda}{n}\frac{\partial}{\partial\lambda}n\right). \tag{4.62}$$

The optical packet group velocity is therefore defined as

$$v_g = \frac{c}{n_g}. \tag{4.63}$$

Leaving a more appropriate discussion of this effect to section 4.10, we note here that one of its macroscopic consequences is that when we set the optical cavity of an FEL-O we should pay much more attention to the relative distance between the cavity mirrors.

To understand this point we should take a step back to the electron-beam structure and note that it is provided by a train of bunches separated by a distance fixed by the RF period. To ensure synchronism between electron and photon bunches, in such a way that after each round trip the photons overlap a fresh bunch

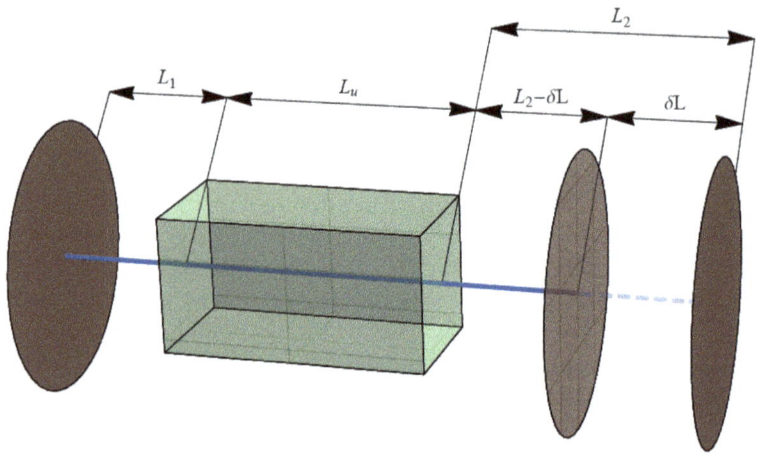

Figure 4.18. FEL-O cavity and mirror mismatch.

of electrons at the undulator entrance, the cavity length should be an integer multiple of the distance between adjacent bunches.

Denoting by T_e the distance between synchronous electron bunches (which are not necessarily adjacent) we must impose the condition that (see figure 4.18)

$$\begin{aligned} &T_e = T_R, \\ &T_R = 2\frac{L_1}{c} + \left(\frac{L_u}{v_g} + \frac{L_u}{c}\right) + 2\frac{L_2 - \delta L^*}{c}, \\ &L_c = L_1 + L_u + L_2, \end{aligned} \tag{4.64}$$

where T_R is the round trip time, evaluated by taking into account the reduced photon packet group velocity inside the undulator, and δL^* is the cavity length reduction necessary to compensate the radiation velocity slowdown. Such a variation is performed with respect to the nominal cavity length set at $L_c = \frac{cT_e}{2}$. In conclusion, we find that the cavity mismatch $\delta\, L^*$, undulator length L_u and group refractive index are linked by the identity

$$n_g = 2\frac{\delta L^*}{L_u} + 1. \tag{4.65}$$

The important point to be stressed is that in the design of an FEL-O, a significant role is played by the cavity mismatch which regulates both the gain and the output gain. In the forthcoming section we will discuss how the thus far developed notions can be embedded in the design of an actual oscillator device.

4.6 Designing an FEL-O device

In this section we will provide the reader with the essential tools for the design of an FEL-O.

Table 4.1. Electron beam, undulator and cavity designing parameters.

Variable	Symbol	Unit
Electron beam		
Energy	E	MeV
Current	$\hat{I}$	ampere
Relative energy spread	σ_ε	none
Emittance radial and vertical	ε_η, $\eta = x, y$	$m \cdot rad$
Longitudinal bunch length	σ_z	meter
Twiss coefficients	β	meter
	γ	meter^{-1}
	α	none
Undulator		
Period length	λ_u	meter
On-axis field	B_0	tesla
Number of periods	N	none
Length	$L_u = N\lambda_u$	meter
Parameter strength	K	none
Cavity		
Cavity length	L_c	meter
Losses		
Active	η_a	
Passive	η_p	
Total	$\eta = \eta_a + \eta_p$	

The first step is to provide a list of parameters we will exploit in the following. They are listed in table 4.1 along with the relevant units.

The peak current reported in table 4.1 refers to the current carried by the electron bunch and related to the longitudinal distribution and bunch charge by

$$I_b(\tau) = \hat{I}\, e^{-\frac{\tau^2}{2\sigma_\tau^2}} \rightarrow Q_b = \int_{-\infty}^{+\infty} I_b(\tau)\, d\tau$$
$$\rightarrow \hat{I} = \frac{Q_b}{\sqrt{2\pi}\sigma_\tau}. \tag{4.66}$$

Note that $\zeta = c\tau$.

The peak current and the undulator parameters allow the evaluation of the gain coefficient

$$\begin{aligned} g_0 &= \frac{16\pi}{\gamma}\frac{\langle J \rangle}{I_0} F\lambda_0 L_u N^2 \xi f_b^2 \\ &= 4\pi\frac{\langle J \rangle}{I_0}\left(\frac{N}{\gamma}\right)^3 (\lambda_u K^* f_b)^2, \\ \xi &= \frac{1}{2}\frac{K^{*2}}{1+K^{*2}}, \end{aligned} \tag{4.67}$$

where

$K^* = k/\sqrt{2}$ for linear undulator,

$K^* = K$ for helical undulator,

$f_b = J_0(\xi) - J_1(\xi)$ for a linear undulator ($J_{0,1}$ cylindrical Bessel function),

$f_b = 1$ for a helical undulator.

In particular the current density J given in A/m^2 is linked to the peak current by

$$\begin{aligned} J &= \frac{\hat{I}}{2\pi\sigma_x\sigma_y}, \\ \sigma_{x,y} &= \sqrt{\beta_{x,y}\varepsilon_{x,y}}\,. \end{aligned} \tag{4.68}$$

The quantity F denotes the so-called filling factor and can be assumed to be 1 if the electron beam in the interaction region (the undulator) is contained within the cavity laser mode waist (see figure 4.19), which, for a confocal cavity is

$$w_0 = \sqrt{\frac{\lambda L_c}{2\pi}}. \tag{4.69}$$

The condition for containing the electron beam in the optical mode is therefore

$$\begin{aligned} \sqrt{\beta_{x,y}\varepsilon_{x,y}} &\leqslant \sqrt{\frac{\lambda L_c}{2\pi}} \\ \rightarrow \beta_{x,y} &\leqslant \frac{\lambda}{2\pi\varepsilon_{x,y}} L_c. \end{aligned} \tag{4.70}$$

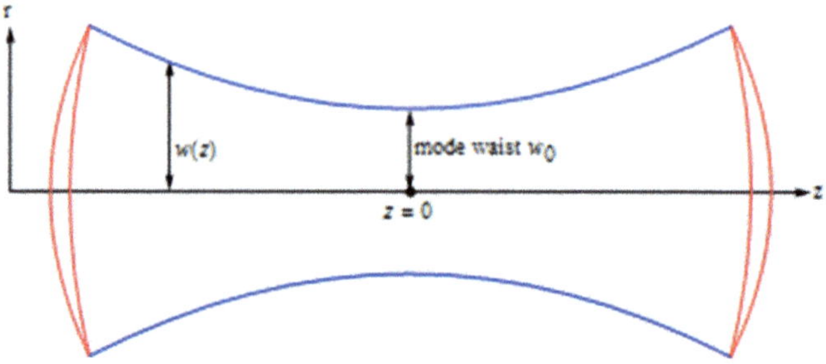

Figure 4.19. Laser mode distribution along the cavity axis.

The electron beam is not mono-energetic and the relevant energy distribution is written

$$f(\varepsilon) = \frac{1}{\sqrt{2\pi}\sigma_\varepsilon} e^{-\frac{\delta_\varepsilon^2}{2\sigma_\varepsilon^2}},$$
$$\delta_\varepsilon = \frac{\gamma - \gamma_0}{\gamma_0}, \tag{4.71}$$

with σ_ε being the relative energy spread. A non-mono-energetic beam causes emission at slightly different wavelengths by the individual electrons inside the bunch, thus determining an inhomogeneous broadening effect, causing a reduction of the gain. The FEL integral equation should accordingly modified as

$$\frac{\partial}{\partial\tau} a(\zeta,\ \tau) = i\pi g_0(\zeta + \Delta\tau)\int_0^\tau \tau' a(\zeta + \Delta\tau',\ \tau - \tau')\, e^{-i\nu\tau'} e^{-\frac{(\pi\mu_\varepsilon t)^2}{2}} d\tau',$$
$$\mu_\varepsilon = 4N\sigma_\varepsilon, \tag{4.72}$$

where the parameter μ_ε determines the gain reduction effect associated with the non-ideal nature of the beam itself. Just to give an idea of the importance of these effects we note that the gain reduction (in the low gain case) is given by

$$G_M(\mu_\varepsilon) \cong \frac{0.85 g_0}{1 + 1.7\mu_\varepsilon^2}. \tag{4.73}$$

(Analogous formulae hold for the high-gain case and they are not reported here for brevity, some examples are reported in section 4.10.)

The gain reduction due to the energy spread of a beam with $\sigma_\varepsilon \cong 10^{-3}$, driving an FEL-O with an undulator having $N = 30$ is negligible (for $\mu_\varepsilon^2 \ll 1$, $\frac{\Delta G}{G} \cong 1.7\,\mu_\varepsilon^2$, thus for this specific example $\frac{\Delta G}{G} \cong 2.4\%$).

An analogous formula includes the effects associated with multimode coupling and slippage, and indeed we find (see section 4.10)

$$G_M(\mu_c) \cong \frac{0.85 g_0}{1 + \frac{1}{3}\mu_c}. \tag{4.74}$$

We have so far learned how to embed both beam and undulator parameters in our design strategy; next is the cavity, by including the effects of the lethargy and of losses.

4.6.1 Lethargy

We have stressed that the relative mismatch of the cavity length with respect to the nominal distance is necessary to compensate the radiation slowdown and allow the synchronism between electron and photons at each round trip.

This can be achieved by either moving the mirrors mechanically or by acting on the RF signal inside the accelerating cavities in order to modulate the distance between electron bunches. It is, however, evident that the mirror distance mismatch $\delta\mathcal{L}$ affects both the gain and the equilibrium intra-cavity intensity.

The pivotal parameter to include in the gain $\delta\mathcal{L}$ (>0 means cavity reduction) is

$$\theta = \frac{4\delta\mathcal{L}}{g_0\Delta}, \tag{4.75}$$

sometimes called cavity *desynchronism* and providing a measure of the mirror mismatch with respect to the slippage length.

We modify the maximum gain formula with the inclusion of θ as follows

$$\begin{gathered} G(\theta, \mu_c) = -0.85 g_0 \frac{\theta}{\theta_s}\left[\ln\left(\frac{\theta}{\theta_s}\gamma(\mu_c)\right) - 1\right], \\ \gamma(\mu_c) = 1 + \frac{\mu_c}{3}, \quad \theta_s \cong 0.456, \quad \theta = \frac{4\delta\mathcal{L}}{g_0\Delta}, \end{gathered} \tag{4.76}$$

(it is to be noted that if we include the effect of the energy spread, $\gamma(\mu_c)$ should be modified as

$$\gamma(\mu_\varepsilon, \mu_c) = \left[\left(1 + \frac{\mu_c}{3}\right)\left(1 + 1.7\mu_\varepsilon^2\right)\right]^{-1}\Bigg),$$

which exhibits the maximum gain (4.74) at

$$\theta^* = \theta_s(\gamma(\mu_c))^{-1}. \tag{4.77}$$

The inclusion of saturation modifies the gain equation as

$$\begin{gathered} G(X, \theta, \mu_c) = -0.85 g_0 \frac{\theta}{\theta_s}\left[\ln\left(\frac{\theta}{\theta^*}F(X)\right) - 1\right], \\ \theta^* = \theta_s(\gamma(\mu_c))^{-1}, \end{gathered} \tag{4.78}$$

and the equilibrium intra-cavity intensity is finally obtained as

$$I_e(\theta, \mu_c) \cong \left(\sqrt{2} + 1\right)\left[\sqrt{\frac{\theta_s}{\theta\gamma(\mu_c)}}\exp\left[\frac{1}{2}\left(1 - \frac{\eta}{1-\eta}\frac{\theta_s}{0.85 g_o \theta}\right)\right] - 1\right] I_s. \tag{4.79}$$

These results which may sound arbitrary have been found using a hybrid procedure, exploiting both analytical and numerical tools; a brief account is given in section 4.10.

In figure 4.20 we show gain and equilibrium intensity versus θ.

In figure 4.21 we have reported the intra-cavity equilibrium intensity computed with PROMETEO and compared with the predictions from (4.79).

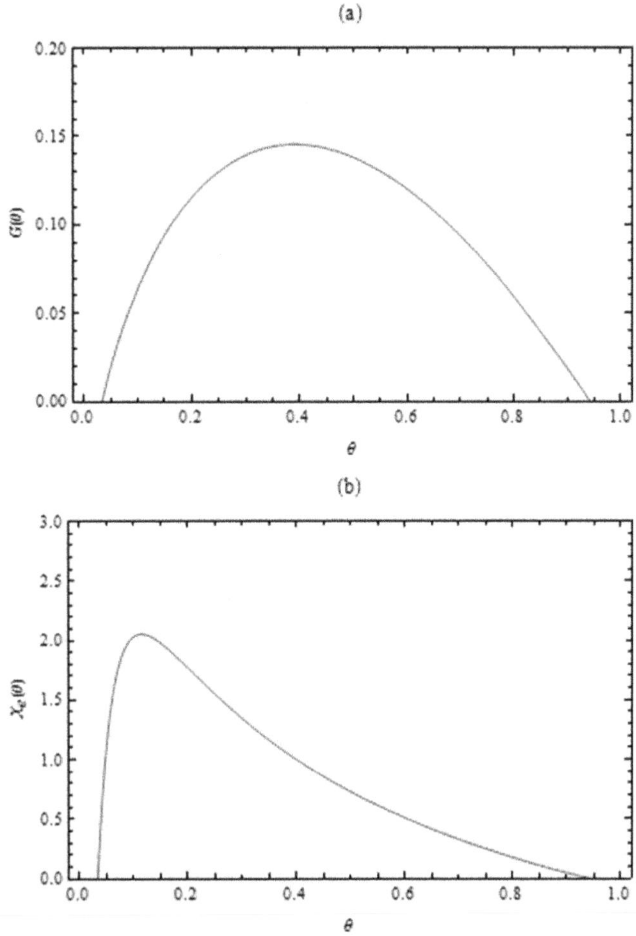

Figure 4.20. Net gain ($G_{\text{net}} = (1 - \eta)\, G_M - \eta$) (a) and intra-cavity equilibrium intensity (b) for $\mu_c = 0.5$, $\eta = 6\%$ and $g_0 = 30\%$.

Two further points are worth adding regarding the previous figures:

a) The value of the cavity desynchronism parameter ensuring the maximum gain is given by (4.77) which yields the following optimum cavity mismatch

$$\delta\mathcal{L}^* = \frac{0.456}{4} \frac{g_0 \Delta}{1 + \frac{\mu_c}{3}}, \tag{4.80}$$

and according to (4.65) we can evaluate the group velocity refractive index as

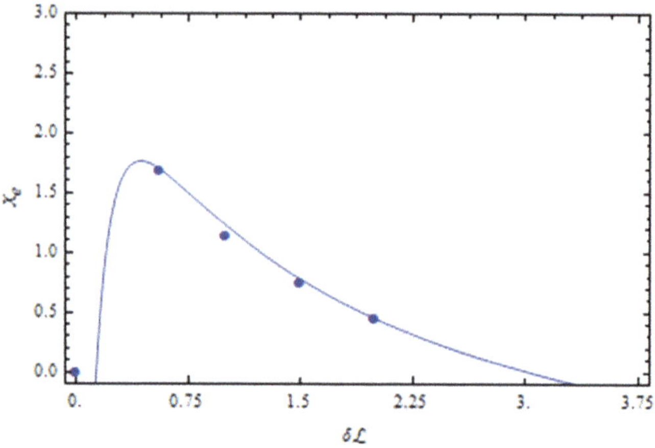

Figure 4.21. Equilibrium intra-cavity intensity; comparison between (4.79) (continuous line) and PROMETEO simulation (dotted line), $\mu_c = 1$, $\eta = 6\%$ and $g_0 = 30\%$.

$$n_g = \frac{0.456}{2}\frac{\Delta}{L_u}\frac{g_0}{1 + \frac{1}{3}\mu_c} \cong 0.63\frac{\Delta}{L_u}G_M(\mu_c). \tag{4.81}$$

b) The maximum of the intra-cavity equilibrium power density is located at a different value of the cavity mismatch yielding the optimum power. We invite the reader to provide a physical argument supporting this statement and consider also the further comments in section 4.10.

4.6.2 Cavity losses

As already stressed the cavity losses consists of two parts, active and passive. The first provides the means to allow the coupling of the intra-cavity radiation outside. The passive losses are associated with real losses, namely radiation absorption, heating, etc.

In the following we will comment on a naïve procedure aimed at optimizing the output power density I_o, linked to the intra-cavity intensity by

$$I_o = \eta_a I_E = \eta_a\left(\sqrt{2} + 1\right)\left(\sqrt{\frac{1 - \eta_a(1 + \xi)}{\eta_a(1 + \xi)}G_M} - 1\right)I_S, \quad \xi = \frac{\eta_p}{\eta_a}. \tag{4.82}$$

An example of the dependence of the extracted power versus the active losses is given in figure 4.22.

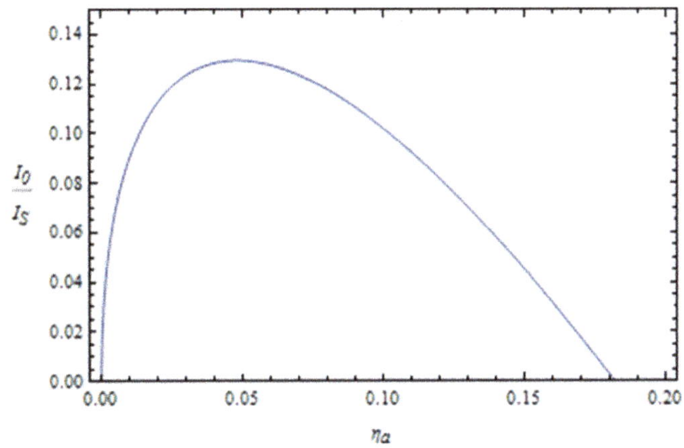

Figure 4.22. Output power density $\frac{I_o}{I_s}$ versus η_a for $\xi = 0.1,\ G_M = 0.25$.

The optimal cavity losses are determined from the condition $\frac{\partial}{\partial \eta_A} I_T = 0$ and, along with the optimum extracted intensity, are approximatively given by

$$\eta_A^* \cong 0.22 \times \frac{G_M}{1+\xi},$$
$$I_o^* \cong 0.5 \times G_M \times I_s. \tag{4.83}$$

In this section we have learned how notions from FEL and conventional laser theory can be embedded to obtain the first elements to design an FEL-O. Further elements more specific in nature, with reference to existing devices, will be found in section 4.10.

4.7 High-gain FEL devices

In the previous section we have attempted a 'practical' treatment of FELs devoted to oscillator devices, and we have stressed the crucial role of the cavity and of cavity losses. In figure 4.23 we show two plots regarding the link between the FEL wavelength and electron-beam energy. In figure 4.23(a) we show: the wavelength region covered by oscillators, where efficient mirrors to confine radiation do exist; the region below 120 nm where the optics have poor performance (large passive losses); and the VUV soft x-ray region where optical cavities are very challenging and where the most probable solution is a device operating without mirrors in a single pass. This option has led to the conception of the fourth generation SR devices and has paved the way for the SASE operating mode.

The dynamics of a single passage FEL device is reported in figure 4.24, describing a beam of electrons entering a sufficiently long undulator to accomplish the process from energy modulation (induced by the spontaneous emission and shot noise) to saturation.

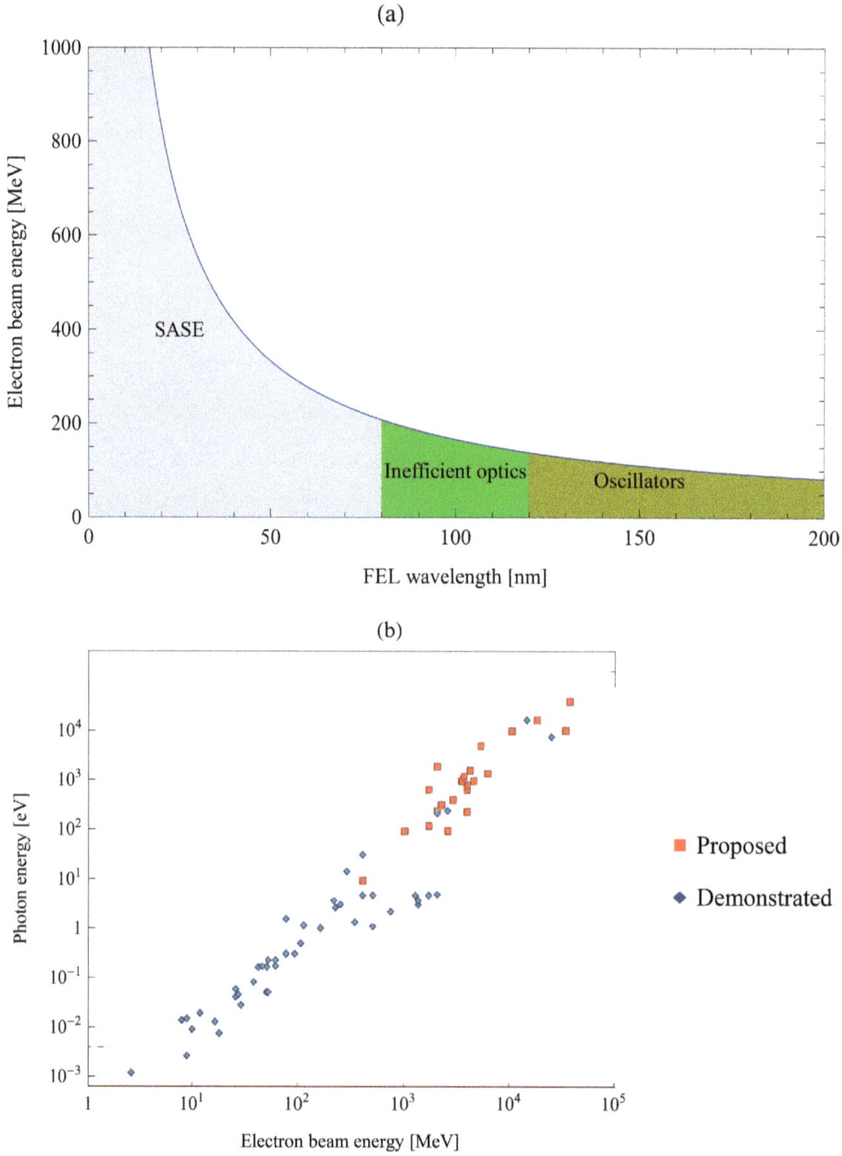

Figure 4.23. (a) A Letardi chart of electron energy versus wavelength. (b) ln–ln plot FEL photon energy versus electron energy.

Assuming that such an option is available and that an FEL device of this sort is characterized by a gain α per unit length, the coherent power growth along the undulator axis should be provided by

$$P(z) = P_n\, e^{\alpha z},$$
$$\alpha = \frac{1}{L_g}, \tag{4.84}$$
$$L_g \equiv \text{gain length},$$

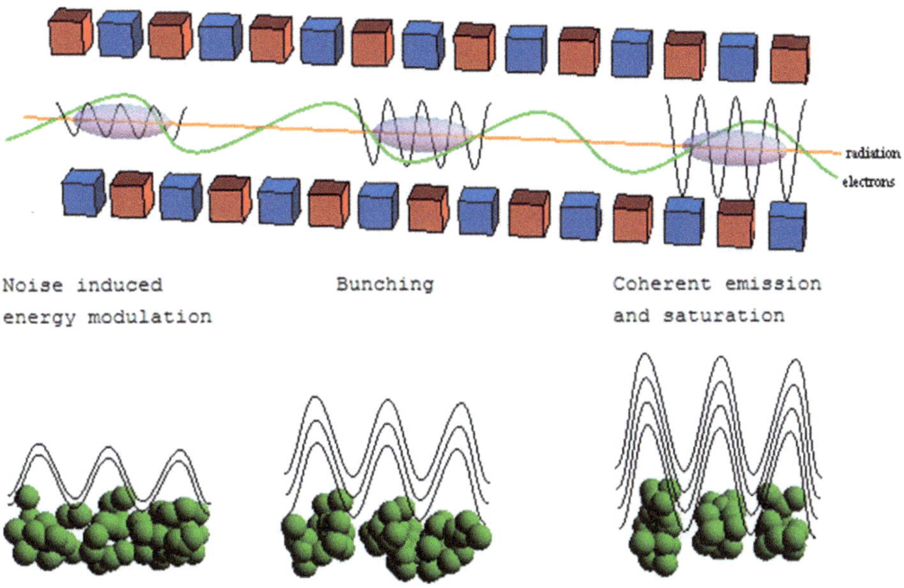

Figure 4.24. FEL single pass (high-gain) from initial noise to saturation.

where P_n is the input noise power. If we then assume that the device undergoes a saturation pattern analogous to that of an FEL-O we write

$$P(z) = P_n \frac{e^{\frac{z}{L_g}}}{1 + \frac{P_n}{P_S}(e^{\frac{z}{z_g}} - 1)}, \tag{4.85}$$

where P_S is the saturated power that we can reasonably assume to be a (small) fraction of the electron-beam power. The length of the undulator allowing the growth of the field from noise level to saturated power, namely the saturation length L_s, is

$$\begin{aligned} &P_n e^{\frac{L_S}{L_g}} \cong P_S \\ &\to L_s \cong L_g \ln\left(\frac{P_S}{P_n}\right), \end{aligned} \tag{4.86}$$

assuming $\frac{P_S}{P_n} \cong 10^8$ we find $L_S \cong 18.5 L_g$, if we impose an undulator length less than 20 m we find $L_g[\mathrm{m}] \leqslant 1.08$.

In the previous qualitative description we have introduced two crucial quantities, namely the gain length and the saturated power. In the following we will obtain the relevant expression by using the elements at our disposal.

We have exploited the integral equation (see (4.43)) to extract from it as much physics as possible. We have found it useful to formulate the problem using a set of dimensionless quantities, convenient for the oscillator configuration. We have chosen to use the dimensionless time $\tau = \frac{z}{L_u}$, namely the longitudinal coordinate normalized to the undulator length and $\nu = 2\pi N \frac{\omega - \omega_0}{\omega_0}$, namely the detuning parameter. The physical quantities determining the normalization are, in the case of τ, the length of the undulator and, regarding ν, twice the number of periods. It should be noted that $2N = (\Delta\omega/\omega)^{-1}$ is the inverse of the FEL bandwith, linked to the time of flight of the radiation inside the undulator. It is evident that the choice of the dimensionless variables is due to the fact that the length of the undulator fixes the gain of the device itself. In the case of single passage FEL devices, the notion of undulator length should be replaced by that of saturation length and therefore we will rewrite (4.43) using explicitly the z-coordinate

$$\frac{d}{dz}a(z) = i\left(\frac{4\pi\rho}{\lambda_u}\right)^3 \int_0^z \chi e^{-i\frac{\nu}{L_u}\chi} a(z - \chi)d\chi,$$
$$\rho = \frac{1}{4\pi N}\sqrt[3]{\pi g_0}. \tag{4.87}$$

The change of variables has naturally put in evidence the parameter ρ, often called the Pierce parameter (PP) which plays a different role than the small signal gain coefficient g_0, which is tailored for oscillator operations. The PP does not depend explicitly on the number of undulator periods (see (4.33)). Let us now define the following 'reference length':

$$L_g = \frac{\lambda_u}{4\pi\sqrt{3}\rho}. \tag{4.88}$$

The reason for the $\sqrt{3}$ in the definition of the gain length will be clear from the following:

$$\frac{d}{d\tilde{z}}a(\tilde{z}) = i\int_0^{\tilde{z}} \tilde{\chi} e^{-i\tilde{\nu}\tilde{\chi}} a(\tilde{z} - \tilde{\chi})d\tilde{\chi},$$
$$\tilde{z} = \frac{z}{L_g}, \quad \tilde{\nu} = \frac{1}{2\rho}\delta_\omega, \quad \delta_\omega = \frac{\omega - \omega_0}{\omega_0}. \tag{4.89}$$

It is now fairly natural to imagine that ρ is associated with the gain length, while the maximum relative bandwidth is

$$(\delta_\omega)_{\mathrm{MAX}} \cong \rho. \tag{4.90}$$

It is evident that the parameter ρ is of fundamental importance since it determines both the gain length (hence the saturation length) and the relative bandwidth (hence the saturated power).

We have so far fixed the main elements of the game, in the following sections we will provide a more substantive analysis.

4.8 High-gain FEL equation and the relevant consequences

In the previous sections we have stretched the FEL high-gain equation to keep out some relevant consequences, but we did not specify anything regarding its mathematical properties. It is an integro-differential equation of Volterra type which can be cast in the form of a third order ordinary differential equation (ODE) if one notes that it can be rewritten as

$$e^{i\tilde{\nu}\tilde{z}}\frac{da}{d\tilde{z}} = i\int_0^{\tilde{z}} [e^{i\,\tilde{\nu}\,(\tilde{z}-\tilde{z}')}\, a(\tilde{z}-\tilde{z}')\tilde{z}']\, d\,\tilde{z}' = i\; D_{\tilde{z}}^{-2}[a(\tilde{z})\, e^{i\tilde{\nu}\tilde{z}}], \tag{4.91}$$

where we have used the symbol $D_{\tilde{z}}^{-2}$ to denote a double integration in the Cauchy sense, according to which the repeated integration is written

$$\hat{D}_x^{-n} f(x) = \frac{1}{(n-1)!}\int_0^x (x-\xi)^{n-1} f(\xi)\, d\xi = \int_0^x d\,x_1 \ldots \int_0^{x_{n-2}} d\,x_{n-1} f(x_{n-1}). \tag{4.92}$$

If we derive twice both sides of (4.91) with respect to $\tilde{z}$ we end up with the following third order ODE with the relevant initial conditions,

$$\begin{gathered}\left(\hat{D}_{\tilde{z}}^{3} + 2i\nu\hat{D}_{\tilde{z}}^{2} - \nu^2\hat{D}_{\tilde{z}}\right) a(\tilde{z}) = ia(\tilde{z}),\\ a|_{\tilde{z}=0} = a_0, \quad \hat{D}_{\tilde{z}}a|_{\tilde{z}=0} = 0, \quad \hat{D}_{\tilde{z}}^{2}\, a|_{\tilde{z}=0} = 0,\end{gathered} \tag{4.93}$$

whose solution can be obtained by standard means.

A particular case of practical interest can be obtained from (4.93) for zero detuning, the equation is drastically simplified and reduces to

$$\hat{D}_{\tilde{z}}^{3} a(\tilde{z}) = ia(\tilde{z}) \tag{4.94}$$

and the relevant solution can be written as

$$a(\tilde{z}) = \sum_{j=0}^{2} a_j e^{\delta_j \tilde{z}}, \tag{4.95}$$

where δ_j are the cubic roots of the imaginary unit and a_j are amplitudes which can be obtained from the initial conditions.

The signal growth is dominated by the solution associated with the real part of the complex roots, which yields

$$a(z) \propto e^{\frac{4\pi\sqrt{3}\rho}{2}\frac{z}{\lambda_u}} = e^{\frac{1}{2}\frac{z}{L_g}}, \tag{4.96}$$

and clarifies the origin of the coefficient $\sqrt{3}$ in the definition of gain length.

The field intensity is associated with $|a(z)|^2$ and including all three roots along with the relevant complex part we end up with the following function accounting for the small signal evolution

$$|a(z)|^2 = A(z)\,|a_0|^2,$$
$$A(z) = \frac{1}{9}\left[3 + 2\cosh\left(\frac{z}{L_g}\right) + 4\cos\left(\frac{\sqrt{3}}{2}\frac{z}{L_g}\right)\cosh\left(\frac{z}{2L_g}\right)\right]. \quad (4.97)$$

The previous curve describes the power growth in the first two regions, modulation bunching and exponential gain.

It is worth noting that

$$A(z)|_{\frac{z}{L_g}\gg 1} \cong \frac{1}{9}e^{\frac{z}{L_g}} \quad (4.98)$$

corresponds to the approximation of the exponential regime.

According to the description of the previous sections the system should achieve a stationary state recognized as the saturation of the device. Leaving the technical

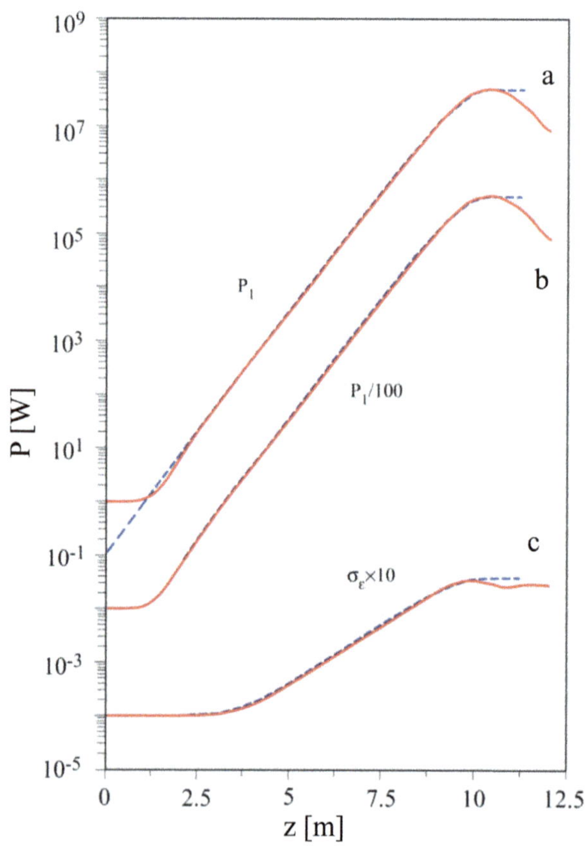

Figure 4.25. Power growth versus longitudinal coordinate, for the following specific parameters $\rho = 2.64 \times 10^{-3}$, $\lambda_u = 0.028$ m, $K = 2.133$ and $P_E = 1.32 \times 10^{10}$ W. (a) Power curve growth numerical simulation (continuous line), comparison with the logistic curve in the exponential regime approximation (dashed line). (b) The same as (a) with a comparison with (4.99) (the factor 100 reduction has been adopted to distinguish between the comparison with the logistic curves). (c) Interaction induced energy spread (magnified by a factor 10).

details to section 4.10, we stress that the 'complete' equation including the physical processes shown in figure 4.25 are summarized in the logistic equation

$$P(z) = P_0 \frac{A(z)}{1 + \frac{P_0}{P_s}(A(z) - 1)},$$
$$P_s = \sqrt{2}\rho P_E, \tag{4.99}$$

which, as shown in figure 4.25, nicely reproduces the numerical data (the further content of the plot will be discussed in this section). We have so far left open our definition of saturated power in (4.99), the square root is indeed justified by the fact that in the high-gain regime, for which in the region of physical interest ($\rho \cong 10^{-3}$) the gain curve is well reproduced by the following Gaussian function ($\delta_{\omega *}$ is a constant close to zero):

$$G(\delta_\omega) = \frac{1}{\sqrt{2\pi}\left(\frac{\rho}{\sqrt{2}}\right)} e^{-\frac{(\delta_\omega - \delta_{\omega *})^2}{2\left(\frac{\rho}{\sqrt{2}}\right)^2}}. \tag{4.100}$$

We have already stressed the importance of the PP ρ which has been exploited to quantify the gain length, the saturated power and the saturation length, which is easily shown to be

$$L_s = 1.066 L_g \ln\left(\frac{9P_F}{P_0}\right). \tag{4.101}$$

The effects of energy spread, emittances and other beam qualities can be included and quantified using this parameter and this aspect of the problem will be discussed later in this chapter.

Before closing this section, we note that the growth of the power is accompanied by a corresponding increase of the energy spread, this is one of the macroscopic effects and causes of the saturation process itself. This and other more technical issues will be discussed in section 4.10, dedicated to complements and exercises.

4.9 Concluding comments

In this section we will provide further elements concerning the FEL high-gain single pass configuration and insist on the pivotal role of the PP, which will be recognized as the crucial design parameter of this type of device.

4.9.1 Inclusion of inhomogeneous broadening effects

To include the effect of the energy spread on the high-gain FEL process, we proceed as before and consider the integral equation (4.72) written in the form

$$\frac{\partial}{\partial \tilde{z}} a(\tilde{z}) = i \int_0^{\tilde{z}} \tilde{z}' a(\tilde{z}, \tilde{z} - \tilde{z}')\, e^{-i\tilde{\nu}\tilde{z}'} e^{-\frac{\left(\frac{\tilde{\mu}_\varepsilon}{4\sqrt{3}}\tilde{z}'\right)^2}{2}} d\tilde{z}',$$
$$\tilde{\mu}_\varepsilon = \frac{2\sigma_\varepsilon}{\rho}, \tag{4.102}$$

within such a new variable definition the parameter controlling the effect of the beam energy spread on the growth process is $\tilde{\mu}_\varepsilon$. We expect that the presence of a non-negligible σ_ε yields a reduction of the gain, with the macroscopic consequence of an increase of the gain length and thus of the saturation length. This effect may be accounted for by the following fairly simple formula (see section 4.10 for further comments):

$$L_g(\tilde{\mu}_\varepsilon) \cong L_g\left(1 + 0.185 \frac{\sqrt{3}}{2} \tilde{\mu}_\varepsilon^2\right). \tag{4.103}$$

4.9.2 Slippage effects

In the previous sections we have shown that the key parameter accounting for short pulse dynamics in an FEL-O is the coupling parameter μ_c, which controls the effects associated with longitudinal mode coupling, mode locking and associated gain reduction effects.

In the case of high gain the role of slippage is more subtle. Indeed, rewriting the integral FEL equation in the form

$$\frac{\partial}{\partial \tilde{z}} a(\zeta, \tilde{z}) = i f_0(\zeta + l_c\tilde{z}) \int_0^{\tilde{z}} \tilde{z}' a(\zeta + l_c\tilde{z}', \tilde{z} - \tilde{z}')\, e^{-i\tilde{\nu}\tilde{z}'} d\tilde{z}',$$
$$l_c = \frac{\lambda}{4\pi\sqrt{3}\rho}, \quad f_0(\zeta) = \sqrt{2\pi} f(\zeta), \tag{4.104}$$

we can recognize the appearance of the quantity l_c known as *coherence length.*

Even though the concept of longitudinal mode coupling is hardly applicable to operation without a radiation confining cavity, nonetheless the coherence length can be viewed as the number of wavelengths within the optical pulses after each undulator period advance, namely

$$l_c = N_c\lambda,$$
$$N_c = \frac{1}{4\pi\sqrt{3}\rho}. \tag{4.105}$$

These wavelengths constituting a kind of a coherent macromode are characterized by a unique phase. If $l_c < \sigma_z$ we have approximatively $N_{sl} \cong \frac{\sigma_z}{l_c}$ of these modes contributing to the FEL evolution, each one with a different phase. If we sample the electron bunch as indicated in figure 4.26, with each slice corresponding to a portion of the packet with length l_c, we obtain accordingly that inside the undulator the field will be built up by the different coherence pulses. These, according to the

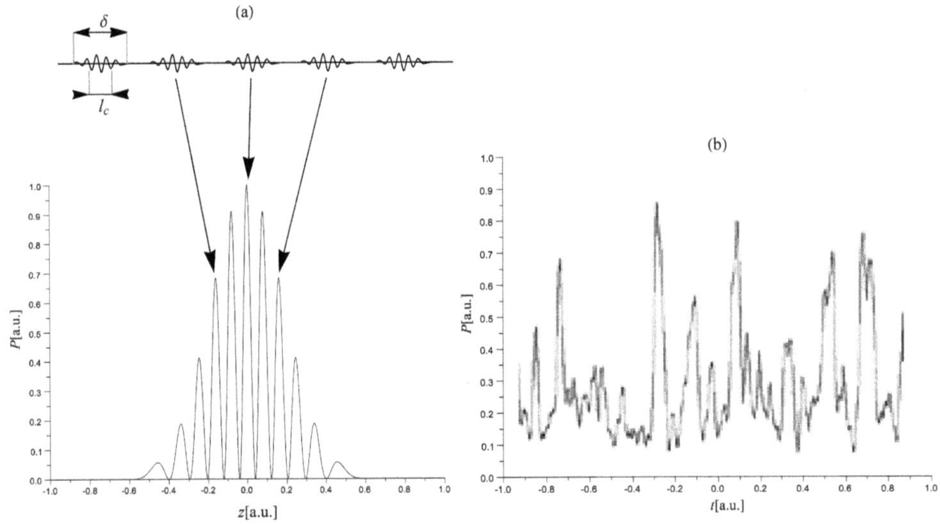

Figure 4.26. (a) Bunch length sampled with the number of coherence length (bottom); wave train from the single slice (up). (b) Temporal structure of the SASE power.

randomness of their initial phases, will determine a corresponding 'chaotic' output signal (figure 4.26) which is the result of the phase interference and slippage mechanisms (for further comments see section 4.10).

4.9.3 Transverse mode dynamics

In the absence of a cavity, another important role is played by the propagation of transverse optical modes and by the associated diffraction. The field may indeed transversally diffuse and become significantly larger than the electron-beam cross section. Therefore, the absence of a proper overlapping between radiation and electrons determines a spoiling of the efficiency of electron–wave transfer energy.

We can use a heuristic argument to understand the importance of these effects and how they can be included in the present analysis. The argument we exploit is the following.

The characteristic quantity of the FEL high gain is the gain length L_g, the relevance of the diffraction is specified by the Rayleigh length, which can be written as

$$
{}_{\eta}Z_R = \frac{4\pi(\varepsilon_\eta \, \beta_\eta)}{\lambda}, \qquad \eta = x, y, \tag{4.106}
$$

where $\sigma_\eta^2 = \varepsilon_\eta \, \beta_\eta$ is the electron-beam transverse section if we assume that the radiation waist coincides with the beam section itself. The effect of the diffraction is negligible if

$$
L_g \ll {}_{\eta}Z_R. \tag{4.107}
$$

We introduce therefore the parameter

$$\mu_\eta^D = \frac{L_g}{{}_\eta Z_R} = \frac{\lambda\lambda_u}{(4\pi)^2\sqrt{3}\,\beta_\eta\varepsilon_\eta\rho}, \tag{4.108}$$

called the high-gain diffraction parameter, as a measure of the importance of the diffraction in the FEL process itself.

Let us now define the characteristic length of the combined process (gain and diffraction), which will be assumed to be the mean of the individual length

$$L_g^{(D)} = \sqrt{L_{g\eta} Z_R}. \tag{4.109}$$

The optical waist measured at that distance is

$$W_\eta = W_0\sqrt{1 + \left(\frac{L_g^{(D)}}{{}_\eta Z_R}\right)^2} = W_0\sqrt{\left(1 + \mu_\eta^D\right)}. \tag{4.110}$$

Finally, recalling the definition of the PP, which is proportional to the cubic root of the current density,

$$\rho \propto \left(\frac{I}{2\,\pi\,\sigma_x\sigma_y}\right)^{\frac{1}{3}}. \tag{4.111}$$

This definition is valid if the radiation beam waist coincides with the electron transverse section; when the increase of the waist due the diffraction is included we end up with the following 'three-dimensional' version of the PP:

$$\begin{aligned} &\rho^D = \mathrm{F}\left(\mu_x^D, \mu_y^D\right)\,\rho, \\ &\mathrm{F}\left(\mu_x^D, \mu_y^D\right) = \left[\left(1 + \mu_x^D\right)\left(1 + \mu_y^D\right)\right]^{-1/6}, \end{aligned} \tag{4.112}$$

which can be exploited to define, e.g., the correction to the saturation length when diffraction may play a role. In the case of an FEL operating with $\rho \cong 10^{-3}$, $\lambda_u = 2.8\text{cm}$, $\lambda = 500\text{nm}$, $\sigma_{x,y} \cong 100\ \mu m$ we obtain an increase of the saturation length of the order of 8% of the diffraction free counterpart.

In this chapter we have touched upon many aspects of the FEL physics and many technical issues. We have only touched upon the surface of a field which is becoming wider and wider, but we will partially fill many of the gaps left open in the main body of this chapter in the forthcoming section dedicated to complements and exercises.

4.10 Complements and exercises

4.10.1 The one atom maser laser system

In the previous parts of this chapter we have described the physics underlying the processes of stimulated emission and absorption in a two-level atomic system. We pointed out that they are the archetype mechanisms of any laser device and we reduced the problem to the study of a physical system with two degrees of freedom (the two-level atom), interacting with an external field. We will show in the following that such a formal description holds for the FEL as well, but the appropriate framing of the problem requires a detour through *micro-maser* (*m*icrowave *a*mplified *s*pontaneous *e*mission (of) *r*adiation) physics and on the associated formalism. The reason for this choice is that in our opinion such a treatment allows a smooth transition between the theory of atom lasers and FELs.

A sketch of a micro-maser device (MMD) is shown in figure 4.27 and can be described as follows:

a) A beam of atoms of rubidium (^{85}Rb, or more in general alkaline atoms) undergoes, after a selection in velocity, a laser excitation to a Rydberg state.
b) The beam is then passed into a high Q cavity, where it interacts with the cavity TE modes, eventually becomes de-excited and pumps the cavity.
c) The cavity outgoing beam is then suitably detected to infer information relevant to the details of the intra-cavity interaction.

Before proceeding further, we should clarify what a Rydberg atom is and why it is exploited to pump the optical cavity. In alkaline metal atoms the valence electron is in an s-state and is excited to a state with a large principal quantum number n. This process is responsible for a transition inducing a large radius of the atom itself and a consequent large electric dipole moment. The transitions involved in the excitation process are shown in figure 4.28. For completeness we recall that atoms are characterized by the quantum numbers n, l, m, namely principal, angular and magnetic. A generic atomic state is denoted by

$$F = nA_{l\pm 1/2},$$

where $A = S,\ P,\ D, \ldots$ denotes the value of the angular momentum $l = 0,\ 1,\ 2\ldots.$

The transition, caused by the absorption of a photon, brings the atom from the fundamental s-level to a p-level with large principal number, namely a Rydberg state. Masing transitions then occur between the $63\mathrm{p}_{3/2}$ and $61\ \mathrm{d}_{5/2}$ states. The α, β states realize the up and down levels of the Maser dynamics and are usually denoted by

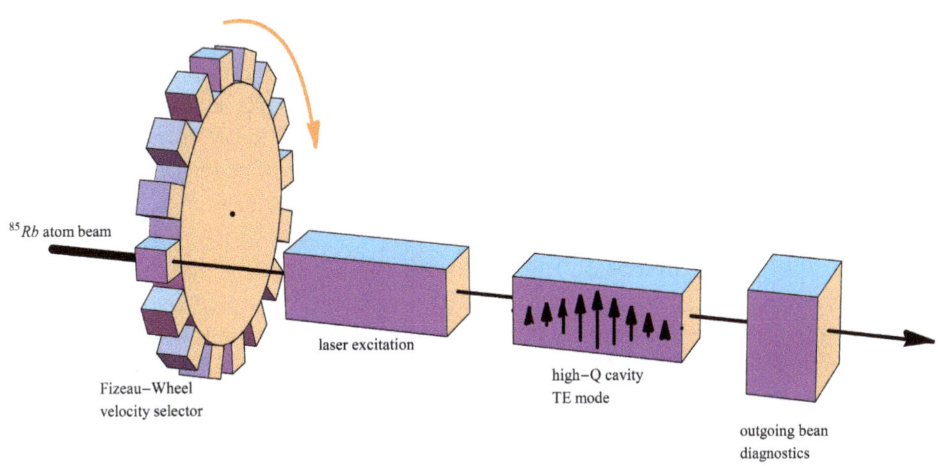

Figure 4.27. Sketch of an MMD.

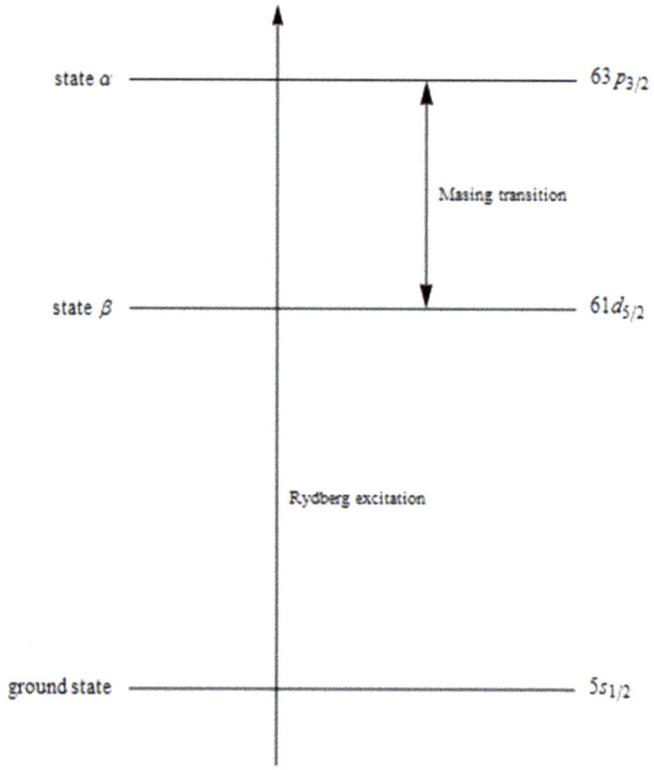

Figure 4.28. Spectroscopy of ^{85}Rb excitation to a Rydberg state and associated masing transitions.

$$\begin{aligned} |\alpha\rangle &= \left|\begin{smallmatrix} -\bullet- \\ -\,-\,- \end{smallmatrix}\right\rangle, \\ |\beta\rangle &= \left|\begin{smallmatrix} -\,-\,- \\ -\bullet- \end{smallmatrix}\right\rangle. \end{aligned} \tag{4.113}$$

The physical properties of Rydberg atoms are briefly summarized below. They are kind of like monsters, if compared to atoms in their ground state or in their first few excited energy states. Indeed we recall that, with the increase of the principal quantum number, the atom radius scales as $R \propto n^2 R_0$ (R_0 is the radius in the fundamental state), since in the case of rubidium $R_0 \cong 4.81 \times 10^{-1}$ nm, we infer that for states with $n = 60$ the atom may reach dimensions of the order of μm. The huge dimensions determine a comparably large dipole moment ($d \propto ZeR$) and a large radiative lifetime which extend to nearly 100 μs.

In the cavity the excited atom interacts with the TE field, which is tuned at the masing transition level and the interaction is dominant, owing to the large electric dipole moment induced by the transition from the ground to the Rydberg state. The energy gaps between the $|\alpha\rangle$, $|\beta\rangle$ levels are within the microwave range and the large lifetime of the associated transitions matches the extremely long lifetime of the high quality of the cavity (super-conducting), characterized by particularly low losses. Having restricted the interaction to that of a two-level system with an

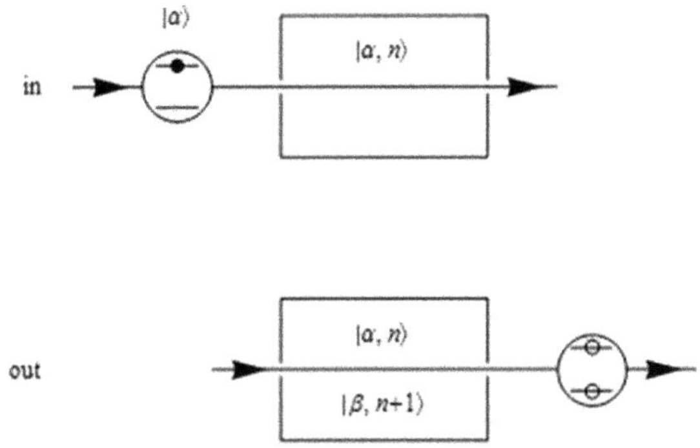

Figure 4.29. The one-atom maser and the relevant intra-cavity states.

electromagnetic field, all the MMD physics is contained in a nut shell and depicted as shown in figure 4.29, which is self- explaining. An excited atom enters inside the cavity and interacts with the field, where the initial state is provided by an entangled combination of excited atoms and photons, namely

$$|\overline{\alpha}\rangle = |\alpha,\, n\rangle. \tag{4.114}$$

The interaction with the field inside the cavity may lead to a $|\overline{\alpha}\rangle$ state or to

$$|\overline{\alpha}\rangle = |\beta,\, n + 1\rangle, \tag{4.115}$$

namely to the de-excited atom with an extra photon (see below for a more detailed understanding of the nature of these quantum states).

The mathematical tools necessary to treat such a problem are a set of operators, acting on the atomic states, such that

$$\begin{aligned} \hat{\sigma}_+ \left| \begin{matrix} -\,- \\ -\bullet- \end{matrix} \right\rangle &= \left| \begin{matrix} -\bullet- \\ -\,- \end{matrix} \right\rangle, \\ \hat{\sigma}_- \left| \begin{matrix} -\bullet- \\ -\,- \end{matrix} \right\rangle &= \left| \begin{matrix} -\,- \\ -\bullet- \end{matrix} \right\rangle, \end{aligned} \tag{4.116}$$

namely providing a transition from the ground state to the upper level and vice versa.

We can mathematically realize these operator devices as follows:

$$\begin{aligned} \hat{\sigma}_+ &= |\beta\rangle\langle\alpha|, \\ \hat{\sigma}_- &= |\alpha\rangle\langle\beta|. \end{aligned} \tag{4.117}$$

Exercise 4.1 Assuming that $\langle\beta|\beta\rangle = \langle\alpha|\alpha\rangle = 1,\ \langle\alpha|\beta\rangle = 0$, show that

$$\begin{aligned} [\hat{\sigma}_+,\ \hat{\sigma}_-] &= 2\,\hat{\sigma}_3, \\ \hat{\sigma}_3 &= \frac{1}{2}(|\beta\rangle\langle\beta| - |\alpha\rangle\langle\alpha|), \\ [\hat{\sigma}_3,\ \hat{\sigma}_\pm] &= \pm\hat{\sigma}_\pm. \end{aligned} \tag{4.118}$$

Hint: note that

$$[\hat{\sigma}_+, \ \hat{\sigma}_-] = |\beta\rangle\langle\alpha|\alpha\rangle\langle\beta| - |\alpha\rangle\langle\beta|\beta\rangle\langle\alpha|.$$

Exercise 4.2

1. Show that the Pauli matrices

$$\hat{\sigma}_3 = \begin{pmatrix} 1 & 0 \\ 0 & -1 \end{pmatrix}, \ \hat{\sigma}_+ = \begin{pmatrix} 0 & 1 \\ 0 & 0 \end{pmatrix}, \ \hat{\sigma}_- = \begin{pmatrix} 0 & 0 \\ 1 & 0 \end{pmatrix} \tag{4.119}$$

can be used as one of the possible (restricted) realization of the operators $\hat{\sigma}_+, \ \hat{\sigma}_-, \ \hat{\sigma}_3$.

Hint: check, e.g., the relevant commutation relations.

2. An analogous set of operators for the electromagnetic field, yielding the creation and the destruction of a photon, namely

$$\begin{aligned} \hat{a}^+ \, |n\rangle &= \sqrt{n+1} \, |n+1\rangle, \\ \hat{a} \, |n\rangle &= \sqrt{n} \, |n-1\rangle. \end{aligned} \tag{4.120}$$

They are the ordinary creation annihilation operators, satisfying the commutation bracket $[\hat{a}, \ \hat{a}^+] = 1$.

Putting everything together, we find that the Hamiltonian ruling the evolution of the system is

$$\begin{aligned} \hat{H} &= \hbar \, \Delta_c \, \hat{a}^+\hat{a} + \hbar\frac{\Delta_a}{2}\hat{\sigma}_3 + \hbar g[\hat{a}^+\hat{\sigma}_- + \hat{a} \, \hat{\sigma}_+], \\ \Delta_c &= \omega - \omega_c, \ \ \Delta_a = \omega - \Omega, \end{aligned} \tag{4.121}$$

where $\omega, \ \omega_a, \ \Omega$ are the frequencies associated with the laser exciting energy, the laser cavity photons and the two-level energy gap, respectively. $\Delta_{c,a}$ are the relevant detunings, the first two terms and last term in (4.121) are the free and interaction parts of the Hamiltonian, with g specifying the coupling constant.

The operator (4.121) is the Jaynes–Cummings Hamiltonian; it has been introduced in a rather heuristic way, for a more appropriate treatment see the lectures by Englert B-G (2002 Elements of Micromaser Physics arxiv: quant-ph/0203052v1), which have largely inspired this section. The correct derivation of Hamiltonian (4.121) requires the use of the rotating wave approximation, which is outside the purposes of this book.

Exercise 4.3 Fix the values of the constants $\omega, \ \omega_a, \ \Omega, \ g$ appearing in (4.121).

Hint: use a cylindrical cavity with radius and length 24 mm and operating in the TE_{121} mode and obtain

$$\omega \cong \Omega = 2\pi \times 21.5 \text{ GHz}.$$

The coupling constant is obtained from the dipole interaction and is, in the case of rubidium and for the considered levels, $g \cong 44$ kHz.

Exercise 4.4 Define the operator number $N = \hat{a}^{+}\hat{a}$ and use the previously quoted commutation rules to prove that the problem under study admits the following conserved quantity:

$$\dot{\hat{N}} + \dot{\hat{\sigma}}_3 = 0. \tag{4.122}$$

Discuss the relevant physical meaning.

Hint: use the Heisenberg equations of motion.

Exercise 4.5 Show that for a fixed number of photons the dynamics is reduced to the exchange of one photon only, $\Delta n = \pm 1$, and that the Hamiltonian can be written as

$$\hat{H} = \hbar \begin{pmatrix} \Delta_c(n-1) + \frac{1}{2}\Delta_a & g\sqrt{n} \\ g\sqrt{n} & \Delta_c n - \frac{1}{2}\Delta \end{pmatrix}. \tag{4.123}$$

Hint: under these conditions the problem is restricted to a 2×2 sub-space since the only transitions it allows are $|\alpha, n-1\rangle \Leftrightarrow |\beta, n\rangle$.

Exercise 4.6 Solve the associated Schrödinger equation.

Hint: first note that

$$i\frac{d}{dt}\underline{\varphi} = \begin{pmatrix} \Delta_c(n-1) + \frac{1}{2}\Delta_a & g\sqrt{n} \\ g\sqrt{n} & \Delta_c n - \frac{1}{2}\Delta_a \end{pmatrix}\underline{\varphi}. \tag{4.124}$$

Solve it in the form

$$\underline{\varphi} = e^{-it\begin{pmatrix} \omega(n-1)+\frac{1}{2}\Omega & g\sqrt{n} \\ g\sqrt{n} & \omega n-\frac{1}{2}\Omega \end{pmatrix}}\varphi_0,$$

and use standard algebraic means to end up with

$$\underline{\varphi} = \begin{pmatrix} \frac{\delta}{\sqrt{\Delta}}\sin\left(\frac{\sqrt{\Delta}}{2}t\right) + \cos\left(\frac{\sqrt{\Delta}}{2}t\right) & \frac{g\sqrt{n}}{\sqrt{\Delta}}\sin\left(\frac{\sqrt{\Delta}}{2}t\right) \\ \frac{g\sqrt{n}}{\sqrt{\Delta}}\sin\left(\frac{\sqrt{\Delta}}{2}t\right) & -\frac{\delta}{\sqrt{\Delta}}\sin\left(\frac{\sqrt{\Delta}}{2}t\right) + \cos\left(\frac{\sqrt{\Delta}}{2}t\right) \end{pmatrix} e^{i\delta t}\varphi_0, \tag{4.125}$$

$\Delta = \delta^2 + 4ng^2, \quad \delta = \Delta_a - \Delta_c.$

Discuss the relevant physical meaning.

Exercise 4.7 Find the eigenvalues and eigenvectors of the Hamiltonian (4.123) and discuss the relevant physical meaning.

Hint: the energy eigenvalues are

$$\frac{E_\pm}{\hbar} = \Delta_c\left(n - \frac{1}{2}\right) \pm \frac{1}{2}\sqrt{\Delta}. \tag{4.126}$$

The associated eigen-states (usually referred as dressed states) are expressed as

$$\begin{aligned} |+, n\rangle &= \sin(\varphi_n)|\,\beta, n-1\rangle + \cos(\varphi_n)|\,\alpha, n+1\rangle, \\ |-, n\rangle &= \cos(\varphi_n)|\,\beta, n-1\rangle - \sin(\varphi_n)|\,\alpha, n+1\rangle, \\ \tan(\varphi_n) &= \frac{\Delta - \delta^2}{\sqrt{\Delta} - \delta}. \end{aligned} \tag{4.127}$$

The transition from bare (atomic and field) to dressed states is shown in figure 4.30. It is worth stressing that the state splitting increase when n increases.

4.10.2 The one electron quantum theory of FEL

We will show below how the FEL quantum theory can be formulated in complete analogy with that of the MMD developed in the previous part of the section.

The quantum theory of FELs is usually formulated in a reference frame in which the undulator field and the laser field are seen as electromagnetic fields with the same

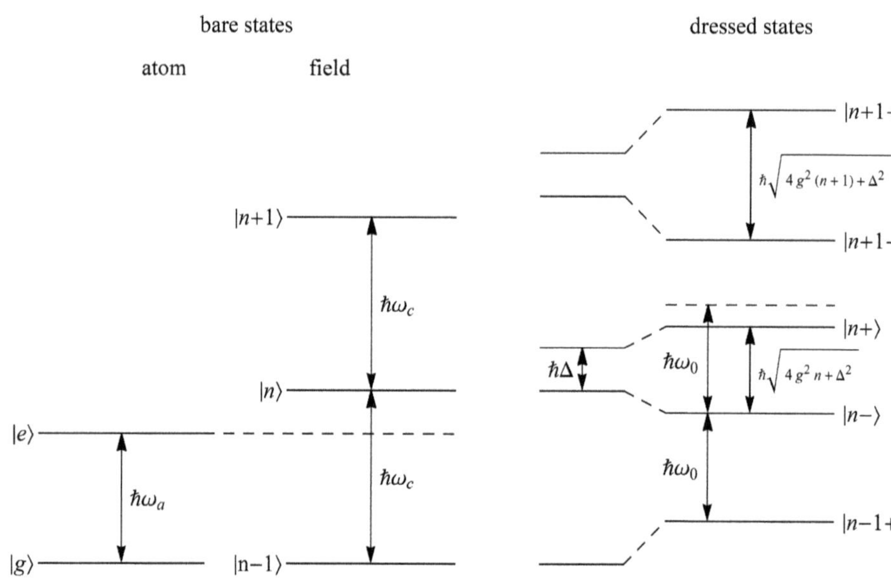

Figure 4.30. From bare to dressed states.

Figure 4.31. Radiation (R) and undulator (U) photons simultaneously interacting with one electron.

wavelength. This is the so-called Bambini–Renieri frame and will be discussed below (Arturo Bambini and Alberto Renieri formulated the theory of FEL in 1978 *Nuovo Cimento Lett.* **21** 399). According to their analysis the FEL dynamics has been viewed in a single particle fashion and they provided the description of the FEL interaction in terms of a simple (non-relativistic) Hamiltonian picture, whose main result has been the derivation of a pendulum-like equation, analogous to the one introduced by W B Colson. Alberto Renieri has been one of the most prominent scientist in FEL physics, introducing new concepts in this research field and providing a significant contribution on both the theoretical and experimental side.

Exercise 4.8 Define the frame where the undulator and radiation field appear to have the same wavelength.

Hereafter we outline a few remarks allowing the solution of the exercise:

a. In this reference frame we can regard the elementary FEL process as described in figure 4.31, according to which an electron interacts with two counter-propagating fields having opposite wave vectors and equal frequencies.
b. We choose a relativistic factor γ^* to specify the reference frame where the laser and 'undulator' wave appear with the same wavelength (for a practical introduction to relativistic kinematics see Artioli M and Dattoli G 2013 *Appunti di Cinematica Relativistica* (Rome: Aracne Editrice)).
c. We first note that in this reference frame the undulator period undergoes the usual Lorentz contraction and therefore

$$\lambda_u^* = \frac{\lambda_u}{\gamma^*},$$

while the laser wavelength is Doppler shifted and

$$\lambda_r^* = \lambda_r \sqrt{\frac{1+\beta^*}{1-\beta^*}}.$$

By assuming $\lambda_r^* = \lambda_u^*$ we end up with

$$\lambda_r \sqrt{\frac{1+\beta^*}{1-\beta^*}} = \frac{\lambda_u}{\gamma^*},$$

which yields

$$\frac{1}{(\gamma^*)^2} \frac{1}{1+\beta^*} = \frac{\lambda_r}{\lambda_u},$$

yielding the kinematic specifications of the desired transformations. If we furthermore assume that

$$\frac{\lambda_r}{\lambda_u} = \frac{1 + \frac{K^2}{2}}{2\gamma^2}$$

we may also conclude that γ^* is almost equal to the electron relativistic factor.

Exercise 4.9 Show that in this reference frame the FEL process can be written in terms of the following Hamiltonian

$$H = m_e c^2 \sqrt{1 + \left[\frac{(\vec{p} - e\vec{A})}{m_e c}\right]^2} + \mathcal{E}_r + \mathcal{E}_u, \qquad \vec{A} = \vec{A}_u + \vec{A}_r, \tag{4.128}$$

where $\mathcal{E}_{r,u}$ are the energies associated with the undulator and radiation fields, furthermore

$$\vec{A}_u = \mathcal{A}_u e^{-i(-k_r z + \omega_r t)} \underline{\kappa} + \text{c. c.} \qquad \vec{A}_r = \mathcal{A}_r e^{-i(k_r z + \omega_r t)} \underline{\kappa} + \text{c. c.} \tag{4.129}$$

($\underline{\kappa}$ is the polarization vector) are the vector fields associated with the radiation and undulator.

The static undulator is treated as an electromagnetic field using a kind of Weiszacker–Williams approximation. (Regarding this point, see Madey J M J 1971 *J. Appl. Phys.* **42** 1906; see also Dattoli G and Renieri A 1985 *Experimental and Theoretical Aspects of Free Electron Laser (Laser Handbook* vol 4) ed M L Stitch and M Bass (Amsterdam: North Holland).)

Exercise 4.10 Use the previous result to show that the Hamiltonian can be reduced, in the non-relativistic limit, to the form (usually called the Bambini–Renieri (B–R) Hamiltonian)

$$H = \frac{p_z^2}{2m_e} + \hbar(\omega + \Omega)(\hat{a}_u^+ \hat{a}_u + \hat{a}_r^+ \hat{a}_r) + \hbar\,\Omega\,(\hat{a}_r^+ \hat{a}_u e^{-2ikz} + \hat{a}_u^+ \hat{a}_r e^{2ikz}), \qquad \Omega = \frac{\alpha \lambda_c \lambda}{4\pi V} c. \tag{4.130}$$

(note that the fine structure constant α, the Compton wavelength $\lambda_c = \frac{h}{m_e c}$, the radiation wavelength λ and the quantization volume V are included in the definition of the parameter),
where the operators $\hat{a}$ are electromagnetic field creation–annihilation operators, satisfying the commutation relations

$$[\hat{a}_l^+, \hat{a}_m] = -\delta_{l,m}. \tag{4.131}$$

Hint:

a. Assume that the field variables are dependent on the longitudinal coordinate only. The canonical momenta $p_{x,y}$ are therefore conserved quantities.
b. Note that

$$\begin{aligned}&(\vec{p} - e\vec{A})^2 = |\vec{p}\,|^2 - 2e\,\vec{p}\cdot\vec{A} + e^2\vec{A}\cdot\vec{A},\\ &\text{Coulomb gauge} \Rightarrow \vec{p}\cdot\vec{A} = 0\,,\\ &\vec{A}\cdot\vec{A} = \vec{A}_u\cdot\vec{A}_u + 2\,\vec{A}_r\cdot\vec{A}_u + \vec{A}_r\cdot\vec{A}_r,\end{aligned} \tag{4.132}$$

thus finding

$$\sqrt{|\vec{p}\,|^2 - 2e\vec{p}\cdot\vec{A} + e^2\vec{A}\cdot\vec{A} + (m_e c)^2} \cong m^* c\left(1 + \frac{|\vec{p}\,|^2}{2(m^* c)^2} + \frac{e^2}{(m^* c)^2}\vec{A}_r\cdot\vec{A}_u\right),$$

$$(m^*)^2 = m_e^2 + e^2(\vec{A}_u\cdot\vec{A}_u + \vec{A}_r\cdot\vec{A}_r).$$

If we average on the oscillating part and define the mass as

$$(m^*)^2 = m_e^2(1 + K_u^2 + K_r^2),$$

where $K_{u,\,r}$ are the strength parameters associated with the undulator and radiation fields, respectively. If we assume that in the B–R frame the FEL dynamics is non-relativistic (which is not true in the case of the linear undulator, explain why) we find

$$H = \frac{p_z^2}{2\,m^*} + e^2\vec{A}_r\cdot\vec{A}_u + e^2\,|\vec{A}_u|^2 + e^2\,|\vec{A}_r|^2,$$

which, using a formalism of second quantization for the fields is written as reported in (4.130) (for the specific technical details see the original paper by Bambini and Renieri).

Exercise 4.11 Show that the following conserved quantity naturally emerges from the B–R formalism

$$\begin{aligned}&\hat{P} = \hat{p}_z + \hbar k(\hat{n}_r - \hat{n}_u),\\ &\hat{N} = \hat{n}_r + \hat{n}_u,\end{aligned} \tag{4.133}$$

where $\hat{n}_{r,u}$ is the operator number of photons associated with the radiation and undulator fields, respectively. Clarify the relevant physical meaning.

Hint: use the Heisenberg equations and compare with (4.122), then think in terms of conservation of momentum (electron and fields) and total number of photons (radiation + undulator).

Exercise 4.12 Define the operators

$$\begin{aligned} \hat{\Sigma}_+ &= \hat{a}_r^+\hat{a}_u, \ \hat{\Sigma}_- = \hat{a}_u^+\hat{a}_r, \\ \hat{\Sigma}_3 &= \frac{1}{2}(\hat{a}_r^+\hat{a}_r - \hat{a}_u^+\hat{a}_u), \end{aligned} \tag{4.134}$$

and show that

$$\begin{aligned} [\hat{\Sigma}_+, \ \hat{\Sigma}_-] &= 2\hat{\Sigma}_3, \\ [\hat{\Sigma}_3, \ \hat{\Sigma}_\pm] &= \pm\hat{\Sigma}_\pm. \end{aligned} \tag{4.135}$$

Write the B–R formalism in the form

$$\begin{aligned} H &= \frac{p_z^2}{2m_e} + \hbar\Omega(\hat{\Sigma}_+\hat{S}_- + \hat{\Sigma}_-\hat{S}_+) + \hbar\omega(a_u^+a_u + a_r^+a_r), \\ \hat{S}_\pm &= e^{\pm 2ikz}, \ \Omega = \frac{\alpha\lambda_c\lambda}{4\pi V}c, \end{aligned} \tag{4.136}$$

and point out the analogies with the Jaynes–Cummings model.

Hint: note that the part of the Hamiltonian containing the total number of photons is a conserved quantity and can be eliminated by a unitary transformation.

Exercise 4.13 Use the conserved quantities in (4.133) to write the Schrödinger equation for the FEL process ruled by the B–R Hamiltonian.

Hint: note that from the conservation of the total momentum and total number of photons it follows that

$$\begin{aligned} n_r &= n_r^0 + l, \\ n_u &= n_u^0 - l, \\ p &= p_0 + 2l\hbar k. \end{aligned} \tag{4.137}$$

The quantum number l accounting for the number of exchanged photons during the process is sufficient to describe the FEL dynamics. The associated wave function can accordingly be written as

$$\Psi(t) = \sum_{l=-n_r^0}^{n_u^0} C_l(t)\left| p_0 - 2l\hbar k, \ n_u^0 - l, \ n_r^0 + l \right\rangle. \tag{4.138}$$

Exercise 4.14 Write the FEL Schrödinger equation and show that the time-dependent $C_l(t)$ coefficients satisfy the equation

$$i\dot{C}_l = (\varepsilon l - w) l C_l + \Omega\left[\sqrt{(n_u^0 - l + 1)(n_r^0 + l)}\, C_{l-1} + \sqrt{(n_u^0 - l)(n_r^0 + l + 1)}\, C_{l+1}\right], \tag{4.139}$$

$$\varepsilon = \frac{2\hbar k^2}{m_e}, \quad w = \frac{2kp_0}{m_e}.$$

Hint: note that

$$i\hbar\, \partial_t \Psi(t) = \left[\frac{1}{2m_e}\hat{p}_z^2 + \hbar\Omega\left(\hat{\Sigma}_+\hat{S}_- + \hat{\Sigma}_-\hat{S}_+\right)\right]\Psi(t),$$

$$\hat{p}_z^2 \Psi(t) = \sum_{l=-n_r^0}^{n_u^0} (p_0 - 2l\hbar k)^2 C_l(t) \left| p_0 - 2l\hbar k,\ n_u^0 - l,\ n_r^0 + l \right\rangle,$$

$$\hat{\Sigma}_+\hat{S}_-\Psi(t) = \sum_{l=-n_r^0}^{n_u^0} C_l(t)\sqrt{\left(n_u^0 - l\right)\left(n_r^0 + l + 1\right)}\ \left| p_0 - 2(l-1)\hbar k,\ n_u^0 - l - 1,\ n_r^0 + l + 1 \right\rangle,$$

$$\hat{\Sigma}_-\hat{S}_+\Psi(t) = \sum_{l=-n_r^0}^{n_u^0} C_l(t)\sqrt{\left(n_u^0 - l + 1\right)\left(n_r^0 + l\right)}\ \left| p_0 - 2(l+1)\hbar k,\ n_u^0 - l + 1,\ n_r^0 + l - 1 \right\rangle.$$

Then use the orthogonality condition

$$\left\langle p_0 - 2l'\hbar k,\ n_u^0 - l',\ n_r^0 + l' \middle| p_0 - 2l\hbar k,\ n_u^0 - l,\ n_r^0 + l \right\rangle = \delta_{l,l'}.$$

Carefully explain why the term $H_0 = \frac{1}{2m_e} p_0^2 + \hbar\omega\,(n_u^0 + n_r^0)$ can be eliminated without any problem.

Exercise 4.15 Explain the physical meaning of the term $(\varepsilon l - w)\, l$.

Hint: note that it is associated with $\Delta E_k = \frac{1}{2m_e}(p_0 - 2l\hbar k)^2 - \frac{1}{2m_e} p_0^2$.

In the previous equation we have treated the undulator field as a real electromagnetic wave and we have associated it with the number of undulator photons n_u^0. This is, however, quite a large quantity and always such that $n_u^0 \gg n_r$. Any photon exchange leaves the number of undulator 'photons' unaffected, therefore we can replace, for any practical purposes, (4.139) with

$$i\dot{C}_l = (\varepsilon l - w)lC_l + \overline{\Omega}\left[\sqrt{\left(n_r^0 + l\right)}\ C_{l-1} + \sqrt{\left(n_r^0 + l + 1\right)}\ C_{l+1}\right],$$
$$\overline{\Omega} = \Omega\sqrt{n_u^0}. \tag{4.140}$$

Exercise 4.16 Equation (4.140) represents the evolution of the probability amplitudes of emitting l-photons at a given time. It couples all the amplitudes, but we can truncate the equation and limit our treatment to the case of the emission and absorption of one photon only. Show that in this case we can cast our equation in the matrix form

$$i\frac{d}{dt}\begin{pmatrix} C_1 \\ C_0 \\ C_{-1} \end{pmatrix} = \begin{pmatrix} \varepsilon - w & \overline{\Omega}\sqrt{n_r^0 + 1} & 0 \\ \overline{\Omega}\sqrt{n_r^0 + 1} & 0 & \overline{\Omega}\sqrt{n_r^0} \\ 0 & \overline{\Omega}\sqrt{n_r^0} & \varepsilon + w \end{pmatrix}\begin{pmatrix} C_1 \\ C_0 \\ C_{-1} \end{pmatrix},$$
$$\begin{pmatrix} C_1 \\ C_0 \\ C_{-1} \end{pmatrix}_{t=0} = \begin{pmatrix} 0 \\ 1 \\ 0 \end{pmatrix}. \tag{4.141}$$

Discuss the meaning of the choice of the initial conditions and use any procedure (analytical, numerical, etc) to obtain the solution for the above set of equations.

Hint: solve the equation as

$$\begin{pmatrix} C_1 \\ C_0 \\ C_{-1} \end{pmatrix} = \hat{U}(t)\begin{pmatrix} 0 \\ 1 \\ 0 \end{pmatrix},$$
$$\hat{U}(t) = \exp\left[-i\hat{M}t\right],$$
$$\hat{M} = \begin{pmatrix} \varepsilon - w & \overline{\Omega}\sqrt{n_r^0 + 1} & 0 \\ \overline{\Omega}\sqrt{n_r^0 + 1} & 0 & \overline{\Omega}\sqrt{n_r^0} \\ 0 & \overline{\Omega}\sqrt{n_r^0} & \varepsilon + w \end{pmatrix}. \tag{4.142}$$

Proceed then as follows:

a. Find the eigenvalues $\lambda_{1,2,3}$ of the matrix $\hat{M}$.
b. Use the Kayley–Hamilton theorem to write

$$\hat{U}(t) = c_0\hat{1} + c_1\hat{M} + c_2\hat{M}^2.$$

c. Find the coefficients $c_{1,2,3}$ by setting

$$e^{-i\lambda_j t} = c_0\hat{1} + c_1\lambda_j + c_2\lambda_j^2,$$
$$j = 1, 2, 3.$$

Compare the solutions using any numerical procedure.

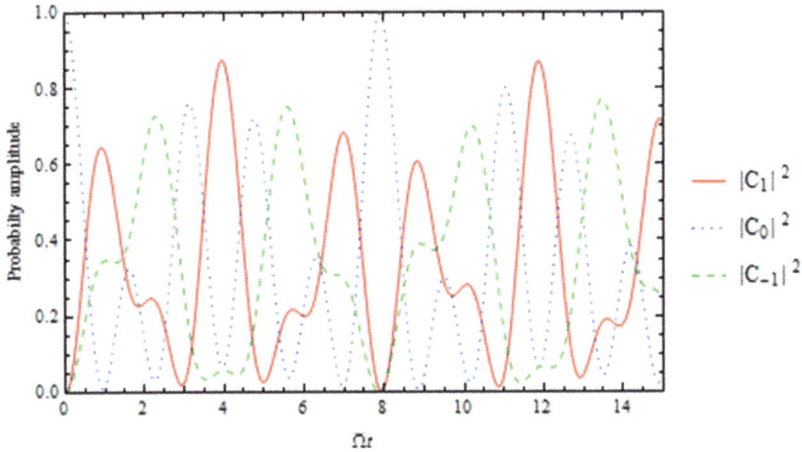

Figure 4.32. Probability versus time of emitting or absorbing a photon in a quantum FEL description, using the approximation in (4.141), ($\varepsilon/\overline{\Omega} = 0.01,\ w/\overline{\Omega} = 1,\ n_r^0 = 1$).

Exercise 4.17 Comment on the physical meaning of figure 4.32, where we have shown the behavior versus time of the probability of emitting zero photons $|C_0|^2$, emitting one $|C_{+1}|^2$ or absorbing one $|C_{-1}|^2$.

Exercise 4.18 Express the matrix $\hat{M}$as

$$\hat{M} = \hat{A} + \hat{B},$$

$$\hat{A} = \begin{pmatrix} \varepsilon - w & 0 & 0 \\ 0 & 0 & 0 \\ 0 & 0 & \varepsilon + w \end{pmatrix},$$

$$\hat{B} = \begin{pmatrix} 0 & \overline{\Omega}\sqrt{n_r^0 + 1} & 0 \\ \overline{\Omega}\sqrt{n_r^0 + 1} & 0 & \overline{\Omega}\sqrt{n_r^0} \\ 0 & \overline{\Omega}\sqrt{n_r^0} & 0 \end{pmatrix}, \tag{4.143}$$

$$\hat{U} = \exp\left[-i\hat{A}\,\frac{t}{2}\right]\exp(-i\hat{B}t)\exp\left[-i\hat{A}\,\frac{t}{2}\right].$$

Then use the symmetric split method to approximate the evolution operator as

$$\hat{U}(\delta) \cong \exp\left[-i\hat{A}\frac{\delta}{2}\right]\exp(-i\hat{B}\delta)\exp\left[-i\hat{A}\,\frac{\delta}{2}\right], \tag{4.144}$$

where δ is a small time step.

Then show that

$$\begin{pmatrix} C_1 \\ C_0 \\ C_{-1} \end{pmatrix} \cong \hat{U}(\delta)^t \begin{pmatrix} 0 \\ 1 \\ 0 \end{pmatrix} \tag{4.145}$$

and compare such an approximate solution with its 'exact' counterpart (4.142).

Hint: note first that

$$\exp\left[-i\hat{A}\,\frac{\delta}{2}\right] = \begin{pmatrix} e^{-i\frac{(\varepsilon-w)}{2}\delta} & 0 & 0 \\ 0 & 1 & 0 \\ 0 & 0 & e^{-i\,\frac{(\varepsilon+w)}{2}\delta} \end{pmatrix},$$

$$\exp(-i\hat{B}\delta) = \hat{1} - i\,\frac{\sin\left(\sqrt{2n_r^0+1}\,\overline{\Omega}\,\delta\right)}{\sqrt{2n_r^0+1}} \begin{pmatrix} 0 & \sqrt{n_r^0+1} & 0 \\ \sqrt{n_r^0+1} & 0 & \sqrt{n_r^0} \\ 0 & \sqrt{n_r^0} & 0 \end{pmatrix}$$

$$- \frac{1-\cos\left(\sqrt{2n_r^0+1}\,\overline{\Omega}\,\delta\right)}{2n_r^0+1} \begin{pmatrix} n_r^0+1 & 0 & \sqrt{n_r^0(n_r^0+1)} \\ 0 & 2n_r^0+1 & 0 \\ \sqrt{n_r^0(n_r^0+1)} & 0 & n_r^0 \end{pmatrix}$$

Then obtain

$$\hat{U}(\delta) \cong \begin{pmatrix} e^{-i\varepsilon_-\delta} & 0 & 0 \\ 0 & 1 & 0 \\ 0 & 0 & e^{-i\varepsilon_+\delta} \end{pmatrix} - i\frac{\sin\left(\sqrt{2n_r^0+1}\,\overline{\Omega}\,\delta\right)}{\sqrt{2n_r^0+1}}$$

$$\times \begin{pmatrix} 0 & \sqrt{n_r^0+1}\,e^{-i\varepsilon_-\frac{\delta}{2}} & 0 \\ \sqrt{n_r^0+1}\,e^{-i\varepsilon_-\frac{\delta}{2}} & 0 & \sqrt{n_r^0}\,e^{-i\varepsilon_+\frac{\delta}{2}} \\ 0 & \sqrt{n_r^0}\,e^{-i\varepsilon_+\frac{\delta}{2}} & 0 \end{pmatrix}$$

$$-\frac{1-\cos\left(\sqrt{2n_r^0+1}\,\overline{\Omega}\,\delta\right)}{2n_r^0+1}$$

$$\times \begin{pmatrix} (n_r^0+1)\,e^{-i\,\varepsilon_-\delta} & 0 & \sqrt{n_r^0(n_r^0+1)}\;e^{-i(\varepsilon_++\varepsilon_-)\frac{\delta}{2}} \\ 0 & 2n_r^0+1 & 0 \\ \sqrt{n_r^0(n_r^0+1)}\;e^{-i(\varepsilon_++\varepsilon_-)\frac{\delta}{2}} & 0 & n_r^0\,e^{-i\varepsilon_+\delta} \end{pmatrix},$$

$$\varepsilon_\pm = \varepsilon \pm w,$$

and

$$\begin{pmatrix} C_1 \\ C_0 \\ C_{-1} \end{pmatrix}(\delta) \cong \begin{pmatrix} -i\dfrac{\sin\left(\sqrt{2n_r^0+1}\,\overline{\Omega}\,\delta\right)}{\sqrt{2n_r^0+1}}\sqrt{n_r^0+1}\,e^{-i\varepsilon-\frac{\delta}{2}} \\ 1-\dfrac{1-\cos\left(\sqrt{2n_r^0+1}\,\overline{\Omega}\,\delta\right)}{2n_r^0+1}\left(2n_r^0+1\right) \\ -i\dfrac{\sin\left(\sqrt{2n_r^0+1}\,\overline{\Omega}\,\delta\right)}{\sqrt{2n_r^0+1}}\sqrt{n_r^0}\,e^{-i\varepsilon+\frac{\delta}{2}} \end{pmatrix}.$$

Exercise 4.19 Use the previous results to derive the FEL gain formula.

Hint: note that according to the above quantum description the average number of exchanged photons is given by

$$\langle l\rangle = \sum_{l=-1}^{1} l\,|C_l|^2 = |C_{+1}|^2 - |C_{-1}|^2$$

and use (4.142) and (4.145) to draw what is shown in figure 4.33. It is furthermore important to stress that:

1. Figure 4.33(a) compares the numerical solution (continuous line) and the curve

$$\langle l\rangle \propto -\frac{\partial}{\partial w}\left[\frac{\sin\left(\dfrac{w\Delta t}{2}\right)}{\dfrac{w\Delta t}{2}}\right]^2$$

.

2. While (a) refers to the case with a sufficiently large photon number of $n_r^0(\cong 60)$, figure 4.33(b) describes the case with $n_r^0 = 1$ (continuous line) and $n_r^0 = 2$ (dotted line).

We have discussed so far the FEL interaction from the quantum point of view, such a treatment, important from the pedagogical point of view, is mainly academic since the associated quantum signatures are far from being detectable from the experimental point of view. The classical physics is indeed largely sufficient to describe the relevant physics. To underscore this statement we note that in figure 4.33 we have reported the FEL gain function, and the deviation from the classical formula occurs for a very low number of photons. This effect is clearly undetectable in any actual experimental device.

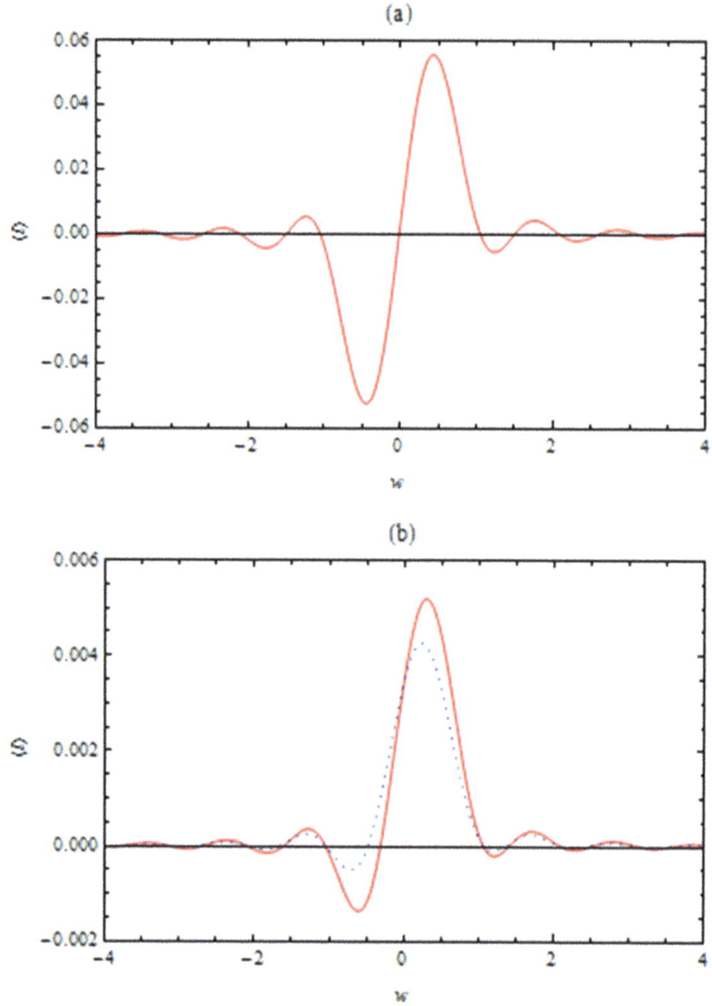

Figure 4.33. Average number of photons versus w, $\Delta t = 6$, $\Omega = 0.01$. (a) $n_r^0 = 60$, numerical (continuous line) and analytical (dotted line). (b) $n_r^0 = 1$ (dotted line), $n_r^0 = 2$ (continuous line).

4.10.3 Optical resonators

4.10.3.1 Optical cavity losses and quality factor

In the main body of the chapter we have considered the theory of FEL oscillators and we have superficially introduced the concept of cavity losses. It is evident that cavity losses are a pivotal element of any laser device and it is useful to understand more carefully the relevant definition and meaning.

We start our discussion by considering a very simple problem, namely that of evaluating the filling of an optical cavity, which is pumped by an external intensity source. According to (4.51) the gain is

$$I_{r+1} = (1 - \eta)\, I_r + I_{in}. \tag{4.146}$$

Assuming that at $r = 0$ the intra-cavity intensity is I_0, we obtain the solution of the difference equation (4.146) in the form

$$\begin{aligned} I_r &= (1 - \eta)^r I_0 + \frac{I_{in}}{\eta}(1 - (1 - \eta)^r), \\ r &\equiv \text{round trip number}. \end{aligned} \tag{4.147}$$

We can replace the discrete number r with time by using the following procedure:

$$\begin{aligned} (1 - \eta)^r &= e^{r \ln(1-\eta)} = e^{-\frac{t}{2L_c}|\ln(1-\eta)|}, \\ r &\cong \frac{ct}{2L_c}, \quad L_c \equiv \text{cavity length}, \end{aligned}$$

thus finding

$$\begin{aligned} I(t) &= I_0 e^{-\frac{t}{\tau_c}} + \frac{I_{in}}{\eta}\left(1 - e^{-\frac{t}{\tau_c}}\right), \\ \tau_c &= \frac{2L_c}{c\,|\ln(1 - \eta)|}, \end{aligned} \tag{4.148}$$

where τ_c defines the photon cavity lifetime.

We can now discuss the same problem using a slightly different language. We assume indeed that the cavity is provided by two mirrors with reflectivity $R_{1,2}$, assuming furthermore that some fractional internal losses (due to diffraction, heating, etc) occur, we can write that, after one round trip, the intensity is provided by, see figure 4.34 (we do not include any cavity filling external source),

$$\begin{aligned} I_1 &= R_1 R_2 (1 - T_i)^2 I_0 \\ i &\equiv \text{internal} \end{aligned}$$

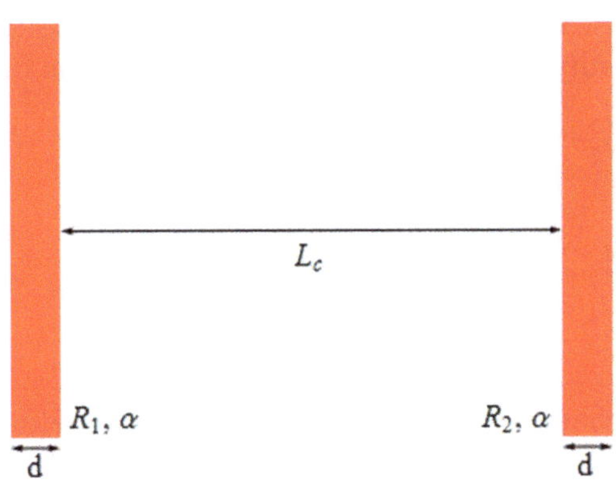

Figure 4.34. A Fabry–Perot resonator. R_1, R_2 are the mirror reflectivities, d the mirror thickness, α the absorption coefficient and L_c the cavity length.

and its extension to the mth iteration simply yields

$$I_m = [R_1R_2(1 - T_i)^2]^m I_0.$$

It is evident, by comparison with (4.147), that the losses η introduced in the previous equations read

$$\eta = 1 - R_1R_2(1 - T_i)^2, \tag{4.149}$$

which is an expected result. Accordingly, we also find

$$\tau_c = \frac{2L_c}{c|\ln[R_1R_2(1 - T_i)^2]|}. \tag{4.150}$$

Let us now introduce the cavity quality factor Q defined as (see also chapter 2)

$$Q = 2\pi\,\frac{\text{energy stored}}{\text{energy lost per cycle}}. \tag{4.151}$$

Accordingly, we write

$$Q = 2\pi\,\frac{\mathcal{E}}{-\left(\dfrac{d}{d(\nu_R t)}\mathcal{E}\right)}, \qquad \nu_R \equiv \text{cycle frequency}, \tag{4.152}$$

which yields

$$\left(\frac{d}{dt}\mathcal{E}\right) = -\frac{2\pi\nu_R}{Q}\mathcal{E},$$

and therefore

$$\mathcal{E} = e^{-\frac{2\pi\nu_R t}{Q}}\mathcal{E}_0.$$

Although we have used the energy and not the field intensity, as in the previous equation we can identify the photon lifetime in terms of the quality factor as

$$\tau_c = \frac{Q}{2\pi\nu_R}, \tag{4.153}$$

thus finally ending up with

$$Q = \frac{4\pi\nu_R L_c}{c|\ln(1 - \eta)|}. \tag{4.154}$$

The frequency ν_R is that of the 'optical' cycle, we find therefore

$$Q = \frac{\omega}{|\ln(1 - \eta)|}\frac{2L_c}{c}.$$

If we assume furthermore that the losses are very small and keep $|\ln(1-\eta)| \cong \eta$, we find

$$Q \cong 2\frac{\omega L_c}{c\eta},$$

namely the Q factor is inversely proportional to the losses of the cavity.

Exercise 4.20 Consider a He–Ne laser oscillating at 633 nm with a cavity length of 0.9 m. Assume that the cavity mirrors are identical, with reflectivity $R = 0.98$ and with no internal losses. Evaluate the cavity lifetime and the associated Q factor.

(*Answer:* $\tau_c \cong \frac{1}{150}$ns, $Q \cong 4.7 \times 10^8$.)

Exercise 4.21 Find a suitable way of introducing the mirror reflectivity and Q for an FEL oscillator cavity

a. Derive from the previous equation the identity

$$1 - \eta = e^{-\frac{\omega}{Q}\frac{2L_c}{c}} = e^{-\frac{8\pi\gamma^2}{Q\left(1+\frac{K^2}{2}\right)}\frac{L_c}{\lambda_u}}.$$

b. Derive the reflectivity as (both mirrors are assumed to be equal)

$$R(1 - T_i) = e^{-\frac{4\pi\gamma^2}{Q\left(1+\frac{K^2}{2}\right)}\frac{L_c}{\lambda_u}}.$$

c. Use the previous equation to write

$$Q = \frac{L_c}{\lambda_u\left(1 + \frac{K^2}{2}\right)}\frac{4\pi\gamma^2}{|\ln[R(1 - T_i)]|}.$$

Discuss the usefulness of the last formula as a design tool (note for example that an FEL-O with $\frac{L_c}{\lambda_u} = 5 \times 10^2$, $\gamma = 200$, $K = 1$, $R \cong 0.9$, $T_i \cong 0$ requires $Q \cong 1.19 \times 10^9$).

The optical cavity is exploited to store the radiation and in such respect the losses are an annoyance since they reduce the gain. On the other side radiation should be extracted from the cavity and they play therefore also a non-passive role. The mirror reflectivity is indeed related to the transmissivity T by the identity

$$R = 1 - T. \tag{4.155}$$

The out-coupled power is therefore associated with the mirror transmissivity. As already noted there are truly passive losses associated with the mirror absorption and heating. So to simplify the problem we assume that the losses can be divided in two parts, namely active η_a losses and passive η_p losses, and we will write

$$\eta = \eta_a + \eta_p, \tag{4.156}$$

which are just phenomenological parameters.

To appreciate their role we go back to (4.147) and (4.148), and note that the equilibrium power is simply

$$I_E = \frac{I_{in}}{\eta}.$$

The out-coupled power density is therefore

$$I_o = \frac{I_{in}}{1 + a},$$
$$a = \frac{\eta_p}{\eta_a}.$$

It is accordingly evident that the power lost in heating is

$$I_l = \eta I_E - I_o = \frac{\eta_p}{1 + a} I_{in}$$

or

$$I_l \cong \eta_p I_{in}, \quad a \ll 1.$$

Before concluding this complement we add a few remarks and an exercise regarding the cavity and the relevant mode confinement.

To this aim we recall that in the case of a stable resonator the modes that fit inside the optical cavity are Gaussian. We will consider a cavity composed of two spherical mirrors with not necessarily identical curvature radii and, given $R_{1,2}$, L, we look for the position of the waist and the waist itself (see figure 4.35).

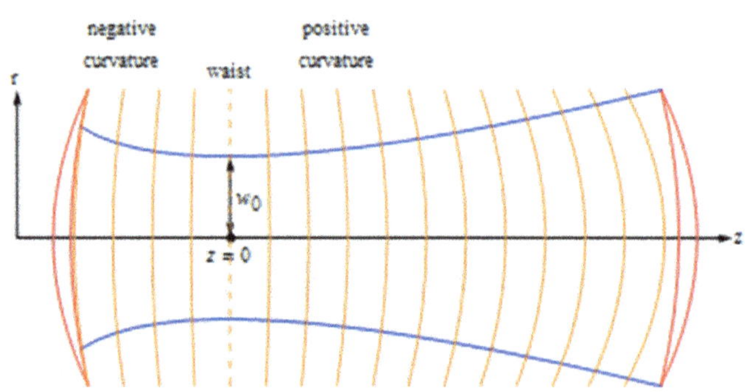

Figure 4.35. Optical resonator and relevant geometrical parameters.

The problem is not dissimilar to what we have learnt for beam propagation. From the optical point of view the mirror can be viewed as a focusing length with focal length $\frac{R}{2}$.

It should be noted that the radius of curvature of the propagating beam changes sign according to the wave front of the mirror where it has been reflected; in the waist the beam has minimum section and infinite curvature radius. The radius is linked to the z-position and to the Rayleigh length by

$$R(z) = \frac{z^2 + z_R^2}{z},$$
$$z_R = \frac{\pi W_0^2}{\lambda}. \tag{4.157}$$

The use of the sign convention reported in figure 4.35 yields

$$-R_1 = \frac{z_1^2 + z_R^2}{z_1},$$
$$R_2 = \frac{z_2^2 + z_R^2}{z_2}, \tag{4.158}$$
$$L = z_2 - z_1.$$

Exercise 4.22 Solve the previous system to obtain $z_{1,2,R}$ and obtain

$$z_R^2 = \frac{g_1 g_2(1 - g_1 g_2)}{(g_1 + g_2 - 2g_1 g_2)^2} L^2,$$

$$z_1 = -\frac{g_2(1 - g_1)}{(g_1 + g_2 - 2g_1 g_2)} L,$$

$$z_2 = \frac{g_1(1 - g_2)}{(g_1 + g_2 - 2g_1 g_2)} L,$$

$$g_\alpha = \left(1 - \frac{L}{R_\alpha}\right),$$

and the waist

$$w_0^2 = \frac{L\lambda}{\pi} \frac{\sqrt{g_1 g_2(1 - g_1 g_2)}}{(g_1 + g_2 - 2g_1 g_2)}.$$

Exercise 4.23 Keeping in account that

$$w(z)^2 = w_0^2\left[1 + \left(\frac{z}{z_R}\right)^2\right],$$

show that the spot size at the mirror location is

$$w_1^2 = \frac{L\lambda}{\pi}\sqrt{\frac{g_2}{g_1(1 - g_1 g_2)}},$$
$$w_2^2 = \frac{L\lambda}{\pi}\sqrt{\frac{g_1}{g_2(1 - g_1 g_2)}}.$$

Use the previous identities to specify the mirror size, to avoid large diffraction losses.

Exercise 4.24 The resonator reported in figure 4.36 is composed of spherical mirrors with radii $R_{1,2}$.

Consider the back and forward propagation of a ray as shown in figure 4.37, and show that it can be represented by the following propagation matrix:

$$\hat{M} = \begin{pmatrix} 1 & L \\ 0 & 1 \end{pmatrix}\begin{pmatrix} 1 & 0 \\ -\frac{1}{f_1} & 1 \end{pmatrix}\begin{pmatrix} 1 & L \\ 0 & 1 \end{pmatrix}\begin{pmatrix} 1 & 0 \\ -\frac{1}{f_2} & 1 \end{pmatrix}. \tag{4.159}$$

Exercise 4.25 Show that the round trip evolution is equivalent to the optical chain reported in figure 4.38 and show that the matrix associated with n round trips is simply $\hat{M}^n$.

Exercise 4.26 Establish an analogy with the electron-beam transport matrix and define the relevant Twiss parameters.

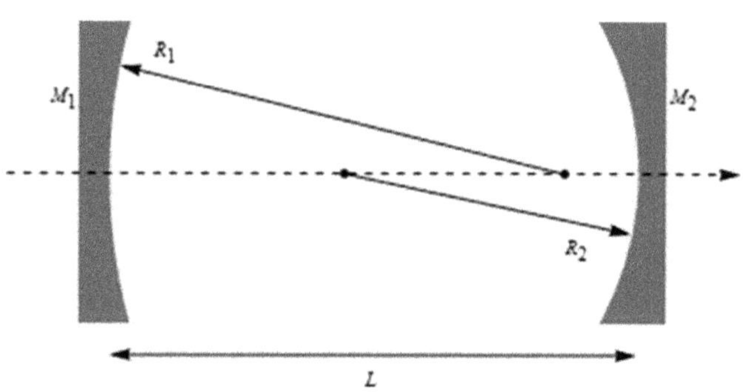

Figure 4.36. Optical resonator composed by spherical mirrors M_1 and M_2 with radii R_1 and R_2, respectively.

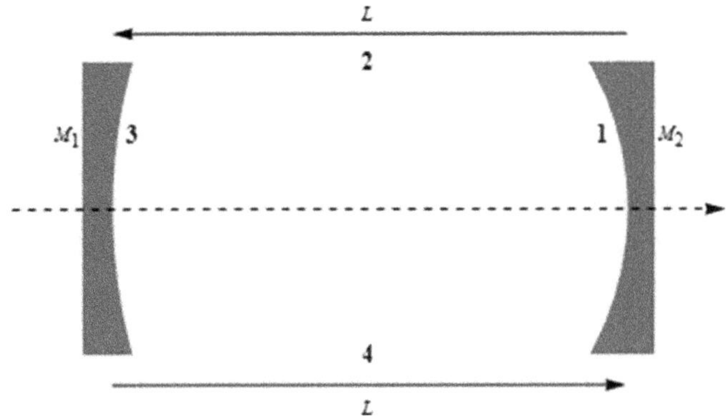

Figure 4.37. A round trip in an optical cavity. The numbers refer to the action of the optical element on the ray: 1 → focusing mirror 2; 2 → straight section; 3 → focusing mirror 1; 4 → straight section.

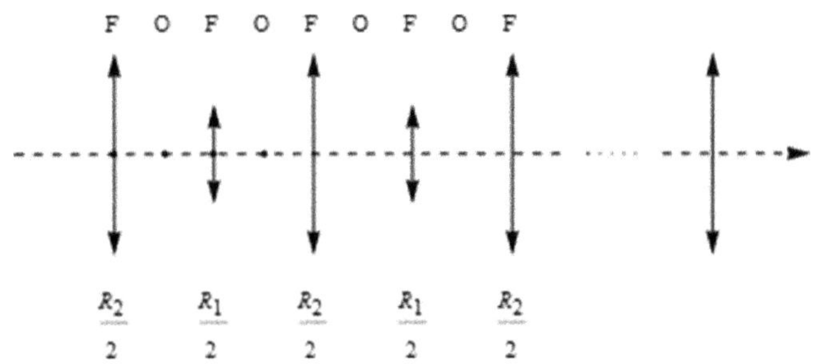

Figure 4.38. Equivalent *FOFOF* lattice for an optical cavity.

Hint: the correspondence is obtained in an almost straightforward manner by noting that in the waist (the minimum of the beam section) we have $\alpha = 0$, since for a diffraction limited beam the radiation emittance is

$$\varepsilon_r = \frac{\lambda}{4\pi}.$$

We end up with

$$\beta_r = 4L\sqrt{\frac{g_2}{g_1(1 - g_1 g_2)}}.$$

Exercise 4.27 Discuss the physical meaning of the cavity stability and show that, writing in general

$$\hat{M} = \begin{pmatrix} a & b \\ c & d \end{pmatrix}, \tag{4.160}$$

the stability requires the condition

$$\left|\frac{a+d}{2}\right| < 1. \tag{4.161}$$

Hint: note that the difference between a stable and unstable resonator is that the ray does not escape the cavity after a finite number of bounces (figure 4.39), recalling furthermore that (see also section 1.14)

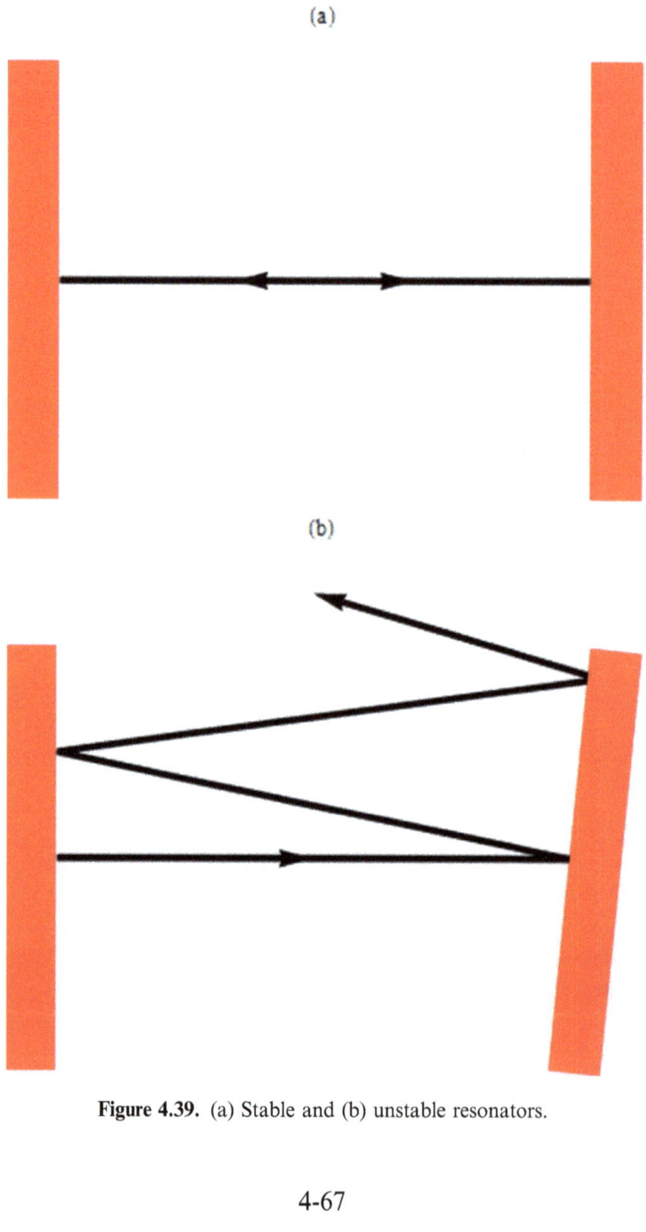

Figure 4.39. (a) Stable and (b) unstable resonators.

$$\hat{M}^n = \frac{\lambda_+^n \lambda_- - \lambda_-^n \lambda_+}{\lambda_- - \lambda_+} \hat{1} + \frac{\lambda_+^n - \lambda_-^n}{\lambda_+ - \lambda_-} \hat{M},$$

where $\lambda_\pm$ are the eigenvalues of the matrix $\hat{M}$ we infer that stability requires $\mathrm{Re}(\lambda_\pm)$.

Exercise 4.28 Show that for the matrix (4.159) such a condition leads to the condition

$$0 < g_1 g_2 < 1. \tag{4.162}$$

Exercise 4.29 Discuss the meaning of the stability diagram in figure 4.40.

The stability diagram of a resonator is geometrically represented as illustrated in figure 4.40, the hyperbola delimit the stability region according to the condition (4.162). In this picture we have

a. A confocal resonator ($R_1 = R_2 = L$) is represented by a point located at the origin ($g_{1,2} = 0$).
b. A plan–plan resonator (infinite curvature radii) is represented by the point (1, 1) lying on the hyperbola.
c. A concentric resonator ($R_1 = R_2 = \frac{L}{2}$) is represented by the point (−1, −1).

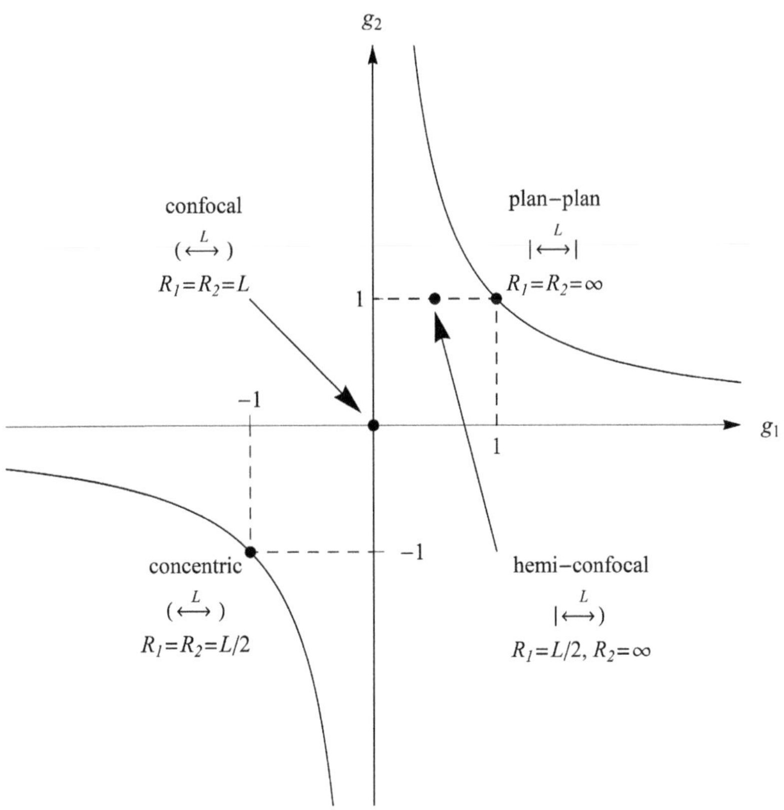

Figure 4.40. Resonator stability diagram.

d. A hemi-confocal resonator ($R_2 = \frac{L}{2}$, $R_1 = \infty$) is represented by the point $(\frac{1}{2}, 1)$.

Exercise 4.30 Show that for a confocal resonator we have

$$w_1^2 = w_2^2 = \frac{\lambda L}{\pi}.$$

We have dwelt on these aspects of the problem to emphasize that the mode section inside the optical cavity does not have a constant profile along the optical axis. Regarding the FEL gain we have already stressed that we should take care of this effect for an appropriate definition of the filling factor, and we will discuss this aspect of the problem further in section 4.10.5.

4.10.4 Colson's pendulum equations, dimensionless variables and non-linear harmonic generation

Exercise 4.31 In the previous chapter we have discussed the derivation of the FEL pendulum equation, we invite the reader to write it in the non-dimensional form

$$\begin{aligned} \frac{d^2}{d\tau^2}\zeta &= |a|\ \cos(\zeta + \varphi), \\ \frac{da}{d\tau} &= -j\langle e^{-i\zeta}\rangle, \\ \tau &= \frac{z}{L}, \end{aligned} \tag{4.163}$$

and prove that the Colson dimensionless field amplitude a and current j are defined as

$$\begin{aligned} |a| &= \frac{k_s}{\gamma^4}K^*(N\lambda_u)^2(1 + K^{*2})f_b e_s = 4\pi\left(\frac{N}{\gamma}\right)^2 K^*\lambda_u\left(1 - \frac{\nu}{2\pi N}\right)f_b e_s, \\ j &= 2\pi g_0, \\ g_0 &= \frac{16\pi}{\gamma}\frac{|J|}{I_0}\lambda_0 L N^2 \xi f_b^2 \\ &= 4\pi\frac{|J|}{I_0}\left(\frac{N}{\gamma}\right)^3(\lambda_u K^* f_b)^2, \end{aligned} \tag{4.164}$$

where g_0 is the small signal gain FEL coefficient.

Exercise 4.32 Derive the FEL integral equation of the small signal regime (see (4.43)).

Hint: set $\zeta \cong \zeta_0 + \nu_0\tau + \delta\zeta$, where $\delta\zeta$ is a perturbation, then use the first equation of (4.163) to obtain (in the small signal approximation)

$\delta\zeta \cong \int_0^\tau d\tau'(\tau - \tau')Re[a(\tau')e^{i(\zeta_0+\nu_0\tau')}]$. Insert this last relation in the second equation of (4.163) and find

$$\frac{da}{d\tau} = -2\pi g_0\langle e^{-i\zeta_0}\rangle_{\zeta_0} + i\pi g_0\int_0^\tau d\tau'(\tau - \tau')\, a(\tau')\, e^{-i\nu_0(\tau-\tau')}$$
$$+\, i\pi g_0\langle e^{-2i\zeta_0}\rangle_{\zeta_0}\int_0^\tau d\tau'(\tau - \tau')\, a^*(\tau')\, e^{-i\nu_0(\tau+\tau')},$$

where $\langle\ldots\rangle_{\zeta_0}$ denotes the average on the initial phase distribution, which is assumed to be of the form

$$f(\zeta_0) = \sum_{n=0}^{\infty} b_n e^{in\zeta_0},$$
$$\int_0^{2\pi} f(\zeta_0)d\zeta_0 = 2\pi,$$

with b_n the bunching coefficients, noting therefore that

$$\langle e^{-i\zeta_0}\rangle_{\zeta_0} = \frac{1}{2\pi}\int_0^{2\pi} f(\zeta_0)e^{-i\zeta_0}d\zeta_0 = b_1,$$
$$\langle e^{-2i\zeta_0}\rangle_{\zeta_0} = \frac{1}{2\pi}\int_0^{2\pi} f(\zeta_0)e^{-2i\zeta_0}d\zeta_0 = b_2.$$

We eventually end up with

$$\frac{da}{d\tau} = -2\pi g_0 b_1 + i\pi g_0\int_0^\tau d\tau'(\tau - \tau')\, a(\tau')e^{-i\nu_0(\tau-\tau')}$$
$$+\, i\pi g_0 b_2\int_0^\tau d\tau'(\tau - \tau')\, a^*(\tau')\, e^{-i\nu_0(\tau+\tau')}. \tag{4.165}$$

Assuming that the beam is not initially bunched ($b_{1,2} = 0$) we obtain the FEL integral equation discussed in the previous part of the chapter.

Exercise 4.33 Assume $b_2 = 0$ in (4.165) and show that the field amplitude a satisfies the same third order equation derived in (4.93) with the initial conditions

$$a(0) = a_0,\ \dot{a}(0) = -2\pi g_0 b_1,\ \ddot{a}(0) = 2\pi i\nu_0 g_0 b_1.$$

Exercise 4.34 Show that the integral equation cannot be reduced to an ODE when the second order bunching coefficient b_2 is included.

Exercise 4.35 Discuss the physical meaning and role of this last term.

Exercise 4.36 Consider the solution of the (4.93) for zero detuning and, assuming the initial conditions

$$a|_{\tau=0} = a_0\,, \qquad \hat{D}_\tau\, a|_{\tau=0} = 0\,, \qquad \hat{D}_\tau^2\, a\Big|_{\tau=0} = 0,$$

show that

$$|a(z)|^2 = B(z)\,|\,a_0|^2,$$
$$B(z) = \frac{2}{9}\left[\cosh\left(\frac{z}{L_{g,1}}\right) - \exp\left(-\frac{z}{2L_{g,1}}\right)\cos\left(\frac{\pi}{3} + \frac{\sqrt{3}\,z}{2L_{g,1}}\right) - \exp\left(\frac{z}{2L_{g,1}}\right)\cos\left(\frac{\pi}{3} - \frac{\sqrt{3}\,z}{2L_{g,1}}\right)\right]. \tag{4.166}$$

Exercise 4.37 Consider the full equation (4.93) and, using the initial conditions

$$a\Big|_{\tau=0} = 0, \qquad \hat{D}_\tau a|_{\tau=0} = -2\pi\, g_0 b_1\,, \qquad \hat{D}_\tau^2\, a\Big|_{\tau=0} = 0,$$

show that the solution reads (this solution was first given by Professor H Fang in an unpublished paper. The same result was later published in Dattoli G, Renieri A and Torre A 1990 *Lectures on Free Electron Laser Theory and Related Topics* (Singapore: World Scientific))

$$a(\tau) = \frac{a_0}{3\,(\nu + p + q)}\, e^{-\frac{2}{3}\, i\nu\tau}\Bigg\{(-\nu + p + q)\, e^{-\frac{i}{3}\,(p+q)\,\tau} + 2(2\,\nu + p + q)e^{\frac{i}{6}(p+q)\tau}\left[\cosh\left(\frac{\sqrt{3}}{6}(p-q)\tau\right) + i\frac{\sqrt{3}\,\nu}{p-q}\sinh\left(\frac{\sqrt{3}}{6}(p-q)\tau\right)\right]\Bigg\}, \tag{4.167}$$
$$p = \left[\frac{1}{2}\left(r + \sqrt{d}\right)\right]^{\frac{1}{3}}, \qquad q = \left[\frac{1}{2}\left(r - \sqrt{d}\right)\right]^{\frac{1}{3}},$$
$$r = 27\pi g_0 - 2\nu^3, \qquad d = 27\pi g_0\left[27\pi g_0 - 4\nu^3\right].$$

Hint: use the Cardano method to obtain the solution of the associated third degree algebraic equation.

Exercise 4.38 Show that the operation at higher harmonics in an FEL device is ruled by the following pendulum equation, explain the relevant physical meaning

$$\begin{aligned}
&\frac{d^2}{d\tau^2}\zeta = \sum_{n=0}^{\infty}|a_n|\cos(\psi_n), \quad \psi_n = n\zeta + \varphi_n, \\
&\frac{d}{d\tau}a_n = -j_n\langle e^{-in\zeta}\rangle, \\
&j_n = 2\pi g_{0,n},
\end{aligned} \tag{4.168}$$

where n is the harmonic number (odd numbers for a linear undulator), a_n is the dimensionless amplitude, φ_n is the field amplitude phase, j_n is the dimensionless current and a_1, φ_1, j_1 correspond to a, φ, j of (4.163). Also,

$$\begin{aligned}
&g_{0,n} = g_0\left(\frac{f_{b,n}}{f_{b,1}}\right)^2 \quad \text{small signal gain coefficient,} \\
&f_{b,n}(\xi) = (-1)^{\frac{n-1}{2}}\left[J_{\frac{n-1}{2}}(n\xi) - J_{\frac{n+1}{2}}(n\xi)\right] \quad \text{Bessel function factor.}
\end{aligned} \tag{4.169}$$

Exercise 4.39 Show that the PP and the gain length associated with the higher order harmonic operation are given by

$$\begin{aligned}
&\rho_n = \frac{1}{4\pi}\left[\frac{\pi g_{0,n}}{N^3}\right]^{\frac{1}{3}} = \rho_1\left[\frac{f_{b,n}}{f_{b,1}}\right]^{\frac{2}{3}}, \\
&L_{g,n} = \frac{\lambda_u}{4\pi\sqrt{3}\rho_n}.
\end{aligned} \tag{4.170}$$

4.10.5 A practical suggestion for FEL-O design

We will dedicate the forthcoming discussion to general considerations regarding the design of an FEL-O. Although we underline that there is not a universal strategy leading to such a design, we note that whole process reduces to a procedure of finding a coherent environment to frame the electron, cavity and undulator parameters.

We have already stressed the existence of a global parameter; in the case of low–medium gain devices the parameter g_0 is one of them. We have also seen that beam parameters, such as current and energy spread, come into play and the latter modifies the gain introducing the inhomogeneous broadening contributions. To this aim we have introduced the parameter $\mu_\varepsilon = 4N\sigma_\varepsilon$. To understand the genesis of this term we go back to the high-gain equation and note that if the beam is not mono-energetic but characterized by a Gaussian distribution we should replace $e^{-i\nu\tau'}$ in (4.72) with

$$e^{-i\nu\tau'} \to \int_{-\infty}^{+\infty} f(\varepsilon)\, e^{-i\nu(\varepsilon)\,\tau'} d\varepsilon.$$

Indeed by noting that

$$\nu(\varepsilon) = \nu + 4\pi N \frac{\delta\gamma}{\gamma}, \quad \varepsilon = \frac{\delta\gamma}{\gamma},$$
$$f(\varepsilon) = \frac{1}{\sqrt{2\pi}\,\sigma_\varepsilon} e^{-\frac{\varepsilon^2}{2\sigma_\varepsilon^2}},$$

we obtain from the previous integral

$$\int_{-\infty}^{+\infty} f(\varepsilon)\, e^{-i\nu(\varepsilon)\tau'} d\varepsilon = e^{-i\nu\tau'} e^{-\frac{(\pi\mu_\varepsilon\tau')^2}{2}}.$$

Furthermore we have made reference to the beam current density, which is defined as the ratio between the current and the transverse beam area. Such a definition must be treated carefully. We require that the integral over the transverse coordinates of the current density provides the current intensity, namely

$$\iint J(x, y) dx\, dy = I\,,$$
$$J(x, y) = J\, e^{-\left(\frac{x^2}{\sigma_x^2} + \frac{y^2}{\sigma_y^2}\right)},$$

thus obtaining

$$J = \frac{I}{2\pi\sigma_x\sigma_y}, \qquad \Sigma_e = 2\pi\sigma_x\sigma_y. \tag{4.171}$$

The link with the beam Twiss parameters is therefore evident since

$$\sigma_\eta = \sqrt{\beta_\eta \varepsilon_\eta}\,.$$

Exercise 4.40 Use the same argument as before to show that bunch charge Q_b and current are linked by

$$I = \frac{1}{\sqrt{2\pi}} \frac{Q_b}{\sigma_\tau}, \qquad \sigma_\tau = \frac{\sigma_z}{c}, \tag{4.172}$$

where σ_z is the electron bunch length,

$$\sigma_\eta = \sqrt{\beta_\eta \varepsilon_\eta}.$$

In FEL theory approximate formulae including effects of various type have been exploited and an extensive compendium can be found in the paper by Dattoli, Ottaviani and Pagnutti (henceforth DOP; Dattoli P, Ottaviani L and Pagnutti P 2007 *ENEA Frascati Internal Report* ENEA-RT/2007/40/FIM). P L Ottaviani (1943–2014) provided a significant contribution to FEL physics. He was the main architect of the code PROMETEO and a very acute contributor to a significant body of scaling 'laws', now largely exploited in the design of FEL devices. We have quoted these sources in previous parts of the chapter and they will be further exploited below in these complements.

We have seen that the effect of the gain reduction due to the energy spread is reproduced (in the case of gain coefficients $g_0 \leqslant 0.2$), see (4.73)). In this section we will use a formula that is also valid in a more significant gain interval ($g_0 \leqslant 10,\ \mu_\varepsilon \leqslant 1$),

$$\begin{aligned} G_M(g_0, \mu_\varepsilon) &= G_M(g_0)\frac{\exp\left(-0.132P(g_0)\mu_\varepsilon^2\right)}{1 + 1.6Q(g_0)\mu_\varepsilon^2}, \\ P(g_0) &= 1 - 3.57 \times 10^{-3}g_0 + 6.83 \times 10^{-2}g_0^2 - 4.79 \times 10^{-3}g_0^3, \\ Q(g_0) &= 1 + 0.142g_0 - 1.1 \times 10^{-2}g_0^2 + 1.8 \times 10^{-4}g_0^3. \end{aligned} \tag{4.173}$$

The forthcoming discussion is aimed at providing an idea of how the previous remarks and the theoretical notions we have acquired in the first part of this chapter can be organized to develop a preliminary design of an FEL-O.

Suppose that we have been requested to specify the parameters for an FEL-O operating in the near infrared (NIR; e.g. 7 μm) with an average power of 10 W; the paradigmatic steps to be followed are reported below.

We write the small signal gain coefficient in MKS units, namely

$$\begin{aligned} g_0 &= 2\pi \frac{J}{I_0}\left(\frac{N}{\gamma}\right)^3 \left[\lambda_u K f_b(K)\right]^2, \\ J &= \frac{I}{\Sigma} \equiv \text{current density}. \end{aligned} \tag{4.174}$$

The explicit form of Σ, in terms of the so-called filling factor, will be specified later.

We invert (4.174) with respect to I

$$I = \frac{g_0 I_0}{2\pi}\frac{\Sigma}{\lambda_u^2}\frac{1}{\left(Kf_b(K)\right)^2}\left(\frac{\gamma}{N}\right)^3, \tag{4.175}$$

and obtain the current necessary to obtain the small signal gain g_0. The use of the inhomogeneous broadening parameter $\mu_\varepsilon = 4N\sigma_\varepsilon$ (see (4.72)) allows us to eliminate the undulator number of periods in (4.175), thus obtaining

$$I = 32\frac{g_0 I_0}{\pi}\frac{\Sigma}{\lambda_u^2}\frac{1}{\left(Kf_b(K)\right)^2}\left(\frac{\gamma\sigma_\varepsilon}{\mu_\varepsilon}\right)^3. \tag{4.176}$$

If we express the energy in terms of the resonance condition (namely $\gamma = \sqrt{\frac{\lambda_u}{2\lambda}(1 + \frac{K^2}{2})}$) we obtain from (4.176),

$$I = 16\frac{g_0 I_0}{\pi\sqrt{2}}\frac{\Sigma}{\lambda_u^{\frac{1}{2}}}\frac{\left(1 + \frac{K^2}{2}\right)^{\frac{3}{2}}}{\left(Kf_b(K)\right)^2}\left(\frac{\sigma_\varepsilon}{\mu_\varepsilon}\right)^3 \lambda^{-\frac{3}{2}}. \tag{4.177}$$

Let us now assume that

$$g_0 = 2.45,$$
$$\mu_\varepsilon = 1,$$
$$\lambda = 7\ \mu\text{m},$$
$$N = 77.$$

It is evident that our requests are compatible with a relative energy spread $\sigma_\varepsilon = \frac{1}{4 \cdot 77} \cong 3.5 \times 10^{-3}$. After inserting the numerical value of the Alfven current (in (4.177), $I_0 \cong 1.7 \times 10^4$), we find for the current

$$I \cong 9.92 \times 10^5 \frac{\Sigma}{\sqrt{\lambda_u}\left(Kf_b(K)\right)^2}\left(1 + \frac{K^2}{2}\right)^{\frac{3}{2}}. \tag{4.178}$$

In figure 4.41 we have reported the maximum gain in terms of the energy spread inhomogeneous broadening parameter for $g_0 = 2.45$. In correspondence to $\mu_\varepsilon = 1$ we can expect a maximum gain of the order of 0.9.

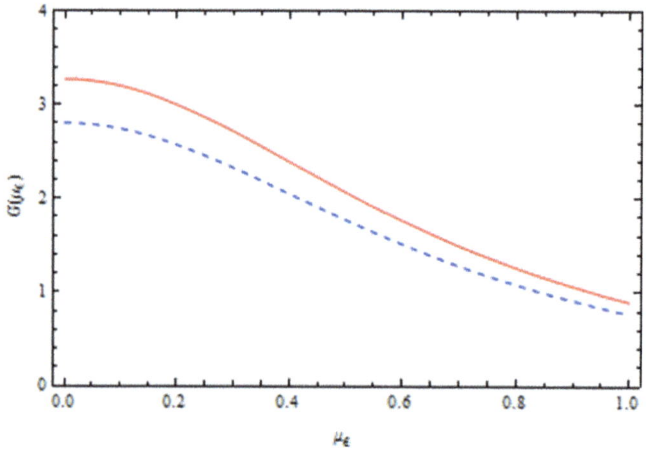

Figure 4.41. Maximum gain versus μ_ε for $g_0 = 2.45$ (see (4.173)). Continuous line = without short bunch correction; dotted line = with bunch corrections $\mu_c = 0.5$.

To avoid further gain dilution due to slippage, we impose the following condition

$$\mu_c = \frac{N\lambda}{\sigma_z} \cong 0.3,$$

amounting to a micro-bunch length $\sigma_z \cong 3N\lambda = 1.6$ mm . A slippage factor of 0.3 is responsible for a gain reduction of about 10%, with respect to the previous quoted value of 0.9. Furthermore, by keeping

$$\begin{aligned} \lambda_u &= 2.8 \text{ cm}, \\ K &= 1, \\ f_b(1) &= 0.91, \end{aligned} \tag{4.179}$$

we eventually end up with

$$I \cong 1.6 \times 10^7 \Sigma. \tag{4.180}$$

The previous conditions (4.179) fix the beam energy to be

$$E \cong 28 \text{ MeV}.$$

If we furthermore assume that the normalized emittance in both directions is

$$\varepsilon_{x,y}^{(n)} = 5 \text{ mm} \cdot \text{mrad}$$

and take a round beam inside the undulator, with Twiss parameter $\beta_{x,y} \cong 2$ m we obtain a beam rms radius given by

$$\sigma_{x,y} = \sqrt{\beta_{x,y}\frac{\varepsilon_{x,y}^{(n)}}{\gamma}} \cong 0.43 \text{ mm}.$$

The optical cavity is assumed to be confocal and 3.55 m long (see figure 4.42); the choice of a confocal cavity stems from the fact that it is very insensitive to small misalignments and because it has the smallest waist.

With a waist at the center given by $\pi W_0^2 = \frac{\lambda L_c}{2} \cong 12.24$ mm^2, the current density can now be specified by keeping

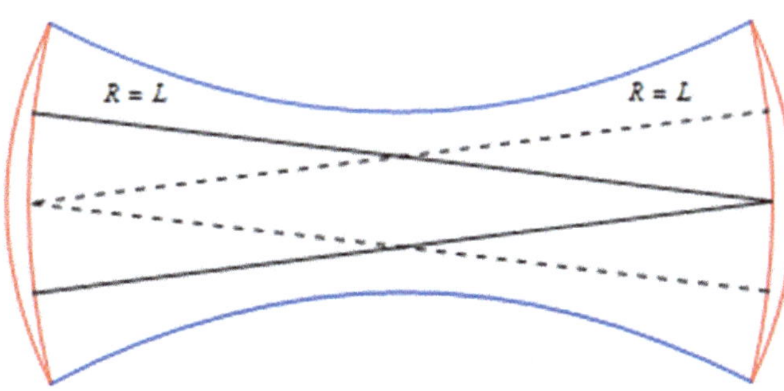

Figure 4.42. Confocal cavity and relevant geometrical features.

$$\Sigma = \Sigma_e \sqrt{\left[1 + \left(\frac{W_0}{\sigma_x\sqrt{2}}\right)^2\right]\left[1 + \left(\frac{W_0}{\sigma_y\sqrt{2}}\right)^2\right]}, \tag{4.181}$$

where $\Sigma_e = 2\pi\sigma_x\sigma_y$ is the electron-beam transverse section, and the term with the square root represents the filling factor correction. If we use the previously quoted values we obtain from (4.180) a current intensity of $I \cong 50$ A.

The cavity filling time is the time which the intra-cavity power density takes to reach the equilibrium condition, it can accordingly be calculated according to the following identities:

$$\begin{aligned} \tau_R &\cong \frac{2}{3}\,10^{-8} L_c \frac{\ln\left(0.1\,\dfrac{I_e}{I_0}\right)}{\ln[(1-\eta)\,G_M + 1]}, \\ I_e &= \left(\sqrt{2} + 1\right)\left(\sqrt{\frac{1-\eta}{\eta}G_M} - 1\right) I_S, \\ I_S[\mathrm{MW/cm^2}] &\cong 6.9132 \times 10^2 \left[\frac{\gamma}{N}\right]^4 \frac{1}{\left(\lambda_u[\mathrm{cm}]\,K\,f_b(K)\right)^2}. \end{aligned} \tag{4.182}$$

We find, for the previously discussed parameters $\tau_R \cong 1$ μs, for a cavity with a length of 3.5 m with total losses around 10%, in correspondence to an intra-cavity equilibrium intensity $I_e \cong 115$ MW cm^{-2}, 4.12 times the saturation intensity I_S and 10^8 above the spontaneous emission term I_0.

In figure 4.43 we have reported the round trip evolution of the intra-cavity dimensionless power $X = \frac{I}{I_S}$, the saturation occurs in about 40 round trips consistent with the estimation of 1 μs for a 3.6 m long cavity.

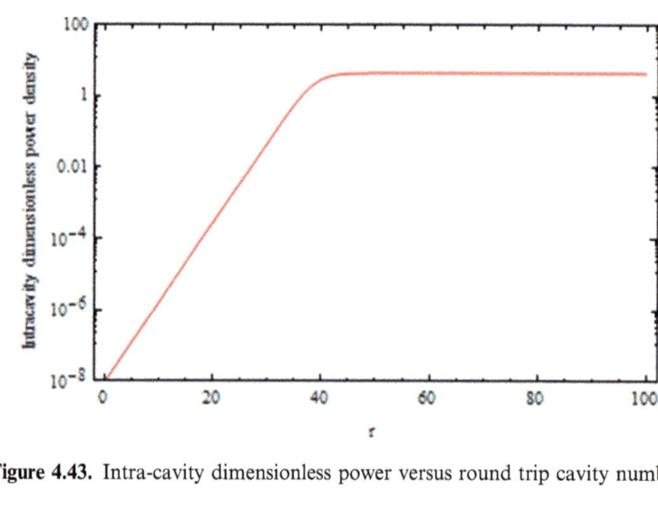

Figure 4.43. Intra-cavity dimensionless power versus round trip cavity number.

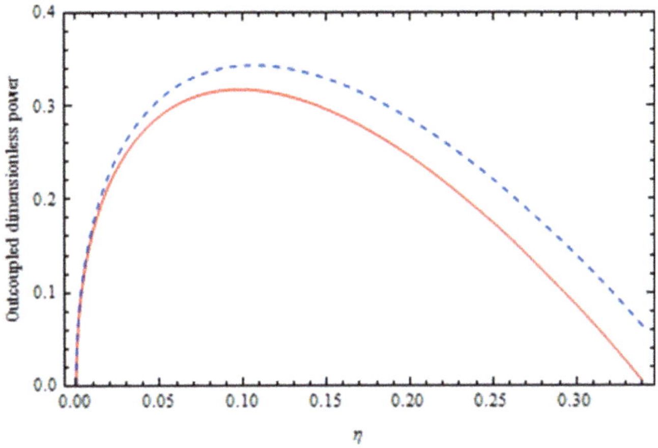

Figure 4.44. Output power versus the cavity losses for passive losses of 2% (dotted line) and 3% (continuous line).

In figure 4.44 we have reported the normalized output power versus the active losses for two different values of $a = \dfrac{\eta_a}{\eta_P}$. The present computation yields an optimum for the active losses around 10%.

The area of the mode on the mirror is $4\pi W_0^2 \cong 0.5$ cm^2, the amount of peak power distributed on the mirror surface is therefore 57.5 MW, a fraction, almost 10%, is transmitted and about 2% is absorbed. The amount of absorbed peak power is around 1.15 MW.

The FEL average absorbed power on the mirror is

$$\begin{aligned} &P_L \cong (\tau_p - \tau_R)\left(\eta_p \hat{P}_e\right)\delta f_r, \\ &\delta \equiv \text{microscopic duty–cycle}, \\ &\tau_{\rm P} \equiv \text{macropulse duration}, \\ &\eta_{\rm p} \equiv \text{passive losses}. \end{aligned} \tag{4.183}$$

If we assume that

$$\begin{aligned} &\tau_p = 10\ \mu\text{s}, \\ &\delta = \frac{5}{330}, \\ &f_r = 10\ \text{Hz}, \end{aligned}$$

we obtain that less than 1 W is dissipated on the mirror surface. The output power is obtained by means of the same procedure by replacing η_p with η_a in (4.183), we find an average power around 10 W.

If we use an undulator with λ_u= 2.8 cm, the gap should be kept around 16.4 mm to obtain $K = 1$. The gap is sufficiently large and the undulator is sufficiently short to support a 3.5 m long confocal cavity with a waist W_0 = 2 mm, sufficiently large to contain the transverse part of the electron-beam. Furthermore, the Rayleigh length $Z_R = \frac{\pi W_0^2}{\lambda} \cong 1.75$ mensures that the laser beam radius at the mirror's output is less than 0.5 cm.

The charge of the bunches is provided by

$$Q = \sqrt{2\pi} I \frac{\sigma_z}{c} \cong 0.7\text{nC}. \tag{4.184}$$

The design criteria we have outlined yield a first insight on how the pivotal parameters specifying an FEL-O should be merged to obtain an efficient and 'fast' design tool. The procedure has been elaborated and has become a fairly mature 'semi-analytical' code called PARSIFEL running with MATHEMATICA (the reader is encouraged to visit the dedicated and continuously upgraded website devoted to FEL physics research and hosted by ENEA at www.fel.enea.it).

4.10.6 A practical suggestion for the high-gain FEL design

The use of resonators for FEL operation in the oscillator configuration is a convenient solution if mirrors, for an efficient confinement of radiation, do exist. The demand for ultra-brilliant, highly coherent, femtosecond x-ray pulses led to the conception of fourth-generation light sources. The x-ray FELs (XFELs) are now running and the radiation they produce is a unique tool for frontier experiments.

We have already stressed that XFEL radiation requires a high-energy linac, producing extremely high-density, high-energy electron beams and long undulators, where electrons radiate, acquire bunching, and drive the FEL power to an exponential growth and eventually to saturation.

Even though we have already stressed the intensity of the technological and financial effort for the realization of these devices, we note that to produce, e.g., a 10 keV FEL intense photon flux, a beam energy of about 8 GeV is necessary, if we use conventional permanent magnet undulator technology.

Regarding the financial efforts, we note that the world's first XFEL facility, the Linac Coherent Light Source (LCLS) at SLAC, would have been unlikely to be constructed if it had not been possible to reuse one-third of the existing S-band linac to accelerate an electron beam up to 14 GeV, with an acceleration gradient of about 17 MV m^{-1}. Such a long linac with a dedicated building would have been difficult to realize as an *ex novo* XFEL facility, because of the prohibitive costs.

We will come later to a deeper discussion of these topics and summarize here the elements of a design tool for high-gain (not necessarily x-ray) FELs. We will follow the same criteria that inspired the previous section. The procedure we follow is the same as before, namely embedding the electron beam, undulator and matching elements to end up with a design allowing the prediction of the performance of the device itself, and its size, cost, etc.

We devote this complement to the pragmatic task of acquiring the essential tools for the design of an FEL in the SASE regime. In the case of the oscillator the key parameter has been the small signal gain coefficient g_0; for the SASE regime an analogous role is played by the PP ρ linked to the small signal gain coefficient by the identity

$$\rho = \frac{1}{4\pi}\left[\frac{\pi g_0}{N^3}\right]^{\frac{1}{3}} = \frac{8.36 \times 10^{-3}}{\gamma}\left[J(\lambda_u K f_b)^2\right]^{\frac{1}{3}}. \tag{4.185}$$

It must be stressed that ρ does not depend explicitly on the number of undulator periods. Although the first equality in (4.185) can be misleading, we have used it to underline the continuity with the description of the case.

We have already noted that ρ defines both the gain length and the saturated power according to the formulae

$$\begin{aligned} L_g &= \frac{\lambda_u}{4\pi\sqrt{3}\rho}, \\ P_s &\cong \sqrt{2}\rho P_e, \end{aligned} \tag{4.186}$$

and it has also been stressed that the power evolution can be reproduced by a logistic-type curve (the factor a can be neglected for any practical purposes),

$$\begin{aligned} P_1(z) &= P_0 \frac{A(z)\exp\left(a\dfrac{z}{Z_{F,1}}\right)}{1 + \dfrac{P_0}{P_{F,1}}[A(z) - 1]}, \\ A(z) &= \frac{1}{9}\left[3 + 2\cosh\left(\frac{z}{L_g}\right) + 4\cos\left(\frac{\sqrt{3}}{2}\frac{z}{L_g}\right)\cosh\left(\frac{z}{2L_g}\right)\right], \\ a &\cong 0.223, \quad z = \text{longitudinal coordinate}, \end{aligned} \tag{4.187}$$

which allows us to define the saturation length as

$$Z_s = 1.066\ L_g \ln\left(\frac{9P_{F,1}}{P_0}\right) \qquad \text{saturation length,}$$

$$P_0 \qquad \text{input seed power.}$$

To better appreciate the role played by the previously introduced parameters, we invite the reader to consider the following exercise

Exercise 4.41 Saturation in high-gain FEL devices is characterized by a power level around 10^8 of the initial seed P_0. Show that this is automatically satisfied if the total number of periods of the undulator is $N \cong \dfrac{1}{\rho}$, in particular verify that the maximum amplification and saturated power density are, respectively,

$$\frac{\Delta P}{P_0} \cong \frac{1}{9} e^{4\pi\sqrt{3}} \cong 3.15 \times 10^8,$$
$$I_s \cong \frac{2\sqrt{2}}{\pi} (4\pi)^3 I_S \cong 1.79 \times 10^3 I_S. \tag{4.188}$$

Hint: use the growth equation using the fastest growth only, namely $P(z) = \frac{P_0}{9} e^{\frac{z}{L_g}}$, and then evaluate $P(\frac{\lambda_u}{\rho})$.

Regarding the second part of the exercise note that (see the first equality in (4.185)) $(N\rho)^3 = (\frac{1}{4\pi})^3 \pi g_0$, which for $N\rho \cong 1$ yields $g_0 \cong \frac{(4\pi)^3}{\pi}$, with $g_0 I_S = \frac{1}{2N} P_e$.

Also, in the present case the beam qualities play a significant role and may be crucial in determining the feasibility of the device under study. The effect of the energy spread can be quantified through the parameter

$$\tilde{\mu}_\varepsilon = \frac{2\sigma_\varepsilon}{\rho}, \tag{4.189}$$

which is the high-gain counterpart of the inhomogeneous broadening parameters, controlling the gain dilution effect in the oscillator regime. Along with energy spread, further sources of inhomogeneous broadening are the beam transverse size and angular divergence, provided by

$$\tilde{\mu}_\eta = \frac{E_\eta}{\gamma_\eta} \gamma_N^2, \quad \tilde{\mu}_{\eta'} = \frac{E_\eta}{\beta_\eta}, \quad \alpha_\eta = 0, \quad \eta = x, y,$$
$$\gamma_N = \frac{1}{\beta_N}, \quad \beta_N = \text{undulator eigen-beta}, \tag{4.190}$$
$$E_\eta = \varepsilon_\eta \frac{\gamma^2}{\rho} \left(1 + \frac{K^2}{2}\right)^{-1},$$
$$\varepsilon_\eta \equiv \text{beam emittance}.$$

The most remarkable consequence of a beam with a finite energy spread and emittances is an increase of the gain length and a decrease of the saturated power. This is not surprising and, for example, the energy spread is responsible for an explicit dependence of the gain length on the energy spread according to the simple formula

$$L_g(\tilde{\mu}_\varepsilon) = l(\tilde{\mu}_\varepsilon)\ L_g,$$
$$l(\tilde{\mu}_\varepsilon) = \left[1 + d\tilde{\mu}_\varepsilon^2\right], \quad d = 0.185\ \frac{\sqrt{3}}{2}. \tag{4.191}$$

It is evident that if $\tilde{\mu}_\varepsilon < 1$ (say 0.5) the increase of the gain length (and of the saturation length as well) is limited to 4% of the ideal case (for further comments see below).

A more complete analysis, including the degrading effects of the emittance, yields the following less straightforward result:

$$
\begin{aligned}
&L_g(\chi) = \chi L_g,\\
&\chi = F_3^{-1}, \qquad F_3 = \frac{F_1}{F_2}\exp\left(c\tilde{\mu}_\varepsilon^2\right),\\
&F_1 = \frac{1 + a\left(\tilde{\mu}_x^2 + \tilde{\mu}_{x\prime}^2 + \tilde{\mu}_y^2 + \tilde{\mu}_{y\prime}^2\right) + b\left(\tilde{\mu}_x + \tilde{\mu}_{x\prime} + \tilde{\mu}_y + \tilde{\mu}_{y\prime}\right)}{\sqrt{\left(1+\tilde{\mu}_x^2\right)\left(1+\tilde{\mu}_{x\prime}^2\right)\left(1+\tilde{\mu}_y^2\right)\left(1+\tilde{\mu}_{y\prime}^2\right)}},\\
&F_2 = 1 + dF_1\tilde{\mu}_\varepsilon^2,\\
&a = 0.159, \quad b = -0.066, \quad c = -0.034, \quad d = 0.185\,\frac{\sqrt{3}}{2},
\end{aligned}
\tag{4.192}
$$

where the function χ summarizes the effect of all the inhomogeneous broadening parameters and if, as happens in actual SASE FEL devices, $\tilde{\mu}_\eta$, $\tilde{\mu}_{\eta\prime}$ are negligible, $F_1 = 1$ and χ reduces to $l(\tilde{\mu}_\varepsilon)$ (the inclusion of the extra term containing the factor c extends the limits of validity of the formula to values of $\tilde{\mu}_\varepsilon$ slightly larger than unity).

The effect on the saturated power is obtained through the same universal function χ,

$$
\begin{aligned}
&P_s(\chi) = \sqrt{2}\;\Phi(\chi)\rho\, P_E,\\
&\Phi(\chi) = \exp\left[-\chi(\chi-1)\right] + \sqrt{2}\;\frac{\chi-1}{\chi^3},
\end{aligned}
\tag{4.193}
$$

and the increase of the saturation length is finally written as

$$
z_s(\chi) = 1.066\;L_g(\chi)\ln\left(\frac{9P_s(\chi)}{P_0}\right). \tag{4.194}
$$

The power growth is reproduced by the same logistic-type equations reported in (4.187), upon replacing

$$
L_g(\chi) \to L_g, \qquad P_s(\chi) \to P_s. \tag{4.195}
$$

The derivation of the previous formula did not require any significant conceptual effort, but a noticeable analysis of numerical data corroborated by wise ansatz of appropriate scaling formulae based on analytical solutions of the FEL integral equation. The details of their derivation are not particularly interesting due to boring technicalities; the reader is, however, invited to consider carefully the relevant physical meaning.

Before proceeding further it is better to stress that the absence of a cavity determines a further problem, associated with the fact that the radiation emitted by the electrons diffracts and the optical mode section may, after a while, significantly exceed the beam section itself, thus reducing the PP and causing an increase of the gain length. To appreciate the importance of this effect we note that the effective laser section is the result of a convolution between the optical mode section and the electron-beam section, namely

$$\Sigma_{r,T} = \sqrt{\Sigma_e^2 + \Sigma_r^2}. \tag{4.196}$$

If we assume that the beam is properly matched, the beam is almost constant inside the undulator

$$\Sigma_e^2 \cong 2\pi\beta_\eta\varepsilon_\eta. \tag{4.197}$$

The radiation after one undulator passage is just given by

$$\Sigma_r^2 = \frac{\lambda}{4\pi}\frac{\lambda_u}{2}. \tag{4.198}$$

We must now take into account that the mode evolution is not free but guided by the FEL interaction itself. The mode evolution is counteracted by the presence of ρ and the effect of the diffraction can be corrected as

$$\Sigma_r^2 = \frac{\lambda}{4\pi}\frac{\lambda_u}{2\rho} \cong \sqrt{3}\,\lambda L_g. \tag{4.199}$$

The effect of the diffraction can be included in our analysis by replacing the current density with its diffraction corrected counterpart

$$\begin{aligned} J^{(D)} &= \left[\left(1 + \mu_x^{(D)}\right)\left(1 + \mu_y^{(D)}\right)\right]^{-1/2} J, \\ \mu_\eta^{(D)} &= \frac{\lambda\lambda_u}{(4\pi)^2\beta_\eta\varepsilon_\eta\rho} = \sqrt{3}\frac{\lambda L_g}{4\beta_\eta\varepsilon_\eta}, \end{aligned} \tag{4.200}$$

thus finally obtaining for the PP the following form, often called three-dimensional ρ:

$$\begin{aligned} \rho^{(D)} &= \mathrm{F}\left(\mu_x^{(D)}, \mu_y^{D}\right)\rho, \\ \mathrm{F}\left(\mu_x^{(D)}, \mu_y^{(D)}\right) &= \left[\left(1 + \mu_x^{(D)}\right)\left(1 + \mu_y^{(D)}\right)\right]^{-1/6}. \end{aligned} \tag{4.201}$$

Previously defined quantities maintain the same expressions provided that
$\rho^{(D)} \to \rho$, obtaining, e.g., a function $\chi^{(D)}$ in place of χ,
except the saturated power which is better reproduced by the formula

$$P_{F,1} = \sqrt{2}\left(\frac{\rho^{(D)}}{\rho}\right)^2 \Phi(\chi^{(D)})\rho^{(D)} P_E. \tag{4.202}$$

It is evident that the diffraction is insignificant if

$$\sqrt{3}\frac{\lambda L_g}{4\beta_\eta\varepsilon_\eta} \ll 1. \tag{4.203}$$

If, as often happens in SASE devices, the beam is produced with a transverse emittance given by $\varepsilon_\eta \cong \frac{\lambda}{4\pi}$, the previous equation yields the following condition between the gain length and beta Twiss parameter,

$$L_g \ll \frac{4}{\sqrt{3}}\beta, \tag{4.204}$$

which is not difficult to satisfy.

We have underlined, in the derivation of the solutions of the FEL integral equation, that the FEL high-gain growth may not be triggered by an input seed but by a bunching contribution induced by the spontaneous emission during the early stage of the interaction of the beam with the undulator. In this case the associated logistic growth can be written as

$$\begin{aligned} B_l(z) &= P_b \frac{B(z)}{1 + \frac{P_b}{P_F} B(z)}, \\ B(z) &= 2\left[\cosh\left(\frac{z}{L_g}\right) - \exp\left(-\frac{z}{2L_g}\right)\cos\left(\frac{\pi}{3} + \frac{\sqrt{3}z}{2L_{g,1}}\right)\right. \\ &\quad \left. - \exp\left(\frac{z}{2L_{g,1}}\right)\cos\left(\frac{\pi}{3} - \frac{\sqrt{3}z}{2L_{g,1}}\right)\right], \end{aligned} \tag{4.205}$$

where P_b is an equivalent initial seed power and will be discussed later.

Exercise 4.42 The reader is invited to discuss the difference between the power growth determined by (4.187) (neglecting the constant a) and (4.205) assuming for $P_{b,0}$ approximately identical values.

Exercise 4.43 Use the data reported in figure 4.45 regarding the experimental data from LCLS and the relevant comparison with GENESIS, use (4.205) to make an analogous comparison with the model discussed in this chapter.

According to figure 4.45(b) the comparison with the semi-analytical model is fairly good and we can safely conclude that the use of the model suggested here is reliable.

We invite the reader to repeat the same check with other experimental results available in literature.

We have so far fixed a general procedure to put together the elements of the game to end up with a fairly simple procedure to result in a preliminary design of an SASE FEL device. An analogous method has been discussed by Ming Xie (Xie M 1996 *IEEE Proc. Particle Accelerator Conference* vol 95 (New York: IEEE) paper N 95CH3584, p 183), the reader is invited to write a program (using MATH-CAD or MATHEMATICA) to study the analogies and point out the discrepancies between these two approaches.

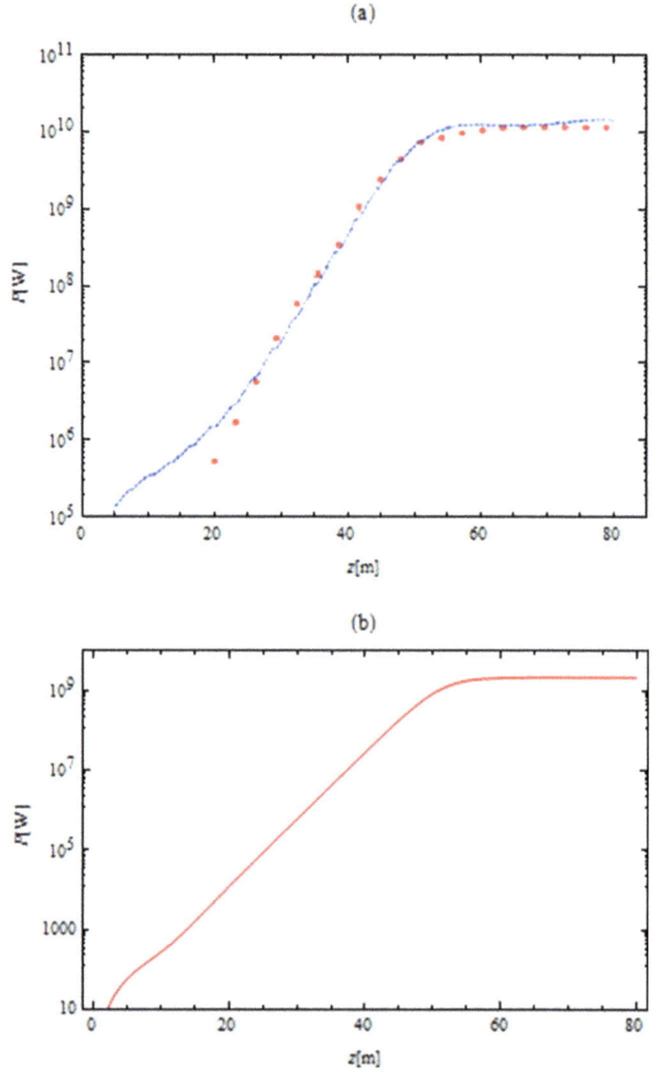

Figure 4.45. (a) LCLS FEL measured performance with $\lambda = 1.5$ Å, $\lambda_u = 3$ cm, $\rho \cong 6 \times 10^{-4}$ and comparison with GENESIS. (b) Comparison with (4.205).

4.10.7 Design of an SASE FEL source

We will now outline the design of a fourth-generation SASE source operating in the nanometer region with a 3 GeV linac producing a bunch of electrons with a charge of 250 pC. We will evaluate the FEL power evolution along the longitudinal coordinate using (4.205), by using as seed power

$$P_b[\mathrm{W}] \cong a \frac{186\, \rho Q[\mathrm{C}]}{(\lambda[\mathrm{m}])^2}.$$

The 'seed' power depends on the packet charge Q, on the operating wavelength λ[m] and on the normalization parameter a, which will be chosen to be around 10^{-3}.

The electron bunch current is linked to its charge by

$$I[\mathrm{A}] = \frac{Q[\mathrm{C}]}{\tau[\mathrm{s}]},$$
$$\tau = \sqrt{2\pi}\,\sigma_\tau.$$

If we assume a current of 1 kA (thus a beam power of 3×10^3 GW at 3 GeV) we will have, for a charge of 250 pC, $\tau[\mathrm{s}] = 2.510^{-13}$.

Furthermore by considering a chain of undulators with $\lambda_u = 2.8$ cm, we can cover the wavelength range shown in figure 4.46 by adjusting the K parameter.

Regarding the tuning of the K parameter, we recall the Halbach formula (see table 4.2), yielding the dependence of the on-axis field on the undulator geometrical parameters (see figure 4.47).

We can obtain an idea of the length of the undulator chain by evaluating the three-dimensional PP (not including inhomogeneous broadening contributions) $\rho \cong 5 \times 10^{-4}$ at 1 nm. We accordingly derive the power evolution along the undulator coordinate of the FEL SASE power, as shown in figure 4.48.

The saturation length, for the operation at 1 nm/3 GeV, is therefore around 55 m.

The operation at 1 nm with 3 GeV and $\lambda_u = 2.8$ cm requires $K \cong 1.71$, the dependence of the PP (in units of 5×10^{-4}) on the operating wavelength is provided in figure 4.49, and therefore the PP for 0.6 nm is $\rho \approx 4.129 \times 10^{-4}$.

The decrease of the PP at shorter wavelengths imposes an increase of the undulator chain length, as shown in figure 4.50, where we have compared the behavior of the power versus z for the 0.6 nm and 1 nm cases. The

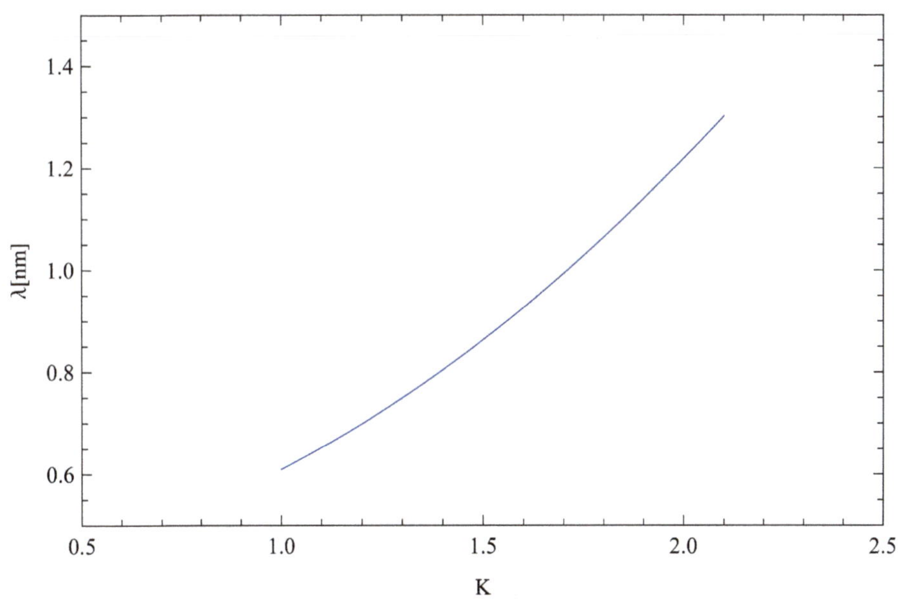

Figure 4.46. Wavelength tunability (in nm) versus K for an operation at 3 GeV.

Table 4.2. Halbach configuration undulator parameters.

$$B_0 = 2B_r \frac{\sin\left(\frac{\pi}{M}\right)}{\left(\frac{\pi}{M}\right)} \left[1 - \exp\left(-\frac{2\pi h}{\lambda_u}\right)\right] \exp\left(-\frac{\pi g}{\lambda_u}\right)$$

M	number of blocks per period
h	magnet height
g	gap
B_r	remanent field

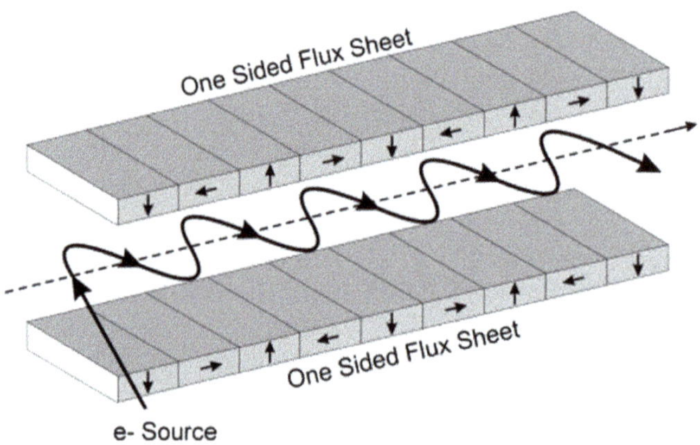

Figure 4.47. Schematic view of a planar magnetic undulator and the associated sinusoidal charged particle trajectory. This image has been obtained by the author(s) from the Wikimedia website where it was made available 'Hiltonj' under a CC BY-SA 3.0 licence. It is included within this article on that basis. It is attributed to 'Hiltonj'. (Image source: https://commons.wikimedia.org/wiki/File:HalbachArrayFEL2.png.)

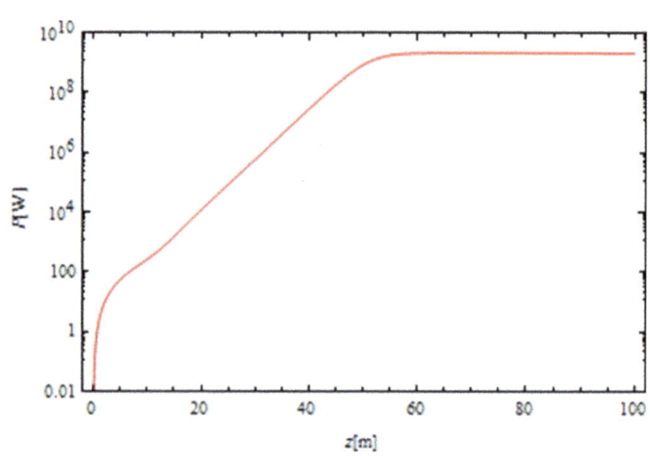

Figure 4.48. SASE FEL power (W) growth versus the undulator coordinate (m) at 1 nm with $\rho = 5 \times 10^{-4}$.

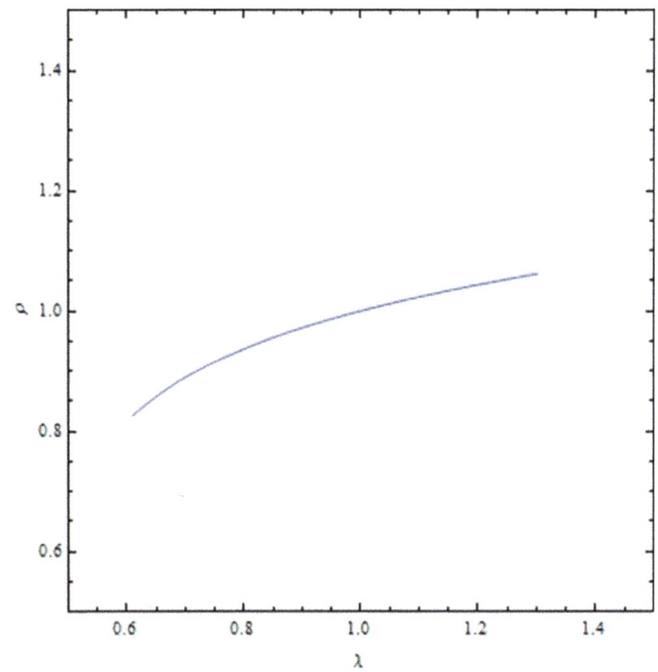

Figure 4.49. Behavior of the PP (in units of 5×10^{-4}) versus the operating wavelength (in nm).

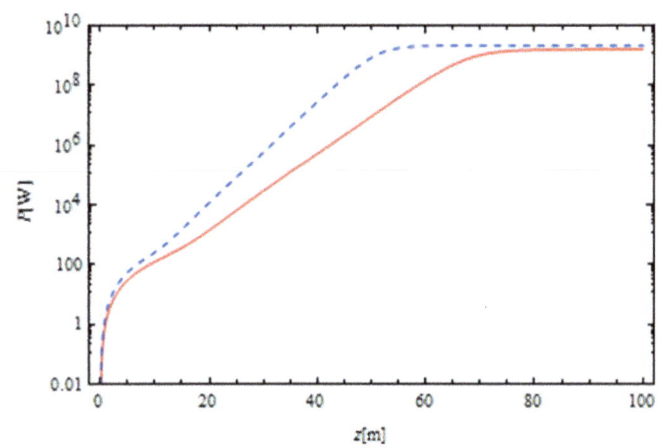

Figure 4.50. FEL SASE at 1 nm (dot line) and at 0.6 nm (continuous line).

operation at shorter wavelengths requires an increase of the saturation length around 10 m.

Let us now discuss whether the values of ρ we have assumed fit with the capabilities of an actual linac. From the definition of the PP

$$\rho = \frac{8.36 \times 10^{-3}}{\gamma}\left[J(\lambda_u K f_b)^2\right]^{\frac{1}{3}},$$

we obtain the following expression for the current density

$$J[\mathrm{A\ m^{-2}}] = 10^9 \frac{(\gamma\rho)^3}{\left(\lambda_u[\mathrm{m}]K f_b\right)^2 (8.36)^3}.$$

According to the previously exploited values ($\rho = 5 \times 10^{-4}$, $\lambda_u = 2.8 \times 10^{-2}$ m, $K = 1.71$, $\gamma = \frac{3 \times 10^3}{0.511}$) we find for the current density $J \cong 2.733 \times 10^{10}$ A m^{-2}, which corresponds to a bunch transverse length inside the undulator of about $\sigma \cong 7.6 \times 10^{-5}$ m, therefore for a normalized emittance of 1 mm · mrad, the corresponding Twiss β parameter can be set around 34.2 m.

The relative energy spread σ_ε compatible with a reasonable length of the saturation length can be specified using a relatively simple argument. To this aim we recall that the inhomogeneous broadening parameter induced by σ_ε is

$$\mu_\varepsilon = \frac{2\sigma_\varepsilon}{\rho},$$

and the associated saturation length increase is written

$$\frac{\Delta L}{L_S} \cong 1 + 0.185\frac{\sqrt{3}}{2}\mu_\varepsilon^2.$$

To keep the length increment within acceptable values, we keep $\mu_\varepsilon \leqslant 1$, which for the upper limit corresponds to $\sigma_\varepsilon \cong \frac{\rho}{2}$. In figure 4.51 it is shown that the operation with these values of the energy spread implies an increase of around 16% for the homogeneous case.

On the other side if $\sigma_\varepsilon \cong \rho$ (see figure 4.52) the increase is significantly larger.

If we reduce the bunch length, we increase the current and thus the PP. We can therefore operate with a larger energy spread, without being obliged to consider a

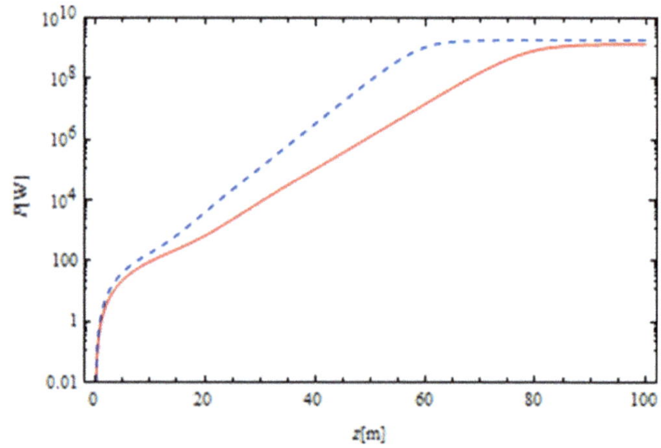

Figure 4.51. The same as figure 4.50 for $\sigma_\varepsilon \cong 2.5 \times 10^{-4}$.

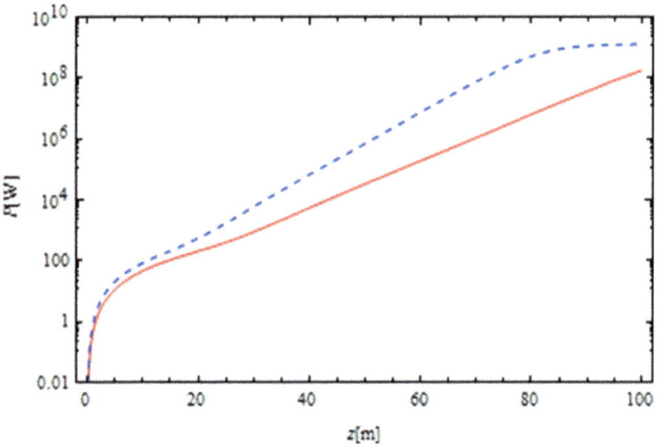

Figure 4.52. The same as figure 4.50, but for $\sigma_\varepsilon \cong 5 \times 10^{-4}$.

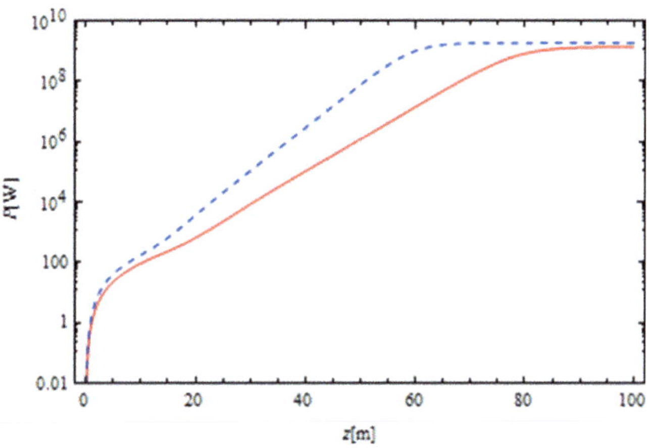

Figure 4.53. The same as figure 4.50 but for $\sigma_\varepsilon \cong 7.2 \times 10^{-4}$.

prohibitively long undulator. SASE operation with a relative energy spread of $\sigma_\varepsilon > 5 \times 10^{-4}$ can be reasonably sustained with a peak current of 3 kA, as shown in figure 4.53

We can therefore summarize the parameters for the sub-nanometer IRIDE SASE FEL operation as follows:

$$\begin{aligned}
&I \cong 3\,\text{kA},\\
&E \cong 3\,\text{GeV},\\
&\sigma_\varepsilon \cong 7.2 \times 10^{-4},\\
&\tau \cong 330\ \text{fs},\\
&\lambda_u = 2.8\,\text{cm},\\
&K \leqslant 1.71,\\
&\varepsilon_n = 1\ \text{mm} \cdot \text{mrad},\\
&\beta \cong 34\,\text{m}.
\end{aligned}$$

The wavelength tunability can be achieved by adjusting the K parameter, which is realized by suitably moving the undulator gap. Further tunability is offered by the non-linear harmonic generation mechanism discussed in the next section.

4.10.8 Non-linear harmonic generation mechanism

The mechanisms underlying the saturation of FEL devices are rather complex, but their macroscopic effects are essentially twofold:

a) A loss of energy of the beam, identified as the power acquired by the laser field.
b) A dilution of the beam qualities, which can be recognized as an increase of the beam energy spread.

The FEL interaction produces therefore both an increase of the laser intensity and an increase of the electron bunch relative energy spread.

Regarding the induced energy spread it is worth considering the following exercise.

Exercise 4.44 The energy spread induced by the FEL interaction can be reproduced by the formula (see DOP) (note that the total energy spread is given by the quadratic composition of induced and natural contributions, namely

$$\begin{aligned} &\sigma_\varepsilon(z) = \left[\sigma_{\varepsilon,0}^2 + \sigma_i^2(z)\right]^{1/2}, \qquad \sigma_{\varepsilon,0} = \sigma_\varepsilon(0) \\ &\sigma_i(z) \cong 3C\sqrt{\frac{A(z)}{1 + 9B[A(z) - 1]}}, \\ &C = \frac{1}{2}\sqrt{\frac{\rho\, P_0}{P_E}}, \qquad B \cong \frac{1.24}{9}\frac{P_0}{P_s}, \\ &\sigma_{i,F} \cong \frac{C}{\sqrt{B}} \cong 1.6\rho. \end{aligned} \tag{4.206}$$

The reader is invited to draw the growth versus z of both laser intensity and $\sigma_i(z)$, and to discuss the relevant physical meaning.

Exercise 4.45 Show that the undulator length necessary to reach an induced energy spread $\sigma_i = \dfrac{\rho}{2}$ is

$$Z^* \cong 0.94 Z_s - 2.44 L_g. \tag{4.207}$$

Explain why (4.206) and (4.207) might be useful in the design of an FEL device.

The induced energy spread is a non-linear by-product of the FEL interaction itself, and can be considered a concomitant cause of the saturation mechanism but, as in any non-linear process, cause and effect are mixed.

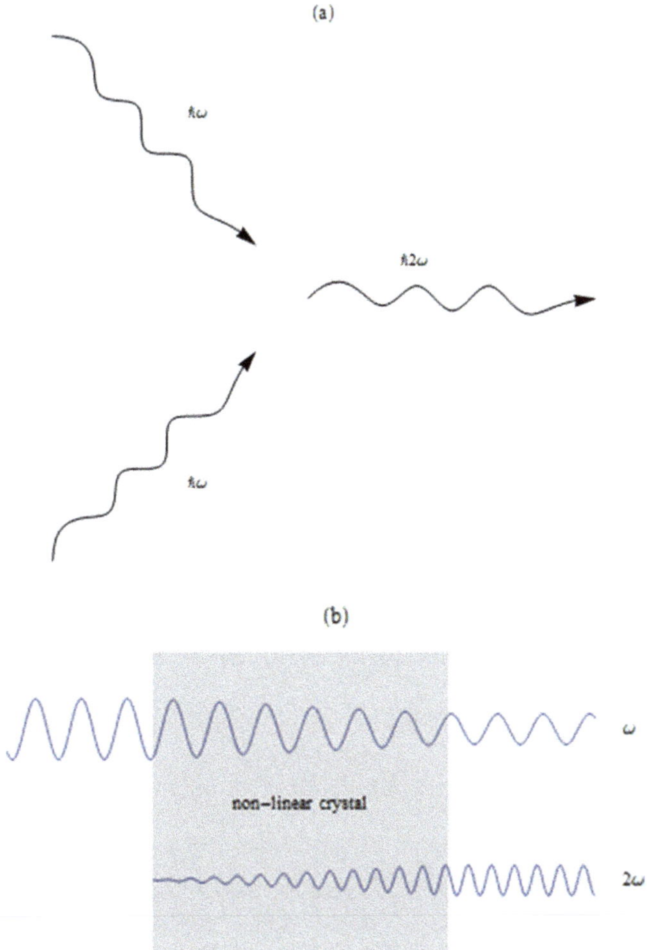

Figure 4.54. (a) Microscopic interpretation of harmonic generation in a non-linear crystal. (b) Macroscopic view, in which the intensity of the field at 2 ω grows at the expense of that oscillating at frequency ω.

We have stressed the presence of a non-linearity to emphasize other mechanisms occurring in optics regarding so-called non-linear harmonic generation. In these processes a laser beam of a given frequency hitting a crystal (see figure 4.54) may generate radiation at the second (or at higher harmonics) of the incident pump.

The microscopic interpretation of the effect is shown in the upper part of figure 4.54, describing two photons of the 'primary wave', which are mixed inside the crystal by producing a photon with double energy, and thus with double frequency with respect to the frequency of the incoming wave. In the lower part of the plot we have reported the macroscopic view, in which the intensity of the field at 2ω grows at the expense of that oscillating at frequency ω.

The process occurs whenever the condition of energy conservation and phase matching are satisfied.

A non-dissimilar mechanism is responsible for third harmonic generation, in which three photons compete in a non-linear crystal and provide outside radiation at the third harmonic of the pump.

Something similar may occur in FELs: along with the resonant frequency higher order harmonics may grow and saturate at much lower power. The behavior is that reported in figure 4.55 and we will comment on it after a few remarks concerning the harmonic generation in a non-linear crystal.

In the weak field limit the power density of the harmonics is linked to that of the fundamental by

$$\begin{aligned} I_2(L) &\propto \frac{4\,(\omega T)^2}{n^2}[\chi^{(2)}]^2 I_1^2, \\ T &= \frac{L}{c}, \quad n = \frac{k_2}{k_1}, \end{aligned} \tag{4.208}$$

and $\chi^{(2)}$ is the second order non-linear polarizability of the medium.

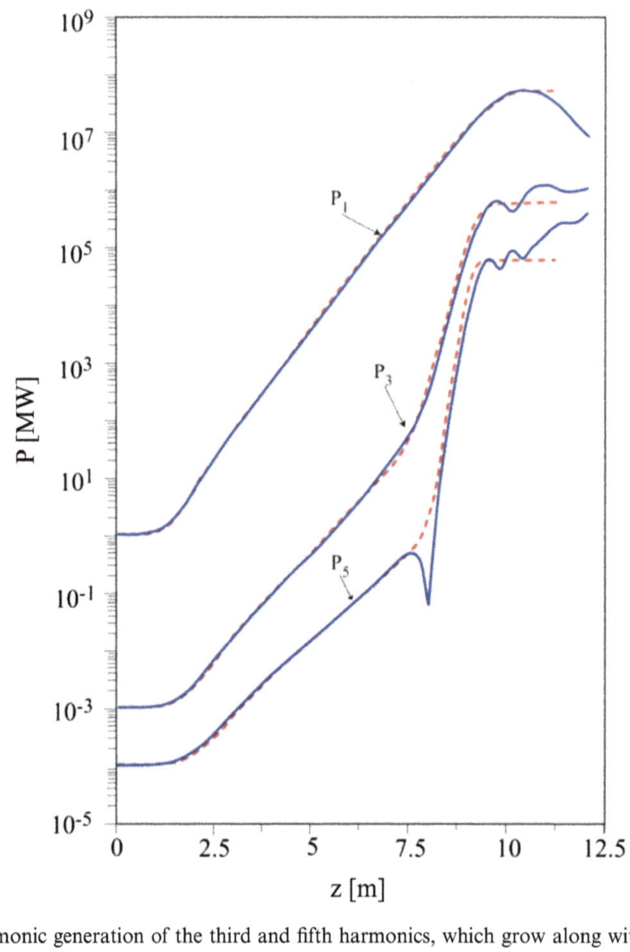

Figure 4.55. Harmonic generation of the third and fifth harmonics, which grow along with the power of the fundamental harmonics.

Apart from the specific meaning of the symbols it is worth stressing that the power of the second harmonic depends on the square of the initial power of the pump. More in general regarding the third harmonic we obtain $I_3(L) \propto I_1^3$.

Let us now go back to figure 4.55, displaying the harmonic generation of the third and fifth harmonics, which grow along with the power of the fundamental. Unlike the previously quoted generation mechanisms of non-linear optics, the harmonics are not depleting the fundamental harmonics. We should furthermore emphasize that the evolution of the harmonics consists of two exponential parts, characterized by two different growth rates.

The first region represents an analogy of the exponential growth of the fundamental harmonics. According to what we have learned in chapter 3, in linearly polarized undulators harmonics are emitted on-axis and SASE radiation can be also generated from these harmonics. The various harmonics grow by following analogous and independent patterns, whose growth in the small signal regime is ruled by the amplitude $A_n(z)$, in which the gain length is provided by $L_{g,n}$.

We have already stressed that the FEL process is characterized by the bunching process. We can expect therefore that with increasing first harmonic power a large amount of bunching is produced and the higher order harmonic is dominated by the non-linear harmonic generation mechanism. The harmonic power growth rate changes slope since the evolution along z is ruled by $I_1^n \propto e^{\frac{n z}{L_g}}$.

In conclusion, denoting by $\Lambda_n(z)$ and $\Pi_n(z)$ the linear and non-linear contributions we have

$$\begin{aligned} P_n(z) &= \Lambda_n(z) + \Pi_n(z) , \qquad n = 3, 5, 7 \ldots \\ \Lambda_n(z) &= P_{0,n} A_n(z) , \end{aligned} \tag{4.209}$$

where

$$L_{g,n}^* = \frac{\lambda_u}{4\pi\sqrt{3}\rho_n^*} = n^{-1/3} L_{g,n} , \qquad \rho_n^* = n^{1/3}\rho_n , \tag{4.210}$$

and $P_{0,n}$ is the input seed power. $A_n(z)$ is given by $A(z)$, the same as for the fundamental harmonics after replacing L_g with $L_{g,n}^*$.

Furthermore, the non-linear term reads

$$\Pi_n(z) = \Pi_{0,n} \frac{\exp\left(\frac{n z}{L_{g,1}}\right)}{1 + \frac{\Pi_{0,n}}{\Pi_{F,n}}\left[\exp\left(\frac{n z}{L_{g,1}}\right) - 1\right]} , \tag{4.211}$$

with

$$\Pi_{0,n} = c_n \left(\frac{P_0}{9\rho_1 P_E}\right)^n \Pi_{F,n} , \qquad c_3 = 8 , \quad c_5 = 116 , \tag{4.212}$$

$$\Pi_{F,n} = \sqrt{n}\left(\frac{\rho_n}{n\,\rho_1}\right)^3 P_{F,n} = \frac{1}{\sqrt{n}}\left(\frac{f_{b,n}}{n\,f_{b,1}}\right)^2 P_F, \tag{4.213}$$

where $\Pi_{F,n}$ is an estimation of the 'saturated' harmonic power.

Exercise 4.46 The reader is invited to evaluate the length of the undulator where the non-linear generation overtakes the linear harmonic growth.

Hint: set $\Lambda_n(z_n) = \Pi_{0,n}$.

To complete the picture of the harmonic generation process we report in figure 4.56 the evolution of the nth bunching coefficient, which is expressible as

$$|b_n(z)|^2 = \alpha_n\left(\frac{P_1(z)}{\rho_1 P_E}\right)^n,$$
$$\alpha_1 = 1, \quad \alpha_2 = \frac{20}{9}, \quad \alpha_3 = 6, \quad \alpha_4 = \frac{56}{3}, \quad \alpha_5 = 60. \tag{4.214}$$

The previous relation yields the behaviour of the bunching coefficients for the nth harmonic versus the longitudinal coordinate and yields an idea of how they are induced by the growth of the fundamental harmonic. The figure reports the case of linear and helical undulators, and it is important to emphasize that the FEL interaction induces higher order bunching in the beam distribution, regardless of the possibility of radiating or not, which is secured only by the existence of a resonance condition at the specified harmonics.

In the case of the linearly polarized undulators the resonance at odd harmonics is guaranteed to be on-axis and the relevant gain coefficient or PP is linked to the Bessel factor $f_{b,n}$, which is zero in the helical case.

It is evident that in the case of an FEL operating with linearly polarized undulators the fundamental harmonic power comes along with that of the higher harmonics at significantly lower power.

Exercise 4.47 Use (4.213) to provide a plot of the ratio of the nth harmonic power to the fundamental harmonics versus K.

Exercise 4.48 Comment on the possibility of realizing a 'segmented' undulator, namely a configuration in which the undulator consists of two sections, and in which the parameters of the second are arranged to be tuned at a harmonic of the first.

We invite the reader to reproduce the results shown in figure 4.57, where we have considered the following parameters.

First section

$$\lambda_u = 2.8\text{ cm}, \quad K = 1.5,\ P_E = 1.32\times 10^4\text{ MW},\ \sigma_{\varepsilon,0} = 10^{-5},\ \rho_1^{I} = 2.303\times 10^{-3},$$
$$N = 300.$$

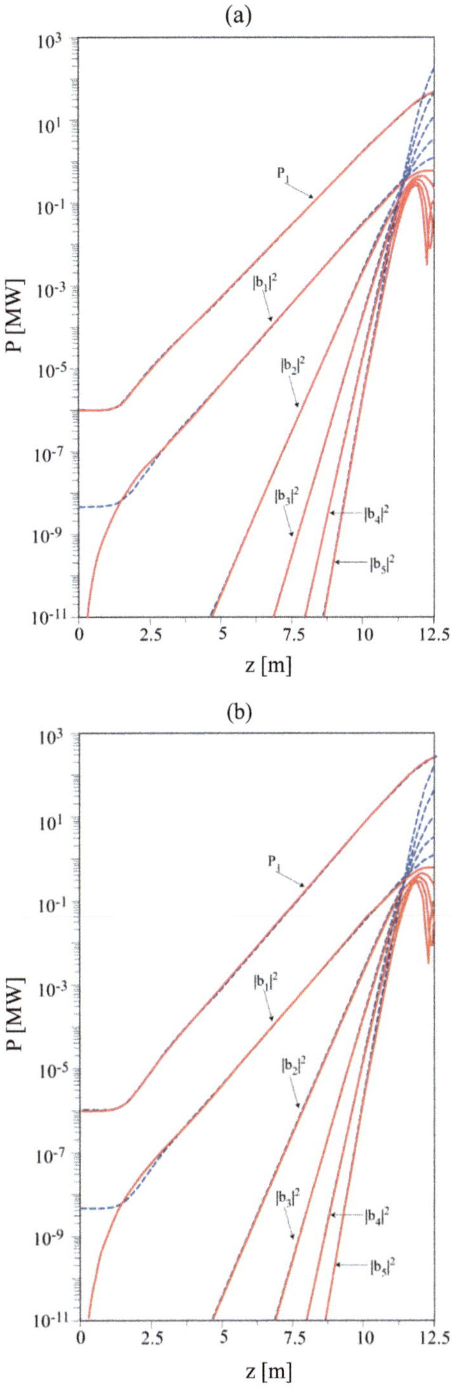

Figure 4.56. Evolution of the power of the fundamental harmonic and of the associated bunching parameters. (a) Linearly and (b) helically polarized undulators. The continuous line refers to the numerical computation and the dashed line to the analytical formula (4.214).

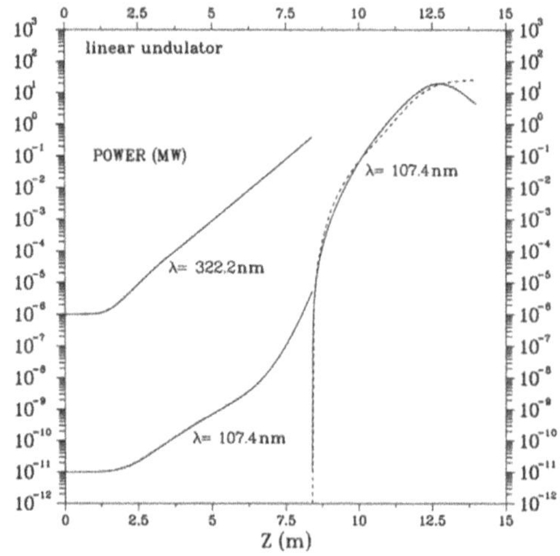

Figure 4.57. Evolution of the FEL power in a segmented undulator configuration where the second section is tuned at the third harmonic of the first section. The continuous line shows a one-dimensional simulation, the dashed line refers to an analytical formula not reported here but available in DOP.

Second section

$$\lambda_u = 0.93 \text{ cm} \ , \quad K = 1.5 \ , \quad \rho_1^{II} = 1.107 \times 10^{-3}.$$

Further reading for chapter 4

Books and review articles

Dattoli G and Renieri A 1985 *Experimental and Theoretical Aspects of Free Electron Laser Physics (Laser Handbook* vol 4) ed Stitch M L and Bass M (Amsterdam: North Holland)

Colson W B, Pellegrini C and Renieri A 1990 (eds) *Free Electron Lasers* (*Laser Handbook* vol 6) (Amsterdam: North Holland)

Colson W B 1990 *Classical Free Electron Laser Theory* (*Laser Handbook* vol 6) (Amsterdam: North Holland)

Dattoli G, Renieri A and Torre A 1993 *Lecture on Free Electron Laser Theory and Related Topics* (Singapore: World Scientific)

Ciocci F, Dattoli G, Torre A and Renieri A 2000 *Insertion Devices for Synchrotron Radiation and Free Electron Lasers* (Singapore: World Scientific)

Dattoli G, Del Franco M, Labat M, Ottaviani P L and Pagnutti S 2012 Introduction to the physics of free electron laser and comparison with conventional laser sources *Free Electron Lasers* ed S Varro (Rijeka: Intech)

Saldin E L, Scneidmiller E A and Yurkov M V 2000 *The Physics of Free Electron Lasers* (Berlin: Springer)

Freund H P and Antonsen T M 1996 *Principles of Free Electron Lasers* (London: Chapman & Hall)

Luchini P and Motz H 1990 *Undulators and Free Electron Lasers* (Oxford: Clarendon Press)
Pellegrini C 2012 The history of x-ray free electron lasers *Eur. Phys. J.* H **37** 659–708

Quantum theory

Dattoli G and Renieri A 1990 *FEL Quantum Aspects* (*Laser Handbook* vol 6) (Amsterdam: North Holland)
Ciocci F, Dattoli G, Renieri A and Torre A 1986 About The mathematical aspects of free electrons and their relationship to quantum optics *Phys. Rep.* **141** 1–50
Preiss P and Schleich W *Theory of the FEL: Classical and Quantum Aspects* www.hzdr.de/db/Cms?pOid=36706
For a complete account of the bibliography, including reference to existing and proposed facilities, see the *World Wide Web Virtual Library: Free Electron Laser Research and Applications* at sbfel3.ucsb.edu/www/vl_fel.html.

IOP Publishing

Charged Beam Dynamics, Particle Accelerators and Free Electron Lasers

Giuseppe Dattoli, Andrea Doria, Elio Sabia and Marcello Artioli

Chapter 5

Compact FEL devices and new acceleration schemes

5.1 Introduction

In the previous chapters we have presented an attempt to describe free electron laser (FEL) devices and dedicated accelerators from a systematic and pragmatic point of view. In most of the chapters (apart from chapter 3), rather than presenting the matter using an academic description, we have chosen an exposition providing the reader with the elements that allow them to understand the subject from the perspective of one who wants to apply, in a reasonably short time, what has been learned.

This chapter follows the same, and even more, unadorned style, since the matter we are going to treat is current and under development. In the following we will describe emerging technologies for undulators and accelerators.

It is apparent after the discussion developed in chapter 4 that an FEL device, operating, e.g. in the x-ray region, is neither cheap nor compact. High-energy linacs require long accelerating structures, operation in the hard x-ray region demands for long saturation length. The solution to obtain at least compact devices could be that of reducing the undulator period and/or using high-gradient accelerators.

In both cases a significant technological effort is needed. Regarding, e.g. the undulators, superconductivity allows the realization of short-period devices (less than 1 cm) with a sufficiently high field to warrant a reasonable (close to unity) parameter strength. The real breakthrough, in this respect, might however be provided by non-static magnetic undulators, exploiting, e.g. intense laser waves, which are in principle capable of reducing the length of the associated period to the order of tens of μm, thus allowing a saturation length below one centimeter.

Regarding the accelerating gradient, the current technology does not allow for more than 100 MeV m^{-1} (X-band accelerating cavities, see section 5.7). An electron

doi:10.1088/978-0-7503-1239-4ch5

beam in the range of energy of GeV therefore requires accelerating structures significantly longer than tens of meters.

Such reference numbers derive from an elementary consideration associated with the breakdown field in metallic materials. One may estimate the corresponding field intensity by saying that the electron in the metal may dislocate to another site by the energy barrier on the order of the atomic energy, say 1 eV. Such a dislocation is over an angström, 10^{-8} cm, which yields a corresponding electric field of 100 MV m^{-1}.

Laser plasma acceleration schemes promise astonishingly large values (100 GeV m^{-1}) and therefore miniaturized accelerating structures. It is evident that targeting these numbers requires an intense program of research, for the reasons we will discuss in the following.

Before entering the technical details regarding new accelerator and FEL architectures, we want to start with a general 'philosophical' discussion of the development, maturity and saturation of technologies. In figure 5.1 we have reported the different phases, regarding the growth and decline of technologies: invention, experimentation, use on a wide scale, saturation and necessity for further innovation.

The curve resembles the logistic behavior discussed in the previous chapters. Such a general remark applies to accelerators and in figure 5.2(a) we have reported the Livingston chart, yielding the evolution versus years, of the electron and hadron colliders.

The plot provides the trend of the technological efforts with 'conventional' accelerating technology, and it is evident that a kind of 'saturation' has occurred. According to what we have learned we can reproduce the trend using the equation

$$\varepsilon = \varepsilon_0 \frac{e^{\alpha(t-t_0)}}{1 + \dfrac{\varepsilon_0}{\varepsilon_S}\left(e^{\alpha(t-t_0)} - 1\right)}, \tag{5.1}$$

which is reported in figure 5.2(b).

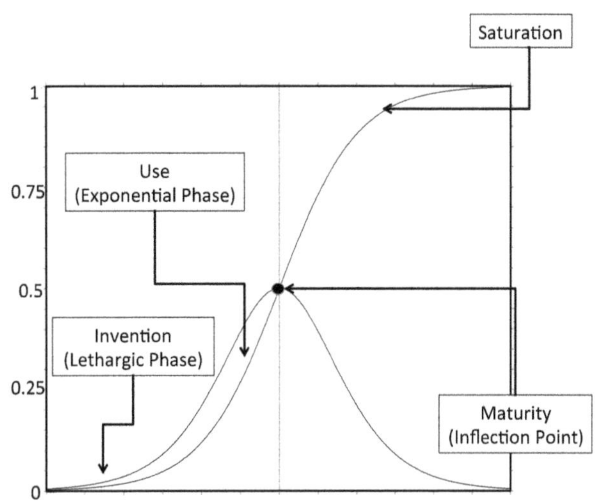

Figure 5.1. Technology from invention to widespread use and necessity for innovation.

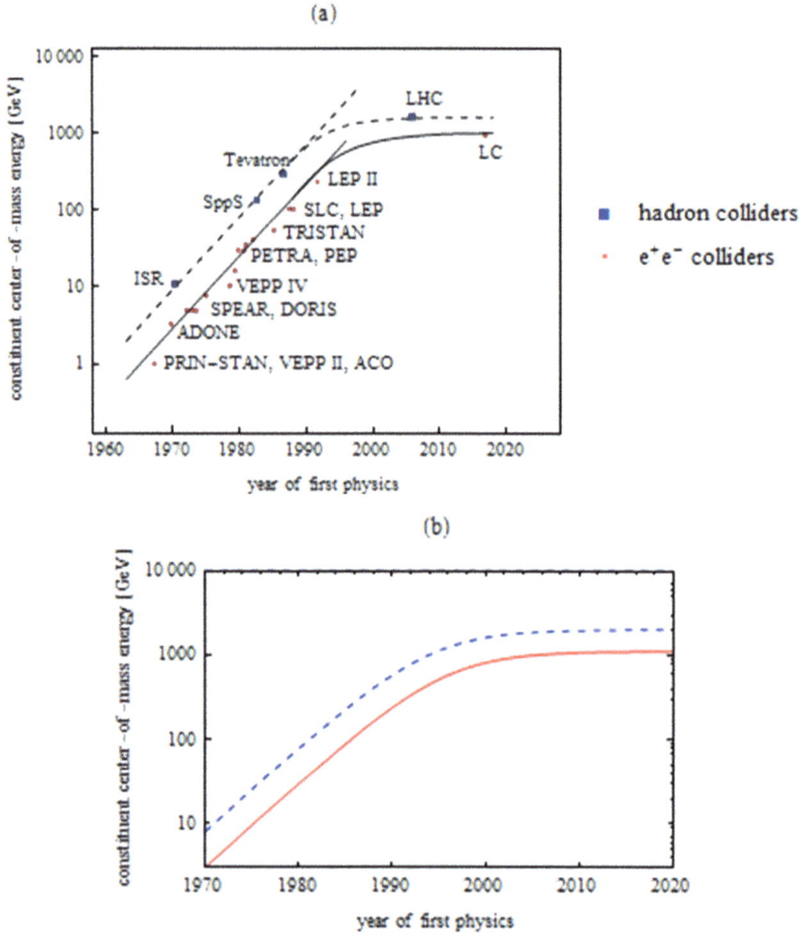

Figure 5.2. (a) A Livingston plot yielding the evolution of the (electron and hadron) colliders. (b) Scaling of the trend with a logistic type behavior (the red continuous line refers to $e^+ e^-$ colliders, the blue dotted line to hadron colliders).

If we trust such a behavior, it means that any, however intense, efforts employing current technologies based on RF acceleration will not provide significant advances in the increase of accelerating gradients in the future. A change in the paradigm is therefore needed if accelerating devices, capable of reaching energies larger than those quoted for CLIC, are required, without involving efforts at a planetary scale.

The plot of figure 5.3 further supports the point of view of technology saturation. On the lower left side of the curve, we have reported the first proposed accelerating structure suitable for providing a beam with energies no larger than few MeV. The real progress was later determined by the use of synchrotron devices, storage rings and high-energy linacs, and finally (lower right) the new technique of plasma acceleration appears to be a promising opportunity, whose evolution exhibits a large positive derivative.

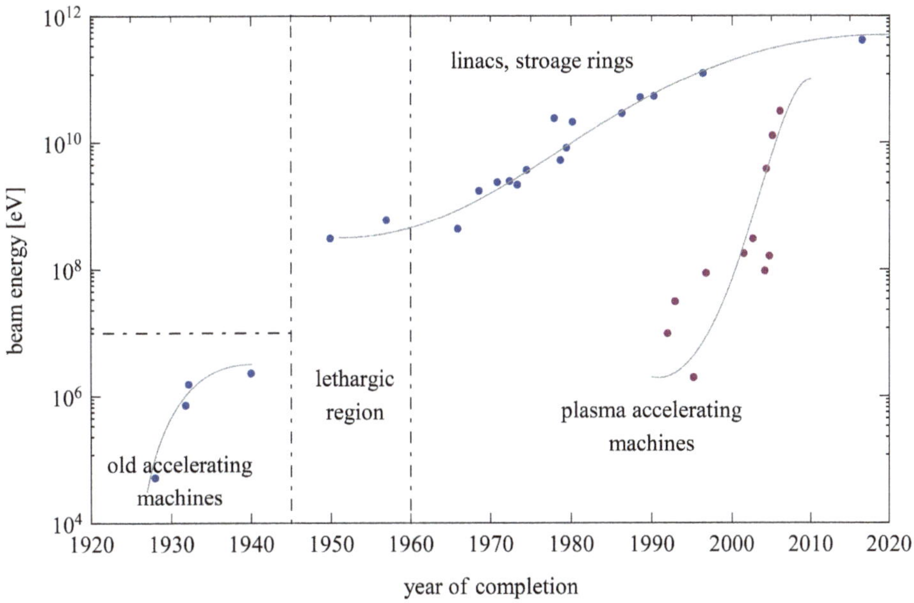

Figure 5.3. Old and new technologies in the same Livingston chart.

In a plasma the collective excitation of the charges generates a field which, unlike the ordinary electromagnetic waves, can be longitudinally polarized. Plasma waves can be excited by a 'very intense laser wave' (see below), propagating in an 'under-dense' plasma (UDP) (see below).

A charged particle, captured in this plasma generated field, may acquire energy in the direction of propagation of the wave and the parameters ruling the acceleration process, which are those associated with the wake field left behind the propagating laser pulse.

The ponderomotive potential of the laser, propagating in a gaseous target, is responsible for the excitation of plasma waves according to the mechanism outlined below (see figure 5.4). The Poynting vector associated with the laser pulse pushes electrons away from the nuclei. The latter, due to their larger masses, are considered to be static in position. The electrostatic forces try to restore equilibrium by attracting the electrons back toward the positively charged ions. This is the mechanism triggering the density oscillations at the 'plasma frequency'. Electrons within the wake field, if in phase with the positive accelerating wave, acquire energy. The electrons may be either injected from outside or self-trapped.

We will provide in the following a more quantitative discussion, but it is worth providing a few reference numbers (which will be commented on more carefully in the following sections of the chapter). We note that plasma frequency ω_p and the associated cold non-relativistic wave breaking electric field E_p are

$$\begin{aligned} \omega_p &= e\sqrt{\frac{n_p}{m\,\varepsilon_0}}, \\ E_p &= \frac{m\,c\,\omega_p}{e}. \end{aligned} \tag{5.2}$$

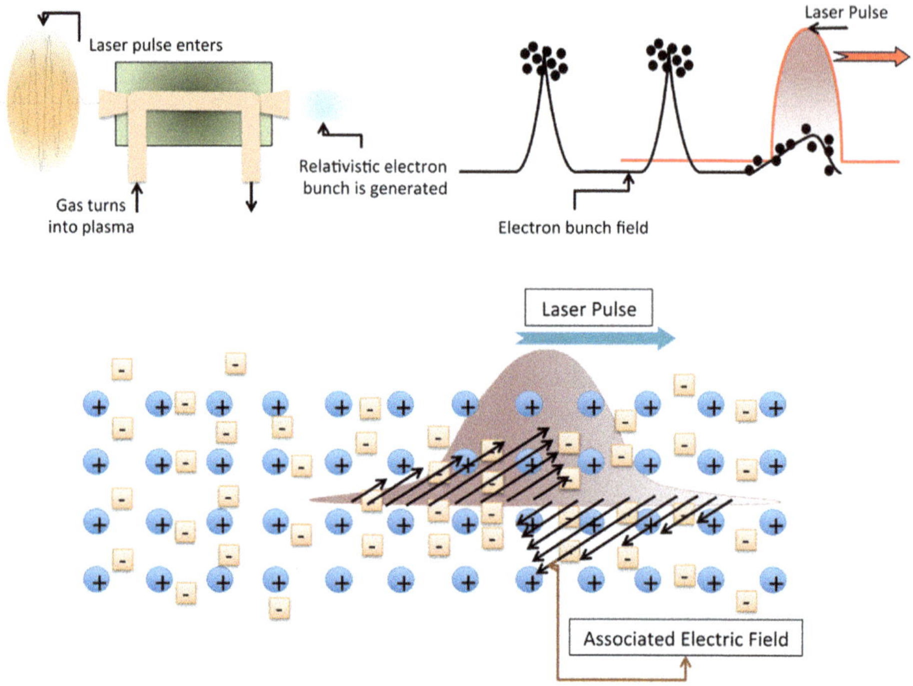

Figure 5.4. Sketch of a laser plasma accelerator. A laser pulse enters a gas target filling plasma and relativistic electrons are repelled outside. Inside the plasma the laser wake generated field is shown on the upper left side. On the right upper side we show self-trapping of electrons from plasma in the non- linear wave regime.

In practical units they read

$$\begin{aligned}\omega_p[\mathrm{Hz}] &\cong 5.6\times 10^{13}\sqrt{\frac{n_e[\mathrm{cm}^{-3}]}{10^{18}}},\\ E_p[\mathrm{GeV/m}] &\cong 96\sqrt{\frac{n_0[\mathrm{cm}^{-3}]}{10^{18}}},\\ \lambda_p[\mu\mathrm{m}] &\cong 33.7\sqrt{\frac{10^{18}}{n_0[\mathrm{cm}^{-3}]}}.\end{aligned} \tag{5.3}$$

Efficient energy transfer between the electromagnetic field and the plasma requires that the laser and plasma waves move at the same speed. In a UDP the phase of the associated wave travels at the laser group velocity, which is close to that of light in a vacuum. The plasma longitudinal electric field can therefore accelerate the relativistic electrons injected into it, which will ride the plasma wave and will be boosted to higher energies in a short length.

The plasma wave phase velocity is linked to the plasma and exciting laser frequencies by (see below)

$$v_p = \frac{\omega}{k} = c\left[1 - \left(\frac{\omega_p}{\omega_l}\right)^2\right]^{\frac{1}{2}}, \tag{5.4}$$

which suggests the introduction of the following relativistic-like factor

$$\gamma_p = \frac{1}{\sqrt{1 - \left(\frac{v_p}{c}\right)^2}} = \frac{\omega_l}{\omega_p}. \tag{5.5}$$

Let us try to put together the above few remarks and consider the energy gain produced on a relativistic electron by the plasma wave field of the order of E_p. Accordingly we will have

$$\Delta\, \mathcal{E}[\mathrm{GeV}] = e\, E_p[\mathrm{GeV/m}]\, L_{\mathrm{eff}}[\mathrm{m}], \tag{5.6}$$

where the effective accelerating length is calculated by taking into account that the electron with velocity approximately equal to that of light moves along the same direction of the plasma wave propagating with velocity v_p. Taking into account that acceleration takes over half a wavelength, we find

$$L_{\mathrm{eff}}[\mathrm{m}] \cong \frac{\lambda_{ph}[\mathrm{m}]}{2\left(1 - \frac{v_p}{c}\right)}. \tag{5.7}$$

If $\frac{v_p}{c} \cong 1$ then $(1 - \frac{v_p}{c}) \cong \frac{1}{2\,\gamma_p^2}$ and we obtain, for the effective interaction length (usually called the dephasing length),

$$L_{\mathrm{eff}}[\mathrm{m}] \cong \gamma_{ph}^2 \lambda_p[\mathrm{m}]. \tag{5.8}$$

Accordingly, we obtain a (maximum) energy gain in one effective length, specified by the simple equation

$$\Delta\, \mathcal{E}[\mathrm{GeV}] \cong 3.23 \times 10^{-3} \gamma_{ph}^2\,. \tag{5.9}$$

A plasma with a volumetric density of 10^{18} cm^{-3} is characterized by a wavelength in the range of tens of micrometers, namely $\lambda_p \cong 3.34 \times 10^{-5}$ m. The associated effective length, if the plasma oscillations are excited by a laser with wavelength $\lambda_l \cong 10^{-6}$ m, is a few centimeters ($L_{\mathrm{eff}} \cong 3.7 \times 10^{-2}$ m) and over this length the energy gain induced by plasma wave–electron interaction may be approximately 4 GeV.

According to the previous statement, the core of the accelerating system is just a few centimeters and not the hundreds of meters necessary to reach these energy gains with conventional technologies. According to these premises it is worth looking more carefully at this type of accelerating mechanism.

5.2 Plasma acceleration: preliminary concepts

Whatever acceleration mechanism we use, the necessary electron-beam characteristics to realize an x-ray FEL are always the same: short bunches, large peak current, small energy spread and emittances. The main issue we will address in this and in the

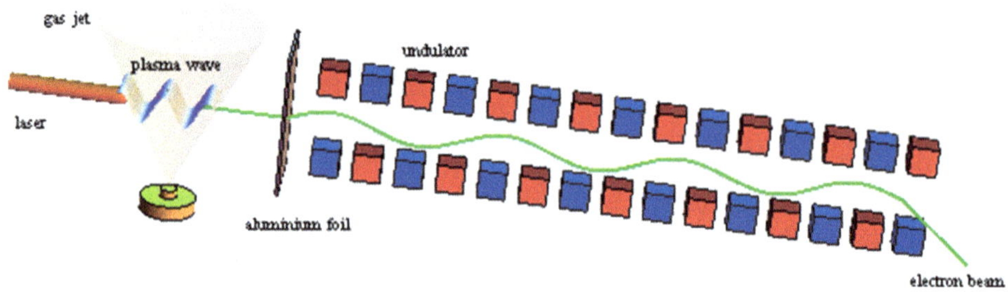

Figure 5.5. A laser plasma accelerated beam.

forthcoming sections is how and whether the plasma accelerating mechanism is able to fulfill these demands.

Let us now consider the layout of an experimental set up (reported in figure 5.5) in which laser–plasma accelerated electrons are injected into an undulator to produce FEL radiation (or at least undulator radiation). The plasma wave is excited by a powerful laser beam (tens of TW, with a duration of a few tens of fs) focused on a gas filled cell. The electrons riding the wave may be either externally injected or are self-trapped in the plasma wave bucket.

There are a few notions which should be premised to appreciate the role of the laser intensity when it interacts with the gas target. First of all we should specify what laser intensities we should consider. We have referred to powerful lasers without specifying what this means. Let us therefore consider a 10 J laser compressed into a pulse of 10 fs, carrying therefore a power of 1 PW (10^{15} W). If the laser is focused on a spot of 50 μm the corresponding intensity is on the order of 10^{19} W cm^{-2}.

The laser power density is defined as

$$\begin{aligned} I_l &= \frac{1}{2}\varepsilon_0 c\, E_l^2 \\ \rightarrow E_l[\mathrm{V/m}] &\cong 27.7\sqrt{I_l[\mathrm{W\,m^{-2}}]}, \end{aligned} \tag{5.10}$$

where E_l is the laser electric field which, for the previously quoted intensity, is extremely large 8.76×10^{12} V m^{-1}, the magnetic field reaches analogous striking values of 2.89×10^4 T and the radiation pressure is of the order of 10^{15} Pa.

According to these numbers, it is natural to ask why one does not use the laser directly as it is to accelerate the electrons; we know, however, that such a possibility is prevented by the Lawson–Woodward theorem (see chapter 2). Within this respect the plasma may provide the symmetry breaking element, allowing a net energy transfer between electrons and radiation.

The electron response to the linearly polarized electric field carried by the laser wave is

$$\begin{aligned} \frac{d}{dt}(\gamma v) &= -\frac{e\,E_0}{m_e}\sin(\omega_l t) \\ \rightarrow \delta(\gamma v) &= \frac{e\,E_0}{m_e\omega_l}\cos(\omega_l t). \end{aligned} \tag{5.11}$$

The dimensionless parameter

$$a_0 = \frac{e\,E_0}{m_e\omega_l c} = \frac{e\,E_0\lambda_l}{2\pi m_e c^2}, \tag{5.12}$$

sometimes called the 'quiver' velocity, is the natural candidate to quantify the amplitudes of the velocity induced oscillations and understand whether the motion, under the influence of the field, becomes relativistic or not. The analogy of such a parameter with the undulator strength is evident.

According to (5.10)–(5.12) the laser intensity, expressed in terms of a_0 reads

$$\begin{aligned} I_l &= \frac{P_{\rm rel}}{\lambda_l^2}a_0^2, \\ P_{\rm rel} &= \frac{m_e c^2}{\tau_e}, \\ \tau_e &= \frac{r_e}{c}, \; r_e \equiv \frac{e^2}{4\pi\varepsilon_0 m_e c^2}. \end{aligned} \tag{5.13}$$

In the previous equation we have introduced the power $P_{\rm rel}$, defined through fundamental quantities only. It indicates a reference (electromagnetic power), interacting with electrons above which the induced motion becomes relativistic.

It is convenient to define the quiver velocity in practical units according to

$$\begin{aligned} a_0 &\cong 0.855\,\lambda_l[\mu\text{m}]\sqrt{\frac{I_l[\text{W cm}^{-2}]}{I_o}}, \\ I_o &= 10^{18}\ \text{W cm}^{-2}. \end{aligned} \tag{5.14}$$

This last equation justifies the alternative denomination of a_0 as the *normalized laser intensity* and its role as a pivotal parameter in our forthcoming discussion. The intensity I_o represents the threshold above which the electrons may acquire relativistic velocities under the influence of the laser field. Taking an average on the laser period, we find indeed from (5.11)

$$\begin{aligned} (\gamma\,\beta)^2 &= \frac{1}{2}a_0^2 \to \beta \propto \frac{a_0}{\sqrt{1+\frac{1}{2}a_0^2}}, \\ \gamma &\propto \sqrt{1+\frac{a_0^2}{2}}. \end{aligned}$$

In figure 5.6 we have collected the laser intensity growth during the last 50 years and the relevant electron response in terms of energy (further comments will be given in section 5.7).

The laser creates a plasma inside the gas target, which is initially quasi-neutral, namely the number densities of electrons and ions with charge state Z are locally balanced ($n_e = Zn_i$). The electrons (the lightest particles) react to any external charge disturbance by restoring the equilibrium. The process is ruled by the plasma frequency, which is the most fundamental time-scale in the physics of plasma

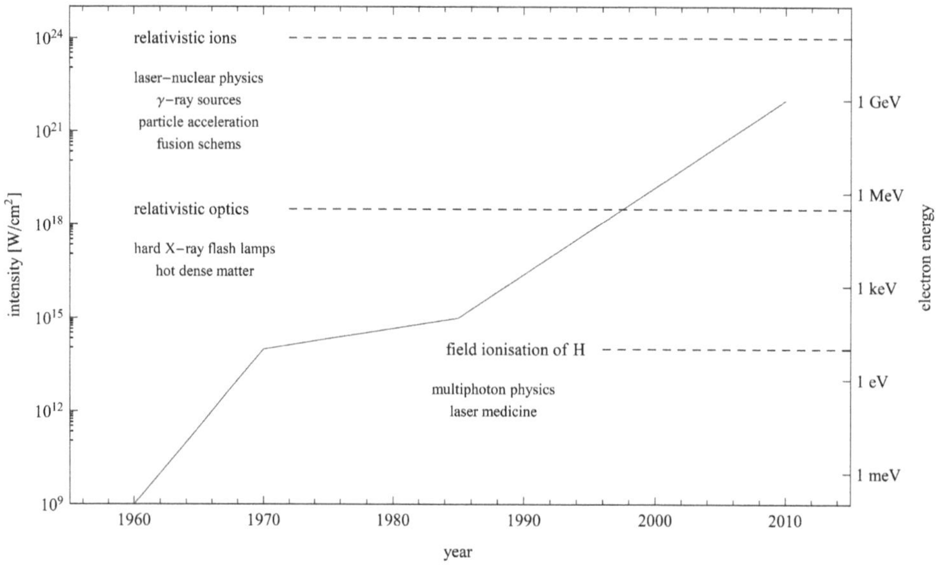

Figure 5.6. Development of laser intensity versus time and electron response.

(there are different plasma frequencies for each species, if not specified, by 'plasma frequency' we invariably mean the *electron* plasma frequency). As already underscored, plasma waves are due to the wake field created by the laser radiation pressure and hence to the associated ponderomotive force, and the physical origin of ω_p is easily understood from elementary considerations.

Let us therefore consider a slab inside the plasma, consisting of one charge species, displaced from its equilibrium position by a small distance δx. The associated charge density on the leading slab surface is $en_0\delta x$ and an equal and opposite charge density develops on the opposite face. The resulting electric field between the two faces is $E = -\frac{\sigma}{\varepsilon_0} = -\frac{en_0}{\varepsilon_0}\delta x$. Any individual charged particle will be then subject to the restoring force

$$m\frac{d^2}{dt^2}\delta x = -\frac{e^2 n_0}{\varepsilon_0}\delta x$$
$$\rightarrow \frac{d^2}{dt^2}\delta x = -\omega_p^2\delta x.$$

The associated oscillation period

$$\tau_p = \frac{1}{\omega_p} \tag{5.15}$$

fixes the time-scale of the plasma phenomena.

To appreciate the role played by ω_p and by the associated 'harmonic potential', we note that, at equilibrium, any particle inside the plasma is characterized by an average kinetic energy linked to its temperature, namely

$$\mathcal{T} \equiv \frac{1}{2} m \langle v \rangle^2 = \frac{1}{2} K_B T, \qquad (5.16)$$
$$K_B \equiv \text{Boltzmann constant.}$$

The potential energy associated with plasma oscillations is, on the other hand, given by

$$\mathcal{U} = \frac{1}{2} m(\omega_p \delta x)^2. \qquad (5.17)$$

If it is negligible with respect to the equilibrium kinetic energy, namely

$$\mathcal{U} \ll \mathcal{T}, \qquad (5.18)$$

the plasma effects do not influence the particle dynamics. The last relation can eventually be cast in the form

$$\delta x \ll \lambda_D = \langle v \rangle \tau_p = \sqrt{\frac{\varepsilon_0 \, K_B T}{e^2 n_0}}, \qquad (5.19)$$

where we have defined the Debye length λ_D, representing a further reference quantity: the maximum amplitude the particle may travel inside before being affected by the interaction with the charges in the plasma. This means that for distances above λ_D the plasma can be viewed as a quasi-neutral system.

A more rigorous definition takes into account the shielding of the Coulomb potential of an ion charge induced by the surrounding electrons, namely

$$\Phi_D = \frac{Q}{4\pi\varepsilon_0 r} e^{-\sqrt{2}\frac{r}{\lambda_D}}. \qquad (5.20)$$

The role of the Debye length is clearly quantified in (5.20) and it is evident that for $\frac{r}{\lambda_D} > 1$ the shielded potential becomes ineffective.

After these remarks we can start the discussion by using a more substantive mathematical description, based on the handling of the Maxwell equations coupled to the electrons' equations of motion, namely

$$\begin{aligned}
&\frac{d}{dt}\vec{p} = - e\,(\vec{E} + \vec{v} \times \vec{B}) \to \text{Lorentz equations},\\
&\vec{p} = m_e \gamma \, \vec{v},\\
&\vec{\nabla} \cdot \vec{E} = -\frac{e}{\varepsilon_0}(n_0 - n_e) \to \text{Gauss law versus plasma charge density deviation from equilibrium},\\
&\vec{\nabla} \times \vec{E} = -\frac{\partial}{\partial t}\vec{B},\\
&c^2 \vec{\nabla} \times \vec{B} = \frac{e}{\varepsilon_0}\vec{J} + \frac{\partial}{\partial t}\vec{E} \to \text{Ampère's law versus plasma charge density},\\
&\vec{J} = -n_e \vec{v} \equiv \text{current density},\\
&\vec{\nabla} \cdot \vec{B} = 0.
\end{aligned} \qquad (5.21)$$

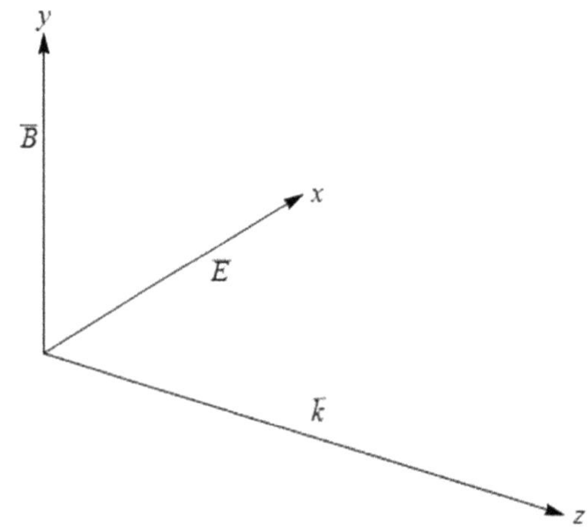

Figure 5.7. Laser wave propagating fields.

The problem is mathematically well defined, but some simplifying assumption could be convenient for further analytical handling.

We essentially assume a plane wave geometry (see figure 5.7) for the laser field, accordingly

$$\vec{B} \equiv \left(0, \ B_y, \ 0\right), \\ \vec{E} \equiv (E_x, \ 0, \ 0). \tag{5.22}$$

Accordingly, introducing vector and scalar potentials,

$$\vec{E} = -\frac{\partial}{\partial t}\vec{A} - \vec{\nabla}\varphi, \\ \vec{B} = \vec{\nabla} \times \vec{A}. \tag{5.23}$$

The first equation in (5.21) yields the canonical momentum conservation as

$$p_x = e\ A_x, \tag{5.24}$$

while the Maxwell equations lead to (see section 5.7 for the details of the derivation)

$$c^2\vec{\nabla} \times (\vec{\nabla} \times \vec{A}) - \frac{\partial^2}{\partial t^2}\vec{A} = \frac{\vec{J}}{\varepsilon_0} - \vec{\nabla}\frac{\partial}{\partial t}\varphi. \tag{5.25}$$

The current density vector can now be split into two parts: the solenoidal (transverse) $J_\perp$ and irrotational (longitudinal) $J_\parallel$ components

$$\vec{J} = \vec{J}_\perp + \vec{J}_\parallel, \tag{5.26}$$

thus obtaining

$$\vec{J}_{||} = \frac{\varepsilon_0}{c^2}\vec{\nabla}\frac{\partial}{\partial t}\varphi, \tag{5.27}$$

and, after using the Coulomb gauge $\vec{\nabla} \cdot \vec{A} = 0$, we end up with (note that $|\vec{J}_{\perp}| = J_x = -ev_x = -\frac{e^2 A_x}{\gamma}$ and for further details see section 5.7)

$$\frac{\partial^2}{\partial t^2}A_x - c^2\nabla^2 A_x = \mu_0 J_x = -\frac{e^2 n_e}{\varepsilon_0 m_e \gamma}A_x. \tag{5.28}$$

Going back to the electron equation of motion we find for the longitudinal component (see section 5.7)

$$\frac{d}{dt}(m_e\gamma\, v_z) = -eE_z - ev_x B_y = -eE_z - \frac{e^2}{2m_e\gamma}\frac{\partial}{\partial z}A_x^2. \tag{5.29}$$

Since from Ampère's and Gauss's laws in (5.21) we find

$$\begin{gathered} \frac{e}{\varepsilon_0}n_e v_z = \frac{\partial}{\partial t}E_z, \\ n_e = n_0 - \frac{\varepsilon_0}{e}\frac{\partial}{\partial z}E_z, \end{gathered} \tag{5.30}$$

we can obtain from (5.29) the differential equations governing the longitudinal electric field.

After a few further manipulations, duly detailed in section 5.7, and passing through an appropriate linearization we end up with

$$\begin{gathered} \left(\frac{\partial^2}{\partial t^2} + \frac{\omega_p^2}{\gamma_0}\right)E_z = -\frac{e^2\tilde{\omega}_p^2}{2m_e\gamma_0^2}\frac{\partial}{\partial z}A_x^2, \\ \gamma_0 = \left[1 + \frac{a_0^2}{2}\right]^{\frac{1}{2}}. \end{gathered} \tag{5.31}$$

The rhs of (5.31) is the ponderomotive contribution, in turn linked to the Lorentz force. This is the mechanism causing the polarization of electrons in the plasma in the longitudinal direction, even though the electric field of the laser is in the transverse direction. The dispersion relation for plasma oscillations is easily obtained from (5.29)–(5.31) as

$$\omega^2 = c^2k^2 + \frac{\omega_p^2}{\gamma_0}, \tag{5.32}$$

thus finding for the associated refractive index (see figure 5.8)

$$\eta = \cos(\vartheta) = \frac{ck}{\omega} = \sqrt{1 - \frac{\omega_p^2}{\gamma_0\omega^2}}. \tag{5.33}$$

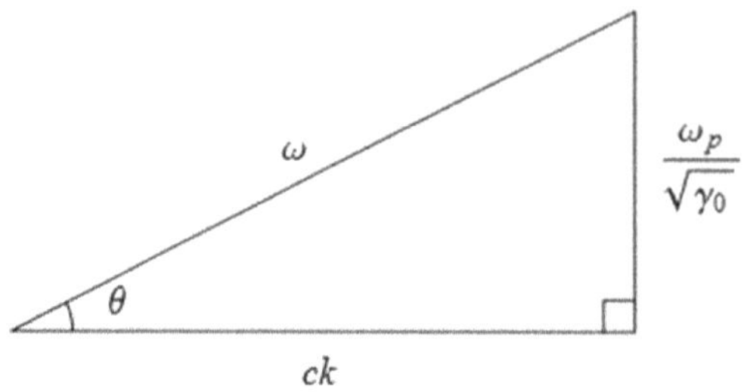

Figure 5.8. Plasma oscillation dispersion triangle.

We can apply the previously introduced diffractive index to understand the previously touched upon notions of under-dense and over-dense plasmas.

Denoting by ω_l the frequency of the laser focused on the plasma, we note that it may be larger or smaller than its plasma frequency. Accordingly the refractive index (5.33) may be imaginary or real.

For $\omega_l > \frac{\omega_p}{\sqrt{\gamma_0}}$ the plasma is said to be *under-dense*, it behaves like a non-linear refractive medium (see figure 5.9a). It can go through the plasma and induce the effects discussed so far. For $\omega_l < \frac{\omega_p}{\sqrt{\gamma_0}}$ the plasma is *over-dense*, the associated refractive index is imaginary, it behaves like a mirror and the laser pulse is reflected (figure 5.9(b)).

The use of the definition of η in terms of the plasma frequency allows the following expression

$$\eta = \sqrt{1 - \frac{n_c}{\gamma_0 n_p}},$$

$$n_c = \frac{\varepsilon_0 m_e \omega_l^2}{e^2},$$

where n_c is the critical density, which in practical units reads

$$n_c = \frac{\varepsilon_0 m_e \omega_l^2}{e^2},$$

$$n_c[\text{cm}^{-3}] \cong 10^{21}(\lambda_l[\mu\text{m}])^{-2}.$$

When $n_c = n_p$ the previous identity defines the critical frequency λ_c, above which laser waves with $\lambda_l > \lambda_c$ are reflected.

The results reported in this section provide the preliminary tools for understanding the basic mechanisms underlying plasma acceleration, discussed in the forthcoming section.

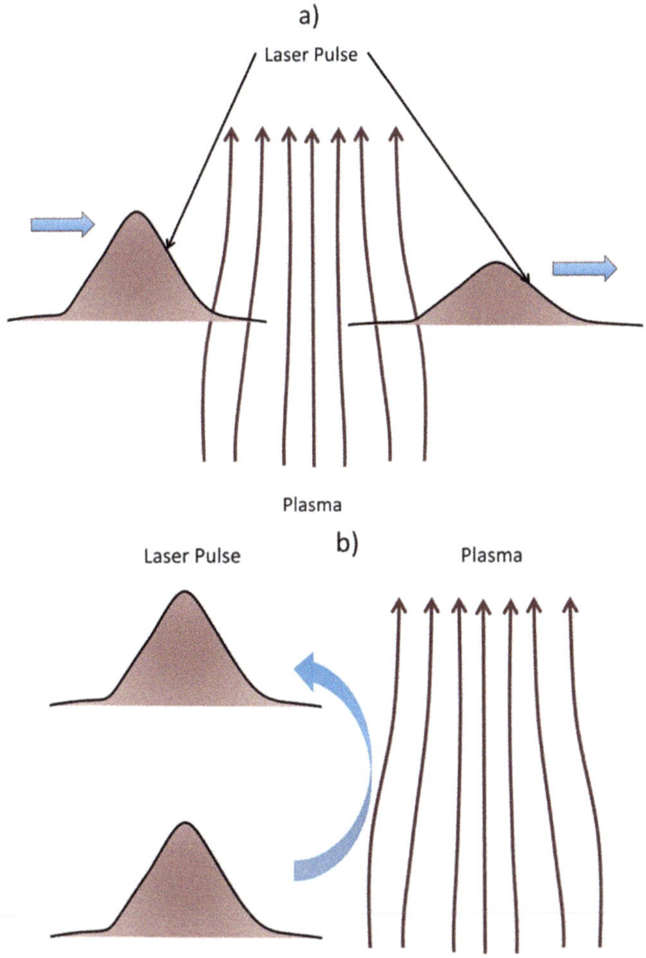

Figure 5.9. Laser pulse and interaction with a plasma: (a) under-dense and (b) over-dense.

5.3 Plasma acceleration: beam production and scaling identities

We will now go deeper into the physical meaning of the equations derived in the previous section, and we will try to derive a set of scaling formulae, useful for the design of an FEL device employing a plasma accelerated beam. We will simply sketch the mathematical details necessary for our purposes (for an exhaustive discussion the reader is addressed to the specialized literature and to a few clarifying remarks in the second part of this chapter).

The field induced by the plasma density oscillations is described by (5.31), which is that of a harmonic oscillator with a source term. The electric field is linked to the scalar potential by $\vec{E} \propto -\vec{\nabla}\varphi$, we make therefore the assumption, to be justified later, that the same equation (5.31) can be adapted to that for the scalar field as

$$\left(\frac{\partial^2}{\partial \zeta^2} + \tilde{k}_p^2\right)\varphi = \frac{\tilde{k}_p^2}{2} m_e c^2 a^2, \tag{5.34}$$
$$\tilde{k}_p = \frac{\omega_p}{c\sqrt{\gamma_0}}, \quad \zeta = z - v_p t, \quad a = \frac{eA}{m_e c}.$$

It should be noted that the φ appearing in (5.34) has the dimensions of an energy divided by a length, namely $e\vec{E} = -\vec{\nabla}\varphi$. The use of the travelling coordinate ζ is obvious and (5.34) follows the plasma oscillation in the co-moving frame.

The relevant solution is therefore found as

$$\varphi = \frac{\tilde{k}_p^2}{2} m_e c^2 \left(\frac{\partial^2}{\partial \zeta^2} + \tilde{k}_p^2\right)^{-1} (a(r, \zeta))^2,$$

where the differential operator on the rhs can be treated by noting that

$$\left(\frac{\partial^2}{\partial \zeta^2} + \tilde{k}_p^2\right)^{-1} = \frac{1}{2\,i\,\tilde{k}_p}\left[\frac{1}{\tilde{k}_p - i\,\dfrac{\partial}{\partial \zeta}} + \frac{1}{\tilde{k}_p + i\,\dfrac{\partial}{\partial \zeta}}\right].$$

Finally using the Laplace transform identity

$$\frac{1}{\tilde{k}_p \mp i\,\dfrac{\partial}{\partial \zeta}}(a(r, \zeta))^2 = \int_0^\infty e^{-sk_p} e^{\mp is\frac{\partial}{\partial \zeta}}(a(r, \zeta))^2 ds = \int_0^\infty e^{-sk_p}(a(r, \zeta \mp is))^2 ds,$$

we end up with

$$\varphi = -\frac{\tilde{k}_p m_e c^2}{2}\int_\zeta^\infty d\zeta' \sin\left[\tilde{k}_p(\zeta - \zeta')\right] a^2(r, \zeta'), \tag{5.35}$$

and we have to use a little math to end up with quantities of practical interest (see section 5.7).

Our next strategy is very elementary, namely we derive an explicit form for the scalar potential and then obtain easily manageable formulae for the radial and longitudinal electric fields. The steps we will follow are sketched below (the details are given in section 5.7).

We write the laser dimensionless field amplitude as

$$a^2(r, \zeta) = U(r) e^{-\frac{\zeta^2}{\sigma_l^2}}, \tag{5.36}$$
$$U(r) = \frac{a_0^2}{2} e^{-\frac{2r^2}{\sigma_r^2}},$$

where σ_l is the longitudinal bunch length and σ_r is the laser section at the entrance of the plasma cell (the reason for neglecting the diffraction effects will be discussed later). Replace it in (5.35) and obtain for the longitudinal and radial electric fields

$$\begin{aligned} e\,E_z &\cong \frac{\sqrt{\pi}}{4} a_0^2 m_e c^2 \tilde{k}_p^2 \sigma_l e^{-\frac{2\,r^2}{\sigma_r^2} - \frac{(\tilde{k}_p \sigma_l)^2}{4}} \cos\left(\tilde{k}_p \zeta\right), \\ e\,E_r &\cong -\sqrt{\pi} a_0^2 m_e c^2 \tilde{k}_p\, \sigma_l \frac{r}{\sigma_r^2} e^{-\frac{2\,r^2}{\sigma_r^2} - \frac{(\tilde{k}_p \sigma_l)^2}{4}} \sin\left(\tilde{k}_p \zeta\right). \end{aligned} \tag{5.37}$$

Adjusting the laser length in such a way that $\tilde{k}_p \sigma_l = 1$ at $r = 0$ we obtain the maximum longitudinal accelerating field as

$$E_z^* \cong 0.35 \frac{a_0^2}{\sqrt{1 + \frac{a_0^2}{2}}} E_p, \tag{5.38}$$

while for the maximum radial field we obtain ($r = \frac{\sigma_r}{2}$ and $\tilde{k}_p \sigma_l = \sqrt{2}$)

$$E_r^* \cong 0.46 \frac{a_0^2 E_p}{\tilde{k}_p \sigma_r}. \tag{5.39}$$

We have fixed so far the essential tools, allowing the understanding of the mechanisms underlying the acceleration process. The derivation of accelerating (transverse and longitudinal) gradients is not sufficient for the design of an FEL device exploiting a plasma accelerated beam. According to the discussion of the previous section, the knowledge of the beam qualities and of the accelerated current are mandatory.

The derivation of the accelerated current passes through the understanding of the amount of charge, which can be 'captured' by the plasma wake bucket. To obtain an idea we use a simple argument.

We consider the volume element

$$V_b = \frac{1}{2}\, \sigma_r^2 \lambda_p. \tag{5.40}$$

In a plasma with volumetric density n_p, we associate to (5.40) the charge

$$\begin{aligned} Q &= eN_b, \\ N_b &= V_b n_p, \end{aligned} \tag{5.41}$$

which is roughly equivalent to the total amount of electron charges displaced by the laser in the excitation of the plasma wave.

We assume that the amount of charge, undergoing the acceleration process, is just proportional to (5.41), namely

$$Q_b = \kappa \frac{e}{2} \sigma_r^2 \lambda_p n_p, \tag{5.42}$$

where κ is left unspecified for the moment. By noting that $\lambda_p n_p = \frac{\pi}{r_e \lambda_p}$ (r_e is the electron classical radius) we find

$$Q_b = \kappa \frac{e}{2}\left(\frac{\sigma_r}{\lambda_p}\right)^2 \frac{\pi}{r_e}\lambda_p. \tag{5.43}$$

Furthermore, since $\lambda_p = \sqrt{\frac{\pi}{n_p r_e}}$ we end up with

$$\begin{aligned} Q_b &= \kappa \frac{e}{4k_l r_e}(\sigma_r k_p)^2 \sqrt{\frac{n_l}{n_p}}, \\ n_l &= \frac{\pi}{\lambda_l^2 r_e}, \quad k_l = \frac{2\pi}{\lambda_l}, \end{aligned} \tag{5.44}$$

which, in practical units, is written

$$Q_b[\text{pC}] \cong \kappa \frac{232}{\sqrt{10}} 10^9 (k_p \sigma_r)^2 \frac{1}{\sqrt{n_p[\text{cm}^{-3}]}}. \tag{5.45}$$

Let us now define the attainable current as

$$\begin{aligned} I_b &= \frac{Q_b}{\tau_b}, \\ \tau_b &= \upsilon \frac{\lambda_p}{2}, \\ \upsilon &< 1, \end{aligned} \tag{5.46}$$

thus, finding for the plasma accelerated current

$$I_b = \frac{\kappa}{\upsilon} e \sigma_r^2 n_p = \frac{\kappa}{\upsilon} I_A \left(\frac{\sigma_r}{\lambda_p}\right)^2, \tag{5.47}$$

where I_A is the Alfvèn current. If we use (5.38) and assume that the energy gain is constant during the whole acceleration length L, we can specify the power of the accelerated beam as

$$P_b[\text{MW}] \cong 0.36 \frac{\kappa}{\upsilon} \frac{a_0^2}{\sqrt{1 + \frac{a_0^2}{2}}} I_A[\text{A}] \left(\frac{\pi\, \sigma_r^2 L}{\lambda_p^3}\right). \tag{5.48}$$

A last effort will be devoted to the understanding of the interaction length.

There are three characteristic lengths which should be understood.

a. *Diffraction length.* The geometry of the laser injection into the plasma target is shown in figure 5.10. The laser is focused at the entrance of the target and then it is expected to diverge in a Rayleigh length. This is, however, not the

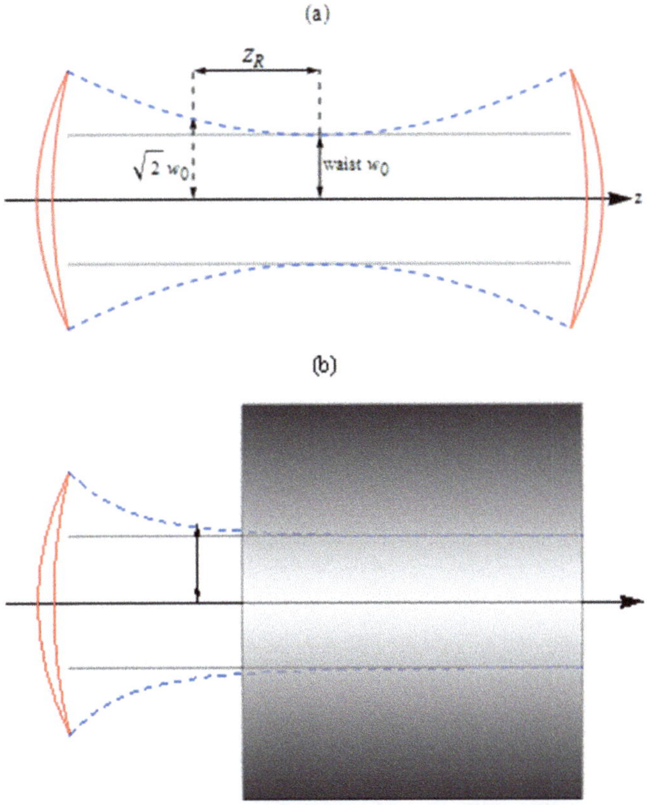

Figure 5.10. Laser beam transport: (a) ordinary propagation with diffraction and (b) guiding inside the plasma.

case, because the self-focusing, induced by the laser plasma interaction, mitigates the radiation spreading. Furthermore, the plasma profile can be suitably tapered to form a channel where the radiation propagates with an almost constant section.

b. *Dephasing length.* We have already stressed that the electrons will increase their energy until they remain in phase with the positive part of the accelerating field, this condition defines the effective length of (5.8), namely

$$L_{\text{eff}}[\text{m}] \cong \gamma_{ph}^2 \lambda_p[\text{m}] = \frac{\lambda_p^3}{\lambda_l^2}. \tag{5.49}$$

Such a condition is valid for a one-dimensional plasma. In the case of a two-dimensional plasma we have not only the accelerating field component, but also the focusing part, so that the two-dimensional dephasing length can be assumed to be half of the value given in (5.49). The three-dimensional case will be touched upon in section 5.7.

c. *Depletion length.* During the propagation of the radiation inside the plasma we can assume that the transverse section is constant with a constant energy loss to the electron bunch. The energy acquired by the N_b electrons while traversing the interaction length L_d is just a fraction of the laser pulse energy

$$\mathcal{E}_l[\mathrm{J}] \propto I_l \pi \sigma_r^2 \frac{\lambda_p}{c}, \tag{5.50}$$

thus finding

$$Q_b E_z^* L_d = \mu\, I_l\left(\pi \sigma_r^2\right)\frac{\lambda_p}{c}. \tag{5.51}$$

In conclusion we end up with

$$L_d = \mu \frac{I_l \pi \sigma_r^2 \frac{\lambda_p}{c}}{Q_b E_z^*} = \frac{\mu}{\kappa} \frac{I_l \frac{\lambda_p}{c}}{0.35\, n_p m_e c^2} \frac{\sqrt{1+\frac{a_0^2}{2}}}{a_0^2} \cong 5.6 \frac{\mu}{\kappa} \frac{\lambda_p^3}{\lambda_l^2} \sqrt{1+\frac{a_0^2}{2}}. \tag{5.52}$$

Even though derived using very naïve arguments, the previous equation contains a grain of truth, it indeed states that the length along which the acceleration occurs scales in terms of plasma density as $L_d \propto n_p^{-\frac{3}{2}}$. Furthermore since $E_z^* \propto n_p^{\frac{1}{2}}$, we can conclude that the final energy behaves as $\Delta\gamma \propto \frac{1}{n}$, which is the experimentally observed behavior (see figure 5.11).

A more accurate expression for the depletion length has been given by Shadwick *et al* (2009 *Phys. Plasmas* **16** 5) and reads

$$L_d \cong 2.8 \frac{\lambda_p^3}{\lambda_l^2} \frac{1+\frac{a_0^2}{2}}{a_0^2}. \tag{5.53}$$

If we limit ourselves to values of the normalized field strength a_0 slightly larger than unity, the depletion and dephasing lengths are almost coincident.

It is evident that a_0, hence the laser intensity, determines the effective height of the electron trapping bucket in the acceleration mode. Combining the previous identities, we find that the force acting on the charge Q_b is

$$Q_b E_z^* = 0.35\, \kappa \frac{e}{k_l\, r_e} \frac{(\sigma_r k_p)^2}{4} \frac{a_0^2}{\sqrt{1+\frac{a_0^2}{2}}} \sqrt{\frac{n_l}{n_p}} E_p. \tag{5.54}$$

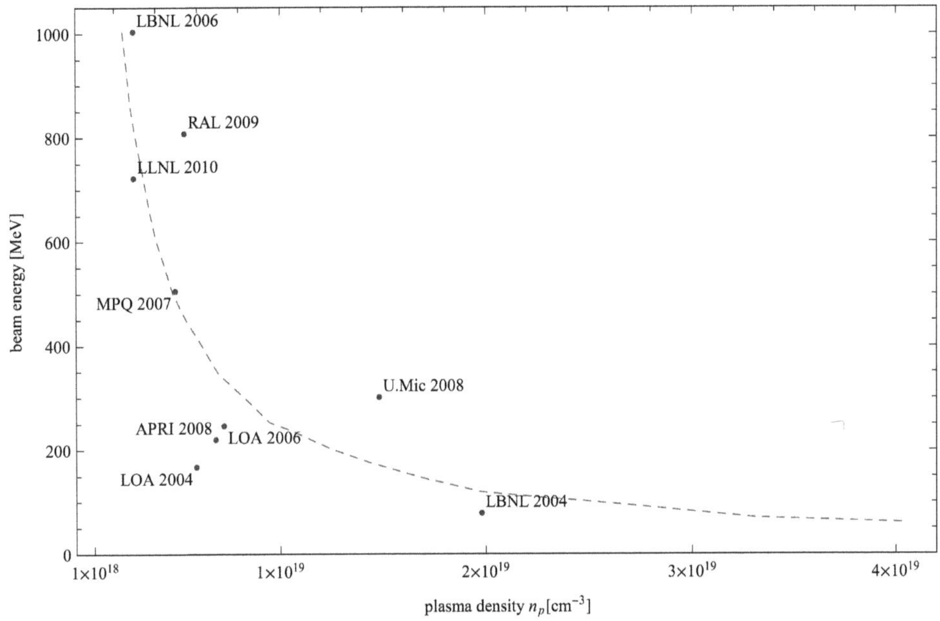

Figure 5.11. Energy versus plasma density scaling and comparison with the experiment.

The *linear regime* (small perturbation) is accordingly defined by

$$\begin{aligned} &a_0 \leqslant 1, \\ &\frac{(\sigma_r k_p)^2}{4} > \frac{a_0^2}{\sqrt{1+\dfrac{a_0^2}{2}}}. \end{aligned} \tag{5.55}$$

The second condition in (5.55) represents a lower bound on the laser section. It ensures that the laser power density does not induce significant distortion in the plasma geometry.

The *non-linear regime* is, in contrast, characterized by (see the concluding section of this chapter)

$$\begin{aligned} &a_0 \gg 1, \\ &\frac{(\sigma_r k_p)}{2} < \sqrt{a_0}. \end{aligned} \tag{5.56}$$

We can now start considering the problem from the point of view of the FEL, and see how to organize the few notions we have acquired to figure out whether a free electron device operating in the x-ray region can be effectively realized using an electron beam obtained from a laser plasma acceleration process.

5.4 FEL and laser plasma accelerated beams: general considerations

We have learned that an FEL operates if very peculiar conditions are satisfied:

1. A sufficiently large current density to ensure a reasonably large value of the ρ parameter.
2. Sufficiently small energy spread and emittances to avoid a significant increase of the gain length and thus of the saturation length.

Let us start our discussion on the effective use of plasma acceleration for an FEL device by considering the experimental set up in figure 5.12, in which a 30 TW, 40 fs $TiAl_2O_3$ laser pulse is focused on a plasma target and produces electrons with 0.5 GeV energy and 20 kA current.

The number reported in the figure allows the evaluation of the parameters relevant to the forthcoming discussion:

$$\begin{aligned} \omega_p[\mathrm{Hz}] &\cong 5.6 \times 10^{13}, \\ \lambda_p[\mu\mathrm{m}] &\cong 33.7, \\ E_p[\mathrm{GeV/m}] &\cong 96\ , \\ E_z^*[\mathrm{GeV/m}] &\cong 28, \\ \gamma_p^2 &\cong 1.8 \cdot 10^3, \\ L_{\mathrm{eff}}[\mathrm{m}] &\cong 0.06, \\ a_0 &\cong 1. \end{aligned}$$

From (5.42)–(5.44) we also infer that to transport a quantity of charge Q_b we need, at the entrance of the gas target, a laser spot size given by

$$\sigma_r \cong \sqrt{\frac{2}{\pi\kappa}\frac{Q_b}{e}\lambda_p r_e}\,. \tag{5.57}$$

Assuming therefore $\kappa \cong 0.5$, we need the previously quoted plasma density $\sigma_r \cong 12\ \mu\mathrm{m}$ to accelerate a charge corresponding to $Q_b \cong 200$ pC, if the length of the bunch is just $\tau_b \cong \frac{1}{2}\frac{\lambda_p}{c}$ we end up with a peak current of 20 kA as reported in figure 5.12. The effective accelerating length L_{eff} is consistent with the length of the

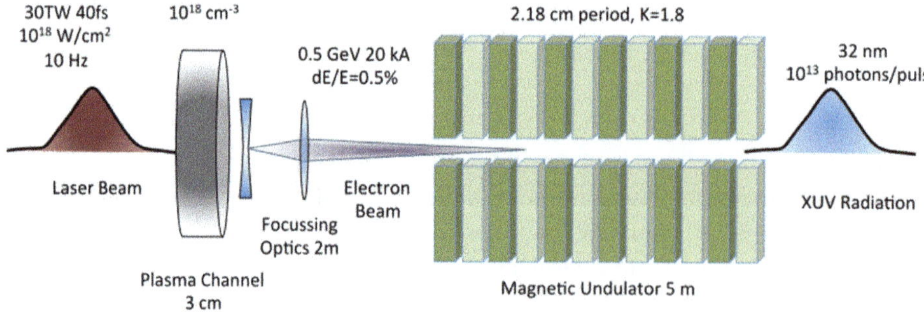

Figure 5.12. Experimental layout of an accelerated plasma operating FEL.

plasma target, so the output energy defined as $eE_z^*[\mathrm{GeV/m}]L_{\mathrm{cell}}$ is within the range of values quoted in the figure.

The plasma accelerated beam extracted from the plasma is adjusted by a focusing channel and is then injected inside an undulator. This is an almost obligatory step because the bunch, at the exit of the 'accelerating section', may be completely unmatched and unsuited for safe FEL operation

The transverse electron dynamics inside the plasma is ruled by the transverse force reported in the second equation of (5.37). For the sake of convenience it will be rewritten as

$$\begin{aligned} r'' &\cong -\omega_\beta^2 r e^{-\frac{2r^2}{\sigma_r^2}} \sin(\psi_p), \\ \psi_p &= \tilde{k}_p \zeta, \\ \omega_\beta^2 &= \sqrt{\pi} a_0^2 \frac{k_p \sigma_l}{\gamma_0^2 \sigma_r^2} e^{-\frac{(\tilde{k}_p \sigma_l)^2}{4}}. \end{aligned} \tag{5.58}$$

The previous equation reduces to that of a harmonic oscillator only for $r < \frac{\sigma_r}{2}$ and for $0 < \psi_p < \pi$. We have also denoted by ω_β a kind of betatron frequency.

It is evident that not all the trapped electrons will be affected by the same quadrupole strength, and that the matching conditions at the exit of the accelerating section become extremely difficult to realize, because outside the plasma the bunch with an extremely small transverse section will undergo a substantial increase of its divergence (see section 5.7).

Assuming, however, that this problem can be overcome and that the relative beam energy spread is limited to 5×10^{-3}, we can use the analytical formulae discussed in the previous chapter to predict (the continuous line in figure 5.13) the effective FEL SASE growth at 32 nm with the parameters and experimental layout of figure 5.12. The comparison with the full numerical simulation (the dotted line in figure 5.13) looks satisfactory (for further comments see section 5.7).

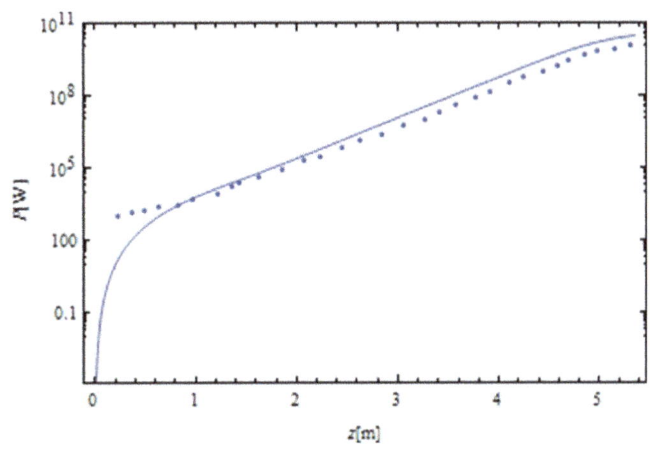

Figure 5.13. FEL growth: semi-analytical formula (continuous line) and numerical simulation (dotted line).

5.5 FEL operating with laser wave undulators: general considerations

In figure 5.14 we have shown the so-called Letardi chart of existing and proposed FEL devices world-wide. Shorter wavelengths demand larger electron-beam energies, and it is evident that a type of technology saturation is occurring for these devices as well. The plot suggests the existence of two different saturation patterns. The left lower side shows 'old style' FELs, operating essentially in the oscillator configuration, whose development toward the VUV–x-ray region has been hampered by the absence of mirrors to confine the radiation. The upper right side shows FELs belonging to the fourth-generation synchrotron radiation source scenario.

In the previous section we have mentioned that the innovation provided by the laser plasma accelerated beam may open new possibilities and allow breakthroughs leading to compact devices, paving the way for fifth-generation synchrotron radiation sources. Even though the electron-beam driver could be amazingly reduced, the undulator length remains on the order of hundreds of meters, if we consider standard magnet technology. A solution, within such a respect, could be provided by the so-called wave undulators (WU), namely by strong electromagnetic waves replacing the undulator itself.

A powerful laser could be a very good candidate, as we will see in this section.

High-power lasers have undergone a spectacular evolution. A record intensity of 2×10^{26} W m^{-2} has been obtained and the upcoming petawatt lasers aim at intensities of the order of 10^{27} W m^{-2}. In the near future, intensities of the order of 10^{28}–10^{30} W m^{-2} are foreseen at the Extreme-Light-Infrastructure (ELI).

The advantages offered by an FEL device operating with a WU are evident. Since lasers with a wavelength in the micrometer range are good candidates, an electron

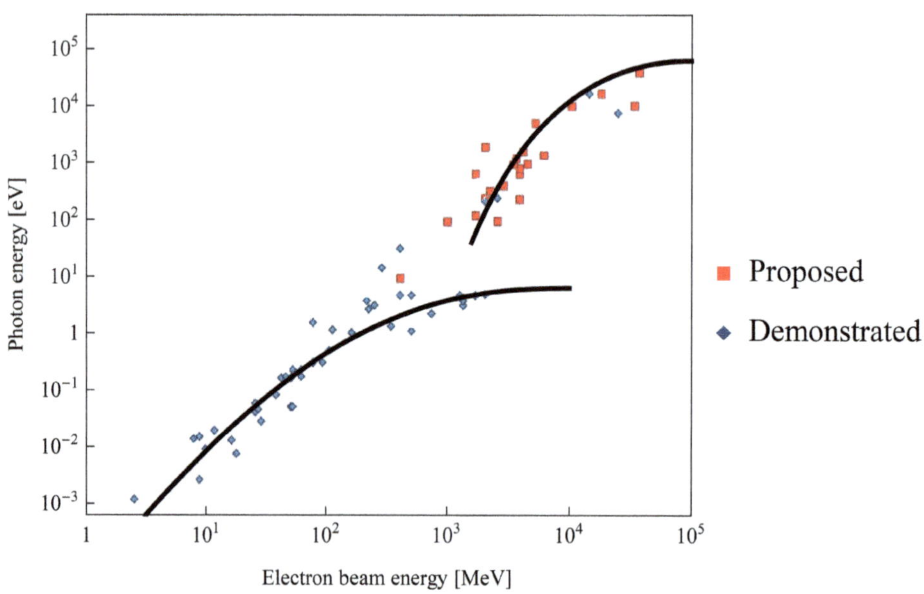

Figure 5.14. Existing and proposed FEL devices.

beam with modest energies could in principle be used to reach the hard x-ray region, and the construction of long undulators could be avoided.

It is, however, important to keep in mind that the underlying high-power Nd–Yag or CO_2 lasers are by no means straightforward devices and require challenging technological efforts to support the growth of a hard x-ray FEL up to saturation.

In this section we will not introduce new concepts and/or different experimental solutions, but we will use the methods discussed in the previous chapters to derive a set of scaling formulae useful for a first design of a linearly polarized WU based SASE FEL. We will study the conditions allowing such an operating regime and check the relevant validity using analytical and numerical analyses.

The fact that an FEL can operate with an electromagnetic undulator is as old as the FEL physics itself. Pantell, Soncini and Puthoff pointed out the possibility of obtaining a laser-like emission by means of the stimulated electron scattering of a CO_2 laser. As shown in chapter 3, Madey described the electron–undulator interaction using the Weiszacker–Williams method of virtual quanta, according to which the radiation emission by the electrons is treated as a Compton back-scattering of the pseudo-photons of the undulator field, which is viewed as an electromagnetic wave carrying a wavelength λ^*, twice the period λ_u of the undulator itself.

From such a point of view the selection of the wavelength λ_s (the resonance condition) of the emitted radiation is the result of a double Doppler shift of the undulator 'photon' wavelength, according to the prescriptions

$$\begin{aligned}
\lambda^* &\to 2\,\lambda_u,\\
\lambda_s &= \frac{1-\beta}{1+\beta}\lambda^* \cong \frac{\lambda_u}{2\,\gamma^2}\left[1+\frac{K^2}{2}\right],\\
\frac{1-\beta}{1+\beta} &\cong \frac{1}{4\,\gamma^2}\left(1+\frac{K^2}{2}\right),\\
K &= \frac{e\,B_0\,\lambda_u}{2\,\pi\,m_0 c^2}.
\end{aligned} \tag{5.59}$$

We recall that the extra-factor in the resonance condition, containing K, is a consequence of the inclusion of the electron transverse motion. An analogous wavelength shift arises in a scattering process involving electrons and an intense laser wave.

We have already stressed, in the previous sections of this chapter, the analogy between the quiver velocity a_0 and K parameters, both due to the motion of the transverse component induced by the interaction itself. In the case of a WU FEL the resonance condition is therefore written

$$\begin{aligned}
\lambda_s &= \frac{1-\beta}{1+\beta}\lambda \cong \frac{\lambda_u}{2\,\gamma^2}\left[1+\frac{K^2}{2}\right],\\
K &\approx 0.855\times 10^{-5}\lambda[\mathrm{m}]\sqrt{(I[\mathrm{W/m^2}])}\,.
\end{aligned} \tag{5.60}$$

In the case of a WU provided by a CO_2 laser, with a 'modest' power density of $I[\mathrm{W\,m^{-2}}] \cong 10^{19}$, the corresponding $K \cong 0.3$ is large enough to support an FEL SASE operation.

The dynamics of FEL operation with magnetic undulators or WUs is essentially the same, if we limit ourselves to modest electron energies and not-too-short laser wavelengths, in order to avoid quantum corrections. The scaling formulae reported in chapter 4 can be therefore used to derive practical formulae yielding the conditions to be satisfied by the wave laser intensity and by the electron-beam qualities to guarantee reliable operation.

In the following we will assume that the Pierce parameter for a WU operating FEL can be written in the same way as for the static magnetic case, furthermore since we will refer to linearly polarized laser fields, we will retain the Bessel factor. For convenience we report the expression of ρ below:

$$\rho \cong \frac{8.36 \times 10^{-3}}{\gamma} \times \left[J[\mathrm{A\,m^{-2}}] \times \left(\lambda[\mathrm{m}] K f_b(\xi)\right)^2\right]^{\frac{1}{3}},$$
$$f_b(\xi) = J_0(\xi) - J_1(\xi), \quad \xi = \frac{1}{4}\frac{K^2}{1 + \frac{K^2}{2}}, \tag{5.61}$$

and note that the current density can be written as

$$J[\mathrm{A\,m^{-2}}] \cong 1.7 \times 10^6 \frac{(\gamma\rho)^3}{\left[K f_b(K) \lambda[\mathrm{m}]\right]^2}. \tag{5.62}$$

Furthermore, as (see (5.60))

$$I[\mathrm{W\,m^{-2}}] \cong 1.37 \times 10^{10} \frac{K^2}{\lambda[\mathrm{m}]^2} \tag{5.63}$$

we combine (5.62) and (5.63) to obtain

$$J[\mathrm{A\,m^{-2}}]\, I[\mathrm{W\,m^{-2}}] f_b^2 \cong 2.33 \times 10^{16} \frac{(\gamma\,\rho)^3}{\lambda[\mathrm{m}]^4}. \tag{5.64}$$

The strategy we had in mind when we started this reshuffling of identities is that of providing simple relations involving current and laser power densities yielding a first glimpse on the conditions necessary to achieve WU FEL operation.

A first indication emerging from the previous analysis is that the condition (5.64) becomes more difficult to satisfy with the decrease of the laser undulator wavelength. Even though this was largely expected, since the corresponding K value is proportional to the WU period, it is worth stressing that the restriction is rather compelling since it scales as λ^{-4}.

The next step is the definition of the gain length, which according to the discussion of the previous chapter reads (note that in the case of a WU the period length λ_u should be replaced by half the laser wavelength)

$$L_g[\mathrm{m}] \cong \frac{\lambda[\mathrm{m}]}{8\,\pi\,\sqrt{3}\rho}. \tag{5.65}$$

The 'full saturation' condition, with an output FEL power of

$$P_F \cong \sqrt{2}\rho\, P_E \tag{5.66}$$

can be achieved in about ten gain lengths, leading to a saturation length given by

$$L_S[\mathrm{m}] \cong \frac{\lambda[\mathrm{m}]}{4\rho}. \tag{5.67}$$

The laser pulse should be at least as long as the saturation length, namely

$$\tau_l[\mathrm{s}] \cong 8.2 \times 10^{-10}\frac{\lambda[\mathrm{m}]}{4\rho}.$$

The laser energy can be therefore specified together with the laser transverse section Σ, according to

$$\begin{aligned} &E[\mathrm{J}] \cong \tau_l[\mathrm{s}]I[\mathrm{W/m^2}]\Sigma[\mathrm{m^2}] \cong 8.2 \times 10^{-10}\frac{\lambda[\mathrm{m}]}{\rho}I[\mathrm{W/m^2}]\,\Sigma[\mathrm{m^2}],\\ &\Sigma = \lambda\, Z_R,\\ &Z_R \equiv \text{Rayleigh length}. \end{aligned} \tag{5.68}$$

We assume, moreover, that the electron-beam is matched to the laser beam section and, therefore, its current is provided by

$$I_E[\mathrm{A}] \cong J[\mathrm{A\ m^{-2}}]\,\Sigma[\mathrm{m^2}]. \tag{5.69}$$

If we fix the laser power density, the following expression for the current can be obtained from (5.64)

$$I_E[\mathrm{A}] \cong \frac{2.33 \times 10^{16}}{f_b^2}\frac{(\gamma\,\rho)^3}{I[\mathrm{W\ m^{-2}}]\,\lambda[\mathrm{m}]^4}\Sigma[\mathrm{m^2}] \tag{5.70}$$

and, in turn, the electron-beam power can be written as

$$P_E[\mathrm{MW}] \cong \frac{2.33 \times 10^{16}}{f_b^2}\frac{(\gamma\,\rho)^3}{I[\mathrm{W\ m^{-2}}]\,\lambda[\mathrm{m}]^4}\Sigma[\mathrm{m^2}]\,E[\mathrm{MeV}]. \tag{5.71}$$

Finally, as a consequence of (5.66), the FEL saturated power reads

$$P_F[\mathrm{MW}] \cong \frac{1.2 \times \sqrt{2} \times 10^{16}}{f_b^2}\frac{(\gamma\,\rho)^4}{I[\mathrm{W\ m^{-2}}]\,\lambda[\mathrm{m}]^4}\Sigma[\mathrm{m^2}]\,. \tag{5.72}$$

We can use the previous relations to infer that the realization of a WU FEL device is not far from present technological capabilities. We will try to be more quantitative and consider the case of a WU provided by a CO_2 laser, with a power density of

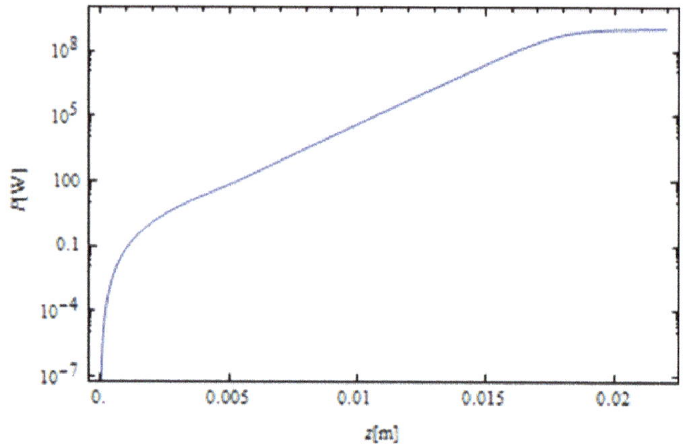

Figure 5.15. FEL intensity growth versus pulse length (in meters), corresponding to power density of $I[\mathrm{W\ m^{-2}}] \cong 8 \times 10^{19}$.

$I[\mathrm{W\ m^{-2}}] \cong 8 \times 10^{19}$(corresponding to $K = 0.76$). The use of an electron beam with an energy of 25 MeV yields an FEL operating wavelength around 1.34 nm. A current density of $10^{11}\ \mathrm{A\ m^{-2}}$ is sufficient to ensure $\rho \cong 3 \times 10^{-4}$ and yields the possibility of reaching full saturation in a laser pulse length of the order of few centimeters. If we further assume that the laser beam is confined in a region with Rayleigh length $Z_R \cong 5 \times 10^{-3}$ m, we obtain a beam current exceeding 5 kA, for a beam with a transverse section around 100 μm. The FEL power density growth versus the pulse length (in meters) with the previously quoted parameters is shown in figure 5.15.

This is just an example aimed at providing general criteria to merge the various elements determining the design of a laser WU-based FEL device and develop an indicative quick design. Further elements of discussion and exercises will be given in section 5.7.

5.6 Concluding comments

In this chapter we have collected a few thoughts providing the elements to assemble a tentative design of FELs operating with powerful lasers, playing the role of accelerators or of undulators. Both techniques require significant improvement to become effective tools and at the moment are interesting possibilities to be pursued with great attention.

The discussion regarding plasma acceleration has been essentially restricted to the quasi-linear regime and we merely mentioned the conditions leading to non-linear operation, whose detailed understanding requires a massive computational effort using particle-in-cell codes. Albeit the analysis of such an operating regime is outside the purposes of this book, we will spend the final section of this chapter discussing some relevant aspects of the use of electron beams for FEL operation.

No analytical treatment is available for this type of regime, a few scaling formulae, emerging from considerations based on similarity criteria (see section 5.7), can be very useful to understand the main parameters of interest for this operation.

Before entering into further details it is worth summarizing what we have discussed so far.

a. A laser pulse, if sufficiently intense, can induce a plasma wave. A more efficient excitation occurs if some conditions on the pulse itself are met, in particular, the excitation is more efficient when the laser pulse duration is of the order of the plasma wavelength. We have seen indeed that the induced wake field reaches its maximum when $\sigma_l \cong \frac{\lambda_p}{2}$.
b. The wake field associated with shorter or longer pulses than the plasma wavelength exhibits significantly different patterns. The long laser pulse is self-modulated with the plasma period, and the resonance between this self-modulation and the plasma frequency leads to effective wake field excitation. The corresponding regime is called self-modulated laser wake field acceleration. Long laser pulses, however, experience not only the one-dimensional self-modulation, but become self-focused and form relativistic channels in the plasma.
c. Trapping of electrons in the plasma waves is a key issue for laser wake field acceleration. Injection and acceleration of external beams has been demonstrated experimentally and various other schemes have been proposed. The wave breaking mechanism may lead to self-trapping and successive acceleration of the electrons.

If the laser pulses are intense enough to break the plasma wave already after the first oscillation, the so-called bubble regime may occur. The transverse electric field associated with a powerful laser, focused on a relatively small region of the plasma, expels the electrons along the radial directions, the field generated by the surrounding plasma charges repels the electrons back. The result is the formation of a bubble containing a large electric field, providing a strong acceleration for the electrons trapped inside.

The features characterizing the bubble regime are as follows (see figure 5.16):

i. A cavity free from cold plasma electrons is formed behind the laser pulse instead of a periodic plasma wave.
ii. A dense bunch of relativistic electrons with a mono-energetic spectrum is self-generated.
iii. The laser pulse propagates many Rayleigh lengths in the homogeneous plasma without a significant diffraction. These features are absent in the ordinary regime of laser wake field acceleration.

Let us now try to understand whether what we have learned in the previous sections can be helpful to understand the features characterizing such an operating regime.

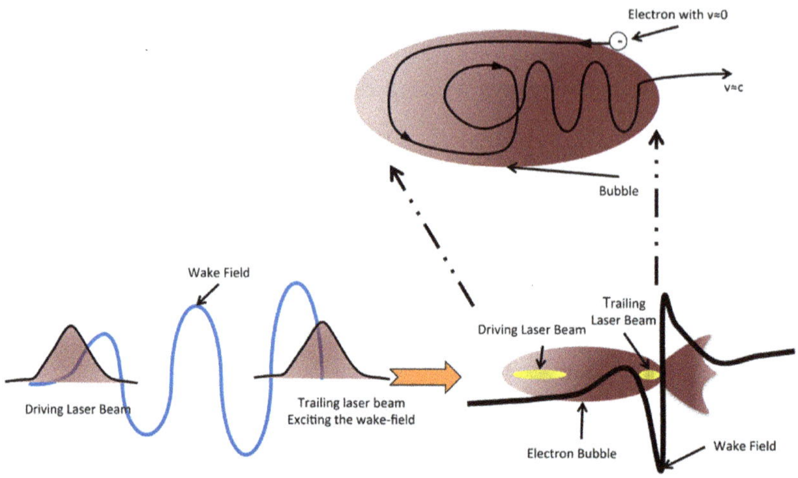

Figure 5.16. Bubble formation in laser plasma acceleration and electron trapping.

We use the following condition on the laser spot size at the entrance of the plasma:

$$\frac{(\sigma_r k_p)^2}{2} \cong a_0, \tag{5.73}$$

which along with (5.43) yields the number of trapped electrons according to the identity

$$N_b = \kappa \frac{1}{2}\left(\frac{\sigma_r}{\lambda_p}\right)^2 \frac{\pi}{r_e}\lambda_p = \kappa \, \frac{1}{4\pi^2} a_o \frac{\pi}{r_e}\lambda_p. \tag{5.74}$$

Using the quiver velocity in the form

$$a_o = \frac{\lambda_l}{\sqrt{\pi}\sigma_r}\sqrt{\frac{P_l}{P_{\text{rel}}}}, \qquad P_l = \pi \, \sigma_r^2 I_l, \tag{5.75}$$

we eventually end up with

$$N_b = \frac{\kappa\sqrt{\pi}}{1.8}\,\frac{1}{\sigma_r k_p} N^*_{\text{Bub}}, \qquad N^*_{\text{Bub}} = 1.8\sqrt{\frac{P}{P_{\text{rel}}}}\frac{1}{r_e k_l}, \tag{5.76}$$

where N_b^* is the number of mono-energetic electrons trapped and accelerated in the bubble regime according to the analysis of Pukhov and Gordienko (2006 Bubble regime of wake field acceleration: similarity theory and optimal scaling

Phil. Trans. R. Soc. **364** 623–33). We therefore obtain complete agreement with the Pukhov–Gordienko formula if we assume that the trapping efficiency is

$$\kappa \cong \frac{1.8}{\sqrt{\pi}}(\sigma_r k_p) \cong \frac{3.6}{\sqrt{\pi}}\sqrt{a_0}. \tag{5.77}$$

An analogous argument can be exploited to discuss the bubble accelerating field, according to (5.38) we find, under the assumption $a_0 \gg 1$ that the accelerating field is

$$\begin{aligned} eE_z^* &\cong 0.495\, a_0 m_e c^2 \frac{2\pi}{\lambda_p} \cong 0.495\, m_e c^2 \frac{\lambda_l}{\sqrt{\pi}} \frac{2\pi}{\lambda_p \sigma_r}\sqrt{\frac{P}{P_{\text{rel}}}} \\ &= \Gamma\,(eE_{\text{Bub}})\,, \\ \Gamma &\cong \frac{0.533}{\sqrt{\pi}} k_p \sigma_r, \end{aligned} \tag{5.78}$$

where E_{Bub} is the Pukhov–Gordienko electric field, namely

$$eE_{\text{Bub}} \cong 0.93\, m_e c^2 \sqrt{\frac{P}{P_{\text{rel}}}} \frac{\lambda_l}{\pi\, \sigma_r^2}. \tag{5.79}$$

It is evident that, at least from the phenomenological point of view, the discussion of this and of the previous section allows us to end up with reliable scaling formulae for the FEL operation. Further comments yielding more 'practical information' can be found in the forthcoming part of the chapter.

5.7 Complements and exercises

In the previous part of the chapter we have used (and abused) notions from plasma physics without the necessary rigor. We have, e.g. touched upon the plasma screening and the associated Debye potential without any justification.

We give an idea of how the different concepts covered in the initial part of the chapter can be embedded to obtain the explicit form of the Debye (Huckel) potential.

It is evident that the form of such a potential is ruled by the Poisson equation, which reads

$$\vec{\nabla}^2 \Phi = \frac{e}{\varepsilon_0}(n_e - n_i),$$

where the electron and ion densities n_e, n_i, respectively, are found from the Boltzmann distribution

$$n_e = n_p e^{e\frac{\Phi}{K_B T}}, \quad n_i = n_p e^{-e\frac{\Phi}{K_B T}}.$$

Thus finding

$$\vec{\nabla}^2 \Phi = 2\frac{e\, n_p}{\varepsilon_0} \sinh\left(\frac{e\,\Phi}{K_B T}\right)$$

by assuming $\frac{e\,\Phi}{K_B T} \ll 1$, we find (the reader is invited to obtain the explicit derivation)

$$\vec{\nabla}^2\Phi = \frac{\Phi}{r_D^2},$$

yielding the Debye potential as well (check that the Debye potential is the solution of the above equation, recall that in spherical coordinates the Laplacian reads $\vec{\nabla}^2 = \frac{1}{r}\partial_r^2 r$).

5.7.1 Ponderomotive forces and plasma waves

In the previous part of this chapter we have often referred to the role of the ponderomotive force that arises in the process of plasma wave interaction and gives rise to plasma wake fields. In this complement we further comment on the relevant meaning and relevance of this phenomenon to the physical processes under study.

In order to address the problem gently, we first discuss the non-relativistic case. We consider the interaction of an electron with an electromagnetic wave, whose field intensities exhibit a smooth dependence on the transverse coordinates or, more precisely, with a gradient derivative smaller than the field wavelength.

The electron will accordingly be ruled by the Lorenz equation

$$\begin{aligned} &m_e\frac{d}{dt}\vec{v} = -e[\vec{E}_0(\vec{r}) + \vec{v} \times \vec{B}_0(\vec{r})]\cos(\omega t) \\ &\vec{E}(\vec{r},\, t) = \vec{E}_0(\vec{r})\cos(\omega t), \\ &\vec{B}(\vec{r},\, t) = \vec{B}_0(\vec{r})\cos(\omega t). \end{aligned} \tag{5.80}$$

We solve the associated motion problem using the assumptions reported below. The particle is initially at rest at position $\vec{r} = \vec{r}_0$, the action of the Lorentz force occurs first through the electric field, which induces a transverse velocity and a coordinate shift specified by

$$\begin{aligned} &m_e\frac{d}{dt}\vec{v}^{(1)} \cong -e\vec{E}_0\cos(\omega t) \\ &\to \vec{v}^{(1)} \cong -\frac{e\vec{E}_0}{m_e\omega}\sin(\omega t),\ \vec{r}^{(1)} - \vec{r}_0 \cong -\frac{e\vec{E}_0}{m_e\omega^2}\cos(\omega t). \end{aligned}$$

The variation in position causes a modification of the electric field, which can be written as

$$\vec{E}_0(\vec{r}_0 + \vec{r}^{(1)}) \cong \vec{E}_0 + (\vec{r}^{(1)}\cdot\vec{\nabla})\vec{E}_0,$$

as a consequence of its smooth transverse dependence.

The induced velocity $\vec{v}^{(1)}$ and position shift $\vec{r}^{(1)}$ are responsible for the onset of a force contribution due to the gradient-dependent term of the electric field and to the magnetic induced deflection; collecting the different contributions and since $\frac{\partial \vec{B}}{\partial t} = -\vec{\nabla} \times \vec{E}$ we can write

$$\vec{F} \cong -e(\vec{r}^{(1)} \cdot \vec{\nabla})\vec{E}_0 \cos(\omega t) - \frac{e}{\omega}\vec{v}^{(1)} \times [\vec{\nabla} \times E_0] \sin(\omega t).$$

Substituting $\vec{v}^{(1)}$ and $\vec{r}^{(1)}$ in the previous equation and taking the average over time, we end up with

$$\vec{F} \cong -\frac{e^2}{2\, m_e \omega^2}\left[(\vec{E}_0 \cdot \vec{\nabla})\, \vec{E}_0 + \vec{E}_0 \times (\vec{\nabla} \times \vec{E}_0)\right],$$

Which, after using the vector identity

$$\vec{E}_0 \times (\vec{\nabla} \times \vec{E}_0) = \frac{1}{2}\vec{\nabla} E_0^2 - (\vec{E}_0 \cdot \vec{\nabla})\, \vec{E}_0,$$

yields the ponderomotive force expressed as

$$\vec{F}_P = -\frac{e^2}{4\, m_e \omega^2}\vec{\nabla}\left(E_0^2(r)\right) = -\frac{m_e c^2}{2\, \pi}\vec{\nabla} a_0^2. \tag{5.81}$$

It is convenient to define $\vec{F}$ through the ponderomotive potential U_p, namely

$$\begin{aligned} \vec{F}_P &= -\, \vec{\nabla}\, U_p, \\ U_p &= \frac{[e\, E_0(r)]^2}{4\, m_e \omega^2} = \frac{m_e c^2}{2\, \pi} a_0^2, \end{aligned} \tag{5.82}$$

which displays the real nature of U_p as that of a kinetic energy, depending on the space coordinate.

A cartoon demonstrating the role of the force in (5.81) is provided in figure 5.17, where it is shown that any spatial variation of the laser field intensity determines a force pushing the electrons (and ions) from the region of higher intensity to the

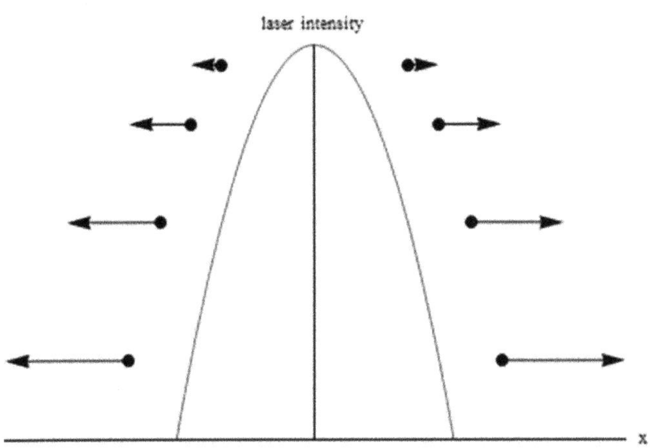

Figure 5.17. Ponderomotive force becomes more intense with decreasing laser intensity.

region of lower intensity through the ponderomotive force. This displacement of electrons creates the large amplitude plasma waves.

Exercise 5.1 Find the relativistic extension of the ponderomotive force.

The generalization to the relativistic case is simply obtained by noting that since the potential U_p in (5.82) is nothing but a spatially dependent kinetic energy, we make the following replacement

$$\frac{m_e c^2}{2} a_0^2 \to m_e c^2 \langle \gamma - 1 \rangle,$$

$$\langle \gamma \rangle = \sqrt{1 + \frac{a_0^2}{2}},$$

thus finally obtaining for the relativistic ponderomotive force the expression

$$\vec{F}_P = -\frac{m_e c^2}{\pi} \vec{\nabla} \left(\sqrt{1 + \frac{a_0^2}{2}} \right). \tag{5.83}$$

Exercise 5.2 Use the definition of quiver velocity as

$$a_0 \cong 8.55 \times 10^{-10} \sqrt{I[\mathrm{W\ cm^{-2}}]\, \lambda[\mu\mathrm{m}]^2},$$

to write the corresponding electric and magnetic field intensities as

$$E_0[\mathrm{GV\ cm^{-2}}] \cong \frac{32.2\, a_0}{\lambda[\mathrm{cm}]},$$

$$B_0[\mathrm{MG}] \cong \frac{107\, a_0}{\lambda[\mathrm{cm}]}.$$

Convert the above quantities into MKSA.

Exercise 5.3 Derive the threshold intensity of a laser inducing atom ionization.

Hint: note that the electric field associated with the Bohr radius $a_B = \frac{\hbar^2}{me^2} \cong 5.3 \times 10^{-9}$ cm is

$$E_A = \frac{e}{4\pi\varepsilon_0 a_B^2} \cong 5.1 \times 10^{-9}\mathrm{cm}$$

$$\to I_A = \frac{\varepsilon_0 c}{2} E_A^2 \cong 3.51 \times 10^{16}\ \mathrm{W\ cm^{-2}}.$$

Exercise 5.4 The electric field of the laser when interacting with an atomic system induces a suppression of the atomic barrier (see figure 5.18), and tunneling ionization may occur. Specify the threshold intensity of the laser necessary to induce such an effect.

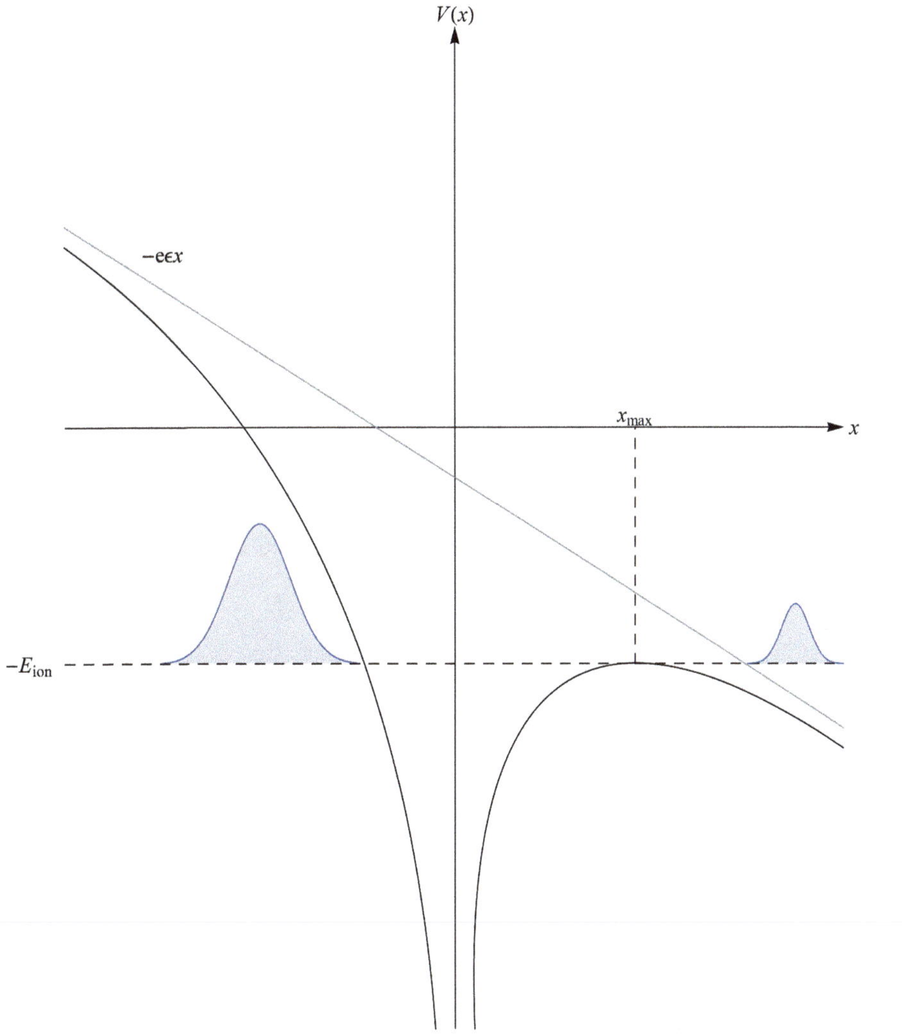

Figure 5.18. Tunneling barrier suppression by an external field.

Hint: the total potential acting on an electron in the atom-laser field is

$$V(x) = -\frac{Ze^2}{4\pi\varepsilon_0 x} - eE_{app}x.$$

The barrier suppression occurs for $x > x_{MAX} = \sqrt{\frac{Ze}{4\,\pi\,\varepsilon_0 E_{app}}}$, setting $V(x_{max}) = E_{Ion}$ we find

$$I_{app} = \frac{c\varepsilon_0 E_{ion}^4}{32\tilde{Z}^2 e^6},$$
$$\tilde{Z} = \frac{Z}{4\pi\varepsilon_0}.$$

5.7.2 Wake fields

In section 5.3 we derived the equations defining the scalar potential associated with the plasma wake fields. But we have left open the algebraic details which will be covered in these complements.

Before providing specific details we invite the reader to consider the following exercise.

Exercise 5.5 Prove the following identity

$$\int_x^{\infty} e^{-a\xi^2 - b\xi} d\xi = \frac{1}{2}\sqrt{\frac{\pi}{a}}\, e^{\frac{b^2}{4a}} erfc\left(\left(x - \frac{b}{2a}\right)\sqrt{a}\right),$$

$$erfc(t) = (1 - erf(t)),$$

$$erf(t) = \frac{2}{\sqrt{\pi}} \int_0^t e^{-\sigma^2} d\sigma.$$

We have shown that the wake scalar potential is defined by the integral transform

$$\varphi = -\frac{\tilde{k}_p m_e c^2}{2} \int_\zeta^{\infty} d\zeta' \sin\left[\tilde{k}_p(\zeta - \zeta')\right] a^2(r, \zeta').$$

The assumption that the dimensionless field amplitude can be cast in the form

$$a^2(r, \zeta) = U(r)\, e^{-\frac{\zeta^2}{\sigma_l^2}},$$

$$U(r) = \frac{a_0^2}{2} e^{-\frac{2 r^2}{\sigma_r^2}},$$

allows the explicit derivation of the transverse and longitudinal electric fields. The mathematical details are summarized below:

$$\int_\zeta^{\infty} d\zeta' \sin\left[\tilde{k}_p(\zeta - \zeta')\right] a^2(r, \zeta') = \frac{a_0^2}{2} e^{-\frac{2 r^2}{\sigma_r^2}} \int_\zeta^{\infty} d\zeta' \sin\left[\tilde{k}_p(\zeta - \zeta')\right] e^{-\frac{\zeta'^2}{2\sigma_l^2}}$$

$$= \frac{a_0^2}{2} e^{-\frac{2 r^2}{\sigma_r^2}} \left[\sin(\tilde{k}_p \zeta)\, C(\zeta) - \cos(\tilde{k}_p \zeta)\, S(\zeta)\right],$$

$$C(\zeta) + i\, S(\zeta) = \int_\zeta^{\infty} d\zeta' e^{-i\,\tilde{k}_p \zeta'}\, e^{-\frac{\zeta'^2}{2\sigma_l^2}}$$

$$= \frac{\sigma_l}{2}\sqrt{2\pi}\, e^{-\frac{(\tilde{k}_p \sigma_l)^2}{2}} (1 - erf)\left(\frac{1}{\sqrt{2}}\left(\frac{\zeta}{\sigma_l} - i\tilde{k}_p \sigma_l\right)\right).$$

We can now look at the behavior of the functions $C(\zeta)$, $S(\zeta)$ reported in figure 5.19 and note that for $\frac{\zeta}{\sigma_l} \ll -1$ (the wake is behind the laser pulse) the functions can be approximated with $C(\zeta) \cong 2,\ S(\zeta) \cong 0$.

The fact that in our computations $U(r)$ does not exhibit any dependence on the ζ variable amounts to the approximation that the laser field does not experience any

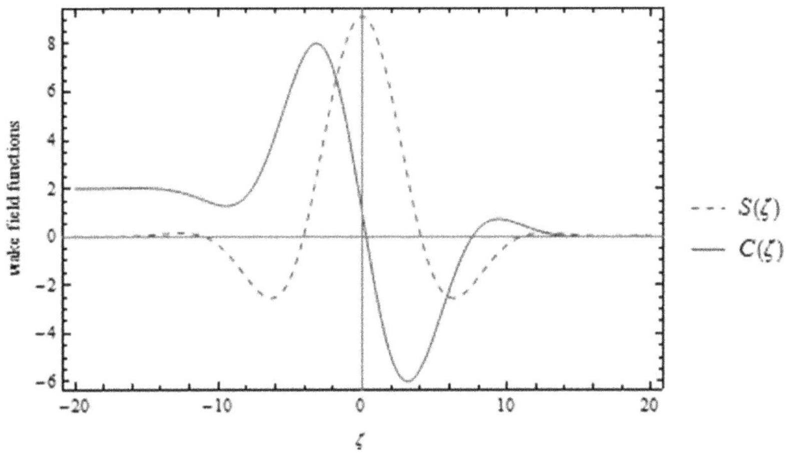

Figure 5.19. Wake field functions versus the co-moving coordinate ζ: (a) $C(\zeta)$, solid line and (b) $S(\zeta)$ dotted line.

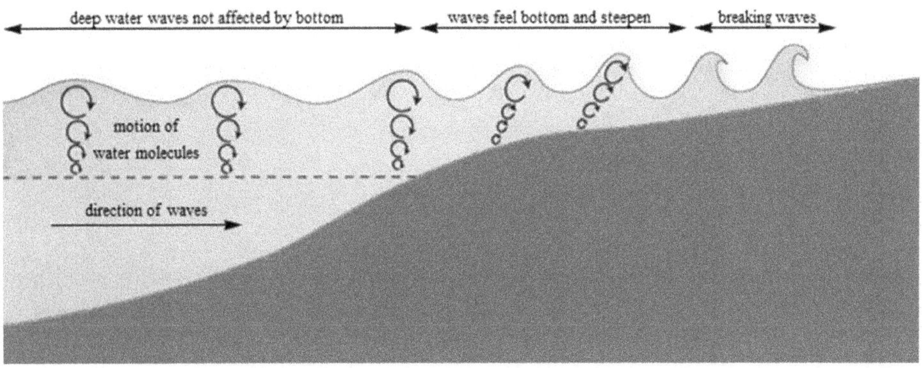

Figure 5.20. Mechanism of a sea wave breaking on the shore.

diffraction effect inside the plasma. We comment further on the validity of such an assumption below.

Exercise 5.6 Find a simple argument defining the maximum intensity of the electric field carried by the plasma wake before the wave breaks.

We have avoided any reference to analogies involving wakes behind a boat when referring to the laser plasma acceleration. However, in talking about wave breaking we take advantage of the analogy of water waves breaking on the shore. When water approaches the shore the density increases and the waves break (see figure 5.20). An analogous effect occurs in the case of plasma waves (see figure 5.21), which undergo a similar effect when the associated electric field overcomes a certain limit. The increase of the electron density is simply due to the fact that electrons acquire energy and their velocity becomes close to the phase velocity of the wave, which is close to the speed of light. We expect therefore that the electronic density increases as $\frac{n}{n_0} = \frac{1}{1 - \frac{\beta}{\beta_p}}$.

The electron and electric field equations, including the effect of density increasing, can be written as (explain why)

$$\left(1 - \frac{\beta}{\beta_p}\right)\frac{\partial E}{\partial \tau} = \beta E_p,$$

$$\frac{1}{1 - \frac{\beta}{\beta_p}}\frac{\partial(\gamma\beta)}{\partial \tau} = -\frac{E}{E_p},$$

$$\tau = -\omega_p \zeta,$$

the first of which is the Poisson equation and the second is the electron relativistic equation of motion.

Combining the two we end up with (why?)

$$\frac{1}{2E_p^2}\frac{\partial E^2}{\partial \tau} = -\frac{\partial}{\partial \tau}\gamma,$$

which once integrated yields (why?)

$$E = \sqrt{2(\gamma_p - 1)}E_p,$$

representing the maximum electric field sustainable by the wave before breaking.

Exercise 5.7 In the previous part of the chapter we have stressed that in the strongly non-linear regime (determined by high-intensity strongly focused lasers) the onset of the so-called bubble regime occurs. In the hypothesis one is interested in a further analogy with ordinary fluids; the closest phenomenon is that of cavitation. In the present case the role of pressure forces is replaced by the ponderomotive force.

The plasma electrons at the front of the pulse and slightly off the laser axis in the plasma channel experience axial and radial ponderomotive and space charge forces, creating an electron evacuated non-uniform ion bubble. The expelled electrons travel along the surface of the bubble and reach the stagnation point (see figure 5.21), where they cumulate until they reach a critical density. In correspondence to which the wave breaks and the electrons are injected into the ion bubble, where they undergo a mono-energetic acceleration process.

Find a physical argument yielding the radius of the ion bubble and the critical size of the electrons determining wave breaking.

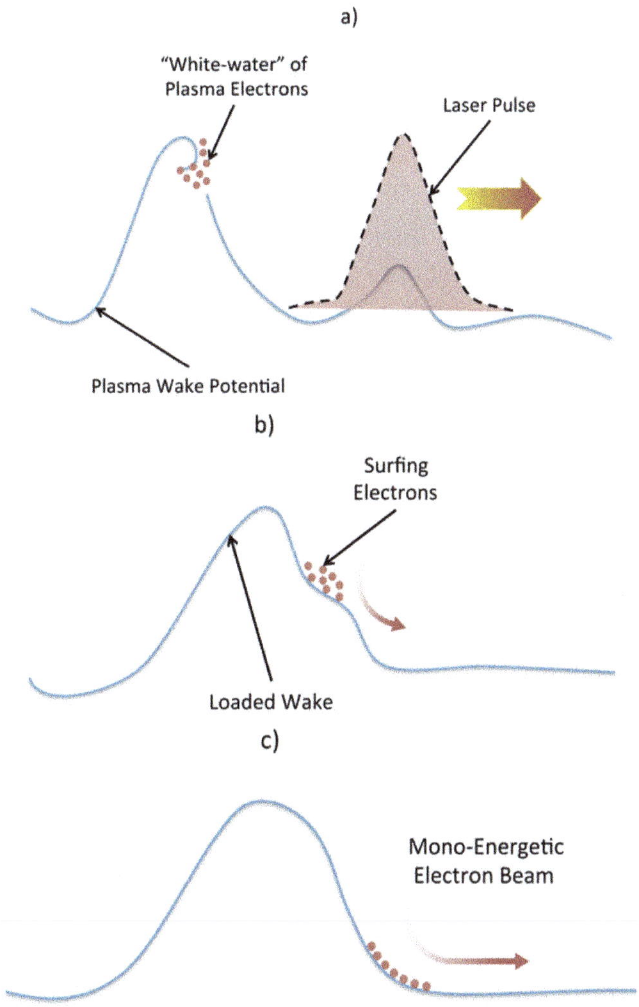

Figure 5.21. Wave breaking in plasma waves.

Hint: recall the definition of ponderomotive potential as kinetic energy and equate it to the ion charge density potential and obtain

$$m_e c^2\left(\sqrt{1+\frac{a^2}{2}}-1\right)=\frac{4}{3}n_p e^2 R^2$$

$$\rightarrow R=\sqrt{3}\frac{c}{\omega_p}\left(\sqrt{1+\frac{a^2}{2}}-1\right).$$

An analogous argument holds for the electron density critical radius, which is reached when the potential energy of the electrons equals their initial kinetic energy, namely

$$m_e c^2(\gamma_{in} - 1) = \frac{4}{3} n_p e^2 r_c^2$$

or (why?)

$$r_c \cong \frac{c}{\omega_p}\sqrt{\frac{n_p}{n_c}}.$$

Exercise 5.8 Find a simple argument explaining the laser focusing mechanisms in plasma

There are different possible sources determining this effect, the plasma wave guiding is due to the plasma density dependence on the radial coordinate. The qualitative explanation of this effect is given in figure 5.22. The supposed plane wave front of the laser field (a) interacts with a region of the plasma with a non-constant density distribution, say parabolic (b). The intensity is minimum along the axis and increases toward the periphery. The refraction index will exhibit the opposite shape (c), thus determining the phase velocity distribution shown in (d), which in turn leads to the focusing mechanism, larger away from the axis than on-axis, thus determining the focusing.

A further source is the so-called self-focusing associate with the r dependence of the laser wave itself.

A more quantitative explanation is given below; to this aim we recall that:

a) Phase and group velocities are given by

$$v_p = \frac{\omega_l}{k} = \sqrt{c^2 + \frac{\omega_p^2}{k^2}},$$

$$v_g = \frac{\partial \omega_l}{\partial k} = \frac{c^2}{v_p}.$$

b) The focusing effects are determined by the refractive index linked to the phase velocity by

$$\eta(r) = \frac{1}{\beta_p(r)} = \sqrt{1 - \frac{\omega_p(r)^2}{\omega_l^2}} \cong 1 - \frac{1}{2}\frac{n_p(r)}{\gamma(r)\, n_c}.$$

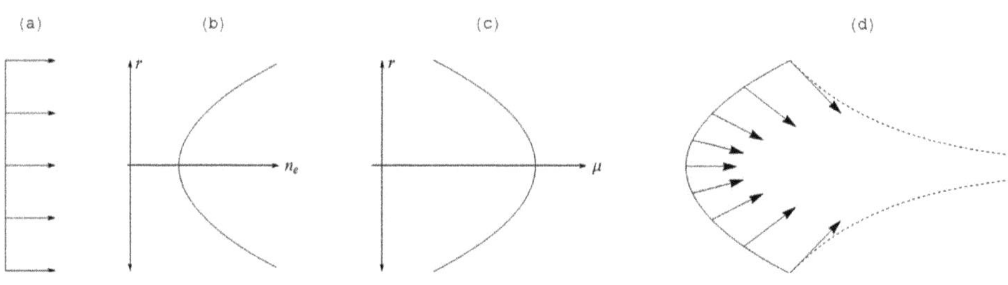

Figure 5.22. Schematics of the self-focusing mechanism in laser plasma interaction.

It should accordingly be stressed that the r dependence of the plasma density induces the previously quoted effect, however, the relativistic factor also depends on r. Neglecting the plasma density dependence and considering the weakly relativistic case, we can write

$$\eta(r) \cong 1 - \frac{1}{2}\frac{n_p}{n_c}\left(1 - \frac{a^2}{2}\right).$$

The condition for refractive guiding is $\frac{\partial}{\partial r}\eta(r) < 0$, we can therefore define a critical value above which self-guiding occurs. The reader is invited to prove that such a threshold is

$$P_c[\mathrm{GW}] \cong 17.5 \times \left(\frac{\omega_l}{\omega_p}\right)^2.$$

Exercise 5.9 Use the laser parameters listed below and a laser cell with a plasma density $n_p = 10^{19}\,/\mathrm{cm}^3$ to design a hard x-ray FEL device,

$$\mathcal{E}_l = 24\ \mathrm{J},\ \tau_l = 43\ \mathrm{fs},\ P_l = 0.56\ \mathrm{PW},\ I_l = 1.1 \times 10^{20}\ \mathrm{W\ cm^{-2}},\ \lambda_l = 0.911\ \mu\mathrm{m}.$$

Exercise 5.10 In the first part of this chapter we have mentioned the Gordienko and Pukhov method to describe the bubble regime and we have mentioned the *scaling theory*. Without entering into the relevant details we note that the whole procedure is based on the definition of one or more dimensionless parameters, constructed by the use of the physical dimensional quantities entering the problem under study, and looking for global dependence in terms of such a parameter.

A simple example of such a procedure is a simple pendulum depending on three physical quantities, time T, length l and acceleration a. It is not difficult to verify that the dimensionless quantity we form using these three quantities is $\sqrt{\frac{l}{a}}\,T$, which sounds very familiar and suggests the existence of a 'law' linking the above three physical quantities (needless to say the most common is that of the isochronism for small oscillations).

A further, less straightforward example, is provided by the Reynold number, which is the similarity parameter regulating the physics of fluids. The dimensional parameters that enter the description of a fluid are a characteristic velocity v, the length l, the dynamics viscosity, dimensionally defined as $[\mu] = [P \times T]$, $P \equiv$ pressure, and the density ρ .

It is evident that one of the possible dimensionless parameters formed through the above quantities is simply

$$\mathrm{Re} = \frac{\rho v l}{\mu},$$

which is the definition of the Reynolds number.

Show that the drag force F_d acting on a uniform sphere moving with constant velocity in an uncompressible fluid is a universal function of the Reynolds number.

Hint: recall that drag force is a kind of friction opposing the sphere's motion. According to the previous remarks F_d depends on the characteristic quantities defining the fluid through a not specified function, namely

$$F_d = f(\rho, v, l, \mu).$$

The argument defining the function can be combined to obtain the following quantity:

$$\Phi = \rho\, v^2\, l^2\, \mathrm{Re}.$$

Since $\rho\, v^2\, l^2$ has the dimension of a force we can safely assume that

$$\frac{F_d}{\rho v^2 l^2} = f(\mathrm{Re}).$$

Exercise 5.11 A single similarity parameter has been introduced by Gordienko and Pukhov, namely $S = \dfrac{n_p}{n_c a_0}$.

Use the argument developed in the previous exercise to justify the following scaling 'laws'

a) *Cavitation condition*

$$\sigma_r k_p = 2\sqrt{a_0}\,.$$

b) *Depletion velocity*

$$v_{\mathrm{dep}} \cong c\left(\frac{\omega_p}{\omega_l}\right)^2.$$

c) *Depletion length*

$$L_{\mathrm{dep}} = c\left(\frac{\omega_p}{\omega_l}\right)^2 \tau_0.$$

d) *Dephasing length*

$$L_d \cong \frac{R}{c - v_p} c \cong \frac{2}{3}\left(\frac{\omega_l}{\omega_p}\right)^2 R.$$

(The condition $L_{\mathrm{dep}} > L_d \rightarrow c\,\tau_0 > \dfrac{2}{3}R$.)

Exercise 5.12 Justify the following scaling for the energy acquired by the electrons at the end of the bubble

$$\Delta\mathcal{E} \cong \frac{2}{3} m_e c^2 \left(\frac{\omega_l}{\omega_p}\right)^2 a_0.$$

Hint: note that the accelerating length is essentially the dephasing length and

$$\frac{e\ E_{\mathrm{acc}}}{m_e c\ \omega_p} \cong \frac{1}{2}\sigma_r k_p \cong \sqrt{a_0}\,.$$

Exercise 5.13 Find an argument supporting the fact that the number of electrons accelerated in a bubble are

$$N_e \cong \kappa \frac{\left(k_p r\right)^3}{k_p r_e},\ \kappa = \frac{1}{30}.$$

(The value of κ derives from the analysis of the numerical simulations.)

Exercise 5.14 Show that in practical units the previous identity reads

$$N_e \cong 2.5 \times 10^7 \frac{\lambda_l[\mu\mathrm{m}]}{0.8}\left(\frac{P}{P_c}\right)^{-\frac{2}{3}},$$

where P[TW].

Exercise 5.15 Obtain an expression for the current and the power of an electron beam obtained via bubble acceleration. In particular, show that

$$P_e \propto \frac{8\pi^2}{3}\kappa \frac{m_e c^2}{e\tau_b}\frac{\sigma_r^3}{r_e \lambda_l^2},$$

where τ_b is the electron bunch length.

Exercise 5.16 Show that the electron pulse cannot be shorter than

$$\tau_b^* = \frac{8\ \pi^2}{3}\kappa \frac{\sigma_r^3}{r_e \lambda_l^2}\frac{m_e c^2}{e\ P_l}$$

and justify why.

5.7.3 Betatron oscillation inside the plasma bubble

The forces acting on a single electron inside the plasma are not particularly complicated, the relevant motion can indeed be reduced to that of a charged particle

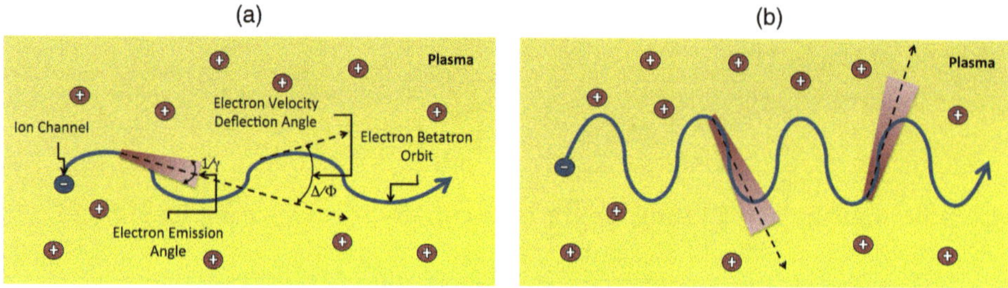

Figure 5.23. (a) Undulator like motion of electrons inside the bubble and (b) wiggler-like motion.

with radial focusing forces and a longitudinal accelerating term. We expect therefore that the electrons undergo betatron oscillations which will be accompanied by an emission of bremsstrahlung radiation associated with the transverse acceleration. The electrons move through the ion channel, executing a wiggling trajectory (see figure 5.23) and radiate according to what has been discussed in chapter 3.

The transverse and longitudinal forces acting on the electrons are expressed in the co-moving frame of the bubble as ($r \leqslant R,\ p_\perp \ll p_z$)

$$F_x \cong -\frac{1}{2} m_e \omega_p^2 x,$$
$$F_y \cong -\frac{1}{2} m_e \omega_p^2 y,$$
$$F_z \cong -\frac{1}{2} m_e \omega_p^2 (\zeta - \beta_x x - \beta_y y).$$

Considering the y-betatron oscillation only we find

$$\frac{d^2}{dt^2} y \cong -\Omega_\beta^2 y,$$
$$\Omega_\beta = \frac{\omega_p}{\sqrt{2}\gamma}.$$

Exercise 5.17 First assume that the longitudinal force is negligible and the energy is constant. Write the transport matrix for the plasma lens, derive the relevant focal length and evaluate the associated Twiss parameter.

Exercise 5.18 Consider the beam at the output of the plasma and study a strategy for the relevant matching to the undulator line.

Exercise 5.19 Include the transverse–longitudinal coupling and discuss the effect on the transport. Write the associated phase-space evolution and discuss the effect on the transverse beam emittance.

Exercise 5.20 Relax the condition of constant energy, include the effect of acceleration and reconsider the problem of the transport.

Exercise 5.21 With reference to figure 5.23(a), show that, inside the bubble and for cone angles $\Delta\vartheta < \frac{1}{\gamma}$, electrons undergo an undulator-like emission providing photons with frequency

$$\omega_B \propto 2\Omega_\beta \gamma^2.$$

Exercise 5.22 With reference to figure 5.23(b), show that inside the bubble electrons for $p_\perp \gg m_c$, $\Delta\vartheta > \frac{1}{\gamma}$ may undergo synchrotron radiation emission, with a spectrum characterized by the critical frequency

$$\omega_c \cong \frac{3\,\Omega_\beta^2 \gamma^3 r_e}{c} \propto \frac{3\,\omega_B r_e}{2\,c}\left(\gamma\,\Omega_\beta\right).$$

Exercise 5.23 Show that the quantity playing the role of undulator period is simply $\lambda_\beta = \sqrt{2\,\gamma}\lambda_p$.

Exercise 5.24 In exercise 5.22 we have not included the parameter strength, show that the emitted frequency is

$$\omega_B = \frac{2\,n\,\Omega_\beta \gamma^2}{1 + \frac{K_\beta^2}{2}},$$

$$K_\beta = \gamma\,\frac{\Omega_\beta y_\beta}{c},$$

$$y_\beta \equiv \text{amplitude of the betatron oscillation.}$$

Hint: recall that in the betatron undulator motion the maximum oscillation is given by $y_{\text{Max}} = \frac{K}{\gamma}\frac{c}{k_u}$ where $k_u = \frac{2\,\pi}{\lambda_u}$, then consider the conclusions of exercise 5.21.

Exercise 5.25 Derive the number of equivalent periods

Hint: note that $N_\beta \cong \frac{L_d}{\lambda_\beta} \cong \frac{2}{3}\left(\frac{\omega_l}{\omega_p}\right)^2 \frac{R}{\lambda_\beta} = \frac{R}{3\pi\sqrt{2\gamma}\,c}\frac{\omega_l^2}{\omega_p}.$

Exercise 5.26 In conventional undulators the electron-beam energy spread induced by the effect of the radiation emission is

$$\sigma_\varepsilon^2 \cong 5.76 \times 10^{-13}\frac{(E[(\mathrm{GeV})])^2 K^3}{(\lambda_u[\mathrm{cm}])^3} L_u[\mathrm{m}]$$

Use the above equation and the previously quoted correspondences to find an analogous expression for the case of radiation emission in the bubble regime.

Hint: write $\sigma_\varepsilon \cong 7.6 \times 10^{-7}\sqrt{N\,K}\dfrac{K\ E[\mathrm{GeV}]}{\lambda_u[\mathrm{cm}]} = 2.75 \times 10^{-10}\sqrt{N_\beta\,K_\beta}\dfrac{K_\beta\,\sqrt{\gamma}}{\lambda_p[\mathrm{cm}]}$.

5.7.4 Laser wave undulators

We have seen that Compton backscattering (CBS) is a mechanism allowing the generation of short wavelength photon beams. The undulator plays a similar role and the key element is the emission of photons with a wavelength reduced by a factor $2\,\gamma^2$ with respect to the undulator period itself.

The emission process by relativistic electrons, propagating in an undulator magnet, can accordingly be viewed as a CBS. The electron in its frame of reference 'sees' the undulator as an intense electromagnetic wave. On the other side the electron perceives, in the laboratory frame, the laser field as an undulator.

The above qualitative statements miss two important differences between the laser and undulator which are linked to the wavelength and magnetic field:

(a) The undulator period is not the same as the laser wavelength.
(b) The undulator field is static in the laboratory frame.

Suppose that we want to define an equivalent undulator period for a laser field with wavelength λ_l.

The difference is easily understood, following the discussion of chapters 3 and 4. We first recall that the wavelength of the radiation emitted in undulator is linked to its period and to the longitudinal velocity of the electrons by

$$(1 - \beta)\lambda_u = \lambda_r.$$

In CBS by a laser with wavelength λ_l, we find

$$\frac{1-\beta}{1+\beta}\lambda_l = \lambda_r.$$

Comparing the lhs of the previous equations, we end up with the conclusion

$$\lambda_u = \frac{\lambda_l}{1+\beta} \cong \frac{\lambda_l}{2},$$

namely the equivalent undulator period is half the laser wavelength.

An analogous reasoning allows us to derive the equivalent undulator field of the laser wave, to this aim we recall that for a plane wave the electric and magnetic fields are linked by

$$\omega\vec{B}_l = \vec{k} \times \vec{E}_l,$$

where $\vec{k}$ is the propagation vector with modulus $|\vec{k}| = \frac{\omega}{c}$. The equivalent magnetic field is therefore

$$B_u = B_l + \frac{E_l}{\beta\, c} = \frac{1+\beta}{\beta} B_l \cong 2B_l.$$

It is therefore evident that the associated undulator strength parameter reads

$$K = \frac{eB_u\lambda_u}{2\pi m_e c} = \frac{eB_l\lambda_l}{2\pi m_e c}.$$

The parameter strength is therefore the same if we replace the equivalent undulator magnetic field B_u and period λ_u with that of the magnetic field B of the plane wave and its wavelength λ_l.

The laser intensity and the magnetic field are linked by the identity

$$B = \frac{\sqrt{Z_0 I_l}}{c}$$

or, in practical units,

$$B[\mathrm{T}] = 6.472 \times 10^{-8}\sqrt{I_l[\mathrm{W\ m^{-2}}]}.$$

A laser with an intensity $I_l = 10^{19}\ \mathrm{W\ m^{-2}}$ is accordingly equivalent to quite a huge magnetic field (204.7 T).

Exercise 5.27 Consider an FEL operating with a CO_2 WU having $I[\mathrm{W\ m^{-2}}] \cong 10^{19}$ and adjust the electron-beam parameters to lase at 1.08 nm. Find the condition to reach saturation in a length of 2 cm and estimate the FEL saturated power.

Exercise 5.28 Discuss the effects of inhomogeneous broadening in a WU FEL and study the relevant consequences on the saturation length for the examples discussed in the previous part of the chapter.

Exercise 5.29 Discuss the possibility of overcoming these effects by increasing the laser intensity and provide some reference identity.

Exercise 5.30 Discuss qualitatively and quantitatively the laser diffraction as a limiting feature of WU operation.

Exercise 5.31 Consider the parameters reported in table 5.1 that are compatible with power and current densities of $I[\mathrm{W\ m^{-2}}] \cong 1.34 \times 10^{19}$, $J[\mathrm{A\ m^{-2}}] \cong 5 \times 10^{11}$ and study the associated FEL performance.

Table 5.1. Electron beam and laser parameters.

Electron beam	
Energy	60 MeV
Length (longitudinal) rms	1 mm
Transverse size rms	25 μm
Relative energy spread	10^{-4}
Emittance	0.6 μm × rad
Gain length	2.83 mm
Pierce parameter	2.81×10^{-4}
Laser pulse	
Wavelength	10 μm
W_0	50 mm
K	0.3

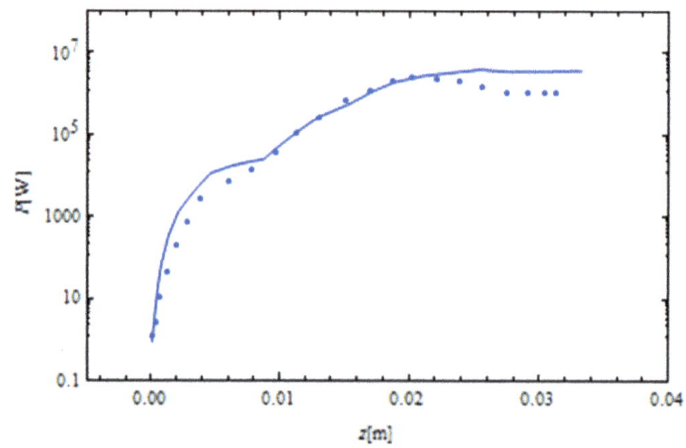

Figure 5.24. Growth of the FEL SASE power (W) versus the laser undulator longitudinal coordinate (m)—analytical (continuous line) and numerical (dotted line)—for the parameters of table 5.1.

Hint: note that the Pierce parameter is $\rho \cong 2.4 \times 10^{-4}$ and, according to the discussion above, we expect a saturation length of the order of 50 ps. In figure 5.24, we provide a comparison between the power growth curve, obtained with an ad hoc developed numerical code, and the analytical formulae of the previous section. The agreement is satisfactory, thus confirming the validity of the scaling relations developed here.

Exercise 5.32 Design an FEL device exploiting a WU and a plasma accelerated beam. Study the conditions for the operation in the hard x-ray region with photon brightness comparable with those of a fourth generation synchrotron radiation source.

5.7.5 Miscellanea on Bessel functions and exotic undulators

In the course of this book we have often referred to Bessel functions which provide a tool of crucial importance to treat processes regarding the emission of radiation by charges executing oscillator motion. The properties of cylindrical Bessel functions can be derived through the relevant integral representation or the generating function. The Bessel cylindrical Bessel function of first kind

$$J_n(x) = \sum_{r=0}^{\left[\frac{n}{2}\right]} \frac{(-1)^r \left(\frac{x}{2}\right)^{n+2r}}{r!\,(n+r)!} \tag{5.84}$$

can be obtained as expansion coefficients of the following function (called the generating function (g.f.)),

$$G(x,t) = e^{\frac{x}{2}\left(t-\frac{1}{t}\right)}, \quad t \neq 0, \ |t| < \infty. \tag{5.85}$$

Exercise 5.33 Derive (5.84) from (5.85).
Hint: expand the exponential as

$$e^{\frac{x}{2}\left(t-\frac{1}{t}\right)} = e^{\frac{x}{2}t} e^{-\frac{x}{2t}} = \sum_{s=0}^{\infty} \frac{t^s}{s!}\left(\frac{x}{2}\right)^s \sum_{q=0}^{\infty} \frac{t^{-q}}{q!}\left(-\frac{x}{2}\right)^q.$$

Then collect parameter t by setting $n = s - q$ and end up with

$$G(x,t) = \sum_{n=-\infty}^{+\infty} t^n J_n(x). \tag{5.86}$$

Exercise 5.34 Use (5.85) to prove that

$$e^{ix\sin(\vartheta)} = \sum_{n=-\infty}^{+\infty} e^{in\vartheta} J_n(x). \tag{5.87}$$

Hint: in (5.85) replace t by $e^{i\vartheta}$.

Exercise 5.35 Consider the radiation emitted by an oscillating charge executing a harmonic dipolar motion with angular frequency ω and amplitude A, use (5.87) to show that the charge radiates at the frequency ω and to its harmonics, and that the radiated intensity scales as $|J_n(\alpha\, A)|^2$ (α is a not specified as constant).

Exercise 5.36 Use (5.87) to derive the integral representation

$$J_m(x) = \frac{1}{\pi}\int_0^{\pi} \cos(m\vartheta - x\sin(\vartheta))\, d\vartheta. \tag{5.88}$$

Hint: multiply both sides of (5.87) by $e^{-im\vartheta}$ then use the orthogonality properties of the trigonometric functions.

We have stressed that oscillating charges emit radiation and that cylindrical Bessel functions of the first kind are useful to understand the power radiated by a charge executing a harmonic dipolar oscillation.

As has been stressed in chapter 3 electrons in undulators execute trajectories more complicated than the simple harmonic motion. For this reason different families of Bessel functions have been introduced, this is indeed the case of two-variable generalized Bessel functions, just touched upon in chapter 3.

They are defined through the generating function

$$G(x, y;\ t) = e^{\frac{x}{2}\left(t-\frac{1}{t}\right)+\frac{y}{2}\left(t^2-\frac{1}{t^2}\right)} = \sum_{n=-\infty}^{+\infty} t^n J_n(x, y), \qquad t \neq 0,\ |t| < \infty. \tag{5.89}$$

Exercise 5.37 Use the previous results to prove that

$$J_n(x, y) = \sum_{l=-\infty}^{+\infty} J_{n-2l}(x) J_l(y). \tag{5.90}$$

Hint: note that $e^{\frac{x}{2}(t-\frac{1}{t})+\frac{y}{2}(t^2-\frac{1}{t^2})} = \sum_{s=0}^{\infty} t^s J_s(x) \sum_{l=0}^{\infty} t^{2l} J_l(x)$, set $s + 2l = n$.

Exercise 5.38 Derive the following integral representation

$$J_m(x, y) = \frac{1}{\pi}\int_0^{\pi} \cos(m\vartheta - x\sin(\vartheta) - y\sin(2\vartheta))\, d\vartheta. \tag{5.91}$$

Exercise 5.39 Discuss the existence of higher order Bessel functions

$$\begin{aligned} {}_mG(x, y; t) &= e^{\frac{x}{2}\left(t-\frac{1}{t}\right)+\frac{y}{2}\left(t^m-\frac{1}{t^m}\right)} = \sum_{n=-\infty}^{+\infty} t^{n(m)} J_n(x, y), \\ {}^{(m)}J_n(x, y) &= \sum_{l=-\infty}^{+\infty} J_{n-ml}(x) J_l(y), \\ {}^{(m)}J_n(x, y) &= \frac{1}{\pi}\int_0^{\pi} \cos(n\vartheta - x\sin(\vartheta) - y\sin(m\vartheta))d\vartheta. \end{aligned} \tag{5.92}$$

Exercise 5.40 Discuss the relevance of the multivariable Bessel function to the emission of radiation by charges executing non-dipolar oscillating motion.

In the past different types of undulators have been proposed to extend, e.g. the emission tunability by increasing the power radiated at higher harmonics. This is indeed the case for bi-harmonic undulators whose on-axis field is specified by

$$\vec{B}_u \equiv [d\ B_0 \sin(h\ k_u z),\ B_0 \sin(k_u z),\ 0]. \tag{5.93}$$

In a structure of this type the FEL will radiate harmonics with polarization along the x- and y-axes.

Exercise 5.41 Show that the FEL pendulum equation has the same structure of that reported in section 4.10. The harmonics have two different indices and the relevant Bessel factor reads

$$\begin{aligned} f_{b,\,n_x} &= (-1)^{\frac{n-1}{2}}\left[{}^{(h)}J_{\frac{n-1}{2}}\left(n\ \xi,\ (-1)^{h+1}\frac{d^2}{h^2}\left(\frac{n\ \xi}{h}\right)\right)\right. \\ &\quad \left. -\ {}^{(h)}J_{\frac{n+1}{2}}\left(n\ \xi,\ (-1)^{h+1}\frac{d^2}{h^2}\left(\frac{n\ \xi}{h}\right)\right)\right], \\ f_{b,\,(h\,n)_y} &= \frac{d}{h}(-1)^{\frac{(n-1)\,h}{2}}\left[{}^{(h)}J_{\frac{h\,(n-1)}{2}}\left(h\ n\ \xi,\ (-1)^{h+1}\frac{d^2}{h^2}(n\ \xi)\right)\right. \\ &\quad \left. +\ (-1)^h\ {}^{(h)}J_{\frac{h\,(n+1)}{2}}\left(h\ n\ \xi,\ (-1)^{h+1}\frac{d^2}{h^2}(n\xi)\right)\right], \end{aligned} \tag{5.94}$$

with

$$\begin{aligned} \xi &= \frac{1}{4}\frac{K^2}{1 + \frac{K^2}{2}\left(1 + \frac{d^2}{h^2}\right)}, \\ \lambda_0 &= \frac{\lambda_u}{2\,\gamma^2}\left[1 + \frac{K^2}{2}\left(1 + \frac{d^2}{h^2}\right)\right]. \end{aligned} \tag{5.95}$$

When $d = h = 3$ one obtains the more simple expressions

$$\begin{aligned} f_{b,1_x} &= {}^{(3)}J_0\left(\xi,\ \frac{\xi}{3}\right) - {}^{(3)}J_1\left(\xi,\ \frac{\xi}{3}\right), \\ f_{b,3_x} &= -\left[{}^{(3)}J_1(3\ \xi,\ \xi) - {}^{(3)}J_2(3\ \xi,\ \xi)\right], \\ f_{b,3_y} &= {}^{(3)}J_0(3\ \xi,\ \xi)\ -\ {}^{(3)}J_3(3\ \xi,\ \xi), \\ \xi &= \frac{1}{4}\frac{K^2}{1 + K^2},\quad \lambda_0 = \frac{1}{2\,\gamma^2}(1 + K^2). \end{aligned}$$

Exercise 5.42 Discuss the effective feasibility of an FEL operating with a bi-harmonic undulator. Provide a computer code to derive the results concerning the power growth reported in figure 5.25.

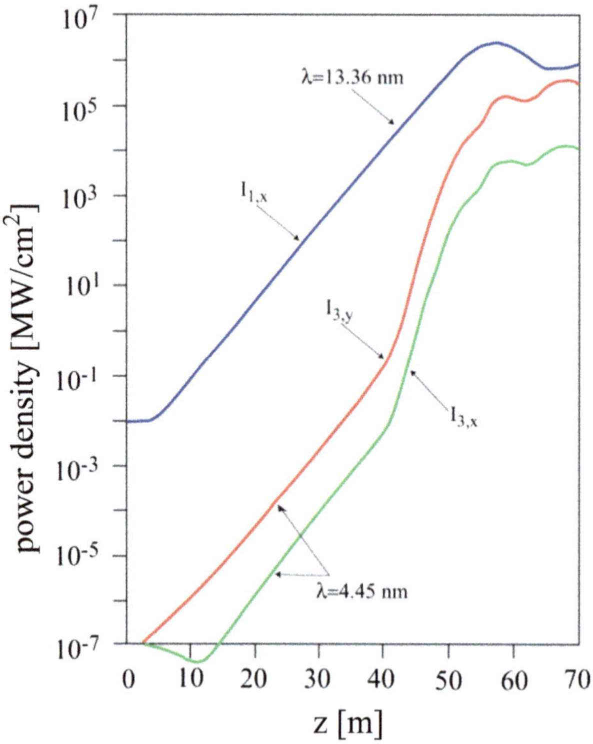

Figure 5.25. Power growth of the main and third harmonics for a bi-harmonic undulator with $d = h = 3$, $E = 1078$ Mev, $\lambda_u = 6$ cm , $K = 0.99$, $\rho = 1.258 \times 10^{-3}$.

The last two exercises are particularly difficult, if the reader succeeds it means that his/her level of understanding of FEL physics is quite good.

5.7.6 High-gradient linac structure

In this chapter we have mentioned high-gradient accelerating structures and we have mainly referred to non-conventional accelerating devices. The possibility of realizing devices with a higher gradient is however offered by conventional linac structures powered by an RF field operating in the so-called X-band. In table 5.2 we have reported the usually adopted frequencies for the cavity accelerating fields and, regarding the X-band, a number of these devices are under construction for FEL SASE operation.

In figure 5.26 we have reported the layout of a 'compact' X-FEL with a linac powered by an S-band cavity at 2.856 GHz, followed by three cavities at 11.424 GHz. In the middle the bunch compressors are used to compress the bunch charge and increase the peak current.

Exercise 5.43 Use the parameters reported in table 5.3 to infer the FEL performance (also make a comparison with the Linac Coherent Light Source (LCLS) case).

Table 5.2. Range of adopted frequencies for the cavity accelerating fields.

1–2 GHz	L-Band
2–4 GHz	S-Band
4–8 GHz	C-Band
8–12 GHz	X-Band

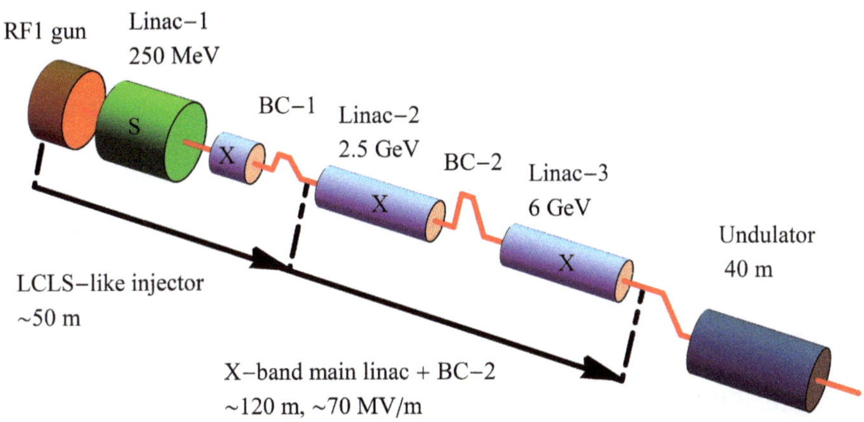

Figure 5.26. A compact X-band linac with two bunch compressors (BC-1 and BC-2) for an X-FEL (see Adolphsen C *et al* 2011 *SLAC National Accelerator Laboratory Report* SLAC-PUB 14276.)

Table 5.3. Electron-beam characteristics at the undulator entrance (Adolphsen C *et al* 2011 *SLAC National Accelerator Laboratory Report* SLAC-PUB 14276).

Parameter	LCLS	X-band FEL	Unit
Bunch charge	250	250	pC
Electron energy	14	6	GeV
Emittance	0.4–0.6	0.4–0.5	μm·rad
Peak current	3.0	3.0	kA
Energy spread	0.01	0.02	%
Undulator period	3	1.5	cm
Undulator parameter	3.5	1.9	
Mean undulator beta	30	8	m
FEL wavelength	1.5	1.5	Å
Saturation length	60	30	m
Saturation power	30	10	GW
FWHM pulse length	80	80	fs
Photons/pulse	2	0.7	10^{12}

Regarding LCLS we note that it was the first X-FEL facility (Emma P *et al* 2010 *Nat. Photonics* **4** 641). The accelerating device pre-existed the X-FEL project and one-third of the existing S-band linac was reused to accelerate an electron beam up to 14 GeV, with an acceleration gradient of about 17 MV m^{-1}.

A replica of LCLS cannot be recommended in view of the high costs for the construction of such a huge accelerating device. The high-gradient C-band linac designed for FEL operation in Japan (Inagaki T *et al* 2014 *Phys. Rev. ST Accel. Beams* RSTAB **17** 080702) has been designed within this framework, starting from general considerations which are quite illuminating and worth being stressed.

According to what we have learned in chapter 2, the accelerating gradient of a constant gradient travelling wave tube structure can be written in terms of the input RF power P_{in}, shunt impedance per unit length R_{sh}, attenuation factor τ and active structure length L,

$$E_{acc} \cong \sqrt{\frac{P_{in}R_{sh}(1 - e^{-\tau})}{L}}.$$

The structure parameters depend on the frequency (f) of the cavity powering field. The shunt impedance scales as $f^{0.5}$ ($R_{sh} = \sqrt{\frac{\mu_0 \pi f}{\sigma_C}}$, $\sigma_C \equiv$ electric conductivity of the material) and the group velocity of the travelling wave as $f^{1.5}$. Keeping therefore the length constant we find that the filling time goes as $f^{-1.5}$. Furthermore, if the pulsed power from the RF source is fixed, a shorter pulse can be used to increase the input peak power (the SLAC Energy Doubler (SLED) technique is used). Combining all these elements we find that the accelerating gradient is proportional to the frequency of the accelerating field. These are the reason why higher frequencies yield larger accelerating fields. The choice of the C-band rather than X has been motivated by the necessity of avoiding cavity break-down.

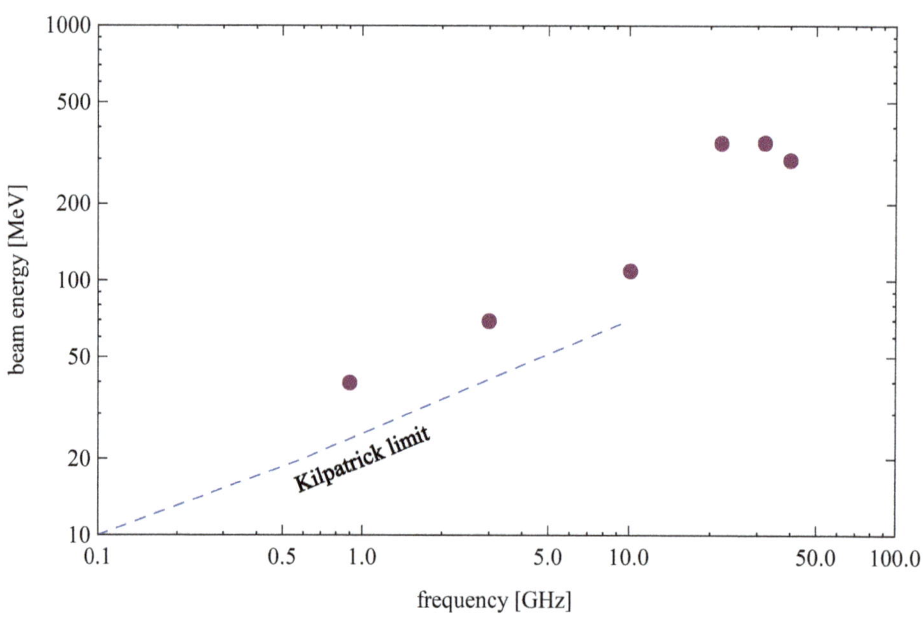

Figure 5.27. Comparison between the Kilpatrick limit and more recent measurements with existing accelerating structures.

To give an idea of the interplay between the accelerating gradient and accelerating field frequency we recall the so-called Kilpatrick criterion according to which safe operation is obtained if

$$f[\text{MHz}] \cong 1.64 \times E[\text{MeV m}^{-1}]e^{-\frac{8.5}{E[\text{MeV m}^{-1}]}}.$$

This is an old empirical criterion which has lost its validity owing to modern cleaning techniques. An idea of the possible improvements to the above Kilpatrick equation is given in figure 5.27.

We hope, at the end of this book, that we have accomplished our goal which has been that of providing sufficient information to start an autonomous work. It is evident that we have just touched upon an immense field; we have treated some topics and others have been left out. We have, for example, made no reference to the physics of the storage ring devoted to FEL operation. The reason is two-fold; these types of problems have been treated in previous books published by some of the current authors, and also because storage ring FELs are not currently a hot topic.

In conclusion, the best we can say is that we hope you enjoyed this book.

Further reading for chapter 5

Plasma physics

Chen F F 2016 *Introduction to Plasma Physics and Controlled Fusion* (Cham: Springer)

Bella P M 2006 *Fundamentals of Plasma Physics* (Cambridge: Cambridge University Press)

Stacey W M 2012 *Fusion Plasma Physics* (Weinheim: Wiley)

Laser-plasma acceleration

Tajima T and Dawson J M 1979 *Phys. Rev. Lett.* **43** 267

Esarey E, Sprangle P, Krall J and Ting A 1996 Overview of plasma-based accelerator concepts *IEEE Trans. Plasma Sci.* **24** 252–88

Kostyukov I, Pukhov A and Kiselev S 2004 Phenomenological theory of laser–plasma interaction in "bubble" regime *Phys. Plasmas* **11** 5256–64

Malka V, Faure J, Gauduel Y A, Lefebvre E, Rousse A and Phuoc K T 2008 Principles and applications of compact laser–plasma accelerators *Nat. Phys* **4** 447–53

Nakajima K 2008 Compact x-ray sources-towards a table-top free-electron laser *Nat. Phys* **4** 92–3

Esarey E, Schroeder C B and Leemans W P 2009 Physics of laser-driven plasma-based electron accelerators *Rev. Mod. Phys.* **81** 1229

Plasma wake acceleration

Holzer B 2016 *Proc. CAS-CERN Accelerator School (Geneva, 23–29 November 2014)* (Geneva: CERN) CERN-2016-001

Nakajima K, Deng Aihua, Yoshitama Hitoshi, Hafz Nasr A M, Lu Haiyang, Shen Baifei, Liu Jiansheng, Li Ruxin and Xu Zhizhan 2012 Laser driven table-top x-ray FEL *Free Electron Lasers* ed S Varro (Rijeka: InTech)

Cros B 2016 Laser-driven plasma wakefield: propagation effects *Proc. CAS-CERN Accelerator School (Geneva, 23–29 November 2014)* ed B Holzer (Geneva: CERN)

Gibbon P 2016 Introduction to plasma physics *Proc. CAS-CERN Accelerator School (Geneva, 23–29 November 2014)* ed B Holzer (Geneva: CERN)

Pukhov A 2016 Particle in cell codes for plasma based acceleration *Proc. CAS-CERN Accelerator School (Geneva, 23–29 November 2014)* ed B Holzer (Geneva: CERN)

Wave undulators

Sengupta N D 1952 *Bull. Math. Soc. (Calcutta)* **44** 175

Brown L S and Kibble T W B 1964 *Phys. Rev* **133** A705

Pantell R H, Soncini G and Puthoff H E 1968 Stimulated amplification of CO_2 laser via stimulated Compton backscattering *IEEE J. Quantum Electron.* **4** 905

Madey J M 1971 Stimulated emission of Bremsstrahlung in a periodic magnetic field *J. Appl. Phys.* **42** 1906

Dattoli G and Renieri A 1985 *Laser Handbook* vol 4 ed M L Stitch and M S Bass (Amsterdam: North Holland)

Dobsiach P, Meystre P and Scully M O 1983 *IEEE J. Quantum Electron.* **19** 1812

Ciocci F, Dattoli G and Walsh J 1985 *Nucl. Instrum. Methods* A **237** 401

Gea Banacloche J, Moore G T, Schlicher R R, Scully M O and Walther H 1987 *IEEE J. Quantum Electron.* **23** 1558

Tang Cha Mei, Hafizi B and Ride S K 1993 *Nucl. Instrum. Methods* A **331** 371

Dattoli G, Letardi T and Vazquez L R 1999 *Phys. Lett.* A **237** 26

Bacci A, Ferrario M, Maroli C, Petrillo V and Serafini L 2006 *Phys. Rev. Spec. Top. (Accel. Beams)* **9** 060704

Polyanskiy M N, Pogorelsky I V and Yakimenko V 2011 *Opt. Express* **19** 7717

Yakimenko V 2011 CO_2 laser based undulator for a compact SASE FEL *Laser and Plasma Accelerators Workshop* 2011 *(Brookhaven, NY, June* 20–21)

Yanovsky V *et al* 2008 *Opt. Express* **16** 2109

Norby J 2005 *Laser Focus World* January 1 http://www.extreme-light-infrastructure.eu/

Exotic undulators and use of generalized Bessel functions

For a list of undulators with non-standard performances see, e.g. http://www.spring8.or.jp/en/about_us/whats_sp8/facilities/bl/light_source_optics/sources/insertion_device/intro_insertion_devices

For the theory of Bi-Harmonic undulators see, e.g.,

Dattoli G, Giannessi L, Ottaviani P L, Freund H P, Biedron S G and Milton S 2002 Two harmonic undulators and harmonic generation in high gain free electron lasers *Nuclear Instrum Methods* A **495** 48–57

Dattoli G and Voykov G K 1993 Spectral properties of two-harmonic undulator radiation *Phys. Rev.* E **48** 3030–9

Ciocci F, Dattoli G, Giannessi L, Ottaviani P L and Renieri A 2001 Biharmonic undulators and FEL dynamics *Proc. 23rd International Free-Electron Laser Conference and 8th FEL Users Workshop (*20–24 *August* 2001, *Darmstadt, Germany)*

Dattoli G, Voykov G K and Carpanese M 1995 Betatron motion and linear undulator brightness: inhomogeneous broadening effects due to electron-beam emittance and Twiss parameters *Phys. Rev.* E **52** 6809–17

Dattoli G and Bucci L 2000 Free electron lasers operating with variably polarization undulators *Nuclear Instrum Methods* A **450** 479–90

www.ingramcontent.com/pod-product-compliance
Ingram Content Group UK Ltd.
Pitfield, Milton Keynes, MK11 3LW, UK
UKHW051907100926
8556IPUK00005B/793

9 780750 312400